Complete Solutions Guide for

CALCULUS

SIXTH EDITION

Larson / Hostetler / Edwards

Volume II
Chapters 6–9

Bruce H. Edwards
University of Florida

Houghton Mifflin Company   Boston   New York

Editor in Chief, Mathematics: Charles Hartford
Managing Editor: Cathy Cantin
Senior Associate Editor: Maureen Brooks
Associate Editor: Michael Richards
Assistant Editor: Carolyn Johnson
Supervising Editor: Karen Carter
Art Supervisor: Gary Crespo
Marketing Manager: Sara Whittern
Associate Marketing Manager: Ros Kane
Marketing Assistant: Carrie Lipscomb
Design: Henry Rachlin
Composition and Art: Meridian Creative Group

Printed in the U.S.A.

ISBN: 0-395-88770-4

5678910-B-09 08 07 06 05

Preface

The *Complete Solutions Guide for Calculus, Sixth Edition* is a supplement to the text by Roland E. Larson, Robert P. Hostetler, and Bruce H. Edwards. Solutions to every exercise in the text are given with all essential algebraic steps included. There are three volumes in the complete set of solutions guides. Volume I contains Chapters P–5, Volume II contains Chapters 6–9, Volume III contains Chapters 10–15, and Appendices.

I have made every effort to see that the solutions are correct. However, I would appreciate hearing about any errors or other suggestions for improvement.

I would like to thank the staff at Larson Texts, Inc. for their help in the production of this guide.

Bruce H. Edwards
University of Florida
Gainesville, Florida 32611
(be@math.ufl.edu)

Contents

CHAPTER 6
Applications of Integration

C H A P T E R 6
Applications of Integration

Section 6.1   Area of a Region Between Two Curves

1. $A = \int_0^6 [0 - (x^2 - 6x)] \, dx = -\int_0^6 (x^2 - 6x) \, dx$

2. $A = \int_{-2}^2 [(2x + 5) - (x^2 + 2x + 1)] \, dx = \int_{-2}^2 (-x^2 + 4) \, dx$

3. $A = \int_0^3 [(-x^2 + 2x + 3) - (x^2 - 4x + 3)] \, dx = \int_0^3 (-2x^2 + 6x) \, dx$

4. $A = \int_0^1 (x^2 - x^3) \, dx$

5. $A = 2\int_{-1}^0 3(x^3 - x) \, dx = 6\int_{-1}^0 (x^3 - x) \, dx \quad$ or $\quad -6\int_0^1 (x^3 - x) \, dx$

6. $A = 2\int_0^1 [(x - 1)^3 - (x - 1)] \, dx$

7. $\int_0^4 \left[(x + 1) - \frac{x}{2} \right] dx$ **8.** $\int_{-1}^1 [(1 - x^2) - (x^2 - 1)] \, dx$ **9.** $\int_0^6 \left[4(2^{-x/3}) - \frac{x}{6} \right] dx$

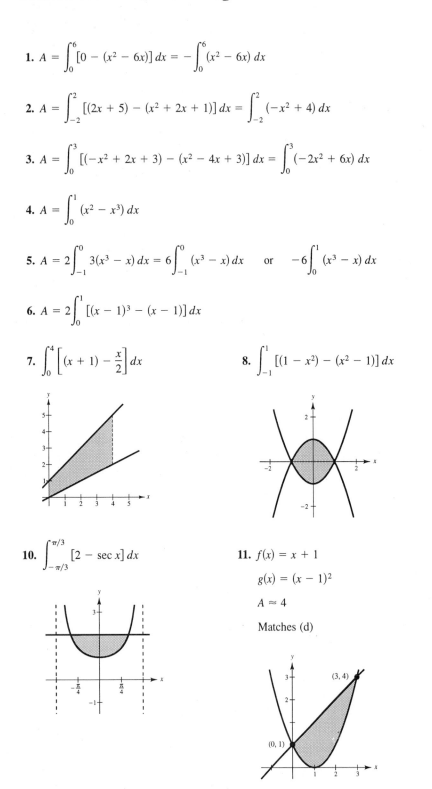

10. $\int_{-\pi/3}^{\pi/3} [2 - \sec x] \, dx$

11. $f(x) = x + 1$

$\quad g(x) = (x - 1)^2$

$\quad A \approx 4$

$\quad$ Matches (d)

12. $f(x) = 2 - \frac{1}{2}x$

$\quad g(x) = 2 - \sqrt{x}$

$\quad A \approx 1$

$\quad$ Matches (a)

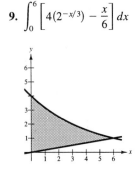

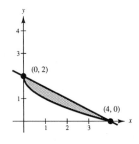

13. The points of intersection are given by:

$$x^2 - 4x = 0$$

$$x(x - 4) = 0 \quad \text{when} \quad x = 0, 4$$

$$A = \int_0^4 [g(x) - f(x)] \, dx$$

$$= -\int_0^4 (x^2 - 4x) \, dx$$

$$= -\left[\frac{x^3}{3} - 2x^2 \right]_0^4$$

$$= \frac{32}{3}$$

14. The points of intersection are given by:

$$3 - 2x - x^2 = 0$$

$$(3 + x)(1 - x) = 0 \quad \text{when} \quad x = -3, 1$$

$$A = \int_{-3}^1 [f(x) - g(x)] \, dx$$

$$= \int_{-3}^1 (3 - 2x - x^2) \, dx$$

$$= \left[3x - x^2 - \frac{x^3}{3} \right]_{-3}^1$$

$$= \frac{32}{3}$$

15. The points of intersection are given by:

$$x^2 + 2x + 1 = 3x + 3$$

$$(x - 2)(x + 1) = 0 \quad \text{when} \quad x = -1, 2$$

$$A = \int_{-1}^2 [g(x) - f(x)] \, dx$$

$$= \int_{-1}^2 [(3x + 3) - (x^2 + 2x + 1)] \, dx$$

$$= \int_{-1}^2 (2 + x - x^2) \, dx$$

$$= \left[2x + \frac{x^2}{2} - \frac{x^3}{3} \right]_{-1}^2 = \frac{9}{2}$$

16. The points of intersection are given by:

$$-x^2 + 4x + 2 = x + 2$$

$$x(3 - x) = 0 \quad \text{when} \quad x = 0, 3$$

$$A = \int_0^3 [f(x) - g(x)] \, dx$$

$$= \int_0^3 [(-x^2 + 4x + 2) - (x + 2)] \, dx$$

$$= \int_0^3 (-x^2 + 3x) \, dx = \left[\frac{-x^3}{3} + \frac{3}{2}x^2 \right]_0^3 = \frac{9}{2}$$

17. The points of intersection are given by:

$$x = 2 - x \quad \text{and} \quad x = 0 \quad \text{and} \quad 2 - x = 0$$

$$x = 1 \qquad\qquad x = 0 \qquad\qquad x = 2$$

$$A = \int_0^1 [(2 - y) - (y)] \, dy = \left[2y - y^2 \right]_0^1 = 1$$

Note that if we integrate with respect to x, we need two integrals. Also, note that the region is a triangle.

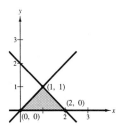

18. $A = \int_1^5 \left(\frac{1}{x^2} - 0 \right) dx = \left[-\frac{1}{x} \right]_1^5 = \frac{4}{5}$

19. The points of intersection are given by:

$$\sqrt{3x} + 1 = x + 1$$

$$\sqrt{3x} = x \quad \text{when} \quad x = 0, 3$$

$$A = \int_0^3 [f(x) - g(x)] \, dx$$

$$= \int_0^3 \left[(\sqrt{3x} + 1) - (x + 1) \right] dx$$

$$= \int_0^3 [(3x)^{1/2} - x] \, dx$$

$$= \left[\frac{2}{9} (3x)^{3/2} - \frac{x^2}{2} \right]_0^3 = \frac{3}{2}$$

20. The points of intersection are given by:

$$\sqrt[3]{x} = x$$

$$x = -1, 0, 1$$

$$A = 2 \int_0^1 [f(x) - g(x)] \, dx$$

$$= 2 \int_0^1 \left(\sqrt[3]{x} - x \right) dx$$

$$= 2 \int_0^1 (x^{1/3} - x) \, dx$$

$$= 2 \left[\frac{3}{4} x^{4/3} - \frac{1}{2} x^2 \right]_0^1 = \frac{1}{2}$$

21. The points of intersection are given by:

$$y^2 = y + 2$$

$$(y - 2)(y + 1) = 0 \quad \text{when} \quad y = -1, 2$$

$$A = \int_{-1}^2 [g(y) - f(y)] \, dy$$

$$= \int_{-1}^2 [(y + 2) - y^2] \, dy$$

$$= \left[2y + \frac{y^2}{2} - \frac{y^3}{3} \right]_{-1}^2 = \frac{9}{2}$$

22. The points of intersection are given by:

$$2y - y^2 = -y$$

$$y(y - 3) = 0 \quad \text{when} \quad y = 0, 3$$

$$A = \int_0^3 [f(y) - g(y)] \, dy$$

$$= \int_0^3 [(2y - y^2) - (-y)] \, dy$$

$$= \int_0^3 (3y - y^2) \, dy = \left[\frac{3}{2} y^2 - \frac{1}{3} y^3 \right]_0^3 = \frac{9}{2}$$

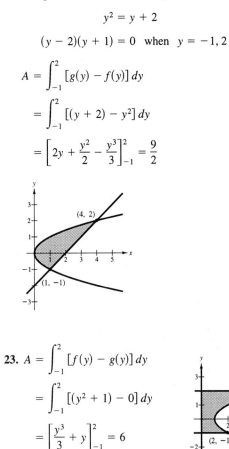

23. $A = \int_{-1}^2 [f(y) - g(y)] \, dy$

$$= \int_{-1}^2 [(y^2 + 1) - 0] \, dy$$

$$= \left[\frac{y^3}{3} + y \right]_{-1}^2 = 6$$

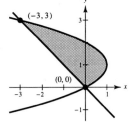

24. $A = \displaystyle\int_0^3 [f(y) - g(y)] \, dy$

$= \displaystyle\int_0^3 \left[\frac{y}{\sqrt{16 - y^2}} - 0 \right] dy$

$= -\dfrac{1}{2} \displaystyle\int_0^3 (16 - y^2)^{-1/2} (-2y) \, dy$

$= \left[-\sqrt{16 - y^2} \right]_0^3 = 4 - \sqrt{7} \approx 1.354$

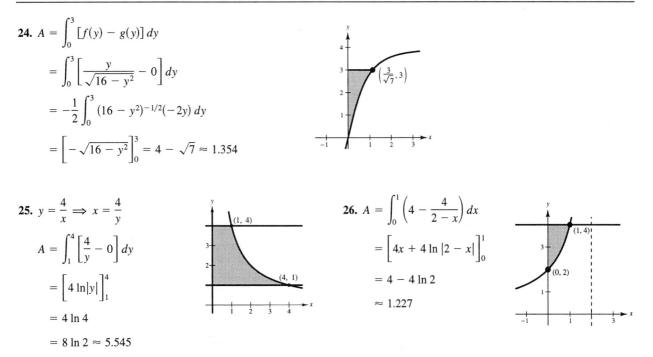

25. $y = \dfrac{4}{x} \implies x = \dfrac{4}{y}$

$A = \displaystyle\int_1^4 \left[\frac{4}{y} - 0 \right] dy$

$= \left[4 \ln|y| \right]_1^4$

$= 4 \ln 4$

$= 8 \ln 2 \approx 5.545$

26. $A = \displaystyle\int_0^1 \left(4 - \frac{4}{2 - x} \right) dx$

$= \left[4x + 4 \ln|2 - x| \right]_0^1$

$= 4 - 4 \ln 2$

≈ 1.227

27. The points of intersection are given by:

$x^3 - 3x^2 + 3x = x^2$

$x(x - 1)(x - 3) = 0$ when $x = 0, 1, 3$

$A = \displaystyle\int_0^1 [f(x) - g(x)] \, dx + \int_1^3 [g(x) - f(x)] \, dx$

$= \displaystyle\int_0^1 [(x^3 - 3x^2 + 3x) - x^2] \, dx + \int_1^3 [x^2 - (x^3 - 3x^2 + 3x)] \, dx$

$= \displaystyle\int_0^1 (x^3 - 4x^2 + 3x) \, dx + \int_1^3 (-x^3 + 4x^2 - 3x) \, dx$

$= \left[\frac{x^4}{4} - \frac{4}{3}x^3 + \frac{3}{2}x^2 \right]_0^1 + \left[\frac{-x^4}{4} + \frac{4}{3}x^3 - \frac{3}{2}x^2 \right]_1^3 = \frac{37}{12}$

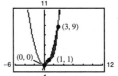

Numerical Approximation: $0.417 + 2.667 \approx 3.083$

28. The point of intersection is given by:

$x^3 - 2x + 1 = -2x$

$x^3 + 1 = 0$ when $x = -1$

$A = \displaystyle\int_{-1}^1 [f(x) - g(x)] \, dx$

$= \displaystyle\int_{-1}^1 [(x^3 - 2x + 1) - (-2x)] \, dx$

$= \displaystyle\int_{-1}^1 (x^3 + 1) \, dx = \left[\frac{x^4}{4} + x \right]_{-1}^1 = 2$

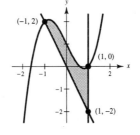

Numerical Approximation: 2.0

29. The points of intersection are given by:

$$x^2 - 4x + 3 = 3 + 4x - x^2$$

$$2x(x - 4) = 0 \quad \text{when} \quad x = 0, 4$$

$$A = \int_0^4 \left[(3 + 4x - x^2) - (x^2 - 4x + 3)\right] dx$$

$$= \int_0^4 (-2x^2 + 8x) \, dx$$

$$= \left[-\frac{2x^3}{3} + 4x^2\right]_0^4 = \frac{64}{3}$$

Numerical Approximation: 21.333

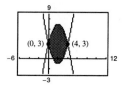

30. The points of intersection are given by:

$$x^4 - 2x^2 = 2x^2$$

$$x^2(x^2 - 4) = 0 \quad \text{when} \quad x = 0, \pm 2$$

$$A = 2\int_0^2 \left[2x^2 - (x^4 - 2x^2)\right] dx$$

$$= 2\int_0^2 (4x^2 - x^4) \, dx$$

$$= 2\left[\frac{4x^3}{3} - \frac{x^5}{5}\right]_0^2 = \frac{128}{15}$$

Numerical Approximation: 8.533

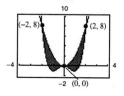

31. $f(x) = x^4 - 4x^2, \quad g(x) = x^2 - 4$

The points of intersection are given by:

$$x^4 - 4x^2 = x^2 - 4$$

$$x^4 - 5x^2 + 4 = 0$$

$$(x^2 - 4)(x^2 - 1) = 0 \quad \text{when} \quad x = \pm 2, \pm 1$$

By symmetry,

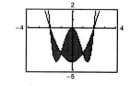

$$A = 2\int_0^1 \left[(x^4 - 4x^2) - (x^2 - 4)\right] dx + 2\int_1^2 \left[(x^2 - 4) - (x^4 - 4x^2)\right] dx$$

$$= 2\int_0^1 (x^4 - 5x^2 + 4) \, dx + 2\int_1^2 (-x^4 + 5x^2 - 4) \, dx$$

$$= 2\left[\frac{x^5}{5} - \frac{5x^3}{3} + 4x\right]_0^1 + 2\left[-\frac{x^5}{5} + \frac{5x^3}{3} - 4x\right]_1^2$$

$$= 2\left[\frac{1}{5} - \frac{5}{3} + 4\right] + 2\left[\left(-\frac{32}{5} + \frac{40}{3} - 8\right) - \left(-\frac{1}{5} + \frac{5}{3} - 4\right)\right] = 8.$$

Numerical Approximation: $5.067 + 2.933 = 8.0$

32. $f(x) = x^4 - 4x^2, \quad g(x) = x^3 - 4x$

The points of intersection are given by:

$$x^4 - 4x^2 = x^3 - 4x$$

$$x^4 - x^3 - 4x^2 + 4x = 0$$

$$x(x - 1)(x + 2)(x - 2) = 0 \quad \text{when} \quad x = -2, 0, 1, 2$$

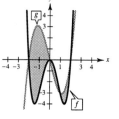

$$A = \int_{-2}^0 \left[(x^3 - 4x) - (x^4 - 4x^2)\right] dx + \int_0^1 \left[(x^4 - 4x^2) - (x^3 - 4x)\right] dx + \int_1^2 \left[(x^3 - 4x) - (x^4 - 4x^2)\right] dx$$

$$= \frac{248}{30} + \frac{37}{60} + \frac{53}{60} = \frac{293}{30}$$

Numerical Approximation: $8.267 + 0.617 + 0.883 \approx 9.767$

33. The points of intersection are given by:

$$\frac{1}{1+x^2} = \frac{x^2}{2}$$

$$x^4 + x^2 - 2 = 0$$

$$(x^2 + 2)(x^2 - 1) = 0$$

$$x = \pm 1$$

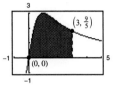

$$A = 2\int_0^1 [f(x) - g(x)]\,dx$$

$$= 2\int_0^1 \left[\frac{1}{1+x^2} - \frac{x^2}{2}\right]dx$$

$$= 2\left[\arctan x - \frac{x^3}{6}\right]_0^1$$

$$= 2\left(\frac{\pi}{4} - \frac{1}{6}\right) = \frac{\pi}{2} - \frac{1}{3} \approx 1.237$$

Numerical Approximation: 1.237

34. $A = \displaystyle\int_0^3 \left[\frac{6x}{x^2+1} - 0\right]dx$

$$= \left[3\ln(x^2+1)\right]_0^3$$

$$= 3\ln 10$$

$$\approx 6.908$$

Numerical Approximation: 6.908

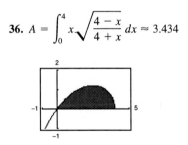

35. $\sqrt{1+x^3} \le \dfrac{1}{2}x + 2$ on $[0, 2]$

Numerical approximation: 1.759

$$A = \int_0^2 \left[\frac{1}{2}x + 2 - \sqrt{1+x^3}\right]dx \approx 1.759$$

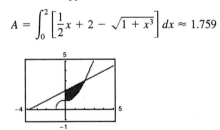

36. $A = \displaystyle\int_0^4 x\sqrt{\frac{4-x}{4+x}}\,dx \approx 3.434$

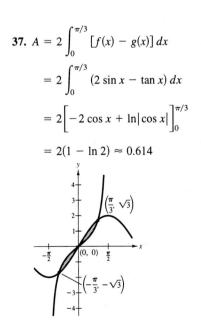

37. $A = 2\displaystyle\int_0^{\pi/3} [f(x) - g(x)]\,dx$

$$= 2\int_0^{\pi/3} (2\sin x - \tan x)\,dx$$

$$= 2\left[-2\cos x + \ln|\cos x|\right]_0^{\pi/3}$$

$$= 2(1 - \ln 2) \approx 0.614$$

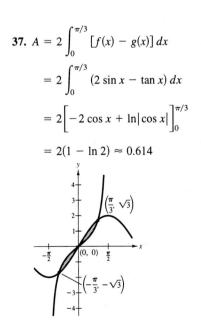

38. $A = 2\displaystyle\int_{\pi/6}^{\pi/2} [f(x) - g(x)]\,dx$

$$= 2\int_{\pi/6}^{\pi/2} [\sin 2x - \cos x]\,dx$$

$$= 2\left[-\frac{1}{2}\cos 2x - \sin x\right]_{\pi/6}^{\pi/2}$$

$$= \frac{1}{2}$$

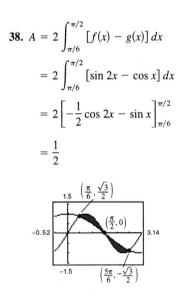

39. $A = \displaystyle\int_0^1 [xe^{-x^2} - 0]\, dx$

$= \left[-\dfrac{1}{2}e^{-x^2} \right]_0^1 = \dfrac{1}{2}\left(1 - \dfrac{1}{e} \right) \approx 0.316$

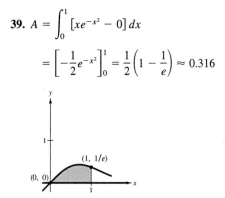

40. From the graph we see that f and g intersect twice at $x = 0$ and $x = 1$.

$A = \displaystyle\int_0^1 [g(x) - f(x)]\, dx$

$= \displaystyle\int_0^1 [(2x + 1) - 3^x]\, dx$

$= \left[x^2 + x - \dfrac{1}{\ln 3}(3^x) \right]_0^1$

$= 2\left(1 - \dfrac{1}{\ln 3} \right) \approx 0.180$

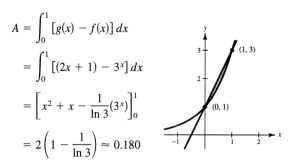

41. $A = \displaystyle\int_0^{\pi} [(2\sin x + \sin 2x) - 0]\, dx$

$= \left[-2\cos x - \dfrac{1}{2}\cos 2x \right]_0^{\pi} = 4.0$

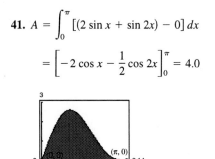

42. $A = \displaystyle\int_0^{\pi} [(2\sin x + \cos 2x) - 0]\, dx$

$= \left[-2\cos x + \dfrac{1}{2}\sin 2x \right]_0^{\pi} = 4$

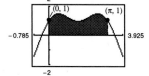

43. $A = \displaystyle\int_1^3 \left[\dfrac{1}{x^2}e^{1/x} - 0 \right] dx$

$= \left[-e^{1/x} \right]_1^3 = e - e^{1/3} \approx 1.323$

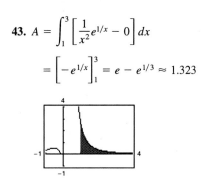

44. $A = \displaystyle\int_1^5 \left[\dfrac{4\ln x}{x} - 0 \right] dx$

$= \left[2(\ln x)^2 \right]_1^5 = 2(\ln 5)^2 \approx 5.181$

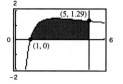

45. (a) $y = \sqrt{\dfrac{x^3}{4 - x}}, \quad y = 0, \quad x = 3$

(b) $A = \displaystyle\int_0^3 \sqrt{\dfrac{x^3}{4 - x}}\, dx,$

No, it cannot be evaluated by hand.

(c) 4.7721

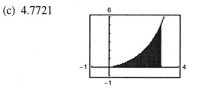

46. (a) $y = \sqrt{x}\, e^x, \quad y = 0, \quad x = 0, \quad x = 1$

(b) $A = \displaystyle\int_0^1 \sqrt{x}\, e^x\, dx.$

No, it cannot be evaluated by hand.

(c) 1.2556

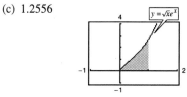

47. $A = \int_0^c \left[\left(\frac{b-a}{c} y + a \right) - \frac{b}{c} y \right] dy$

$= \int_0^c \left(-\frac{a}{c} y + a \right) dy$

$= \left[-\frac{a}{2c} y^2 + ay \right]_0^c$

$= -\frac{ac}{2} + ac = \frac{ac}{2} \quad \left(= \frac{1}{2} (\text{base})(\text{height}) \right)$

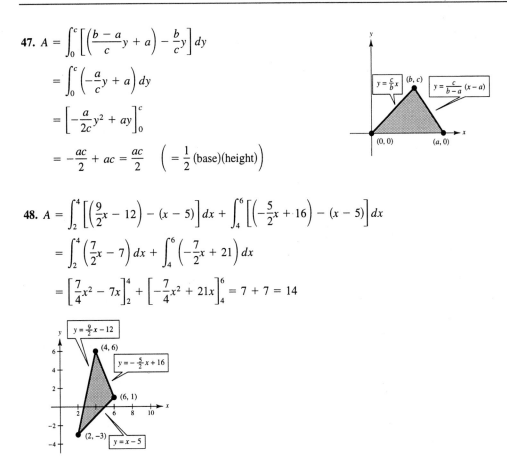

48. $A = \int_2^4 \left[\left(\frac{9}{2} x - 12 \right) - (x - 5) \right] dx + \int_4^6 \left[\left(-\frac{5}{2} x + 16 \right) - (x - 5) \right] dx$

$= \int_2^4 \left(\frac{7}{2} x - 7 \right) dx + \int_4^6 \left(-\frac{7}{2} x + 21 \right) dx$

$= \left[\frac{7}{4} x^2 - 7x \right]_2^4 + \left[-\frac{7}{4} x^2 + 21x \right]_4^6 = 7 + 7 = 14$

49. $f(x) = x^3$

$f'(x) = 3x^2$

At $(1, 1)$, $f'(1) = 3$.

Tangent line:

$\quad y - 1 = 3(x - 1)$ or $y = 3x - 2$

The tangent line intersects $f(x) = x^3$ at $x = -2$.

$A = \int_{-2}^1 [x^3 - (3x - 2)] \, dx = \left[\frac{x^4}{4} - \frac{3x^2}{2} + 2x \right]_{-2}^1 = \frac{27}{4}$

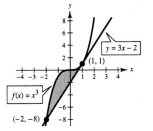

50. $f(x) = \frac{1}{x^2 + 1}$

$f'(x) = -\frac{2x}{(x^2 + 1)^2}$

At $\left(1, \frac{1}{2} \right)$, $f'(1) = -\frac{1}{2}$.

Tangent line:

$\quad y - \frac{1}{2} = -\frac{1}{2}(x - 1)$ or $y = -\frac{1}{2} x + 1$

The tangent line intersects $f(x) = \frac{1}{x^2 + 1}$ at $x = 0$.

$A = \int_0^1 \left[\frac{1}{x^2 + 1} - \left(-\frac{1}{2} x + 1 \right) \right] dx = \left[\arctan x + \frac{x^2}{4} - x \right]_0^1 = \frac{\pi - 3}{4} \approx 0.0354$

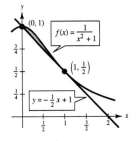

51. $x^4 - 2x^2 + 1 \le 1 - x^2$ on $[-1, 1]$

$$A = \int_{-1}^{1} [(1 - x^2) - (x^4 - 2x^2 + 1)] \, dx$$

$$= \int_{-1}^{1} (x^2 - x^4) \, dx$$

$$= \left[\frac{x^3}{3} - \frac{x^5}{5} \right]_{-1}^{1} = \frac{4}{15}$$

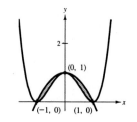

You can use a single integral because $x^4 - 2x^2 + 1 \le 1 - x^2$ on $[-1, 1]$.

52. $x^3 \ge x$ on $[-1, 0]$

$x^3 \le x$ on $[0, 1]$

Both functions symmetric to origin

$$\int_{-1}^{0} (x^3 - x) \, dx = -\int_{0}^{1} (x^3 - x) \, dx.$$

Thus, $\int_{-1}^{1} (x^3 - x) \, dx = 0$.

$$A = 2 \int_{0}^{1} (x - x^3) \, dx = 2 \left[\frac{x^2}{2} - \frac{x^4}{4} \right]_{0}^{1} = \frac{1}{2}$$

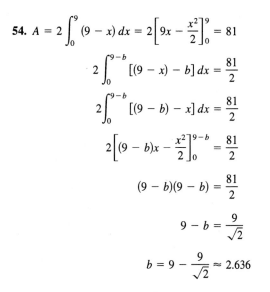

53.

$$A = \int_{-3}^{3} (9 - x^2) \, dx = 36$$

$$\int_{-\sqrt{9-b}}^{\sqrt{9-b}} [(9 - x^2) - b] \, dx = 18$$

$$\int_{0}^{\sqrt{9-b}} [(9 - b) - x^2] \, dx = 9$$

$$\left[(9 - b)x - \frac{x^3}{3} \right]_{0}^{\sqrt{9-b}} = 9$$

$$\frac{2}{3} (9 - b)^{3/2} = 9$$

$$(9 - b)^{3/2} = \frac{27}{2}$$

$$9 - b = \frac{9}{\sqrt[3]{4}}$$

$$b = 9 - \frac{9}{\sqrt[3]{4}} \approx 3.330$$

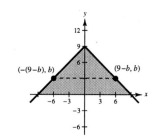

54. $A = 2 \int_{0}^{9} (9 - x) \, dx = 2 \left[9x - \frac{x^2}{2} \right]_{0}^{9} = 81$

$$2 \int_{0}^{9-b} [(9 - x) - b] \, dx = \frac{81}{2}$$

$$2 \int_{0}^{9-b} [(9 - b) - x] \, dx = \frac{81}{2}$$

$$2 \left[(9 - b)x - \frac{x^2}{2} \right]_{0}^{9-b} = \frac{81}{2}$$

$$(9 - b)(9 - b) = \frac{81}{2}$$

$$9 - b = \frac{9}{\sqrt{2}}$$

$$b = 9 - \frac{9}{\sqrt{2}} \approx 2.636$$

55. $\displaystyle\lim_{\|\Delta\|\to 0}\sum_{i=1}^{n}(x_i - x_i^2)\,\Delta x$

where $x_i = \dfrac{i}{n}$ and $\Delta x = \dfrac{1}{n}$ is the same as

$$\int_0^1 (x - x^2)\,dx = \left[\frac{x^2}{2} - \frac{x^3}{3}\right]_0^1 = \frac{1}{6}.$$

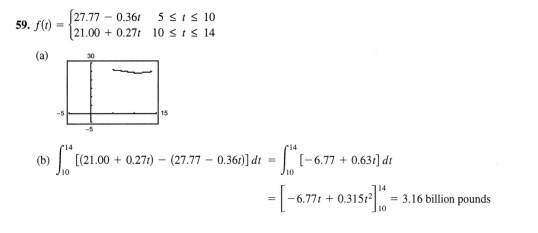

56. $\displaystyle\lim_{\|\Delta\|\to 0}\sum_{i=1}^{n}(4 - x_i^2)\,\Delta x$

where $x_i = -2 + \dfrac{4i}{n}$ and $\Delta x = \dfrac{4}{n}$ is the same as

$$\int_{-2}^{2} (4 - x^2)\,dx = \left[4x - \frac{x^3}{3}\right]_{-2}^{2} = \frac{32}{3}.$$

57. $\displaystyle\int_0^5 \left[(7.21 + 0.58t) - (7.21 + 0.45t)\right]dt = \int_0^5 0.13t\,dt = \left[\frac{0.13t^2}{2}\right]_0^5 = \$1.625 \text{ billion}$

58. $\displaystyle\int_0^5 \left[(7.21 + 0.26t + 0.02t^2) - (7.21 + 0.1t + 0.01t^2)\right]dt = \int_0^5 (0.01t^2 + 0.16t)\,dt$

$$= \left[\frac{0.01t^3}{3} + \frac{0.16t^2}{2}\right]_0^5$$

$$= \frac{29}{12}\text{ billion} \approx \$2.417 \text{ billion}$$

59. $f(t) = \begin{cases} 27.77 - 0.36t & 5 \le t \le 10 \\ 21.00 + 0.27t & 10 \le t \le 14 \end{cases}$

(a)

```
        30




   -5          15

        -5
```

(b) $\displaystyle\int_{10}^{14}\left[(21.00 + 0.27t) - (27.77 - 0.36t)\right]dt = \int_{10}^{14}\left[-6.77 + 0.63t\right]dt$

$$= \left[-6.77t + 0.315t^2\right]_{10}^{14} = 3.16 \text{ billion pounds}$$

60. $\displaystyle\int_0^{10}\left[(568.50 + 7.15t) - (525.60 + 6.43t)\right]dt = \int_0^{10}(42.90 + 0.72t)\,dt = \left[42.90t + 0.36t^2\right]_0^{10} = \$465 \text{ million}$

61. 5% : $P_1 = 893{,}000\,e^{(0.05)t}$

$3\frac{1}{2}\%$: $P_2 = 893{,}000\,e^{(0.035)t}$

Difference in profits over 5 years:

$$\int_0^5 \left[893{,}000e^{0.05t} - 893{,}000e^{0.035t}\right]dt = 893{,}000\left[\frac{e^{0.05t}}{0.05} - \frac{e^{0.035t}}{0.035}\right]_0^5$$

$$\approx 893{,}000\left[(25.6805 - 34.0356) - (20 - 28.5714)\right]$$

$$\approx 893{,}000(0.2163) \approx \$193{,}156$$

Note: Using a graphing utility you obtain \$193,183.

62. Proposal 2 is better, since the cummulative deficit (the area under the curve) is less.

63. The total area is 8 times the area of the shaded region to the right. A point (x, y) is on the upper boundary of the region if

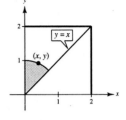

$$\sqrt{x^2 + y^2} = 2 - y$$
$$x^2 + y^2 = 4 - 4y + y^2$$
$$x^2 = 4 - 4y$$
$$4y = 4 - x^2$$
$$y = 1 - \frac{x^2}{4}.$$

We now determine where this curve intersects the line $y = x$.

$$x = 1 - \frac{x^2}{4}$$
$$x^2 + 4x - 4 = 0$$
$$x = \frac{-4 \pm \sqrt{16 + 16}}{2} = -2 \pm 2\sqrt{2} \implies x = -2 + 2\sqrt{2}$$

$$\text{Total area} = 8 \int_0^{-2+2\sqrt{2}} \left(1 - \frac{x^2}{4} - x\right) dx$$
$$= 8 \left[x - \frac{x^3}{12} - \frac{x^2}{2} \right]_0^{-2+2\sqrt{2}} \approx 8(0.4379) = 3.503$$

64. The curves intersect at the point where the slope of y_2 equals that of y_1, 1.

$$y_2 = 0.08x^2 + k \implies y'_2 = 0.16x = 1 \implies x = \frac{1}{.16} = 6.25$$

(a) The value of k is given by

$$y_1 = y_2$$
$$6.25 = (0.08)(6.25)^2 + k$$
$$k = 3.125.$$

(b) $\text{Area} = 2 \int_0^{6.25} (y_2 - y_1) \, dx$

$$= 2 \int_0^{6.25} (0.08x^2 + 3.125 - x) \, dx$$
$$= 2 \left[\frac{0.08x^3}{3} + 3.125x - \frac{x^2}{2} \right]_0^{6.25}$$
$$= 2(6.510417) \approx 13.02083$$

65. (a) $A = 2 \left[\int_0^5 \left(1 - \frac{1}{3}\sqrt{5 - x} \right) dx + \int_5^{5.5} (1 - 0) \, dx \right]$

$$= 2 \left(\left[x + \frac{2}{9}(5 - x)^{3/2} \right]_0^5 + \left[x \right]_5^{5.5} \right) = 2 \left(5 - \frac{10\sqrt{5}}{9} + 5.5 - 5 \right) \approx 6.031 \text{ m}^2$$

(b) $V = 2A \approx 2(6.031) \approx 12.062 \text{ m}^3$

(c) $5000 \, V \approx 5000(12.062) = 60,310 \text{ pounds}$

66. (a) $A \approx 6.031 - 2 \left[\pi \left(\frac{1}{16}\right)^2 \right] - 2 \left[\pi \left(\frac{1}{8}\right)^2 \right] \approx 5.908$

(b) $V = 2A \approx 2(5.908) \approx 11.816 \text{ m}^3$

(c) $5000V \approx 5000(11.816) = 59,082 \text{ pounds}$

67. $50 - 0.5x = 0.125x$

$$x = 80$$

$p_1(80) = p_2(80) = 10$

Point of equilibrium: (80, 10)

$$CS = \int_0^{80} [(50 - 0.5x) - 10] \, dx$$

$$= \left[-\frac{0.5x^2}{2} + 40x \right]_0^{80} = 1600$$

$$PS = \int_0^{80} [10 - 0.125x] \, dx$$

$$= \left[10x - \frac{0.125x^2}{2} \right]_0^{80} = 400$$

68. $1000 - 0.4x^2 = 42x$

$$x = 20$$

$p_1(20) = p_2(20) = 840$

Point of equilibrium: (20, 840)

$$CS = \int_0^{20} [(1000 - 0.4x^2) - 840] \, dx$$

$$= \left[160x - \frac{0.4x^3}{3} \right]_0^{20} \approx 2133.33$$

$$PS = \int_0^{20} [840 - 42x] \, dx$$

$$= \left[840x - 21x^2 \right]_0^{20} = 8400$$

69. True

70. True

Section 6.2   Volume: The Disc Method

1. $V = \pi \int_0^1 (-x + 1)^2 \, dx = \pi \int_0^1 (x^2 - 2x + 1) \, dx = \pi \left[\frac{x^3}{3} - x^2 + x \right]_0^1 = \frac{\pi}{3}$

2. $V = \pi \int_0^2 (4 - x^2)^2 \, dx = \pi \int_0^2 (x^4 - 8x^2 + 16) \, dx = \pi \left[\frac{x^5}{5} - \frac{8x^3}{3} + 16x \right]_0^2 = \frac{256\pi}{15}$

3. $V = \pi \int_1^4 (\sqrt{x})^2 \, dx = \pi \int_1^4 x \, dx = \pi \left[\frac{x^2}{2} \right]_1^4 = \frac{15\pi}{2}$

4. $V = \pi \int_0^2 (\sqrt{4 - x^2})^2 \, dx = \pi \int_0^2 (4 - x^2) \, dx = \pi \left[4x - \frac{x^3}{3} \right]_0^2 = \frac{16\pi}{3}$

5. $V = \pi \int_0^1 [(x^2)^2 - (x^3)^2] \, dx = \pi \int_0^1 (x^4 - x^6) \, dx = \pi \left[\frac{x^5}{5} - \frac{x^7}{7} \right]_0^1 = \frac{2\pi}{35}$

6. $2 = 4 - \dfrac{x^2}{4}$

$8 = 16 - x^2$

$x^2 = 8$

$x = \pm 2\sqrt{2}$

$$V = \pi \int_{-2\sqrt{2}}^{2\sqrt{2}} \left[\left(4 - \frac{x^2}{4} \right)^2 - (2)^2 \right] dx$$

$$= 2\pi \int_0^{2\sqrt{2}} \left[\frac{x^4}{16} - 2x^2 + 12 \right] dx$$

$$= 2\pi \left[\frac{x^5}{80} - \frac{2x^3}{3} + 12x \right]_0^{2\sqrt{2}}$$

$$= 2\pi \left[\frac{128\sqrt{2}}{80} - \frac{32\sqrt{2}}{3} + 24\sqrt{2} \right]$$

$$= \frac{448\sqrt{2}}{15} \pi \approx 132.69$$

7. $y = x^2 \implies x = \sqrt{y}$

$$V = \pi \int_0^4 \left(\sqrt{y}\right)^2 dy = \pi \int_0^4 y \, dy$$

$$= \pi \left[\frac{y^2}{2}\right]_0^4 = 8\pi$$

8. $y = \sqrt{16 - x^2} \implies x = \sqrt{16 - y^2}$

$$V = \pi \int_0^4 \left(\sqrt{16 - y^2}\right)^2 dy = \pi \int_0^4 (16 - y^2) \, dy$$

$$= \pi \left[16y - \frac{y^3}{3}\right]_0^4 = \frac{128\pi}{3}$$

9. $y = x^{2/3} \implies x = y^{3/2}$

$$V = \pi \int_0^1 (y^{3/2})^2 \, dy = \pi \int_0^1 y^3 \, dy = \pi \left[\frac{y^4}{4}\right]_0^1 = \frac{\pi}{4}$$

10. $V = \pi \int_1^4 (-y^2 + 4y)^2 \, dy = \pi \int_1^4 (y^4 - 8y^3 + 16y^2) \, dy$

$$= \pi \left[\frac{y^5}{5} - 2y^4 + \frac{16y^3}{3}\right]_1^4$$

$$= \frac{459\pi}{15} = \frac{153\pi}{5}$$

11. $y = \sqrt{x}, \ y = 0, \ x = 4$

(a) $R(x) = \sqrt{x}, \ r(x) = 0$

$$V = \pi \int_0^4 \left(\sqrt{x}\right)^2 dx$$

$$= \pi \int_0^4 x \, dx = \left[\frac{\pi}{2}x^2\right]_0^4 = 8\pi$$

(b) $R(y) = 4, \ r(y) = y^2$

$$V = \pi \int_0^2 (16 - y^4) \, dy$$

$$= \pi \left[16y - \frac{1}{5}y^5\right]_0^2 = \frac{128\pi}{5}$$

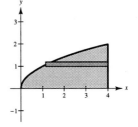

(c) $R(y) = 4 - y^2, \ r(y) = 0$

$$V = \pi \int_0^2 (4 - y^2)^2 \, dy$$

$$= \pi \int_0^2 (16 - 8y^2 + y^4) \, dy$$

$$= \pi \left[16y - \frac{8}{3}y^3 + \frac{1}{5}y^5\right]_0^2 = \frac{256\pi}{15}$$

(d) $R(y) = 6 - y^2, \ r(y) = 2$

$$V = \pi \int_0^2 \left[(6 - y^2)^2 - 4\right] dy$$

$$= \pi \int_0^2 (32 - 12y^2 + y^4) \, dy$$

$$= \pi \left[32y - 4y^3 + \frac{1}{5}y^5\right]_0^2 = \frac{192\pi}{5}$$

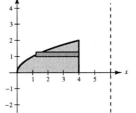

12. $y = 2x^2$, $y = 0$, $x = 2$

(a) $R(y) = 2$, $r(y) = \sqrt{y/2}$

$$V = \pi \int_0^8 \left(4 - \frac{y}{2}\right) dy = \pi \left[4y - \frac{y^2}{4}\right]_0^8 = 16\pi$$

(b) $R(x) = 2x^2$, $r(x) = 0$

$$V = \pi \int_0^2 4x^4 \, dx = \pi \left[\frac{4x^5}{5}\right]_0^2 = \frac{128\pi}{5}$$

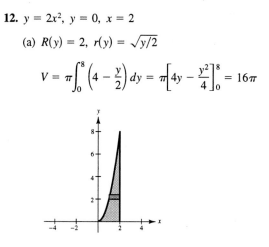

(c) $R(x) = 8$, $r(x) = 8 - 2x^2$

$$V = \pi \int_0^2 \left[64 - (64 - 32x^2 + 4x^4)\right] dx$$

$$= \pi \int_0^2 (32x^2 - 4x^4) \, dx = 4\pi \int_0^2 (8x^2 - x^4) \, dx$$

$$= 4\pi \left[\frac{8}{3}x^3 - \frac{1}{5}x^5\right]_0^2 = \frac{896\pi}{15}$$

(d) $R(y) = 2 - \sqrt{y/2}$, $r(y) = 0$

$$V = \pi \int_0^8 \left(2 - \sqrt{\frac{y}{2}}\right)^2 dy$$

$$= \pi \int_0^8 \left(4 - 4\sqrt{\frac{y}{2}} + \frac{y}{2}\right) dy$$

$$= \pi \left[4y - \frac{4\sqrt{2}}{3}y^{3/2} + \frac{y^2}{4}\right]_0^8 = \frac{16\pi}{3}$$

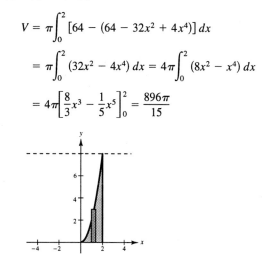

13. $y = x^2$, $y = 4x - x^2$ intersect at $(0, 0)$ and $(2, 4)$.

(a) $R(x) = 4x - x^2$ $r(x) = x^2$

$$V = \pi \int_0^2 \left[(4x - x^2)^2 - x^4\right] dx$$

$$= \pi \int_0^2 (16x^2 - 8x^3) \, dx$$

$$= \pi \left[\frac{16}{3}x^3 - 2x^4\right]_0^2 = \frac{32\pi}{3}$$

(b) $R(x) = 6 - x^2$, $r(x) = 6 - (4x - x^2)$

$$V = \pi \int_0^2 \left[(6 - x^2)^2 - (6 - 4x + x^2)^2\right] dx$$

$$= 8\pi \int_0^2 (x^3 - 5x^2 + 6x) \, dx$$

$$= 8\pi \left[\frac{x^4}{4} - \frac{5}{3}x^3 + 3x^2\right]_0^2 = \frac{64\pi}{3}$$

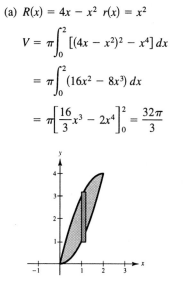

14. $y = 6 - 2x - x^2$, $y = x + 6$ intersect at $(-3, 3)$ and $(0, 6)$.

(a) $R(x) = 6 - 2x - x^2$, $r(x) = x + 6$

$$V = \pi \int_{-3}^{0} [(6 - 2x - x^2)^2 - (x + 6)^2] \, dx$$

$$= \pi \int_{-3}^{0} (x^4 + 4x^3 - 9x^2 - 36x) \, dx$$

$$= \pi \left[\frac{1}{5}x^5 + x^4 - 3x^3 - 18x^2 \right]_{-3}^{0} = \frac{243\pi}{5}$$

(b) $R(x) = (6 - 2x - x^2) - 3$, $r(x) = (x + 6) - 3$

$$V = \pi \int_{-3}^{0} [(3 - 2x - x^2)^2 - (x + 3)^2] \, dx$$

$$= \pi \int_{-3}^{0} (x^4 + 4x^3 - 3x^2 - 18x) \, dx$$

$$= \pi \left[\frac{1}{5}x^5 + x^4 - x^3 - 9x^2 \right]_{-3}^{0} = \frac{108\pi}{5}$$

15. $R(x) = 4 - x$, $r(x) = 1$

$$V = \pi \int_{0}^{3} [(4 - x)^2 - (1)^2] \, dx$$

$$= \pi \int_{0}^{3} (x^2 - 8x + 15) \, dx$$

$$= \pi \left[\frac{x^3}{3} - 4x^2 + 15x \right]_{0}^{3} = 18\pi$$

16. $R(x) = 4 - x^2$, $r(x) = 0$

$$V = 2\pi \int_{0}^{2} (4 - x^2)^2 \, dx$$

$$= 2\pi \int_{0}^{2} (x^4 - 8x^2 + 16) \, dx$$

$$= 2\pi \left[\frac{x^5}{5} - \frac{8x^3}{3} + 16x \right]_{0}^{2}$$

$$= \frac{512\pi}{15}$$

17. $R(x) = 4$, $r(x) = 4 - \dfrac{1}{x}$

$$V = \pi \int_{1}^{4} \left[(4)^2 - \left(4 - \frac{1}{x} \right)^2 \right] dx$$

$$= \pi \int_{1}^{4} \left(\frac{8}{x} - \frac{1}{x^2} \right) dx$$

$$= \pi \left[8 \ln|x| + \frac{1}{x} \right]_{1}^{4}$$

$$= \pi \left(8 \ln 4 - \frac{3}{4} \right) \approx 32.49$$

18. $R(x) = 4$, $r(x) = 4 - \sec x$

$$V = \pi \int_{0}^{\pi/3} [(4)^2 - (4 - \sec x)^2] \, dx$$

$$= \pi \int_{0}^{\pi/3} (8 \sec x - \sec^2 x) \, dx$$

$$= \pi \left[8 \ln|\sec x + \tan x| - \tan x \right]_{0}^{\pi/3}$$

$$= \pi \left[(8 \ln|2 + \sqrt{3}| - \sqrt{3}) - (8 \ln|1 + 0| - 0) \right]$$

$$= \pi \left[8 \ln(2 + \sqrt{3}) - \sqrt{3} \right] \approx 27.66$$

19. $R(y) = 6 - y$, $r(y) = 0$

$$V = \pi \int_0^4 (6 - y)^2 \, dy$$

$$= \pi \int_0^4 (y^2 - 12y + 36) \, dy$$

$$= \pi \left[\frac{y^3}{3} - 6y^2 + 36y \right]_0^4$$

$$= \frac{208\pi}{3}$$

20. $R(y) = 6$, $r(y) = 6 - (6 - y) = y$

$$V = \pi \int_0^4 [(6)^2 - (y)^2] \, dy$$

$$= \pi \left[36y - \frac{y^3}{3} \right]_0^4 = \frac{368\pi}{3}$$

21. $R(y) = 6 - y^2$, $r(y) = 2$

$$V = \pi \int_{-2}^2 [(6 - y^2)^2 - (2)^2] \, dy$$

$$= 2\pi \int_0^2 (y^4 - 12y^2 + 32) \, dy$$

$$= 2\pi \left[\frac{y^5}{5} - 4y^3 + 32y \right]_0^2$$

$$= \frac{384\pi}{5}$$

22. $R(y) = 6 - \dfrac{6}{y}$, $r(y) = 0$

$$V = \pi \int_2^6 \left(6 - \frac{6}{y} \right)^2 dy$$

$$= 36\pi \int_2^6 \left(1 - \frac{2}{y} + \frac{1}{y^2} \right) dy$$

$$= 36\pi \left[y - 2 \ln|y| - \frac{1}{y} \right]_2^6$$

$$= 36\pi \left[\left(\frac{35}{6} - 2 \ln 6 \right) - \left(\frac{3}{2} - 2 \ln 2 \right) \right]$$

$$= 36\pi \left(\frac{13}{3} + 2 \ln \frac{1}{3} \right) = 12\pi(13 - 6 \ln 3) \approx 241.59$$

23. $R(x) = \dfrac{1}{\sqrt{x + 1}}$, $r(x) = 0$

$$V = \pi \int_0^3 \left(\frac{1}{\sqrt{x + 1}} \right)^2 dx$$

$$= \pi \int_0^3 \frac{1}{x + 1} \, dx$$

$$= \left[\pi \ln|x + 1| \right]_0^3 = \pi \ln 4$$

24. $R(x) = x\sqrt{4 - x^2}$, $r(x) = 0$

$$V = 2\pi \int_0^2 \left[x\sqrt{4 - x^2} \right]^2 dx$$

$$= 2\pi \int_0^2 (4x^2 - x^4) \, dx$$

$$= 2\pi \left[\frac{4x^3}{3} - \frac{x^5}{5} \right]_0^2$$

$$= \frac{128\pi}{15}$$

25. $R(x) = \dfrac{1}{x}$, $r(x) = 0$

$$V = \pi \int_1^4 \left(\frac{1}{x} \right)^2 dx$$

$$= \pi \left[-\frac{1}{x} \right]_1^4$$

$$= \frac{3\pi}{4}$$

26. $R(x) = \dfrac{3}{x+1},\ r(x) = 0$

$$V = \pi \int_0^8 \left(\frac{3}{x+1}\right)^2 dx$$

$$= 9\pi \int_0^8 (x+1)^{-2}\, dx$$

$$= 9\pi\left[-\frac{1}{x+1}\right]_0^8 = 8\pi$$

27. $R(x) = e^{-x},\ r(x) = 0$

$$V = \pi \int_0^1 (e^{-x})^2\, dx$$

$$= \pi \int_0^1 e^{-2x}\, dx$$

$$= \left[-\frac{\pi}{2}e^{-2x}\right]_0^1$$

$$= \frac{\pi}{2}(1 - e^{-2}) \approx 1.358$$

28. $R(x) = e^{x/2},\ r(x) = 0$

$$V = \pi \int_0^4 (e^{x/2})^2\, dx$$

$$= \pi \int_0^4 e^x\, dx$$

$$= \left[\pi e^x\right]_0^4$$

$$= \pi(e^4 - 1) \approx 168.38$$

29. $y = 6 - 3x \Rightarrow x = \dfrac{1}{3}(6-y)$

$$V = \pi \int_0^6 \left[\frac{1}{3}(6-y)\right]^2 dy$$

$$= \frac{\pi}{9}\int_0^6 [36 - 12y + y^2]\, dy$$

$$= \frac{\pi}{9}\left[36y - 6y^2 + \frac{y^3}{3}\right]_0^6$$

$$= \frac{\pi}{9}\left[216 - 216 + \frac{216}{3}\right]$$

$$= 8\pi$$

30. $y = 9 - x^2,\ y = 0,\ x = 2,\ x = 3$

$$x = \sqrt{9 - y}$$

$$V = \int_2^5 \left[\left(\sqrt{9-y}\right)^2 - 2^2\right] dy$$

$$= \int_2^5 (5 - y)\, dy$$

$$= \left[5y - \frac{y^2}{2}\right]_2^5$$

$$= \left(25 - \frac{25}{2}\right) - (10 - 2) = \frac{9}{2}$$

31. $V = \pi \displaystyle\int_0^\pi [\sin x]^2\, dx \approx 4.9348$

32. $V = \pi \displaystyle\int_0^{\pi/2} [\cos x]^2\, dx \approx 2.4674$

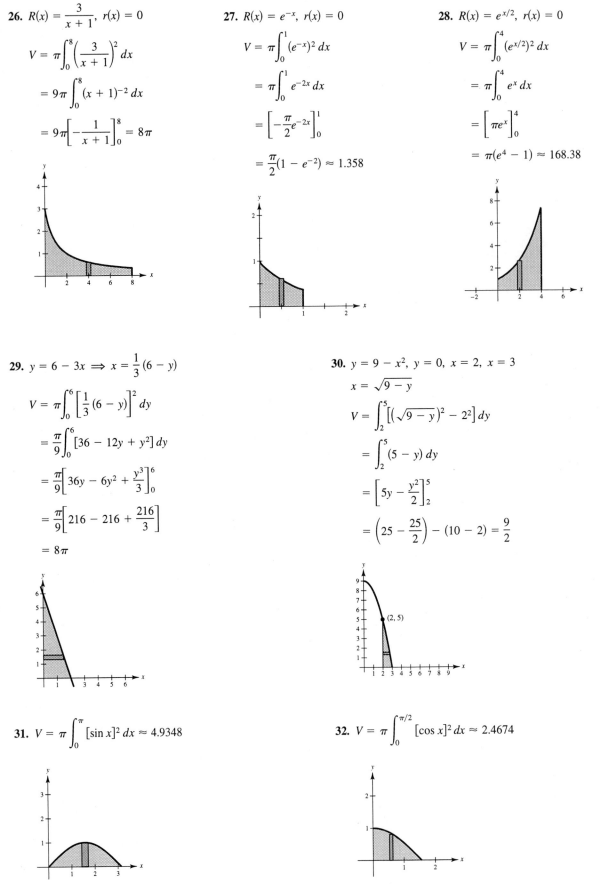

33. $V = \pi \displaystyle\int_0^2 [e^{-x^2}]^2 \, dx \approx 1.9686$

34. $V = \pi \displaystyle\int_1^3 [\ln x]^2 \, dx \approx 3.2332$

35. $V = \pi \displaystyle\int_{-1}^2 [e^{x/2} + e^{-x/2}]^2 \, dx \approx 49.0218$

36. $V = \pi \displaystyle\int_0^5 [2 \arctan (0.2x)]^2 \, dx \approx 15.4115$

37. $A \approx 3$

Matches (a)

38. $A \approx \frac{3}{4}$

Matches (b)

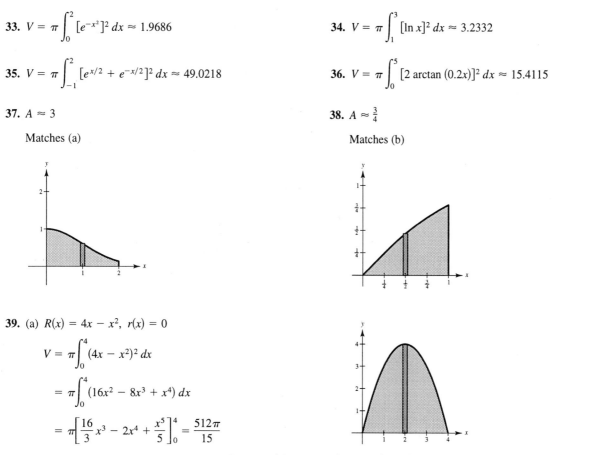

39. (a) $R(x) = 4x - x^2$, $r(x) = 0$

$$V = \pi \int_0^4 (4x - x^2)^2 \, dx$$

$$= \pi \int_0^4 (16x^2 - 8x^3 + x^4) \, dx$$

$$= \pi \left[\frac{16}{3}x^3 - 2x^4 + \frac{x^5}{5} \right]_0^4 = \frac{512\pi}{15}$$

(b) Completing the square we have $4x - x^2 = 4 - (x^2 - 4x + 4) = 4 - (x - 2)^2$. Thus, $y = 4 - x^2$ has the same volume as in part (a) since the solid has been translated only horizontally.

40. (a)   (b)   (c)

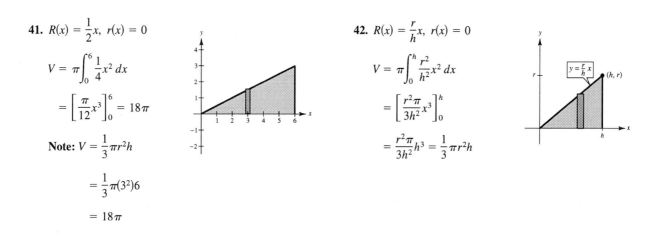

$a < c < b.$

41. $R(x) = \dfrac{1}{2}x$, $r(x) = 0$

$$V = \pi \int_0^6 \frac{1}{4}x^2 \, dx$$

$$= \left[\frac{\pi}{12}x^3 \right]_0^6 = 18\pi$$

Note: $V = \dfrac{1}{3}\pi r^2 h$

$$= \frac{1}{3}\pi(3^2)6$$

$$= 18\pi$$

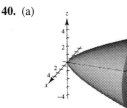

42. $R(x) = \dfrac{r}{h}x$, $r(x) = 0$

$$V = \pi \int_0^h \frac{r^2}{h^2}x^2 \, dx$$

$$= \left[\frac{r^2\pi}{3h^2}x^3 \right]_0^h$$

$$= \frac{r^2\pi}{3h^2}h^3 = \frac{1}{3}\pi r^2 h$$

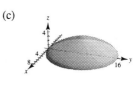

43. $R(x) = \sqrt{r^2 - x^2}, \ r(x) = 0$

$$V = \pi \int_{-r}^{r} (r^2 - x^2) \, dx$$

$$= 2\pi \int_{0}^{r} (r^2 - x^2) \, dx$$

$$= 2\pi \left[r^2 x - \frac{1}{3} x^3 \right]_0^r$$

$$= 2\pi \left(r^3 - \frac{1}{3} r^3 \right) = \frac{4}{3} \pi r^3$$

$y = \sqrt{r^2 - x^2}$

$(-r, 0)$ $(r, 0)$

44. $x = \sqrt{r^2 - y^2}, \ R(y) = \sqrt{r^2 - y^2}, \ r(y) = 0$

$$V = \pi \int_{h}^{r} \left(\sqrt{r^2 - y^2} \right)^2 dy$$

$$= \pi \int_{h}^{r} (r^2 - y^2) \, dy$$

$$= \pi \left[r^2 y - \frac{y^3}{3} \right]_h^r$$

$$= \pi \left[\left(r^3 - \frac{r^3}{3} \right) - \left(r^2 h - \frac{h^3}{3} \right) \right]$$

$$= \pi \left(\frac{2r^3}{3} - r^2 h + \frac{h^3}{3} \right)$$

$$= \frac{\pi}{3} (2r^3 - 3r^2 h + h^3)$$

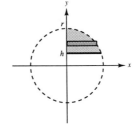

45. $x = r - \frac{r}{H} y = r \left(1 - \frac{y}{H} \right), \ R(y) = r \left(1 - \frac{y}{H} \right), \ r(y) = 0$

$$V = \pi \int_{0}^{h} \left[r \left(1 - \frac{y}{H} \right) \right]^2 dy = \pi r^2 \int_{0}^{h} \left(1 - \frac{2}{H} y + \frac{1}{H^2} y^2 \right) dy$$

$$= \pi r^2 \left[y - \frac{1}{H} y^2 + \frac{1}{3H^2} y^3 \right]_0^h$$

$$= \pi r^2 \left(h - \frac{h^2}{H} + \frac{h^3}{3H^2} \right)$$

$$= \pi r^2 h \left(1 - \frac{h}{H} + \frac{h^2}{3H^2} \right)$$

46. (a) $V = \pi \int_{0}^{4} \left(\sqrt{x} \right)^2 dx = \pi \int_{0}^{4} x \, dx = \left[\frac{\pi x^2}{2} \right]_0^4 = 8\pi$

Let $0 < c < 4$ and set

$$\pi \int_{0}^{c} x \, dx = \left[\frac{\pi x^2}{2} \right]_0^c = \frac{\pi c^2}{2} = 4\pi.$$

$$c^2 = 8$$

$$c = \sqrt{8} = 2\sqrt{2}$$

Thus, when $x = 2\sqrt{2}$, the solid is divided into two parts of equal volume.

(b) Set $\pi \int_{0}^{c} x \, dx = \frac{8\pi}{3}$ (one third of the volume). Then

$$\frac{\pi c^2}{2} = \frac{8\pi}{3}, \ c^2 = \frac{16}{3}, \ c = \frac{4}{\sqrt{3}} = \frac{4\sqrt{3}}{3}.$$

To find the other value, set $\pi \int_{0}^{d} x \, dx = \frac{16\pi}{3}$ (two thirds of the volume). Then

$$\frac{\pi d^2}{2} = \frac{16\pi}{3}, \ d^2 = \frac{32}{3}, \ d = \frac{\sqrt{32}}{\sqrt{3}} = \frac{4\sqrt{6}}{3}.$$

The x-values that divide the solid into three parts of equal volume are $x = \left(4\sqrt{3} \right)/3$ and $x = \left(4\sqrt{6} \right)/3$.

47. $V = \pi \int_0^2 \left(\frac{1}{8} x^2 \sqrt{2-x}\right)^2 dx = \frac{\pi}{64} \int_0^2 x^4 (2-x)\, dx = \frac{\pi}{64} \left[\frac{2x^5}{5} - \frac{x^6}{6}\right]_0^2 = \frac{\pi}{30}$

48. $y = \begin{cases} \sqrt{0.1x^3 - 2.2x^2 + 10.9x + 22.2}, & 0 \le x \le 11.5 \\ 2.95, & 11.5 < x \le 15 \end{cases}$

$V = \pi \int_0^{11.5} \left(\sqrt{0.1x^3 - 2.2x^2 + 10.9x + 22.2}\right)^2 dx + \pi \int_{11.5}^{15} 2.95^2\, dx$

$= \pi \left[\frac{0.1x^4}{4} - \frac{2.2x^3}{3} + \frac{10.9x^2}{2} + 22.2x\right]_0^{11.5} + \pi \left[2.95^2 x\right]_{11.5}^{15}$

≈ 1031.9016 cubic centimeters

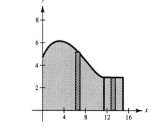

49. (a) $R(x) = \frac{3}{5}\sqrt{25 - x^2}, \ r(x) = 0$

$V = \frac{9\pi}{25} \int_{-5}^{5} (25 - x^2)\, dx$

$= \frac{18\pi}{25} \int_0^5 (25 - x^2)\, dx$

$= \frac{18\pi}{25} \left[25x - \frac{x^3}{3}\right]_0^5 = 60\pi$

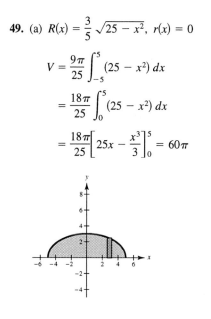

(b) $R(y) = \frac{5}{3}\sqrt{9 - y^2}, \ r(y) = 0, \ x \ge 0$

$V = \frac{25\pi}{9} \int_0^3 (9 - y^2)\, dy$

$= \frac{25\pi}{9} \left[9y - \frac{y^3}{3}\right]_0^3 = 50\pi$

50. (a) First find where $y = b$ intersects the parabola:

$b = 4 - \frac{x^2}{4}$

$x^2 = 16 - 4b = 4(4 - b)$

$x = 2\sqrt{4 - b}$

$V = \int_0^{2\sqrt{4-b}} \pi \left[4 - \frac{x^2}{4} - b\right]^2 dx + \int_{2\sqrt{4-b}}^{4} \pi \left[b - 4 + \frac{x^2}{4}\right]^2 dx$

$= \int_0^4 \pi \left[4 - \frac{x^2}{4} - b\right]^2 dx$

$= \pi \int_0^4 \left[\frac{x^4}{16} - 2x^2 + \frac{bx^2}{2} + b^2 - 8b + 16\right] dx$

$= \pi \left[\frac{x^5}{80} - \frac{2x^3}{3} + \frac{bx^3}{6} + b^2 x - 8bx + 16x\right]_0^4$

$= \pi \left[\frac{64}{5} - \frac{128}{3} + \frac{32}{3} b + 4b^2 - 32b + 64\right]$

$= \pi \left[4b^2 - \frac{64}{3}b + \frac{512}{15}\right]$

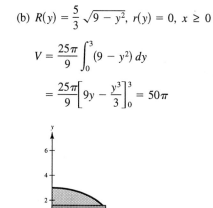

—CONTINUED—

50. —CONTINUED—

(b) graph of $V(b) = \pi\left[4b^2 - \dfrac{64}{3}b + \dfrac{512}{15}\right]$

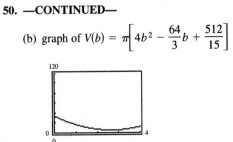

Minimum Volume is 17.87 for $b = 2.67$

(c) $V'(b) = \pi\left[8b - \dfrac{64}{3}\right] = 0 \implies b = \dfrac{64/3}{8} = \dfrac{8}{3} = 2\frac{2}{3}$

$V''(b) = 8\pi > 0 \implies b = \dfrac{8}{3}$ is a relative minimum.

51. Total volume: $V = \dfrac{4\pi(50)^3}{3} = \dfrac{500{,}000\,\pi}{3}\ \text{ft}^3$

Volume of water in the tank:

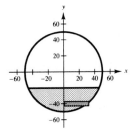

$$\pi\int_{-50}^{y_0}\left(\sqrt{2500 - y^2}\right)^2 dy = \pi\int_{-50}^{y_0}(2500 - y^2)\,dy$$

$$= \pi\left[2500y - \frac{y^3}{3}\right]_{-50}^{y_0}$$

$$= \pi\left(2500y_0 - \frac{y_0^3}{3} + \frac{250{,}000}{3}\right)$$

When the tank is one-fourth of its capacity:

$$\frac{1}{4}\left(\frac{500{,}000\pi}{3}\right) = \pi\left(2500y_0 - \frac{y_0^3}{3} + \frac{250{,}000}{3}\right)$$
$$125{,}000 = 7500y_0 - y_0^3 + 250{,}000$$

$$y_0^3 - 7500y_0 - 125{,}000 = 0$$

$$y_0 \approx -17.36$$

Depth: $-17.36 - (-50) = 32.64$ feet

When the tank is three-fourths of its capacity the depth is $100 - 32.64 = 67.36$ feet.

52. (a) $V = \displaystyle\int_0^{10} \pi[f(x)]^2\,dx$

Simpson's Rule: $b - a = 10 - 0 = 10$, $\quad n = 10$

$V \approx \dfrac{\pi}{3}[(2.1)^2 + 4(1.9)^2 + 2(2.1)^2 + 4(2.35)^2 + 2(2.6)^2 + 4(2.85)^2 + 2(2.9)^2 + 4(2.7)^2 + 2(2.45)^2 + 4(2.2)^2 + (2.3)^2]$

$\approx \dfrac{\pi}{3}[178.405] \approx 186.83\ \text{cm}^3$

(b) $f(x) = 0.00249x^4 - 0.0529x^3 + 0.3314x^2 - 0.4999x + 2.112$

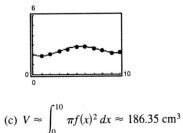

(c) $V \approx \displaystyle\int_0^{10} \pi f(x)^2\,dx \approx 186.35\ \text{cm}^3$

53. (a) $\pi \displaystyle\int_0^h r^2 \, dx$ **(ii)**

is the volume of a right circular cylinder with radius r and height h.

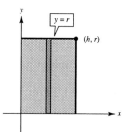

(b) $\pi \displaystyle\int_{-b}^b \left(a\sqrt{1 - \dfrac{x^2}{b^2}}\right)^2 dx$ **(iv)**

is the volume of an ellipsoid with axes $2a$ and $2b$.

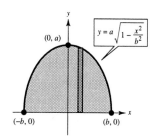

(c) $\pi \displaystyle\int_{-r}^r \left(\sqrt{r^2 - x^2}\right)^2 dx$ **(iii)**

is the volume of a sphere with radius r.

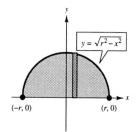

(d) $\pi \displaystyle\int_0^h \left(\dfrac{rx}{h}\right)^2 dx$ **(i)**

is the volume of a right circular cone with the radius of the base as r and height h.

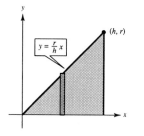

(e) $\pi \displaystyle\int_{-r}^r \left[\left(R + \sqrt{r^2 - x^2}\right)^2 - \left(R - \sqrt{r^2 - x^2}\right)^2\right] dx$ **(v)**

is the volume of a torus with the radius of its circular cross section as r and the distance from the axis of the torus to the center of its cross section as R.

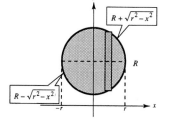

54. $V = \frac{1}{2}(8)(1)(2) = 8 \text{ m}^3$

55.

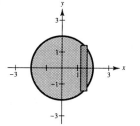

Base of Cross Section $= 2\sqrt{4 - x^2}$

(a) $A(x) = b^2 = \left(2\sqrt{4 - x^2}\right)^2$

$$V = \int_{-2}^2 4(4 - x^2) \, dx$$

$$= 4\left[4x - \frac{x^3}{3}\right]_{-2}^2 = \frac{128}{3}$$

(b) $A(x) = \dfrac{1}{2}bh = \dfrac{1}{2}\left(2\sqrt{4 - x^2}\right)\left(\sqrt{3}\sqrt{4 - x^2}\right)$

$$= \sqrt{3}\,(4 - x^2)$$

$$V = \sqrt{3}\int_{-2}^2 (4 - x^2) \, dx$$

$$= \sqrt{3}\left[4x - \frac{x^3}{3}\right]_{-2}^2 = \frac{32\sqrt{3}}{3}$$

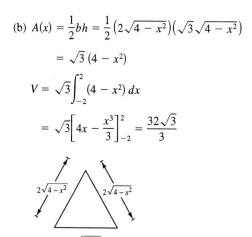

—CONTINUED—

55. —CONTINUED—

(c) $A(x) = \dfrac{1}{2}\pi r^2$

$$= \dfrac{\pi}{2}\left(\sqrt{4-x^2}\right)^2 = \dfrac{\pi}{2}(4-x^2)$$

$$V = \dfrac{\pi}{2}\int_{-2}^{2}(4-x^2)\,dx = \dfrac{\pi}{2}\left[4x - \dfrac{x^3}{3}\right]_{-2}^{2} = \dfrac{16\pi}{3}$$

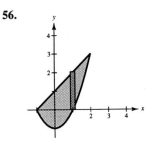

$2\sqrt{4-x^2}$

(d) $A(x) = \dfrac{1}{2}bh$

$$= \dfrac{1}{2}\left(2\sqrt{4-x^2}\right)\left(\sqrt{4-x^2}\right) = 4 - x^2$$

$$V = \int_{-2}^{2}(4-x^2)\,dx = \left[4x - \dfrac{x^3}{3}\right]_{-2}^{2} = \dfrac{32}{3}$$

$\sqrt{4-x^2}$

$2\sqrt{4-x^2}$

56.

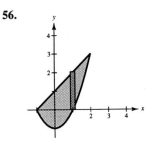

Base of Cross Section $= (x+1) - (x^2 - 1) = 2 + x - x^2$

(a) $A(x) = b^2 = (2 + x - x^2)^2$

$$= 4 + 4x - 3x^2 - 2x^3 + x^4$$

$$V = \int_{-1}^{2}(4 + 4x - 3x^2 - 2x^3 + x^4)\,dx$$

$$= \left[4x + 2x^3 - x^3 - \dfrac{1}{2}x^4 + \dfrac{1}{5}x^5\right]_{-1}^{2} = \dfrac{81}{10}$$

$2+x-x^2$

$2+x-x^2$

(b) $A(x) = bh = (2 + x - x^2)1$

$$V = \int_{-1}^{2}(2 + x - x^2)\,dx = \left[2x + \dfrac{x^2}{2} - \dfrac{x^3}{3}\right]_{-1}^{2} = \dfrac{9}{2}$$

1

$2+x-x^2$

57.

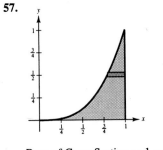

Base of Cross Section $= 1 - \sqrt[3]{y}$

—CONTINUED—

(a) $A(y) = b^2 = \left(1 - \sqrt[3]{y}\right)^2$

$$V = \int_{0}^{1}\left(1 - \sqrt[3]{y}\right)^2\,dy$$

$$= \int_{0}^{1}(1 - 2y^{1/3} + y^{2/3})\,dy$$

$$= \left[y - \dfrac{3}{2}y^{4/3} + \dfrac{3}{5}y^{5/3}\right]_{0}^{1} = \dfrac{1}{10}$$

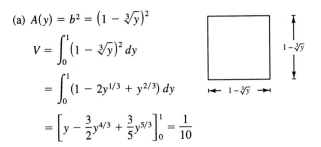

$1-\sqrt[3]{y}$

$1-\sqrt[3]{y}$

57. —CONTINUED—

(b) $A(y) = \frac{1}{2}\pi r^2 = \frac{1}{2}\pi\left(\frac{1 - \sqrt[3]{y}}{2}\right)^2 = \frac{1}{8}\pi\left(1 - \sqrt[3]{y}\right)^2$

$V = \frac{1}{8}\pi\int_0^1 \left(1 - \sqrt[3]{y}\right)^2 dy = \frac{\pi}{8}\left(\frac{1}{10}\right) = \frac{\pi}{80}$

(c) $A(y) = \frac{1}{2}bh = \frac{1}{2}(1 - \sqrt[3]{y})\left(\frac{\sqrt{3}}{2}\right)(1 - \sqrt[3]{y})$

$\qquad = \frac{\sqrt{3}}{4}(1 - \sqrt[3]{y})^2$

$V = \frac{\sqrt{3}}{4}\int_0^1 (1 - \sqrt[3]{y})^2 dy = \frac{\sqrt{3}}{4}\left(\frac{1}{10}\right) = \frac{\sqrt{3}}{40}$

(d) $A(y) = \frac{1}{2}\pi ab = \frac{\pi}{2}(2)(1 - \sqrt[3]{y})\frac{1 - \sqrt[3]{y}}{2}$

$\qquad = \frac{\pi}{2}(1 - \sqrt[3]{y})^2$

$V = \frac{\pi}{2}\int_0^1 (1 - \sqrt[3]{y})^2 dy = \frac{\pi}{2}\left(\frac{1}{10}\right) = \frac{\pi}{20}$

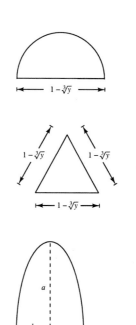

58. The cross sections are squares. By symmetry, we can set up an integral for an eighth of the volume and multiply by 8.

$A(y) = b^2 = \left(\sqrt{r^2 - y^2}\right)^2$

$V = 8\int_0^r (r^2 - y^2)\, dy$

$\quad = 8\left[r^2 y - \frac{1}{3}y^3\right]_0^r$

$\quad = \frac{16}{3}r^3$

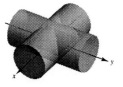

59. Assume that the oil just hits the top of the cylinder, which means that the volume of oil is one-half of the volume of the cylinder. Then we have

$\sin 20° = \frac{d \sin 70°}{h}$

$h = \frac{d \sin 70°}{\sin 20°}$

$\text{Volume} = \frac{1}{2}(\pi r^2 h)$

$\qquad = \frac{\pi}{2}\left(\frac{d}{2}\right)^2\left(\frac{d \sin 70°}{\sin 20°}\right)$

$\qquad = \frac{\pi d^3 \sin 70°}{8 \sin 20°} = \pi d^3 \cot 20°.$

60. Let $A_1(x)$ and $A_2(x)$ equal the areas of the cross sections of the two solids for $a \le x \le b$. Since $A_1(x) = A_2(x)$, we have

$V_1 = \int_a^b A_1(x)\, dx = \int_a^b A_2(x)\, dx = V_2$

Thus, the volumes are the same.

61. (a) Since the cross sections are isosceles right triangles:

$$A(x) = \frac{1}{2}bh = \frac{1}{2}\left(\sqrt{r^2 - y^2}\right)\left(\sqrt{r^2 - y^2}\right) = \frac{1}{2}(r^2 - y^2)$$

$$V = \frac{1}{2}\int_{-r}^{r}(r^2 - y^2)\,dy = \int_{0}^{r}(r^2 - y^2)\,dy = \left[r^2 y - \frac{y^3}{3}\right]_{0}^{r} = \frac{2}{3}r^3$$

(b) $A(x) = \frac{1}{2}bh = \frac{1}{2}\sqrt{r^2 - y^2}\left(\sqrt{r^2 - y^2}\tan\theta\right) = \frac{\tan\theta}{2}(r^2 - y^2)$

$$V = \frac{\tan\theta}{2}\int_{-r}^{r}(r^2 - y^2)\,dy = \tan\theta\int_{0}^{r}(r^2 - y^2)\,dy = \tan\theta\left[r^2 y - \frac{y^3}{3}\right]_{0}^{r} = \frac{2}{3}r^3\tan\theta$$

As $\theta \to 90°$, $V \to \infty$.

62. $V = \pi\int_{-\sqrt{R^2-r^2}}^{\sqrt{R^2-r^2}}\left[\left(\sqrt{R^2 - x^2}\right)^2 - r^2\right]dx$

$= 2\pi\int_{0}^{\sqrt{R^2-r^2}}(R^2 - r^2 - x^2)\,dx$

$= 2\pi\left[(R^2 - r^2)x - \frac{x^3}{3}\right]_{0}^{\sqrt{R^2-r^2}}$

$= 2\pi\left[(R^2 - r^2)^{3/2} - \frac{(R^2 - r^2)^{3/2}}{3}\right]$

$= \frac{4}{3}\pi(R^2 - r^2)^{3/2}$

63. $\frac{4}{3}\pi(25 - r^2)^{3/2} = \frac{1}{2}\left(\frac{4}{3}\right)\pi(125)$

$(25 - r^2)^{3/2} = \frac{125}{2}$

$25 - r^2 = \left(\frac{125}{2}\right)^{2/3}$

$25 - \frac{25}{(2^{2/3})} = r^2$

$25(1 - 2^{-2/3}) = r^2$

$r = 5\sqrt{1 - 2^{-2/3}} \approx 3.0415$

64. (a) When $a = 1$: $|x| + |y| = 1$ represents a square.

When $a = 2$: $|x|^2 + |y|^2 = 1$ represents a circle.

(b) $|y| = (1 - |x|^a)^{1/a}$

$$A = 2\int_{-1}^{1}(1 - |x|^a)^{1/a}\,dx = 4\int_{0}^{1}(1 - x^a)^{1/a}\,dx$$

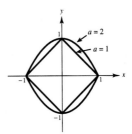

To approximate the volume of the solid, form n slices, each of whose area is approximated by the integral above. Then sum the volumes of these n slices.

Section 6.3   Volume: The Shell Method

1. $p(x) = x$

$h(x) = x$

$$V = 2\pi\int_{0}^{2}x(x)\,dx = \left[\frac{2\pi x^3}{3}\right]_{0}^{2} = \frac{16\pi}{3}$$

2. $p(x) = x$

$h(x) = 1 - x$

$$V = 2\pi\int_{0}^{1}x(1 - x)\,dx$$

$$= 2\pi\int_{0}^{1}(x - x^2)\,dx = 2\pi\left[\frac{x^2}{2} - \frac{x^3}{3}\right]_{0}^{1} = \frac{\pi}{3}$$

3. $p(x) = x$

$h(x) = \sqrt{x}$

$V = 2\pi \int_0^4 x\sqrt{x}\, dx$

$= 2\pi \int_0^4 x^{3/2}\, dx$

$= \left[\dfrac{4\pi}{5} x^{5/2}\right]_0^4 = \dfrac{128\pi}{5}$

4. $p(x) = x$

$h(x) = 8 - (x^2 + 4) = 4 - x^2$

$V = 2\pi \int_0^2 x(4 - x^2)\, dx$

$= 2\pi \int_0^2 (4x - x^3)\, dx$

$= 2\pi \left[2x^2 - \dfrac{x^4}{4}\right]_0^2 = 8\pi$

5. $p(x) = x$

$h(x) = x^2$

$V = 2\pi \int_0^2 x^3\, dx$

$= \left[\dfrac{\pi}{2} x^4\right]_0^2 = 8\pi$

6. $p(x) = x$

$h(x) = x^2$

$V = 2\pi \int_0^4 x^3\, dx$

$= \left[\dfrac{\pi}{2} x^4\right]_0^4 = 128\pi$

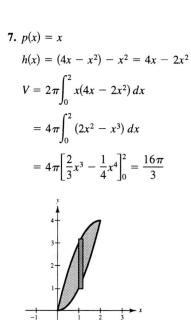

7. $p(x) = x$

$h(x) = (4x - x^2) - x^2 = 4x - 2x^2$

$V = 2\pi \int_0^2 x(4x - 2x^2)\, dx$

$= 4\pi \int_0^2 (2x^2 - x^3)\, dx$

$= 4\pi \left[\dfrac{2}{3} x^3 - \dfrac{1}{4} x^4\right]_0^2 = \dfrac{16\pi}{3}$

8. $p(x) = x$

$h(x) = 4 - x^2$

$V = 2\pi \int_0^2 (4x - x^3)\, dx$

$= 2\pi \left[2x^2 - \dfrac{1}{4} x^4\right]_0^2 = 8\pi$

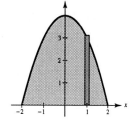

9. $p(x) = x$

$h(x) = 4 - (4x - x^2) = x^2 - 4x + 4$

$V = 2\pi \int_0^2 (x^3 - 4x^2 + 4x)\, dx$

$= 2\pi \left[\dfrac{x^4}{4} - \dfrac{4}{3}x^3 + 2x^2 \right]_0^2 = \dfrac{8\pi}{3}$

10. $p(x) = x$

$h(x) = 4 - 2x$

$V = 2\pi \int_0^2 x(4 - 2x)\, dx$

$= 2\pi \int_0^2 (4x - 2x^2)\, dx$

$= 2\pi \left[2x^2 - \dfrac{2}{3}x^3 \right]_0^2 = \dfrac{16\pi}{3}$

11. $p(x) = x$

$h(x) = \dfrac{1}{\sqrt{2\pi}} e^{-x^2/2}$

$V = 2\pi \int_0^1 x\left(\dfrac{1}{\sqrt{2\pi}} e^{-x^2/2} \right) dx$

$= \sqrt{2\pi} \int_0^1 e^{-x^2/2} x\, dx$

$= \left[-\sqrt{2\pi}\, e^{-x^2/2} \right]_0^1 = \sqrt{2\pi}\left(1 - \dfrac{1}{\sqrt{e}} \right) \approx 0.986$

12. $p(x) = x$

$h(x) = \dfrac{\sin x}{x}$

$V = 2\pi \int_0^\pi x\left[\dfrac{\sin x}{x} \right] dx$

$= 2\pi \int_0^\pi \sin x\, dx = \left[-2\pi \cos x \right]_0^\pi = 4\pi$

13. $p(y) = y$

$h(y) = 2 - y$

$V = 2\pi \int_0^2 y(2 - y)\, dy$

$= 2\pi \int_0^2 (2y - y^2)\, dy$

$= 2\pi \left[y^2 - \dfrac{y^3}{3} \right]_0^2 = \dfrac{8\pi}{3}$

14. $p(y) = -y \quad (p(y) \geq 0 \text{ on } [-2, 0])$

$h(y) = 4 - (2 - y) = 2 + y$

$V = 2\pi \int_{-2}^0 (-y)(2 + y)\, dy$

$= 2\pi \int_{-2}^0 (-2y - y^2)\, dy$

$= 2\pi \left[-y^2 - \dfrac{y^3}{3} \right]_{-2}^0 = \dfrac{8\pi}{3}$

15. $p(y) = y$ and $h(y) = 1$ if $0 \le y < \dfrac{1}{2}$.

$p(y) = y$ and $h(y) = \dfrac{1}{y} - 1$ if $\dfrac{1}{2} \le y \le 1$.

$$V = 2\pi \int_0^{1/2} y\,dy + 2\pi \int_{1/2}^1 (1 - y)\,dy$$

$$= 2\pi \left[\frac{y^2}{2}\right]_0^{1/2} + 2\pi \left[y - \frac{y^2}{2}\right]_{1/2}^1 = \frac{\pi}{4} + \frac{\pi}{4} = \frac{\pi}{2}$$

16. $p(y) = y$

$h(y) = 9 - y^2$

$$V = 2\pi \int_0^3 y(9 - y^2)\,dy$$

$$= 2\pi \int_0^3 (9y - y^3)\,dy$$

$$= 2\pi \left[\frac{9}{2}y^2 - \frac{1}{4}y^4\right]_0^3 = \frac{81\pi}{2}$$

17. $p(x) = 4 - x$

$h(x) = 4x - x^2 - x^2 = 4x - 2x^2$

$$V = 2\pi \int_0^2 (4 - x)(4x - 2x^2)\,dx$$

$$= 2\pi(2) \int_0^2 (x^3 - 6x^2 + 8x)\,dx$$

$$= 4\pi \left[\frac{x^4}{4} - 2x^3 + 4x^2\right]_0^2 = 16\pi$$

18. $p(x) = 2 - x$

$h(x) = 4x - x^2 - x^2 = 4x - 2x^2$

$$V = 2\pi \int_0^2 (2 - x)(4x - 2x^2)\,dx$$

$$= 2\pi \int_0^2 (8x - 8x^2 + 2x^3)\,dx$$

$$= 2\pi \left[4x^2 - \frac{8}{3}x^3 + \frac{1}{2}x^4\right]_0^2 = \frac{16\pi}{3}$$

19. $p(x) = 5 - x$

$h(x) = 4x - x^2$

$$V = 2\pi \int_0^4 (5 - x)(4x - x^2)\,dx$$

$$= 2\pi \int_0^4 (x^3 - 9x^2 + 20x)\,dx$$

$$= 2\pi \left[\frac{x^4}{4} - 3x^3 + 10x^2\right]_0^4 = 64\pi$$

20. $p(x) = 6 - x$

$h(x) = \sqrt{x}$

$$V = 2\pi \int_0^4 (6 - x)\sqrt{x}\,dx$$

$$= 2\pi \int_0^4 (6x^{1/2} - x^{3/2})\,dx$$

$$= 2\pi \left[4x^{3/2} - \frac{2}{5}x^{5/2}\right]_0^4 = \frac{192\pi}{5}$$

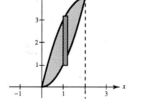

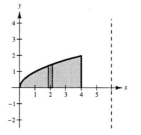

21. (a) **Disc**

$$R(x) = x^3$$

$$r(x) = 0$$

$$V = \pi \int_0^2 x^6 \, dx = \pi \left[\frac{x^7}{7}\right]_0^2 = \frac{128\pi}{7}$$

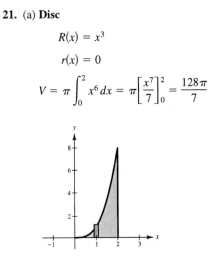

(b) **Shell**

$$p(x) = x$$

$$h(x) = x^3$$

$$V = 2\pi \int_0^2 x^4 \, dx = 2\pi \left[\frac{x^5}{5}\right]_0^2 = \frac{64\pi}{5}$$

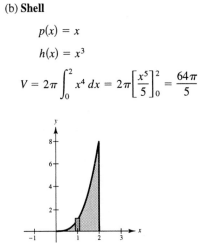

(c) **Shell**

$$p(x) = 4 - x$$

$$h(x) = x^3$$

$$V = 2\pi \int_0^2 (4 - x)x^3 \, dx$$

$$= 2\pi \int_0^2 (4x^3 - x^4) \, dx$$

$$= 2\pi \left[x^4 - \frac{1}{5}x^5\right]_0^2 = \frac{96\pi}{5}$$

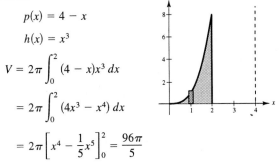

22. (a) **Disc**

$$R(x) = \frac{1}{x^2}$$

$$r(x) = 0$$

$$V = \pi \int_1^4 \left(\frac{1}{x^2}\right)^2 \, dx$$

$$= \pi \left[-\frac{1}{3x^3}\right]_1^4 = \frac{21\pi}{64}$$

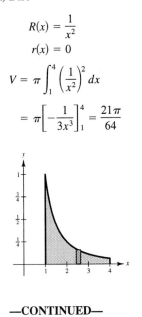

—CONTINUED—

(b) **Shell**

$$p(x) = x$$

$$h(x) = \frac{1}{x^2}$$

$$V = 2\pi \int_1^4 x\left(\frac{1}{x^2}\right) \, dx$$

$$= 2\pi \int_1^4 \frac{1}{x} \, dx$$

$$= 2\pi \left[\ln|x|\right]_1^4 = 2\pi \ln 4$$

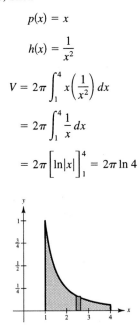

22. **—CONTINUED—**

(c) **Disc**

$$R(x) = 1$$

$$r(x) = 1 - \frac{1}{x^2}$$

$$V = \pi \int_1^4 \left[1 - \left(1 - \frac{1}{x^2} \right)^2 \right] dx$$

$$= \pi \int_1^4 \left(\frac{2}{x^2} - \frac{1}{x^4} \right) dx$$

$$= \pi \left[-\frac{2}{x} + \frac{1}{3x^3} \right]_1^4 = \frac{225\pi}{192}$$

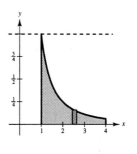

23. (a) **Shell**

$$p(y) = y$$

$$h(y) = (a^{1/2} - y^{1/2})^2$$

$$V = 2\pi \int_0^a y(a - 2a^{1/2}y^{1/2} + y) \, dy$$

$$= 2\pi \int_0^a (ay - 2a^{1/2}y^{3/2} + y^2) \, dy$$

$$= 2\pi \left[\frac{a}{2}y^2 - \frac{4a^{1/2}}{5}y^{5/2} + \frac{y^3}{3} \right]_0^a$$

$$= 2\pi \left[\frac{a^3}{2} - \frac{4a^3}{5} + \frac{a^3}{3} \right] = \frac{\pi a^3}{15}$$

(b) Same as part a by symmetry

(c) **Shell**

$$p(x) = a - x$$

$$h(x) = (a^{1/2} - x^{1/2})^2$$

$$V = 2\pi \int_0^a (a - x)(a^{1/2} - x^{1/2})^2 \, dx$$

$$= 2\pi \int_0^a (a^2 - 2a^{3/2}x^{1/2} + 2a^{1/2}x^{3/2} - x^2) \, dx$$

$$= 2\pi \left[a^2x - \frac{4}{3}a^{3/2}x^{3/2} + \frac{4}{5}a^{1/2}x^{5/2} - \frac{1}{3}x^3 \right]_0^a = \frac{4\pi a^3}{15}$$

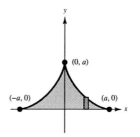

24. (a) **Disc**

$$R(x) = (a^{2/3} - x^{2/3})^{3/2}$$

$$r(x) = 0$$

$$V = \pi \int_{-a}^a (a^{2/3} - x^{2/3})^3 \, dx$$

$$= 2\pi \int_0^a (a^2 - 3a^{4/3}x^{2/3} + 3a^{2/3}x^{4/3} - x^2) \, dx$$

$$= 2\pi \left[a^2x - \frac{9}{5}a^{4/3}x^{5/3} + \frac{9}{7}a^{2/3}x^{7/3} - \frac{1}{3}x^3 \right]_0^a$$

$$= 2\pi \left(a^3 - \frac{9}{5}a^3 + \frac{9}{7}a^3 - \frac{1}{3}a^3 \right) = \frac{32\pi a^3}{105}$$

—CONTINUED—

24. —CONTINUED—

(b) Same as part a by symmetry

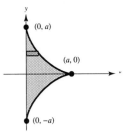

25. (a)

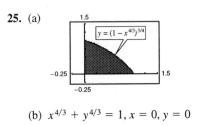

(b) $x^{4/3} + y^{4/3} = 1$, $x = 0$, $y = 0$

$$y = (1 - x^{4/3})^{3/4}$$

$$V = 2\pi \int_0^1 x(1 - x^{4/3})^{3/4} \, dx \approx 1.5056$$

26. (a)

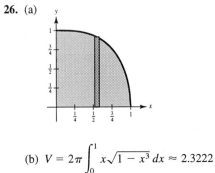

(b) $V = 2\pi \int_0^1 x\sqrt{1 - x^3} \, dx \approx 2.3222$

27. (a)

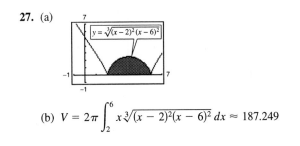

(b) $V = 2\pi \int_2^6 x\sqrt[3]{(x - 2)^2(x - 6)^2} \, dx \approx 187.249$

28. (a)

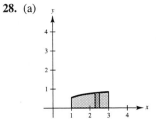

(b) $V = 2\pi \int_1^3 \dfrac{2x}{1 + e^{1/x}} \, dx \approx 19.0162$

29. (a)

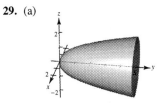

(b)

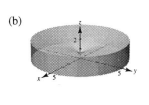

(c)

$a < c < b$

30. (a) $V = \displaystyle\int_0^5 \pi(x^{2/5})^2 \, dx = \left[\dfrac{5\pi x^{9/5}}{9}\right]_0^5 \approx 31.6$

(b) $V = \displaystyle\int_0^5 2\pi x(x^{2/5}) \, dx = \int_0^5 2\pi x^{7/5} \, dx = \left[2\pi x^{12/5}\left(\dfrac{5}{12}\right)\right]_0^5 = \dfrac{5\pi}{6}x^{12/5} \approx 124.6$

(c) $V = \displaystyle\int_0^5 2\pi(5 - x)x^{2/5} \, dx = \int_0^5 2\pi(5x^{2/5} - x^{7/5}) \, dx = 2\pi\left[5x^{7/5}\left(\dfrac{5}{7}\right) - x^{12/5}\left(\dfrac{5}{12}\right)\right]_0^5 \approx 89.0$

31. $y = 2e^{-x}$, $y = 0$, $x = 0$, $x = 2$

Volume ≈ 7.5

Matches (d)

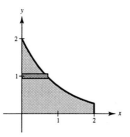

32. $y = \tan x$, $y = 0$, $x = 0$, $x = \dfrac{\pi}{4}$

Volume ≈ 1

Matches (e)

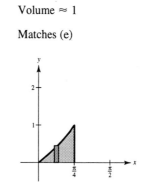

33. $p(x) = x$

$h(x) = 2 - \dfrac{1}{2}x^2$

$$V = 2\pi \int_0^2 x\left(2 - \frac{1}{2}x^2\right) dx = 2\pi \int_0^2 \left(2x - \frac{1}{2}x^3\right) dx = 2\pi\left[x^2 - \frac{1}{8}x^4\right]_0^2 = 4\pi \text{ (total volume)}$$

Now find x_0 such that

$$\pi = 2\pi \int_0^{x_0} \left(2x - \frac{1}{2}x^3\right) dx$$

$$1 = 2\left[x^2 - \frac{1}{8}x^4\right]_0^{x_0}$$

$$1 = 2x_0^2 - \frac{1}{4}x_0^4$$

$$x_0^4 - 8x_0^2 + 4 = 0$$

$$x_0^2 = 4 \pm 2\sqrt{3} \qquad \text{(Quadratic Formula)}$$

Take $x_0 = \sqrt{4 - 2\sqrt{3}}$ since the other root is too large.

Diameter: $2\sqrt{4 - 2\sqrt{3}} \approx 1.464$

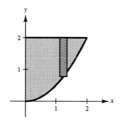

34. Total volume of the hemisphere is $\frac{1}{2}\left(\frac{4}{3}\right)\pi r^3 = \frac{2}{3}\pi(3)^3 = 18\pi$. By the Shell Method, $p(x) = x$, $h(x) = \sqrt{9 - x^2}$. Find x_0 such that

$$6\pi = 2\pi \int_0^{x_0} x\sqrt{9 - x^2} \, dx$$

$$6 = -\int_0^{x_0} (9 - x^2)^{1/2}(-2x) \, dx$$

$$= \left[-\frac{2}{3}(9 - x^2)^{3/2}\right]_0^{x_0} = 18 - \frac{2}{3}(9 - x_0^2)^{3/2}$$

$$(9 - x_0^2)^{3/2} = 18$$

$$x_0 = \sqrt{9 - 18^{2/3}} \approx 1.460.$$

Diameter: $2\sqrt{9 - 18^{2/3}} \approx 2.920$

35. $x = \sqrt{r^2 - \left(\dfrac{h}{2}\right)^2} = \dfrac{\sqrt{4r^2 - h^2}}{2}$

$V = 4\pi \displaystyle\int_{\sqrt{4r^2 - h^2}/2}^{r} x\sqrt{r^2 - x^2}\,dx$

$\quad = \left[-2\pi\left(\dfrac{2}{3}\right)(r^2 - x^2)^{3/2}\right]_{\sqrt{4r^2 - h^2}/2}^{r}$

$\quad = 0 + \dfrac{4\pi}{3}\left[r^2 - \dfrac{4r^2 - h^2}{4}\right]^{3/2}$

$\quad = \dfrac{4\pi}{3}\left(\dfrac{h^2}{4}\right)^{3/2}$

$\quad = \dfrac{4\pi}{3}\cdot\dfrac{h^3}{8} = \dfrac{\pi h^3}{6}$ (independent of r!)

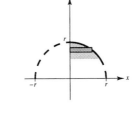

36. $V = 4\pi \displaystyle\int_{-1}^{1} (2 - x)\sqrt{1 - x^2}\,dx$

$\quad = 8\pi \displaystyle\int_{-1}^{1} \sqrt{1 - x^2}\,dx - 4\pi \displaystyle\int_{-1}^{1} x\sqrt{1 - x^2}\,dx$

$\quad = 8\pi\left(\dfrac{\pi}{2}\right) + 2\pi \displaystyle\int_{-1}^{1} x(1 - x^2)^{1/2}(-2)\,dx$

$\quad = 4\pi^2 + \left[2\pi\left(\dfrac{2}{3}\right)(1 - x^2)^{3/2}\right]_{-1}^{1} = 4\pi^2$

37. $V = 4\pi \displaystyle\int_{-r}^{r} (R - x)\sqrt{r^2 - x^2}\,dx$

$\quad = 4\pi R \displaystyle\int_{-r}^{r} \sqrt{r^2 - x^2}\,dx - 4\pi \displaystyle\int_{-r}^{r} x\sqrt{r^2 - x^2}\,dx$

$\quad = 4\pi R\left(\dfrac{\pi r^2}{2}\right) + \left[2\pi\left(\dfrac{2}{3}\right)(r^2 - x^2)^{3/2}\right]_{-r}^{r}$

$\quad = 2\pi^2 r^2 R$

38. Disc

$R(y) = \sqrt{r^2 - y^2}$

$r(y) = 0$

$V = \pi \displaystyle\int_{r-h}^{r} (r^2 - y^2)\,dy$

$\quad = \pi\left[r^2 y - \dfrac{y^3}{3}\right]_{r-h}^{r} = \dfrac{1}{3}\pi h^2(3r - h)$

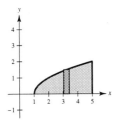

39. $\pi \displaystyle\int_{1}^{5} (x - 1)\,dx = \pi \displaystyle\int_{1}^{5} \left(\sqrt{x - 1}\right)^2 dx$

This integral represents the volume of the solid generated by revolving the region bounded by $y = \sqrt{x - 1}$, $y = 0$, and $x = 5$ about the x-axis by using the Disc Method.

$2\pi \displaystyle\int_{0}^{2} y[5 - (y^2 + 1)]\,dy$

represents this same volume by using the Shell Method.

Disc Method

40. $2\pi \displaystyle\int_{0}^{4} x\left(\dfrac{x}{2}\right) dx$

represents the volume of the solid generated by revolving the region bounded by $y = x/2$, $y = 0$, and $x = 4$ about the y-axis by using the Shell Method.

$\pi \displaystyle\int_{0}^{2} [16 - (2y)^2]\,dy = \pi \displaystyle\int_{0}^{2} [(4)^2 - (2y)^2]\,dy$

represents this same volume by using the Disc Method.

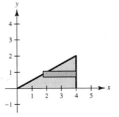

Disc Method

41. (a) $2\pi \int_0^r hx\left(1 - \dfrac{x}{r}\right)dx$ (ii)

is the volume of a right circular cone with the radius of the base as r and height h.

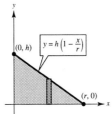

(b) $2\pi \int_{-r}^r (R - x)\left(2\sqrt{r^2 - x^2}\right)dx$ (v)

is the volume of a torus with the radius of its circular cross section as r and the distance from the axis of the torus to the center of its cross section as R.

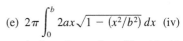

(c) $2\pi \int_0^r 2x\sqrt{r^2 - x^2}\,dx$ (iii) is the volume of a sphere with radius r.

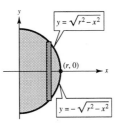

(d) $2\pi \int_0^r hx\,dx$ (i) is the volume of a right circular cylinder with a radius of r and a height of h.

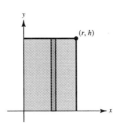

(e) $2\pi \int_0^b 2ax\sqrt{1 - (x^2/b^2)}\,dx$ (iv)

is the volume of an ellipsoid with axes $2a$ and $2b$.

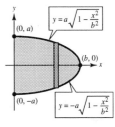

42. (a) $V = 2\pi \int_0^4 xf(x)\,dx$

$$= \dfrac{2\pi(40)}{3(4)}\left[0 + 4(10)(45) + 2(20)(40) + 4(30)(20) + 0\right]$$

$$= \dfrac{20\pi}{3}[5800] \approx 121{,}475 \text{ cubic feet}$$

(b) Top line: $y - 50 = \dfrac{40 - 50}{20 - 0}(x - 0) = -\dfrac{1}{2}x \Rightarrow y = -\dfrac{1}{2}x + 50$

Bottom line: $y - 40 = \dfrac{0 - 40}{40 - 20}(x - 20) = -2(x - 20) \Rightarrow y = -2x + 80$

$$V = 2\pi \int_0^{20} x\left(-\dfrac{1}{2}x + 50\right)dx + 2\pi \int_{20}^{40} x(-2x + 80)\,dx$$

$$= 2\pi \int_0^{20}\left(-\dfrac{1}{2}x^2 + 50x\right)dx + 2\pi \int_{20}^{40}(-2x^2 + 80x)\,dx$$

$$= 2\pi\left[-\dfrac{x^3}{6} + 25x^2\right]_0^{20} + 2\pi\left[-\dfrac{2x^3}{3} + 40x^2\right]_{20}^{40}$$

$$= 2\pi\left[\dfrac{26{,}000}{3}\right] + 2\pi\left[\dfrac{32{,}000}{3}\right]$$

$$\approx 121{,}475 \text{ cubic feet}$$

(Note that Simpson's Rule is exact for this problem.)

43. (a) $V = 2\pi \int_0^{200} x f(x)\, dx$

$\approx \dfrac{2\pi(200)}{3(8)}[0 + 4(25)(19) + 2(50)(19) + 4(75)(17) + 2(100)15 + 4(125)(14) + 2(150)(10) + 4(175)(6) + 0]$

$\approx 1{,}366{,}593$ cubic feet

(b) $d = -0.000561x^2 + 0.0189x + 19.39$

(c) $V \approx 2\pi \int_0^{200} x d(x)\, dx \approx 2\pi(213{,}800) = 1{,}343{,}345$ cubic feet

(d) Number gallons $\approx V(7.48) = 10{,}048{,}221$ gallons

Section 6.4   Arc Length and Surfaces of Revolution

1. $(0, 0), (5, 12)$

(a) $d = \sqrt{(5-0)^2 + (12-0)^2} = 13$

(b) $y = \dfrac{12}{5}x$

$y' = \dfrac{12}{5}$

$s = \int_0^5 \sqrt{1 + \left(\dfrac{12}{5}\right)^2}\, dx = \left[\dfrac{13}{5}x\right]_0^5 = 13$

2. $(1, 2), (7, 10)$

(a) $d = \sqrt{(7-1)^2 + (10-2)^2} = 10$

(b) $y = \dfrac{4}{3}x + \dfrac{2}{3}$

$y' = \dfrac{4}{3}$

$s = \int_1^7 \sqrt{1 + \left(\dfrac{4}{3}\right)^2}\, dx = \left[\dfrac{5}{3}x\right]_1^7 = 10$

3. $y = \dfrac{2}{3}x^{3/2} + 1$

$y' = x^{1/2},\ [0, 1]$

$s = \int_0^1 \sqrt{1 + x}\, dx$

$= \left[\dfrac{2}{3}(1 + x)^{3/2}\right]_0^1$

$= \dfrac{2}{3}\left(\sqrt{8} - 1\right) \approx 1.219$

4. $y = x^{3/2} - 1$

$y' = \dfrac{3}{2}x^{1/2},\ [0, 4]$

$s = \int_0^4 \sqrt{1 + \dfrac{9}{4}x}\, dx$

$= \dfrac{4}{9}\int_0^4 \left(1 + \dfrac{9}{4}x\right)^{1/2}\left(\dfrac{9}{4}\right) dx$

$= \left[\dfrac{8}{27}\left(1 + \dfrac{9}{4}x\right)^{3/2}\right]_0^4$

$= \dfrac{8}{27}(10^{3/2} - 1) \approx 9.073$

5. $y = \dfrac{3}{2}x^{2/3}$

$y' = \dfrac{1}{x^{1/3}}, [1, 8]$

$s = \displaystyle\int_1^8 \sqrt{1 + \left(\dfrac{1}{x^{1/3}}\right)^2}\, dx$

$\quad = \displaystyle\int_1^8 \sqrt{\dfrac{x^{2/3} + 1}{x^{2/3}}}\, dx$

$\quad = \dfrac{3}{2}\displaystyle\int_1^8 \sqrt{x^{2/3} + 1}\left(\dfrac{2}{3x^{1/3}}\right) dx$

$\quad = \dfrac{3}{2}\left[\dfrac{2}{3}(x^{2/3} + 1)^{3/2}\right]_1^8$

$\quad = 5\sqrt{5} - 2\sqrt{2} \approx 8.352$

6. $y = \dfrac{x^5}{10} + \dfrac{1}{6x^3}$

$y' = \dfrac{1}{2}x^4 - \dfrac{1}{2x^4}$

$1 + (y')^2 = \left(\dfrac{1}{2}x^4 + \dfrac{1}{2x^4}\right)^2, [1, 2]$

$s = \displaystyle\int_a^b \sqrt{1 + (y')^2}\, dx$

$\quad = \displaystyle\int_1^2 \sqrt{\left(\dfrac{1}{2}x^4 + \dfrac{1}{2x^4}\right)^2}\, dx$

$\quad = \displaystyle\int_1^2 \left(\dfrac{1}{2}x^4 + \dfrac{1}{2x^4}\right) dx$

$\quad = \left[\dfrac{1}{10}x^5 - \dfrac{1}{6x^3}\right]_1^2 = \dfrac{779}{240} \approx 3.246$

7. $y = \dfrac{x^4}{8} + \dfrac{1}{4x^2}$

$y' = \dfrac{1}{2}x^3 - \dfrac{1}{2x^3}$

$1 + (y')^2 = \left(\dfrac{1}{2}x^3 + \dfrac{1}{2x^3}\right)^2, [1, 2]$

$s = \displaystyle\int_a^b \sqrt{1 + (y')^2}\, dx$

$\quad = \displaystyle\int_1^2 \left(\dfrac{1}{2}x^3 + \dfrac{1}{2x^3}\right) dx$

$\quad = \left[\dfrac{1}{8}x^4 - \dfrac{1}{4x^2}\right]_1^2 = \dfrac{33}{16} \approx 2.063$

8. $y = \dfrac{1}{2}(e^x + e^{-x})$

$y' = \dfrac{1}{2}(e^x - e^{-x})$

$1 + (y')^2 = \left[\dfrac{1}{2}(e^x + e^{-x})\right]^2, [0, 2]$

$s = \displaystyle\int_0^2 \sqrt{\left[\dfrac{1}{2}(e^x + e^{-x})\right]^2}\, dx$

$\quad = \dfrac{1}{2}\displaystyle\int_0^2 (e^x + e^{-x})\, dx$

$\quad = \dfrac{1}{2}\left[e^x - e^{-x}\right]_0^2 = \dfrac{1}{2}\left(e^2 - \dfrac{1}{e^2}\right) \approx 3.627$

9. (a) $y = 4 - x^2, 0 \le x \le 2$

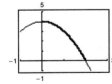

(b) $y' = -2x$

$1 + (y')^2 = 1 + 4x^2$

$L = \displaystyle\int_0^2 \sqrt{1 + 4x^2}\, dx$

(c) $L \approx 4.647$

10. (a) $y = x^2 + x - 2, -2 \le x \le 1$

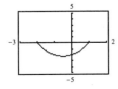

(b) $y' = 2x + 1$

$1 + (y')^2 = 1 + 4x^2 + 4x + 1$

$L = \displaystyle\int_{-2}^1 \sqrt{2 + 4x + 4x^2}\, dx$

(c) $L \approx 5.653$

11. (a) $y = \dfrac{1}{x}, 1 \leq x \leq 3$

(b) $y' = -\dfrac{1}{x^2}$

(c) $L \approx 2.147$

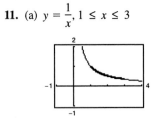

$1 + (y')^2 = 1 + \dfrac{1}{x^4}$

$L = \displaystyle\int_1^3 \sqrt{1 + \dfrac{1}{x^4}}\, dx$

12. (a) $y = \dfrac{1}{1 + x}, 0 \leq x \leq 1$

(b) $y' = -\dfrac{1}{(1 + x)^2}$

(c) $L \approx 1.132$

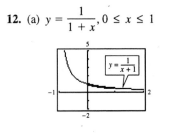

$1 + (y')^2 = 1 + \dfrac{1}{(1 + x)^4}$

$L = \displaystyle\int_0^1 \sqrt{1 + \dfrac{1}{(1 + x)^4}}\, dx$

13. (a) $y = \sin x, 0 \leq x \leq \pi$

(b) $y' = \cos x$

(c) $L \approx 3.820$

$1 + (y')^2 = 1 + \cos^2 x$

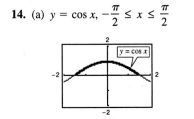

$L = \displaystyle\int_0^\pi \sqrt{1 + \cos^2 x}\, dx$

14. (a) $y = \cos x, -\dfrac{\pi}{2} \leq x \leq \dfrac{\pi}{2}$

(b) $y' = -\sin x$

(c) 3.820

$1 + (y')^2 = 1 + \sin^2 x$

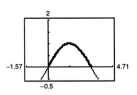

$L = \displaystyle\int_{-\pi/2}^{\pi/2} \sqrt{1 + \sin^2 x}\, dx$

15. (a) $x = e^{-y}, 0 \leq y \leq 2$

(b) $y' = -\dfrac{1}{x}$

(c) $L \approx 2.221$

$y = -\ln x$

$1 \geq x \geq e^{-2} \approx 0.135$

$1 + (y')^2 = 1 + \dfrac{1}{x^2}$

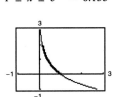

$L = \displaystyle\int_{e^{-2}}^1 \sqrt{1 + \dfrac{1}{x^2}}\, dx$

Alternatively, you can do all the computations with respect to y.

(a) $x = e^{-y}\; 0 \leq y \leq 2$

(b) $\dfrac{dx}{dy} = -e^{-y}$

(c) $L \approx 2.221$

$1 + \left(\dfrac{dx}{dy}\right)^2 = 1 + e^{-2y}$

$L = \displaystyle\int_0^2 \sqrt{1 + e^{-2y}}\, dy$

16. (a) $y = \ln x, 1 \le x \le 5$

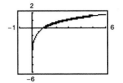

(b) $y' = \dfrac{1}{x}$

$$1 + (y')^2 = 1 + \frac{1}{x^2}$$

$$L = \int_1^5 \sqrt{1 + \frac{1}{x^2}}\, dx$$

(c) $L \approx 4.367$

17. (a) $y = 2\arctan x, 0 \le x \le 1$

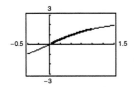

(b) $y' = \dfrac{2}{1 + x^2}$

$$L = \int_0^1 \sqrt{1 + \frac{4}{(1 + x^2)^2}}\, dx$$

(c) $L \approx 1.871$

18. (a) $x = \sqrt{36 - y^2}, 0 \le y \le 3$

$y = \sqrt{36 - x^2}, 3\sqrt{3} \le x \le 6$

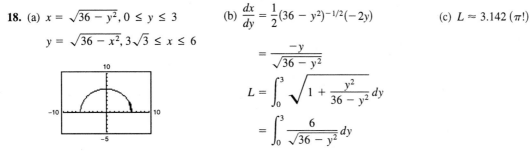

(b) $\dfrac{dx}{dy} = \dfrac{1}{2}(36 - y^2)^{-1/2}(-2y)$

$$= \frac{-y}{\sqrt{36 - y^2}}$$

$$L = \int_0^3 \sqrt{1 + \frac{y^2}{36 - y^2}}\, dy$$

$$= \int_0^3 \frac{6}{\sqrt{36 - y^2}}\, dy$$

(c) $L \approx 3.142\ (\pi!)$

Alternatively, you can convert to a function of x.

$$y = \sqrt{36 - x^2}$$

$$y' = \frac{dy}{dx} = -\frac{x}{\sqrt{36 - x^2}}$$

$$L = \int_{3\sqrt{3}}^6 \sqrt{1 + \frac{x^2}{36 - x^2}}\, dx = \int_{3\sqrt{3}}^6 \frac{6}{\sqrt{36 - x^2}}\, dx$$

Although this integral is undefined at $x = 0$, a graphing utility still gives $L \approx 3.142$.

19. $\displaystyle\int_0^2 \sqrt{1 + \left[\frac{d}{dx}\left(\frac{5}{x^2 + 1}\right)\right]^2}\, dx$

$s \approx 5$

Matches (b)

20. $\displaystyle\int_0^{\pi/4} \sqrt{1 + \left[\frac{d}{dx}(\tan x)\right]^2}\, dx$

$s \approx 1$

Matches (e)

21. $y = x^3, [0, 4]$

(a) $d = \sqrt{(4-0)^2 + (64-0)^2} \approx 64.125$

(b) $d = \sqrt{(1-0)^2 + (1-0)^2} + \sqrt{(2-1)^2 + (8-1)^2} + \sqrt{(3-2)^2 + (27-8)^2} + \sqrt{(4-3)^2 + (64-27)^2}$

   ≈ 64.525

(c) $s = \int_0^4 \sqrt{1 + (3x^2)^2}\, dx = \int_0^4 \sqrt{1 + 9x^4}\, dx \approx 64.666$

(d) 64.672

22. $f(x) = (x^2 - 4)^2, [0, 4]$

(a) $d = \sqrt{(4-0)^2 + (144-16)^2} \approx 128.062$

(b) $d = \sqrt{(1-0)^2 + (9-16)^2} + \sqrt{(2-1)^2 + (0-9)^2} + \sqrt{(3-2)^2 + (25-0)^2} + \sqrt{(4-3)^2 + (144-25)^2}$

   ≈ 160.151

(c) $s = \int_0^4 \sqrt{1 + [4x(x^2-4)]^2}\, dx \approx 159.087$

(d) 160.287

23. (a)

(c) $y_1' = 1, \ L_1 = \int_0^4 \sqrt{2}\, dx \approx 5.657$

$y_2' = \dfrac{3}{4}x^{1/2}, \ L_2 = \int_0^4 \sqrt{1 + \dfrac{9x}{16}}\, dx \approx 5.759$

$y_3' = \dfrac{1}{2}x, \ L_3 = \int_0^4 \sqrt{1 + \dfrac{x^2}{4}}\, dx \approx 5.916$

$y_4' = \dfrac{5}{16}x^{3/2}, \ L_4 = \int_0^4 \sqrt{1 + \dfrac{25}{256}x^3}\, dx \approx 6.063$

(b) y_1, y_2, y_3, y_4

24. Let $y = \ln x, \ 1 \le x \le e, \ y' = \dfrac{1}{x}$ and $L_1 = \int_1^e \sqrt{1 + \dfrac{1}{x^2}}\, dx$.

Equivalently, $x = e^y, \ 0 \le y \le 1, \ \dfrac{dx}{dy} = e^y$, and $L_2 = \int_0^1 \sqrt{1 + e^{2y}}\, dy = \int_0^1 \sqrt{1 + e^{2x}}\, dx$.

Numerically, both integrals yield $L = 2.0035$

25. $y = \dfrac{1}{3}[x^{3/2} - 3x^{1/2} + 2]$

When $x = 0, \ y = \dfrac{2}{3}$. Thus, the fleeting object has traveled $\dfrac{2}{3}$ units when it is caught.

$$y' = \frac{1}{3}\left[\frac{3}{2}x^{1/2} - \frac{3}{2}x^{-1/2}\right] = \left(\frac{1}{2}\right)\frac{x-1}{x^{1/2}}$$

$$1 + (y')^2 = 1 + \frac{(x-1)^2}{4x} = \frac{(x+1)^2}{4x}$$

$$s = \int_0^1 \frac{x+1}{2x^{1/2}}\, dx = \frac{1}{2}\int_0^1 (x^{1/2} + x^{-1/2})\, dx = \frac{1}{2}\left[\frac{2}{3}x^{3/2} + 2x^{1/2}\right]_0^1 = \frac{4}{3} = 2\left(\frac{2}{3}\right)$$

The pursuer has traveled twice the distance that the fleeing object has traveled when it is caught.

26. $y = 31 - 10(e^{x/20} + e^{-x/20})$

$$y' = -\frac{1}{2}(e^{x/20} - e^{-x/20})$$

$$1 + (y')^2 = 1 + \frac{1}{4}(e^{x/10} - 2 + e^{-x/10}) = \left[\frac{1}{2}(e^{x/20} + e^{-x/20})\right]^2$$

$$s = \int_{-20}^{20} \sqrt{\left[\frac{1}{2}(e^{x/20} + e^{-x/20})\right]^2}\, dx$$

$$= \frac{1}{2}\int_{-20}^{20} (e^{x/20} + e^{-x/20})\, dx = \left[10(e^{x/20} - e^{-x/20})\right]_{-20}^{20} = 20\left(e - \frac{1}{e}\right) \approx 47 \text{ ft}$$

Thus, there are $100(47) = 4700$ square feet of roofing on the barn.

27. $y = 20 \cosh\dfrac{x}{20}, \; -20 \le x \le 20$

$$y' = \sinh\frac{x}{20}$$

$$1 + (y')^2 = 1 + \sinh^2\frac{x}{20} = \cosh^2\frac{x}{20}$$

$$L = \int_{-20}^{20} \cosh\frac{x}{20}\, dx = 2\int_0^{20} \cosh\frac{x}{20}\, dx = 2(20) \sinh\frac{x}{20}\Big]_0^{20}$$

$$= 40\sinh(1) \approx 47.008 \text{ m.}$$

28. $y = 693.8597 - 68.7672 \cosh 0.0100333x$

$y' = -0.6899619478 \sinh 0.0100333x$

$$s = \int_{-299.2239}^{299.2239} \sqrt{1 + (-0.6899619478 \sinh 0.0100333x)^2}\, dx \approx 1480$$

(Use Simpson's Rule with $n = 100$ or a graphing utility.)

29. $y = \sqrt{9 - x^2}$

$$y' = \frac{-x}{\sqrt{9 - x^2}}$$

$$1 + (y')^2 = \frac{9}{9 - x^2}$$

$$s = \int_0^2 \sqrt{\frac{9}{9 - x^2}}\, dx$$

$$= \int_0^2 \frac{3}{\sqrt{9 - x^2}}\, dx$$

$$= \left[3\arcsin\frac{x}{3}\right]_0^2$$

$$= 3\left(\arcsin\frac{2}{3} - \arcsin 0\right)$$

$$= 3\arcsin\frac{2}{3} \approx 2.1892$$

30. $y = \sqrt{25 - x^2}$

$$y' = \frac{-x}{\sqrt{25 - x^2}}$$

$$1 + (y')^2 = \frac{25}{25 - x^2}$$

$$s = \int_{-3}^4 \sqrt{\frac{25}{25 - x^2}}\, dx$$

$$= \int_{-3}^4 \frac{5}{\sqrt{25 - x^2}}\, dx$$

$$= \left[5\arcsin\frac{x}{5}\right]_{-3}^4$$

$$= 5\left[\arcsin\frac{4}{5} - \arcsin\left(-\frac{3}{5}\right)\right] \approx 7.8540$$

$$\frac{1}{4}[2\pi(5)] \approx 7.8540 = s$$

31. $y = \dfrac{x^3}{3}$

$y' = x^2, [0, 3]$

$S = 2\pi \displaystyle\int_0^3 \dfrac{x^3}{3} \sqrt{1 + x^4}\, dx$

$\quad = \dfrac{\pi}{6} \displaystyle\int_0^3 (1 + x^4)^{1/2}(4x^3)\, dx$

$\quad = \left[\dfrac{\pi}{9}(1 + x^4)^{3/2} \right]_0^3$

$\quad = \dfrac{\pi}{9}\left(82\sqrt{82} - 1\right) \approx 258.85$

32. $y = \sqrt{x}$

$y' = \dfrac{1}{2\sqrt{x}}, [1, 4]$

$S = 2\pi \displaystyle\int_1^4 \sqrt{x} \sqrt{1 + \dfrac{1}{4x}}\, dx$

$\quad = 2\pi \displaystyle\int_1^4 \dfrac{\sqrt{x}}{2\sqrt{x}} \sqrt{4x + 1}\, dx$

$\quad = \pi \displaystyle\int_1^4 \sqrt{4x + 1}\, dx$

$\quad = \dfrac{\pi}{4} \displaystyle\int_1^4 (4x + 1)^{1/2}(4)\, dx$

$\quad = \left[\dfrac{\pi}{6}(4x + 1)^{3/2} \right]_1^4$

$\quad = \dfrac{\pi}{6}\left(17\sqrt{17} - 5\sqrt{5}\right) \approx 30.85$

33. $\qquad y = \dfrac{x^3}{6} + \dfrac{1}{2x}$

$\qquad y' = \dfrac{x^2}{2} - \dfrac{1}{2x^2}$

$1 + (y')^2 = \left(\dfrac{x^2}{2} + \dfrac{1}{2x^2} \right)^2, [1, 2]$

$\qquad S = 2\pi \displaystyle\int_1^2 \left(\dfrac{x^3}{6} + \dfrac{1}{2x} \right)\left(\dfrac{x^2}{2} + \dfrac{1}{2x^2} \right) dx$

$\qquad = 2\pi \displaystyle\int_1^2 \left(\dfrac{x^5}{12} + \dfrac{x}{3} + \dfrac{1}{4x^3} \right) dx$

$\qquad = 2\pi \left[\dfrac{x^6}{72} + \dfrac{x^2}{6} - \dfrac{1}{8x^2} \right]_1^2 = \dfrac{47\pi}{16}$

34. $\qquad y = \dfrac{x}{2}$

$\qquad y' = \dfrac{1}{2}$

$1 + (y')^2 = \dfrac{5}{4}, [0, 6]$

$\qquad S = 2\pi \displaystyle\int_0^6 \dfrac{x}{2} \sqrt{\dfrac{5}{4}}\, dx$

$\qquad = \left[\dfrac{2\pi\sqrt{5}}{8} x^2 \right]_0^6 = 9\sqrt{5}\,\pi$

35. $y = \sqrt[3]{x} + 2$

$y' = \dfrac{1}{3x^{2/3}}, [1, 8]$

$S = 2\pi \displaystyle\int_1^8 x \sqrt{1 + \dfrac{1}{9x^{4/3}}}\, dx$

$\quad = \dfrac{2\pi}{3} \displaystyle\int_1^8 x^{1/3} \sqrt{9x^{4/3} + 1}\, dx$

$\quad = \dfrac{\pi}{18} \displaystyle\int_1^8 (9x^{4/3} + 1)^{1/2}(12x^{1/3})\, dx$

$\quad = \left[\dfrac{\pi}{27}(9x^{4/3} + 1)^{3/2} \right]_1^8$

$\quad = \dfrac{\pi}{27}\left(145\sqrt{145} - 10\sqrt{10}\right) \approx 199.48$

36. $y = 4 - x^2$

$y' = -2x, [0, 2]$

$S = 2\pi \displaystyle\int_0^2 x\sqrt{1 + 4x^2}\, dx$

$\quad = \dfrac{\pi}{4} \displaystyle\int_0^2 (1 + 4x^2)^{1/2}(8x)\, dx$

$\quad = \left[\dfrac{\pi}{6}(1 + 4x^2)^{3/2} \right]_0^2$

$\quad = \dfrac{\pi}{6}\left(17\sqrt{17} - 1\right) \approx 36.18$

37. $y = \sin x$

$y' = \cos x,\ [0, \pi]$

$S = 2\pi \int_0^{\pi} \sin x \sqrt{1 + \cos^2 x}\ dx$

≈ 14.4236

38. $y = \ln x$

$y' = \dfrac{1}{x}$

$1 + (y')^2 = \dfrac{x^2 + 1}{x^2},\ [1, e]$

$S = 2\pi \int_1^{e} x \sqrt{\dfrac{x^2 + 1}{x^2}}\ dx$

$= 2\pi \int_1^{e} \sqrt{x^2 + 1}\ dx \approx 22.943$

39. $y = \dfrac{hx}{r}$

$y' = \dfrac{h}{r}$

$1 + (y')^2 = \dfrac{r^2 + h^2}{r^2}$

$S = 2\pi \int_0^{r} x \sqrt{\dfrac{r^2 + h^2}{r^2}}\ dx$

$= \left[\dfrac{2\pi\sqrt{r^2 + h^2}}{r}\left(\dfrac{x^2}{2}\right) \right]_0^{r} = \pi r \sqrt{r^2 + h^2}$

40. $y = \sqrt{r^2 - x^2}$

$y' = \dfrac{-x}{\sqrt{r^2 - x^2}}$

$1 + (y')^2 = \dfrac{r^2}{r^2 - x^2}$

$S = 2\pi \int_{-r}^{r} \sqrt{r^2 - x^2} \sqrt{\dfrac{r^2}{r^2 - x^2}}\ dx$

$= 2\pi \int_{-r}^{r} r\ dx = \Big[2\pi r x \Big]_{-r}^{r} = 4\pi r^2$

41. $y = \sqrt{9 - x^2}$

$y' = \dfrac{-x}{\sqrt{9 - x^2}}$

$\sqrt{1 + (y')^2} = \dfrac{3}{\sqrt{9 - x^2}}$

$S = 2\pi \int_0^{2} \dfrac{3x}{\sqrt{9 - x^2}}\ dx$

$= -3\pi \int_0^{2} \dfrac{-2x}{\sqrt{9 - x^2}}\ dx$

$= \left[-6\pi \sqrt{9 - x^2} \right]_0^{2}$

$= 6\pi \left(3 - \sqrt{5} \right) \approx 14.40$

See figure in Exercise 42.

42. From Exercise 41 we have:

$S = 2\pi \int_0^{a} \dfrac{rx}{\sqrt{r^2 - x^2}}\ dx$

$= -r\pi \int_0^{a} \dfrac{-2x\ dx}{\sqrt{r^2 - x^2}}$

$= \left[-2r\pi \sqrt{r^2 - x^2} \right]_0^{a}$

$= 2r^2\pi - 2r\pi\sqrt{r^2 - a^2}$

$= 2r\pi\left(r - \sqrt{r^2 - a^2} \right)$

$= 2\pi r h$ (where h is the height of the zone)

43. $y = \dfrac{1}{3}x^{1/2} - x^{3/2}$

$y' = \dfrac{1}{6}x^{-1/2} - \dfrac{3}{2}x^{1/2} = \dfrac{1}{6}\left(x^{-1/2} - 9x^{1/2}\right)$

$1 + (y')^2 = 1 + \dfrac{1}{36}\left(x^{-1} - 18 + 81x\right) = \dfrac{1}{36}\left(x^{-1/2} + 9x^{1/2}\right)^2$

$S = 2\pi \int_0^{1/3} \left(\dfrac{1}{3}x^{1/2} - x^{3/2}\right) \sqrt{\dfrac{1}{36}\left(x^{-1/2} + 9^{1/2}\right)^2}\ dx = \dfrac{2\pi}{6} \int_0^{1/3} \left(\dfrac{1}{3}x^{1/2} - x^{3/2}\right)\left(x^{-1/2} + 9x^{1/2}\right) dx$

$= \dfrac{\pi}{3} \int_0^{1/3} \left(\dfrac{1}{3} + 2x - 9x^2\right) dx = \dfrac{\pi}{3}\left[\dfrac{1}{3}x + x^2 - 3x^3\right]_0^{1/3} = \dfrac{\pi}{27}\ \text{ft}^2 \approx 0.1164\ \text{ft}^2 \approx 16.8\ \text{in}^2$

Amount of glass needed: $V = \dfrac{\pi}{27}\left(\dfrac{0.015}{12}\right) \approx 0.00015\ \text{ft}^3 \approx 0.25\ \text{in}^3$

44. (a) We approximate the volume by summing 6 discs of thickness 3 and circumference C_i equal to the average of the given circumferences:

$$V \approx \sum_{i=1}^{6} \pi r_i{}^2(3) = \sum_{i=1}^{6} \pi \left(\frac{C_i}{2\pi}\right)^2 (3) = \frac{3}{4\pi} \sum_{i=1}^{6} C_i{}^2$$

$$= \frac{3}{4\pi}\left[\left(\frac{50 + 65.5}{2}\right)^2 + \left(\frac{65.5 + 70}{2}\right)^2 + \left(\frac{70 + 66}{2}\right)^2 + \left(\frac{66 + 58}{2}\right)^2 + \left(\frac{58 + 51}{2}\right)^2 + \left(\frac{51 + 48}{2}\right)^2\right]$$

$$= \frac{3}{4\pi}[57.75^2 + 67.75^2 + 68^2 + 62^2 + 54.5^2 + 49.5^2]$$

$$= \frac{3}{4\pi}[21813.625] = 5207.62 \text{ cubic inches}$$

(b) The lateral surface area of a frustum of a right circular cone is $\pi s(R + r)$. For the first frustum,

$$S_1 \approx \pi\left[3^2 + \left(\frac{65.5 - 50}{2\pi}\right)^2\right]^{1/2}\left[\frac{50}{2\pi} + \frac{65.5}{2\pi}\right]$$

$$= \left(\frac{50 + 65.5}{2}\right)\left[9 + \left(\frac{65.5 - 50}{2\pi}\right)^2\right]^{1/2}.$$

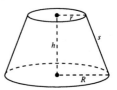

Adding the six frustums together,

$$S \approx \left(\frac{50 + 65.5}{2}\right)\left[9 + \left(\frac{15.5}{2\pi}\right)^2\right]^{1/2} + \left(\frac{65.5 + 70}{2}\right)\left[9 + \left(\frac{4.5}{2\pi}\right)^2\right]^{1/2} +$$

$$\left(\frac{70 + 66}{2}\right)\left[9 + \left(\frac{4}{2\pi}\right)^2\right]^{1/2} + \left(\frac{66 + 58}{2}\right)\left[9 + \left(\frac{8}{2\pi}\right)^2\right]^{1/2}$$

$$\left(\frac{58 + 51}{2}\right)\left[9 + \left(\frac{7}{2\pi}\right)^2\right]^{1/2} + \left(\frac{51 + 48}{2}\right)\left[9 + \left(\frac{3}{2\pi}\right)^2\right]^{1/2}$$

$$\approx 224.30 + 208.96 + 208.54 + 202.06 + 174.41 + 150.37$$

$$= 1168.64$$

(c) $r = 0.00401 y^3 - 0.1416 y^2 + 1.232 y + 7.943$

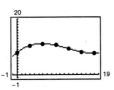

(d) $V = \displaystyle\int_0^{18} \pi r^2 \, dy \approx 5275.9$ cubic inches

$$S = \int_0^{18} 2\pi r(y)\sqrt{1 + r'(y)^2} \, dy$$

$$\approx 1179.5 \text{ square inches}$$

45. Individual project, see Exercise 44.

46. (a) $V = \pi \displaystyle\int_1^b \frac{1}{x^2} \, dx = \left[-\frac{\pi}{x}\right]_1^b = \pi\left(1 - \frac{1}{b}\right)$

(c) $\displaystyle\lim_{b\to\infty} V = \lim_{b\to\infty} \pi\left(1 - \frac{1}{b}\right) = \pi$

(d) Since

$$\frac{\sqrt{x^4 + 1}}{x^3} > \frac{\sqrt{x^4}}{x^3} = \frac{1}{x} > 0 \text{ on } [1, b]$$

we have

$$\int_1^b \frac{\sqrt{x^4 + 1}}{x^3} \, dx > \int_1^b \frac{1}{x} \, dx = \left[\ln x\right]_1^b = \ln b$$

and $\displaystyle\lim_{b\to\infty} \ln b \to \infty$. Thus,

$$\lim_{b\to\infty} 2\pi \int_1^b \frac{\sqrt{x^4 + 1}}{x^3} \, dx = \infty.$$

(b) $S = 2\pi \displaystyle\int_1^b \frac{1}{x}\sqrt{1 + \left(-\frac{1}{x^2}\right)^2} \, dx$

$$= 2\pi \int_1^b \frac{1}{x}\sqrt{1 + \frac{1}{x^4}} \, dx$$

$$= 2\pi \int_1^b \frac{\sqrt{x^4 + 1}}{x^3} \, dx$$

47. (a) $\dfrac{ds}{dx} = \sqrt{1 + [f'(x)]^2}$ (Second Fundamental Theorem of Calculus)

(b) $ds = \sqrt{1 + [f'(x)]^2}\, dx$

$$(ds)^2 = \left[1 + [f'(x)]^2\right](dx)^2 = \left[1 + \left(\frac{dy}{dx}\right)^2\right](dx)^2 = (dx)^2 + (dy)^2$$

(c) $s(x) = \displaystyle\int_1^x \sqrt{1 + \left(\frac{3}{2}t^{1/2}\right)^2}\, dt = \int_1^x \sqrt{1 + \frac{9}{4}t}\, dt = \left[\frac{4}{9}\left(\frac{2}{3}\right)\left(1 + \frac{9}{4}t\right)^{3/2}\right]_1^x$

$\qquad = \dfrac{8}{27}\left[\left(1 + \frac{9}{4}x\right)^{3/2} - \left(\frac{13}{4}\right)^{3/2}\right] = \dfrac{8}{27}\left[\dfrac{(4 + 9x)^{3/2}}{8} - \dfrac{13\sqrt{13}}{8}\right] = \dfrac{1}{27}\left[(4 + 9x)^{3/2} - 13\sqrt{13}\right]$

$s(2) = \dfrac{1}{27}\left[22\sqrt{22} - 13\sqrt{13}\right] = \dfrac{22}{27}\sqrt{22} - \dfrac{13}{27}\sqrt{13} \approx 2.086$

48. (a) $\dfrac{x^2}{9} + \dfrac{y^2}{4} = 1$

Ellipse: $y_1 = 2\sqrt{1 - \dfrac{x^2}{9}}$

$\qquad y_2 = -2\sqrt{1 - \dfrac{x^2}{9}}$

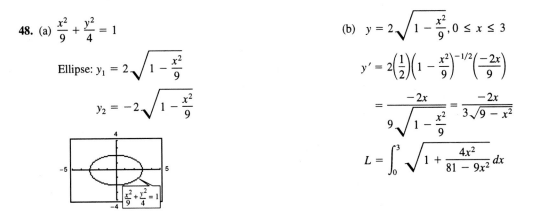

(b) $y = 2\sqrt{1 - \dfrac{x^2}{9}},\ 0 \le x \le 3$

$y' = 2\left(\dfrac{1}{2}\right)\left(1 - \dfrac{x^2}{9}\right)^{-1/2}\left(\dfrac{-2x}{9}\right)$

$\qquad = \dfrac{-2x}{9\sqrt{1 - \dfrac{x^2}{9}}} = \dfrac{-2x}{3\sqrt{9 - x^2}}$

$L = \displaystyle\int_0^3 \sqrt{1 + \dfrac{4x^2}{81 - 9x^2}}\, dx$

(c) You cannot evaluate this definite integral, since the integrand is not defined at $x = 3$. Simpson's Rule will not work for the same reason. Also, the integrand does not have an elementary antiderivative.

49. (a) Area of circle with radius L: $A = \pi L^2$

Area of sector with central angle θ (in radians)

$S = \dfrac{\theta}{2\pi}A = \dfrac{\theta}{2\pi}(\pi L^2) = \dfrac{1}{2}L^2\theta$

(b) Let s be the arc length of the sector, which is the circumference of the base of the cone. Here, $s = L\theta = 2\pi r$, and you have

$S = \dfrac{1}{2}L^2\theta = \dfrac{1}{2}L^2\left(\dfrac{s}{L}\right) = \dfrac{1}{2}Ls = \dfrac{1}{2}L(2\pi r) = \pi rL$

(c) The lateral surface area of the frustum is the difference of the large cone and the small one.

$S = \pi r_2(L + L_1) - \pi r_1 L_1$

$\qquad = \pi r_2 L + \pi L_1(r_2 - r_1)$

By similar triangles, $\dfrac{L + L_1}{r_2} = \dfrac{L_1}{r_1} \Rightarrow Lr_1 = L_1(r_2 - r_1)$

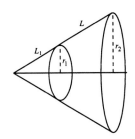

Hence,

$S = \pi r_2 L + \pi L_1(r_2 - r_1) = \pi r_2 L + \pi Lr_1$

$\qquad = \pi L(r_1 + r_2)$.

50. Essay

Section 6.5   Work

1. $W = Fd = (100)(10) = 1000 \text{ ft} \cdot \text{lb}$

2. $W = Fd = (2400)(6) = 14{,}400 \text{ ft} \cdot \text{lb}$

3. $W = Fd = (112)(4) = 448 \text{ joules}$ (newton-meters)

4. $W = Fd = [9(2000)]\left[\frac{1}{2}(5280)\right] = 47{,}520{,}000 \text{ ft} \cdot \text{lb}$

5. Since the work equals the area under the force function, you have $(c) < (d) < (a) < (b)$.

6. (a) $W = \displaystyle\int_0^9 6 \, dx = 54 \text{ ft} \cdot \text{lbs}$

   (b) $W = \displaystyle\int_0^7 20 \, dx + \int_7^9 (-10x + 90) \, dx = 140 + 20$

   $\quad = 160 \text{ ft} \cdot \text{lbs}$

   (c) $W = \displaystyle\int_0^9 \frac{1}{27} x^2 \, dx = \frac{x^3}{81}\Big]_0^9 = 9 \text{ ft} \cdot \text{lbs}$

   (d) $W = \displaystyle\int_0^9 \sqrt{x} \, dx = \frac{2}{3} x^{3/2}\Big]_0^9 = \frac{2}{3}(27) = 18 \text{ ft} \cdot \text{lbs}$

7. $F(x) = kx$

   $5 = k(4)$

   $k = \dfrac{5}{4}$

   $W = \displaystyle\int_0^7 \frac{5}{4} x \, dx = \left[\frac{5}{8} x^2\right]_0^7$

   $\quad = \dfrac{245}{8} \text{ in} \cdot \text{lb}$

   $\quad = 30.625 \text{ in} \cdot \text{lb} \approx 2.55 \text{ ft} \cdot \text{lb}$

8. $W = \displaystyle\int_5^9 \frac{5}{4} x \, dx = \left[\frac{5}{8} x^2\right]_5^9$

   $\quad = 35 \text{ in} \cdot \text{lb} \approx 2.92 \text{ ft} \cdot \text{lb}$

9. $F(x) = kx$

   $250 = k(30) \implies k = \dfrac{25}{3}$

   $W = \displaystyle\int_{20}^{50} F(x) \, dx = \int_{20}^{50} \frac{25}{3} x \, dx = \frac{25x^2}{6}\bigg]_{20}^{50}$

   $\quad = 8750 \text{ n} \cdot \text{cm} = 87.5 \text{ joules or Nm}$

10. $F(x) = kx$

   $800 = k(70) \implies k = \dfrac{80}{7}$

   $W = \displaystyle\int_0^{70} F(x) \, dx = \int_0^{70} \frac{80}{7} x \, dx = \frac{40x^2}{7}\bigg]_0^{70}$

   $\quad = 28000 \text{ n} \cdot \text{cm} = 280 \text{ Nm}$

11. $F(x) = kx$

   $15 = 6k$

   $k = \dfrac{5}{2}$

   $W = \displaystyle\int_0^{12} \frac{5}{2} x \, dx = \left[\frac{5}{4} x^2\right]_0^{12}$

   $\quad = 180 \text{ in} \cdot \text{lb} = 15 \text{ ft} \cdot \text{lb}$

12. $F(x) = kx$

   $15 = k(1) = k$

   $W = 2 \displaystyle\int_0^4 15x \, dx = \left[15x^2\right]_0^4$

   $\quad = 240 \text{ ft} \cdot \text{lb}$

13. $W = 18 = \displaystyle\int_0^{1/3} kx \, dx = \frac{kx^2}{2}\bigg]_0^{1/3} = \frac{k}{18} \implies k = 324$

   $W = \displaystyle\int_{1/3}^{7/12} 324x \, dx = 162x^2\bigg]_{1/3}^{7/12} = 37.125 \text{ ft} \cdot \text{lbs}$

   $\left[\textbf{Note: } 4 \text{ inches} = \frac{1}{3} \text{ foot}\right]$

14. $W = 7.5 = \displaystyle\int_0^{1/6} kx \, dx = \frac{kx^2}{2}\bigg]_0^{1/6} = \frac{k}{72} \implies k = 540$

   $W = \displaystyle\int_{1/6}^{5/24} 540x \, dx = 270x^2\bigg]_{1/6}^{5/24} = 4.21875 \text{ ft} \cdot \text{lbs}$

15. Assume that the earth has a radius of 4000 miles.

$$F(x) = \frac{k}{x^2}$$

(a) $W = \int_{4000}^{4200} \frac{64{,}000{,}000}{x^2}\,dx = \left[-\frac{64{,}000{,}000}{x}\right]_{4000}^{4200} \approx -15{,}238.095 + 16{,}000$

$$4 = \frac{k}{(4000)^2}$$

$$= 761.905 \text{ mi} \cdot \text{ton} \approx 8.05 \times 10^9 \text{ ft} \cdot \text{lb}$$

$$k = 64{,}000{,}000$$

(b) $W = \int_{4000}^{4400} \frac{64{,}000{,}000}{x^2}\,dx = \left[-\frac{64{,}000{,}000}{x}\right]_{4000}^{4400} \approx -14{,}545.455 + 16{,}000$

$$F(x) = \frac{64{,}000{,}000}{x^2}$$

$$= 1454.545 \text{ mi} \cdot \text{ton} \approx 1.54 \times 10^{10} \text{ ft} \cdot \text{lb}$$

16. $\quad W = \int_{4000}^{h} \frac{64{,}000{,}000}{x^2}\,dx = \left[-\frac{64{,}000{,}000}{x}\right]_{4000}^{h} = -\frac{64{,}000{,}000}{h} + 16{,}000$

$\lim\limits_{h\to\infty} W = 16{,}000 \text{ mi} \cdot \text{ton} \approx 1.690 \times 10^{11} \text{ ft} \cdot \text{lb}$

17. Assume that the earth has a radius of 4000 miles.

$$F(x) = \frac{k}{x^2}$$

(a) $W = \int_{4000}^{15{,}000} \frac{160{,}000{,}000}{x^2}\,dx = \left[-\frac{160{,}000{,}000}{x}\right]_{4000}^{15{,}000} \approx -10{,}666.667 + 40{,}000$

$$10 = \frac{k}{(4000)^2}$$

$$= 29{,}333.333 \text{ mi} \cdot \text{ton}$$

$$k = 160{,}000{,}000$$

$$\approx 2.93 \times 10^4 \text{ mi} \cdot \text{ton}$$

$$\approx 3.10 \times 10^{11} \text{ ft} \cdot \text{lb}$$

$$F(x) = \frac{160{,}000{,}000}{x^2}$$

(b) $W = \int_{4000}^{26{,}000} \frac{160{,}000{,}000}{x^2}\,dx = \left[-\frac{160{,}000{,}000}{x}\right]_{4000}^{26{,}000} \approx -6{,}153.846 + 40{,}000$

$$= 33{,}846.154 \text{ mi} \cdot \text{ton}$$

$$\approx 3.38 \times 10^4 \text{ mi} \cdot \text{ton}$$

$$\approx 3.57 \times 10^{11} \text{ ft} \cdot \text{lb}$$

18. Weight on surface of moon: $\frac{1}{6}(12) = 2$ tons

Weight varies inversely as the square of distance from the center of the moon. Therefore,

$$F(x) = \frac{k}{x^2}$$

$$2 = \frac{k}{(1100)^2}$$

$$k = 2.42 \times 10^6$$

$$W = \int_{1100}^{1150} \frac{2.42 \times 10^6}{x^2}\,dx = \left[\frac{-2.42 \times 10^6}{x}\right]_{1100}^{1150} = 2.42 \times 10^6\left(\frac{1}{1100} - \frac{1}{1150}\right)$$

$$\approx 95.652 \text{ mi} \cdot \text{ton} \approx 1.01 \times 10^9 \text{ ft} \cdot \text{lb}$$

19. Weight of each layer: $62.4(20)\,\Delta y$

Distance: $4 - y$

(a) $W = \int_{2}^{4} 62.4(20)(4 - y)\,dy = \left[4992y - 624y^2\right]_{2}^{4} = 2496 \text{ ft} \cdot \text{lb}$

(b) $W = \int_{0}^{4} 62.4(20)(4 - y)\,dy = \left[4992y - 624y^2\right]_{0}^{4} = 9984 \text{ ft} \cdot \text{lb}$

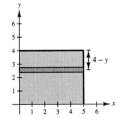

20. The bottom half had to be pumped a greater distance then the top half.

21. Volume of disc of water: $\pi(2)^2 \Delta y = 4\pi \Delta y$

Weight of disc of water: $(1000)(4\pi \Delta y)$

Distance the disc of water is moved: $5 - y$

$$W = \int_0^4 (5 - y)(4000\pi)\, dy = 4000\pi \int_0^4 (5 - y)\, dy$$

$$= 4000\pi \left[5y - \frac{y^2}{2} \right]_0^4$$

$$= 4000\pi[12] = 48000\pi \text{ kg} \cdot \text{m}$$

22. Volume of disc of water: $4\pi \Delta y$

Weight of disc of water: $4000\pi \Delta y$

Distance the disc of water is moved: y

$$W = \int_{10}^{12} y(4000\pi)\, dy = \left[2000\pi y^2 \right]_{10}^{12}$$

$$= 88000\pi \text{ kg} \cdot \text{m}$$

23. Volume of disc: $\pi\left(\frac{2}{3}y\right)^2 \Delta y$

Weight of disc: $62.4\pi\left(\frac{2}{3}y\right)^2 \Delta y$

Distance: $6 - y$

$$W = \frac{4(62.4)\pi}{9} \int_0^6 (6 - y)y^2\, dy = \frac{4}{9}(62.4)\pi \left[2y^3 - \frac{1}{4}y^4 \right]_0^6 = 2995.2\pi \text{ ft} \cdot \text{lb}$$

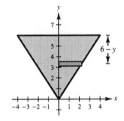

24. Volume of disc: $\pi\left(\frac{2}{3}y\right)^2 \Delta y$

Weight of disc: $62.4\pi\left(\frac{2}{3}y\right)^2 \Delta y$

Distance: y

(a) $W = \frac{4}{9}(62.4)\pi \int_0^2 y^3\, dy = \left[\frac{4}{9}(62.4)\pi\left(\frac{1}{4}y^4\right) \right]_0^2 \approx 110.9\pi \text{ ft} \cdot \text{lb}$

(b) $W = \frac{4}{9}(62.4)\pi \int_4^6 y^3\, dy = \left[\frac{4}{9}(62.4)\pi\left(\frac{1}{4}y^4\right) \right]_4^6 \approx 7210.7\pi \text{ ft} \cdot \text{lb}$

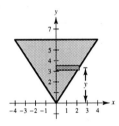

25. Volume of disc: $\pi\left(\sqrt{36 - y^2}\right)^2 \Delta y$

Weight of disc: $62.4\pi(36 - y^2) \Delta y$

Distance: y

$$W = 62.4\pi \int_0^6 y(36 - y^2)\, dy$$

$$= 62.4\pi \int_0^6 (36y - y^3)\, dy = 62.4\pi \left[18y^2 - \frac{1}{4}y^4 \right]_0^6$$

$$= 20{,}217.6\pi \text{ ft} \cdot \text{lb}$$

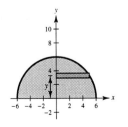

26. Volume of each layer: $\left(\frac{y + 3}{3}\right)(3)\Delta y = (y + 3)\, dy$

Weight of each layer: $55.6(y + 3)\, dy$

Distance: $5 - y$

$$W = 55.6 \int_0^3 (5 - y)(y + 3)\, dy$$

$$= 55.6 \int_0^3 (15 + 2y - y^2)\, dy$$

$$= 55.6 \left[15y + y^2 - \frac{y^3}{3} \right]_0^3 = 2502 \text{ ft} \cdot \text{lb}$$

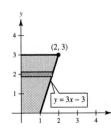

27. Volume of layer: $V = lwh = 4(2)\sqrt{(9/4) - y^2} \, \Delta y$

     Weight of layer: $W = 42(8)\sqrt{(9/4) - y^2} \, \Delta y$

     Distance: $\dfrac{13}{2} - y$

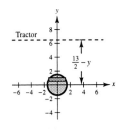

$$W = \int_{-1.5}^{1.5} 42(8)\sqrt{(9/4) - y^2}\left(\frac{13}{2} - y\right) dy$$

$$= 336\left[\frac{13}{2} \int_{-1.5}^{1.5} \sqrt{(9/4) - y^2} \, dy - \int_{-1.5}^{1.5} \sqrt{(9/4) - y^2} \, y \, dy\right]$$

The second integral is zero since the integrand is odd and the limits of integration are symmetric to the origin. The first integral represents the area of a semicircle of radius $\frac{3}{2}$. Thus, the work is

$$W = 336\left(\frac{13}{2}\right)\pi\left(\frac{3}{2}\right)^2\left(\frac{1}{2}\right) = 2457\pi \text{ ft} \cdot \text{lb}$$

28. Volume of layer: $V = 12(2)\sqrt{(25/4) - y^2} \, \Delta y$

     Weight of layer: $W = 42(24)\sqrt{(25/4) - y^2} \, \Delta y$

     Distance: $\dfrac{19}{2} - y$

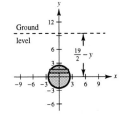

$$W = \int_{-2.5}^{2.5} 42(24)\sqrt{\frac{25}{4} - y^2}\left(\frac{19}{2} - y\right) dy$$

$$= 1008\left[\frac{19}{2} \int_{-2.5}^{2.5} \sqrt{\frac{25}{4} - y^2} \, dy + \int_{-2.5}^{2.5} \sqrt{\frac{25}{4} - y^2}(-y) \, dy\right]$$

The second integral is zero since the integrand is odd and the limits of integration are symmetric to the origin. The first integral represents the area of a semicircle of radius $\frac{5}{2}$. Thus, the work is

$$W = 1008\left(\frac{19}{2}\right)\pi\left(\frac{5}{2}\right)^2\left(\frac{1}{2}\right) = 29{,}925\pi \text{ ft} \cdot \text{lb} \approx 94{,}012.16 \text{ ft} \cdot \text{lb}.$$

29. Weight of section of chain: $3 \, \Delta y$

     Distance: $15 - y$

$$W = 3\int_0^{15} (15 - y) \, dy$$

$$= \left[-\frac{3}{2}(15 - y)^2\right]_0^{15}$$

$$= 337.5 \text{ ft} \cdot \text{lb}$$

30. The lower 10 feet of chain are raised 5 feet with a constant force.

$$W_1 = 3(10)5 = 150 \text{ ft} \cdot \text{lb}$$

The top 5 feet will be raised with variable force.

Weight of section: $3 \, \Delta y$

Distance: $5 - y$

$$W_2 = 3\int_0^5 (5 - y) \, dy = \left[-\frac{3}{2}(5 - y)^2\right]_0^5 = \frac{75}{2} \text{ ft} \cdot \text{lb}$$

$$W = W_1 + W_2 = 150 + \frac{75}{2} = \frac{375}{2} \text{ ft} \cdot \text{lb}$$

31. The lower 5 feet of chain are raised 10 feet with a constant force.

$$W_1 = 3(5)(10) = 150 \text{ ft} \cdot \text{lb}$$

The top 10 feet of chain are raised with a variable force.

Weight per section: $3 \, \Delta y$

Distance: $10 - y$

$$W_2 = 3\int_0^{10} (10 - y) \, dy = \left[-\frac{3}{2}(10 - y)^2\right]_0^{10} = 150 \text{ ft} \cdot \text{lb}$$

$$W = W_1 + W_2 = 300 \text{ ft} \cdot \text{lb}$$

32. The work required to lift the chain is 337.5 ft · lb (from Exercise 29). The work required to lift the 100-pound load is $W = (100)(15) = 1500$. The work required to lift the chain with a 100-pound load attached is

$$W = 337.5 + 1500 = 1837.5 \text{ ft} \cdot \text{lb}.$$

33. Weight of section of chain: $3\,\Delta y$

Distance: $15 - 2y$

$$W = 3\int_0^{7.5} (15 - 2y)\, dy = \left[-\frac{3}{4}(15 - 2y)^2 \right]_0^{7.5} = \frac{3}{4}(15)^2 = 168.75 \text{ ft} \cdot \text{lb}$$

34. $W = 3\int_0^6 (12 - 2y)\, dy = \left[-\frac{3}{4}(12 - 2y)^2 \right]_0^6 = \frac{3}{4}(12)^2 = 108 \text{ ft} \cdot \text{lb}$

35. Work to pull up the ball: $W_1 = 500(15) = 7500 \text{ ft} \cdot \text{lb}$

Work to wind up the top 15 feet of cable: force is variable

Weight per section: $1\,\Delta y$

Distance: $15 - x$

$$W_2 = \int_0^{15} (15 - x)\, dx = \left[-\frac{1}{2}(15 - x)^2 \right]_0^{15}$$

$$= 112.5 \text{ ft} \cdot \text{lb}$$

Work to lift the lower 25 feet of cable with a constant force:

$$W_3 = (1)(25)(15) = 375 \text{ ft} \cdot \text{lb}$$

$$W = W_1 + W_2 + W_3 = 7500 + 112.5 + 375$$

$$= 7987.5 \text{ ft} \cdot \text{lb}$$

36. Work to pull up the ball: $W_1 = 500(40) = 20{,}000 \text{ ft} \cdot \text{lb}$

Work to pull up the cable: force is variable

Weight per section: $1\,\Delta y$

Distance: $40 - x$

$$W_2 = \int_0^{40} (40 - x)\, dx = \left[-\frac{1}{2}(40 - x)^2 \right]_0^{40}$$

$$= 800 \text{ ft} \cdot \text{lb}$$

$$W = W_1 + W_2 = 20{,}000 + 800 = 20{,}800 \text{ ft} \cdot \text{lb}$$

37. $p = \dfrac{k}{V}$

$1000 = \dfrac{k}{2}$

$k = 2000$

$$W = \int_2^3 \frac{2000}{V}\, dV = \left[2000 \ln |V| \right]_2^3$$

$$= 2000 \ln\left(\frac{3}{2} \right) \approx 810.93 \text{ ft} \cdot \text{lb}$$

38. $p = \dfrac{k}{V}$

$2000 = \dfrac{k}{1}$

$k = 2000$

$$W = \int_1^4 \frac{2000}{V}\, dV = \left[2000 \ln |V| \right]_1^4$$

$$= 2000 \ln 4 \approx 2772.59 \text{ ft} \cdot \text{lb}$$

39. $F(x) = \dfrac{k}{(2 - x)^2}$

$$W = \int_{-2}^1 \frac{k}{(2 - x)^2}\, dx = \left[\frac{k}{2 - x} \right]_{-2}^1 = k\left(1 - \frac{1}{4} \right) = \frac{3k}{4} \text{ (units of work)}$$

40. (a) $W = FD = (8000\pi)(2) = 16{,}000\pi \text{ ft} \cdot \text{lbs}$

(b) $W \approx \dfrac{2 - 0}{3(6)}[0 + 4(20{,}000) + 2(22{,}000) + 4(15{,}000) + 2(10{,}000) + 4(5000) + 0]$

$\approx 24{,}88.889 \text{ ft} \cdot \text{lb}$

—CONTINUED—

40. —CONTINUED—

(c) $F(x) = -16,261.36x^4 + 85,295.45x^3 - 157,738.64x^2 + 104,386.36x - 32.4675$

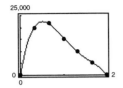

(d) $F(x) = 0$ when $x \approx 0.524$ feet. $F(x)$ is a maximum when $x \approx 0.524$ feet.

(e) $W = \int_0^2 F(x)\, dx \approx 25,180.5$ ft · lbs

41. $W = \int_0^5 1000[1.8 - \ln(x+1)]\, dx \approx 3249.44$

42. $W = \int_0^4 \left(\frac{e^{x^2} - 1}{100}\right) dx \approx 11,494$ ft · lb

43. $W = \int_0^5 100x\sqrt{125 - x^3}\, dx \approx 10,330.3$ ft · lb

44. $W = \int_0^2 1000 \sinh x\, dx \approx 2762.2$ ft · lb

Section 6.6   Moments, Centers of Mass, and Centroids

1. $\bar{x} = \dfrac{6(-5) + 3(1) + 5(3)}{6 + 3 + 5} = -\dfrac{6}{7}$

2. $\bar{x} = \dfrac{7(-3) + 4(-2) + 3(5) + 8(6)}{7 + 4 + 3 + 8} = \dfrac{17}{11}$

3. $\bar{x} = \dfrac{1(7) + 1(8) + 1(12) + 1(15) + 1(18)}{1 + 1 + 1 + 1 + 1} = 12$

4. $\bar{x} = \dfrac{12(-3) + 1(-2) + 6(-1) + 3(0) + 11(4)}{12 + 1 + 6 + 3 + 11} = 0$

5. (a) $\bar{x} = \dfrac{(7+5) + (8+5) + (12+5) + (15+5) + (18+5)}{5} = 17 = 12 + 5$

(b) $\bar{x} = \dfrac{12(-3-3) + 1(-2-3) + 6(-1-3) + 3(0-3) + 11(4-3)}{12 + 1 + 6 + 3 + 11} = -3 = 0 - 3$

6. The center of mass is translated k units as well.

7. $50x = 75(L - x) = 75(10 - x)$

$50x = 750 - 75x$

$125x = 750$

$x = 6$ feet

8. $200x = 550(5 - x)$ (Person on left)

$200x = 2750 - 550x$

$750x = 2750$

$x = 3\frac{2}{3}$ feet

9. $\bar{x} = \dfrac{5(2) + 1(-3) + 3(1)}{5 + 1 + 3} = \dfrac{10}{9}$

$\bar{y} = \dfrac{5(2) + 1(1) + 3(-4)}{5 + 1 + 3} = -\dfrac{1}{9}$

$(\bar{x}, \bar{y}) = \left(\dfrac{10}{9}, -\dfrac{1}{9}\right)$

10. $\bar{x} = \dfrac{10(1) + 2(5) + 5(-4)}{10 + 2 + 5} = 0$

$\bar{y} = \dfrac{10(-1) + 2(5) + 5(0)}{10 + 2 + 5} = 0$

$(\bar{x}, \bar{y}) = (0, 0)$

11. $\bar{x} = \dfrac{3(-2) + 4(-1) + 2(7) + 1(0) + 6(-3)}{3 + 4 + 2 + 1 + 6} = -\dfrac{7}{8}$

$\bar{y} = \dfrac{3(-3) + 4(0) + 2(1) + 1(0) + 6(0)}{3 + 4 + 2 + 1 + 6} = -\dfrac{7}{16}$

$(\bar{x}, \bar{y}) = \left(-\dfrac{7}{8}, -\dfrac{7}{16}\right)$

12. $\bar{x} = \dfrac{4(2) + 2(-1) + 2.5(6) + 5(2)}{4 + 2 + 2.5 + 5} = \dfrac{31}{13.5} = \dfrac{62}{27}$

$\bar{y} = \dfrac{4(3) + 2(5) + 2.5(8) + 5(-2)}{4 + 2 + 2.5 + 5} = \dfrac{32}{13.5} = \dfrac{64}{27}$

$(\bar{x}, \bar{y}) = \left(\dfrac{62}{27}, \dfrac{64}{27}\right)$

13. $m = \rho \displaystyle\int_0^4 \sqrt{x}\, dx = \left[\dfrac{2\rho}{3}x^{3/2}\right]_0^4 = \dfrac{16\rho}{3}$

$M_x = \rho \displaystyle\int_0^4 \dfrac{\sqrt{x}}{2}\left(\sqrt{x}\right) dx = \left[\rho\dfrac{x^2}{4}\right]_0^4 = 4\rho$

$\bar{y} = \dfrac{M_x}{m} = 4\rho\left(\dfrac{3}{16\rho}\right) = \dfrac{3}{4}$

$M_y = \rho \displaystyle\int_0^4 x\sqrt{x}\, dx = \left[\rho\dfrac{2}{5}x^{5/2}\right]_0^4 = \dfrac{64\rho}{5}$

$\bar{x} = \dfrac{M_y}{m} = \dfrac{64\rho}{5}\left(\dfrac{3}{16\rho}\right) = \dfrac{12}{5}$

$(\bar{x}, \bar{y}) = \left(\dfrac{12}{5}, \dfrac{3}{4}\right)$

14. $m = \rho \displaystyle\int_0^4 x^2\, dx = \left[\rho\dfrac{x^3}{3}\right]_0^4 = \dfrac{64}{3}\rho$

$M_x = \rho \displaystyle\int_0^4 \dfrac{x^2}{2}(x^2)\, dx = \left[\rho\dfrac{x^5}{10}\right]_0^4 = \dfrac{512\rho}{5}$

$\bar{y} = \dfrac{M_x}{m} = \dfrac{512\rho}{5}\left(\dfrac{3}{64\rho}\right) = \dfrac{24}{5}$

$M_y = \rho \displaystyle\int_0^4 x(x^2)\, dx = \left[\rho\dfrac{x^4}{4}\right]_0^4 = 64\rho$

$\bar{x} = \dfrac{M_y}{m} = 64\rho\left(\dfrac{3}{64\rho}\right) = 3$

$(\bar{x}, \bar{y}) = \left(3, \dfrac{24}{5}\right)$

15. $m = \rho \displaystyle\int_0^1 (x^2 - x^3)\, dx = \rho\left[\dfrac{x^3}{3} - \dfrac{x^4}{4}\right]_0^1 = \dfrac{\rho}{12}$

$M_x = \rho \displaystyle\int_0^1 \dfrac{(x^2 + x^3)}{2}(x^2 - x^3)\, dx = \dfrac{\rho}{2}\displaystyle\int_0^1 (x^4 - x^6)\, dx = \dfrac{\rho}{2}\left[\dfrac{x^5}{5} - \dfrac{x^7}{7}\right]_0^1 = \dfrac{\rho}{35}$

$\bar{y} = \dfrac{M_x}{m} = \dfrac{\rho}{35}\left(\dfrac{12}{\rho}\right) = \dfrac{12}{35}$

$M_y = \rho \displaystyle\int_0^1 x(x^2 - x^3)\, dx = \rho\displaystyle\int_0^1 (x^3 - x^4)\, dx = \rho\left[\dfrac{x^4}{4} - \dfrac{x^5}{5}\right]_0^1 = \dfrac{\rho}{20}$

$\bar{x} = \dfrac{M_y}{m} = \dfrac{\rho}{20}\left(\dfrac{12}{\rho}\right) = \dfrac{3}{5}$

$(\bar{x}, \bar{y}) = \left(\dfrac{3}{5}, \dfrac{12}{35}\right)$

16. $m = \rho \int_0^1 (\sqrt{x} - x)\, dx = \rho \left[\frac{2}{3}x^{3/2} - \frac{x^2}{2} \right]_0^1 = \frac{\rho}{6}$

$M_x = \rho \int_0^1 \frac{(\sqrt{x} + x)}{2}(\sqrt{x} - x)\, dx = \frac{\rho}{2} \int_0^1 (x - x^2)\, dx = \frac{\rho}{2}\left[\frac{x^2}{2} - \frac{x^3}{3} \right]_0^1 = \frac{\rho}{12}$

$\bar{y} = \frac{M_x}{m} = \frac{\rho}{12}\left(\frac{6}{\rho} \right) = \frac{1}{2}$

$M_y = \rho \int_0^1 x(\sqrt{x} - x)\, dx = \rho \int_0^1 (x^{3/2} - x^2)\, dx = \rho \left[\frac{2}{5}x^{5/2} - \frac{x^3}{3} \right]_0^1 = \frac{\rho}{15}$

$\bar{x} = \frac{M_y}{m} = \frac{\rho}{15}\left(\frac{6}{\rho} \right) = \frac{2}{5}$

$(\bar{x}, \bar{y}) = \left(\frac{2}{5}, \frac{1}{2} \right)$

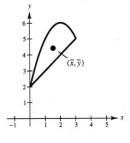

17. $m = \rho \int_0^3 [(-x^2 + 4x + 2) - (x + 2)]\, dx = -\rho \left[\frac{x^3}{3} + \frac{3x^2}{2} \right]_0^3 = \frac{9\rho}{2}$

$M_x = \rho \int_0^3 \left[\frac{(-x^2 + 4x + 2) + (x + 2)}{2} \right][(-x^2 + 4x + 2) - (x + 2)]\, dx$

$= \frac{\rho}{2} \int_0^3 (-x^2 + 5x + 4)(-x^2 + 3x)\, dx = \frac{\rho}{2} \int_0^3 (x^4 - 8x^3 + 11x^2 + 12x)\, dx$

$= \frac{\rho}{2} \left[\frac{x^5}{5} - 2x^4 + \frac{11x^3}{3} + 6x^2 \right]_0^3 = \frac{99\rho}{5}$

$\bar{y} = \frac{M_x}{m} = \frac{99\rho}{5}\left(\frac{2}{9\rho} \right) = \frac{22}{5}$

$M_y = \rho \int_0^3 x[(-x^2 + 4x - 2) - (x + 2)]\, dx = \rho \int_0^3 (-x^3 + 3x^2)\, dx = \rho \left[-\frac{x^4}{4} + x^3 \right]_0^3 = \frac{27\rho}{4}$

$\bar{x} = \frac{M_y}{m} = \frac{27\rho}{4}\left(\frac{2}{9\rho} \right) = \frac{3}{2}$

$(\bar{x}, \bar{y}) = \left(\frac{3}{2}, \frac{22}{5} \right)$

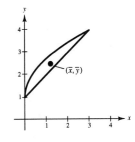

18. $m = \rho \int_0^3 [(\sqrt{3x} + 1) - (x + 1)]\, dx = \rho \left[\frac{2}{9}(3x)^{3/2} - \frac{x^2}{2} \right]_0^3 = \frac{3\rho}{2}$

$M_x = \rho \int_0^3 \frac{[(\sqrt{3x} + 1) + (x + 1)]}{2}[(\sqrt{3x} + 1) - (x + 1)]\, dx$

$= \frac{\rho}{2} \int_0^3 (\sqrt{3x} + x + 2)(\sqrt{3x} - x)\, dx$

$= \frac{\rho}{2} \int_0^3 (2\sqrt{3x} + x - x^2)\, dx$

$= \frac{\rho}{2} \left[\frac{4}{9}(3x)^{3/2} + \frac{x^2}{2} - \frac{x^3}{3} \right]_0^3 = \frac{15\rho}{4}$

$\bar{y} = \frac{M_x}{m} = \frac{15\rho}{4}\left(\frac{2}{3\rho} \right) = \frac{5}{2}$

$M_y = \rho \int_0^3 x[(\sqrt{3x} + 1) - (x + 1)]\, dx = \rho \int_0^3 (x\sqrt{3x} - x^2)\, dx = \rho \left[\frac{2\sqrt{3}}{5}x^{5/2} - \frac{x^3}{3} \right]_0^3 = \frac{9\rho}{5}$

$\bar{x} = \frac{M_y}{m} = \frac{9\rho}{5}\left(\frac{2}{3\rho} \right) = \frac{6}{5}$

$(\bar{x}, \bar{y}) = \left(\frac{6}{5}, \frac{5}{2} \right)$

19.
$$m = \rho \int_0^8 x^{2/3}\, dx = \rho \left[\frac{3}{5} x^{5/3} \right]_0^8 = \frac{96\rho}{5}$$

$$M_x = \rho \int_0^8 \frac{x^{2/3}}{2}(x^{2/3})\, dx = \frac{\rho}{2}\left[\frac{3}{7} x^{7/3} \right]_0^8 = \frac{192\rho}{7}$$

$$\bar{y} = \frac{M_x}{m} = \frac{192\rho}{7}\left(\frac{5}{96\rho}\right) = \frac{10}{7}$$

$$M_y = \rho \int_0^8 x(x^{2/3})\, dx = \rho \left[\frac{3}{8} x^{8/3} \right]_0^8 = 96\rho$$

$$\bar{x} = \frac{M_y}{m} = 96\rho \left(\frac{5}{96\rho}\right) = 5$$

$$(\bar{x}, \bar{y}) = \left(5, \frac{10}{7} \right)$$

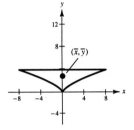

20.
$$m = 2\rho \int_0^8 (4 - x^{2/3})\, dx = 2\rho \left[4x - \frac{3}{5} x^{5/3} \right]_0^8 = \frac{128\rho}{5}$$

By symmetry, M_y and $\bar{x} = 0$.

$$M_x = 2\rho \int_0^8 \left(\frac{4 + x^{2/3}}{2} \right)(4 - x^{2/3})\, dx = \rho \left[16x - \frac{3}{7} x^{7/3} \right]_0^8 = \frac{512\rho}{7}$$

$$\bar{y} = \frac{512\rho}{7}\left(\frac{5}{128\rho}\right) = \frac{20}{7}$$

$$(\bar{x}, \bar{y}) = \left(0, \frac{20}{7} \right)$$

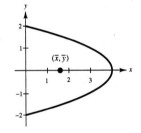

21.
$$m = 2\rho \int_0^2 (4 - y^2)\, dy = 2\rho \left[4y - \frac{y^3}{3} \right]_0^2 = \frac{32\rho}{3}$$

$$M_y = 2\rho \int_0^2 \left(\frac{4 - y^2}{2} \right)(4 - y^2)\, dy = \rho \left[16y - \frac{8}{3} y^3 + \frac{y^5}{5} \right]_0^2 = \frac{256\rho}{15}$$

$$\bar{x} = \frac{M_y}{m} = \frac{256\rho}{15}\left(\frac{3}{32\rho}\right) = \frac{8}{5}$$

By symmetry, M_x and $\bar{y} = 0$.

$$(\bar{x}, \bar{y}) = \left(\frac{8}{5}, 0 \right)$$

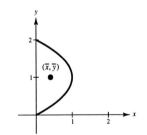

22.
$$m = \rho \int_0^2 (2y - y^2)\, dy = \rho \left[y^2 - \frac{y^3}{3} \right]_0^2 = \frac{4\rho}{3}$$

$$M_y = \rho \int_0^2 \left(\frac{2y - y^2}{2} \right)(2y - y^2)\, dy = \frac{\rho}{2}\left[\frac{4y^3}{3} - y^4 + \frac{y^5}{5} \right]_0^2 = \frac{8\rho}{15}$$

$$\bar{x} = \frac{M_y}{m} = \frac{8\rho}{15}\left(\frac{3}{4\rho}\right) = \frac{2}{5}$$

$$M_x = \rho \int_0^2 y(2y - y^2)\, dy = \rho \left[\frac{2y^3}{3} - \frac{y^4}{4} \right]_0^2 = \frac{4\rho}{3}$$

$$\bar{y} = \frac{M_x}{m} = \frac{4\rho}{3}\left(\frac{3}{4\rho}\right) = 1$$

$$(\bar{x}, \bar{y}) = \left(\frac{2}{5}, 1 \right)$$

23. $m = \rho \int_0^3 [(2y - y^2) - (-y)] \, dy = \rho \left[\dfrac{3y^2}{2} - \dfrac{y^3}{3} \right]_0^3 = \dfrac{9\rho}{2}$

$M_y = \rho \int_0^3 \dfrac{[(2y - y^2) + (-y)]}{2} [(2y - y^2) - (-y)] \, dy = \dfrac{\rho}{2} \int_0^3 (y - y^2)(3y - y^2) \, dy$

$= \dfrac{\rho}{2} \int_0^3 (y^4 - 4y^3 + 3y^2) \, dy = \dfrac{\rho}{2} \left[\dfrac{y^5}{5} - y^4 + y^3 \right]_0^3 = -\dfrac{27\rho}{10}$

$\bar{x} = \dfrac{M_y}{m} = -\dfrac{27\rho}{10} \left(\dfrac{2}{9\rho} \right) = -\dfrac{3}{5}$

$M_x = \rho \int_0^3 y[(2y - y^2) - (-y)] \, dy = \rho \int_0^3 (3y^2 - y^3) \, dy = \rho \left[y^3 - \dfrac{y^4}{4} \right]_0^3 = \dfrac{27\rho}{4}$

$\bar{y} = \dfrac{M_x}{m} = \dfrac{27\rho}{4} \left(\dfrac{2}{9\rho} \right) = \dfrac{3}{2}$

$(\bar{x}, \bar{y}) = \left(-\dfrac{3}{5}, \dfrac{3}{2} \right)$

24. $m = \rho \int_{-1}^2 [(y + 2) - y^2] \, dy = \rho \left[\dfrac{y^2}{2} + 2y - \dfrac{y^3}{3} \right]_{-1}^2 = \dfrac{9\rho}{2}$

$M_y = \rho \int_{-1}^2 \dfrac{[(y + 2) + y^2]}{2} [(y + 2) - y^2] \, dy$

$= \dfrac{\rho}{2} \int_{-1}^2 [(y + 2)^2 - y^4] \, dy = \dfrac{\rho}{2} \left[\dfrac{(y + 2)^3}{3} - \dfrac{y^5}{5} \right]_{-1}^2 = \dfrac{36\rho}{5}$

$\bar{x} = \dfrac{M_y}{m} = \dfrac{36\rho}{5} \left(\dfrac{2}{9\rho} \right) = \dfrac{8}{5}$

$M_x = \rho \int_{-1}^2 y[(y + 2) - y^2] \, dy$

$= \rho \int_{-1}^2 (2y + y^2 - y^3) \, dy = \rho \left[y^2 + \dfrac{y^3}{3} - \dfrac{y^4}{4} \right]_{-1}^2 = \dfrac{9\rho}{4}$

$\bar{y} = \dfrac{M_x}{m} = \dfrac{9\rho}{4} \left(\dfrac{2}{9\rho} \right) = \dfrac{1}{2}$

$(\bar{x}, \bar{y}) = \left(\dfrac{8}{5}, \dfrac{1}{2} \right)$

25. $A = \int_0^1 (x - x^2) \, dx = \left[\dfrac{1}{2}x^2 - \dfrac{x^3}{3} \right]_0^1 = \dfrac{1}{6}$

$M_x = \dfrac{1}{2} \int_0^1 (x^2 - x^4) \, dx = \dfrac{1}{2} \left[\dfrac{x^3}{3} - \dfrac{x^5}{5} \right]_0^1 = \dfrac{1}{2} \left(\dfrac{1}{3} - \dfrac{1}{5} \right) = \dfrac{1}{15}$

$M_y = \int_0^1 (x^2 - x^3) \, dx = \left[\dfrac{x^3}{3} - \dfrac{x^4}{4} \right]_0^1 = \left(\dfrac{1}{3} - \dfrac{1}{4} \right) = \dfrac{1}{12}$

26. $A = \int_1^4 \dfrac{1}{x} \, dx = \left[\ln|x| \right]_1^4 = \ln 4$

$M_x = \dfrac{1}{2} \int_1^4 \dfrac{1}{x^2} \, dx = \left[\dfrac{1}{2} \left(-\dfrac{1}{x} \right) \right]_1^4 = \left(-\dfrac{1}{8} + \dfrac{1}{2} \right) = \dfrac{3}{8}$

$M_y = \int_1^4 x \left(\dfrac{1}{x} \right) \, dx = \left[x \right]_1^4 = 3$

27. $A = \int_0^3 (2x + 4)\,dx = \left[x^2 + 4x\right]_0^3 = 9 + 12 = 21$

$M_x = \dfrac{1}{2}\int_0^3 (2x + 4)^2\,dx = \int_0^3 (2x^2 + 8x + 8)\,dx = \left[\dfrac{2x^3}{3} + 4x^2 + 8x\right]_0^3 = 18 + 36 + 24 = 78$

$M_y = \int_0^3 (2x^2 + 4x)\,dx = \left[\dfrac{2x^3}{3} + 2x^2\right]_0^3 = 18 + 18 = 36$

28. $A = \int_{-2}^2 -(x^2 - 4)\,dx = 2\int_0^2 (4 - x^2)\,dx = \left[8x - \dfrac{2x^3}{3}\right]_0^2 = 16 - \dfrac{16}{3} = \dfrac{32}{3}$

$M_x = \dfrac{1}{2}\int_{-2}^2 (x^2 - 4)(4 - x^2)\,dx = -\dfrac{1}{2}\int_{-2}^2 (x^4 - 8x^2 + 16)\,dx$

$\quad = -\dfrac{1}{2}\left[\dfrac{x^5}{5} - \dfrac{8x^3}{3} + 16x\right]_{-2}^2 = -\left[\dfrac{32}{5} - \dfrac{64}{3} + 32\right] = -\dfrac{256}{15}$

$M_y = 0$ by symmetry.

29. $m = \rho\int_0^5 10x\sqrt{125 - x^3}\,dx \approx 1033.0\rho$

$M_x = \rho\int_0^5 \left(\dfrac{10x\sqrt{125 - x^3}}{2}\right)\left(10x\sqrt{125 - x^3}\right)dx = 50\rho\int_0^5 x^2(125 - x^3)\,dx = \dfrac{3{,}124{,}375\rho}{24} \approx 130{,}208\rho$

$M_y = \rho\int_0^5 10x^2\sqrt{125 - x^3}\,dx = -\dfrac{10\rho}{3}\int_0^5 \sqrt{125 - x^3}(-3x^2)\,dx = \dfrac{12{,}500\sqrt{5}\rho}{9} \approx 3105.6\rho$

$\bar{x} = \dfrac{M_y}{m} \approx 3.0$

$\bar{y} = \dfrac{M_x}{m} \approx 126.0$

Therefore, the centroid is $(3.0, 126.0)$.

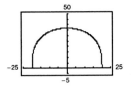

30. $m = \rho\int_0^4 xe^{-x/2}\,dx \approx 2.3760\rho$

$M_x = \rho\int_0^4 \left(\dfrac{xe^{-x/2}}{2}\right)(xe^{-x/2})\,dx = \dfrac{\rho}{2}\int_0^4 x^2e^{-x}\,dx \approx 0.7619\rho$

$M_y = \rho\int_0^4 x^2e^{-x/2}\,dx \approx 5.1732\rho$

$\bar{x} = \dfrac{M_y}{m} \approx 2.2$

$\bar{y} = \dfrac{M_x}{m} \approx 0.3$

Therefore, the centroid is $(2.2, 0.3)$.

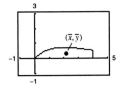

31. $m = \rho\int_{-20}^{20} 5\sqrt[3]{400 - x^2}\,dx \approx 1239.76\rho$

$M_x = \rho\int_{-20}^{20} \dfrac{5\sqrt[3]{400 - x^2}}{2}\left(5\sqrt[3]{400 - x^2}\right)dx$

$\quad = \dfrac{25\rho}{2}\int_{-20}^{20} (400 - x^2)^{2/3}\,dx \approx 20064.27$

$\bar{y} = \dfrac{M_x}{m} \approx 16.18$

$\bar{x} = 0$ by symmetry. Therefore, the centroid is $(0, 16.2)$.

32. $m = \rho \displaystyle\int_{-2}^{2} \dfrac{8}{x^2 + 4} \, dx \approx 6.2832\rho$

$M_x = \rho \displaystyle\int_{-2}^{2} \dfrac{1}{2}\left(\dfrac{8}{x^2 + 4}\right)\left(\dfrac{8}{x^2 + 4}\right) dx = 32\rho \displaystyle\int_{-2}^{2} \dfrac{1}{(x^2 + 4)^2} \, dx \approx 5.14149\rho$

$\bar{y} = \dfrac{M_x}{m} \approx 0.8$

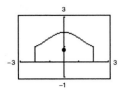

$\bar{x} = 0$ by symmetry. Therefore, the centroid is $(0, 0.8)$.

33. $A = \dfrac{1}{2}(2a)c = ac$

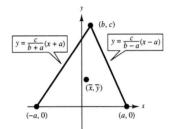

$\dfrac{1}{A} = \dfrac{1}{ac}$

$\bar{x} = \left(\dfrac{1}{ac}\right)\dfrac{1}{2}\displaystyle\int_0^c \left[\left(\dfrac{b-a}{c}y + a\right)^2 - \left(\dfrac{b+a}{c}y - a\right)^2\right] dy$

$\quad = \dfrac{1}{2ac}\displaystyle\int_0^c \left[\dfrac{4ab}{c}y - \dfrac{4ab}{c^2}y^2\right] dy$

$\quad = \dfrac{1}{2ac}\left[\dfrac{2ab}{c}y^2 - \dfrac{4ab}{3c^2}y^3\right]_0^c = \dfrac{1}{2ac}\left(\dfrac{2}{3}abc\right) = \dfrac{b}{3}$

$\bar{y} = \dfrac{1}{ac}\displaystyle\int_0^c y\left[\left(\dfrac{b-a}{c}y + a\right) - \left(\dfrac{b+a}{c}y - a\right)\right] dy$

$\quad = \dfrac{1}{ac}\displaystyle\int_0^c y\left(-\dfrac{2a}{c}y + 2a\right) dy = \dfrac{2}{c}\displaystyle\int_0^c \left(y - \dfrac{y^2}{c}\right) dy$

$\quad = \dfrac{2}{c}\left[\dfrac{y^2}{2} - \dfrac{y^3}{3c}\right]_0^c = \dfrac{c}{3}$

$(\bar{x}, \bar{y}) = \left(\dfrac{b}{3}, \dfrac{c}{3}\right)$

In Exercise 566 of Section P.2, you found that $(b/3, c/3)$ is the point of intersection of the medians.

34. $A = bh = ac$

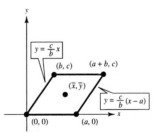

$\dfrac{1}{A} = \dfrac{1}{ac}$

$\bar{x} = \dfrac{1}{ac}\dfrac{1}{2}\displaystyle\int_0^c \left[\left(\dfrac{b}{c}y + a\right)^2 - \left(\dfrac{b}{c}y\right)^2\right] dy$

$\quad = \dfrac{1}{2ac}\displaystyle\int_0^c \left(\dfrac{2ab}{c}y + a^2\right) dy$

$\quad = \dfrac{1}{2ac}\left[\dfrac{ab}{c}y^2 + a^2y\right]_0^c = \dfrac{1}{2ac}[abc + a^2c] = \dfrac{1}{2}(b + a)$

$\bar{y} = \dfrac{1}{ac}\displaystyle\int_0^c y\left[\left(\dfrac{b}{c}y + a\right) - \left(\dfrac{b}{c}y\right)\right] dy = \left[\dfrac{1}{c}\dfrac{y^2}{2}\right]_0^c = \dfrac{c}{2}$

$(\bar{x}, \bar{y}) = \left(\dfrac{b+a}{2}, \dfrac{c}{2}\right)$

This is the point of intersection of the diagonals.

35. $A = \dfrac{c}{2}(a + b)$

$$\frac{1}{A} = \frac{2}{c(a + b)}$$

$$\bar{x} = \frac{2}{c(a + b)} \int_0^c x\left(\frac{b - a}{c}x + a\right) dx = \frac{2}{c(a + b)} \int_0^c \left(\frac{b - a}{c}x^2 + ax\right) dx = \frac{2}{c(a + b)}\left[\frac{b - a}{c}\frac{x^3}{3} + \frac{ax^2}{2}\right]_0^c$$

$$= \frac{2}{c(a + b)}\left[\frac{(b - a)c^2}{3} + \frac{ac^2}{2}\right] = \frac{2}{c(a + b)}\left[\frac{2bc^2 - 2ac^2 + 3ac^2}{6}\right] = \frac{c(2b + a)}{3(a + b)} = \frac{(a + 2b)c}{3(a + b)}$$

$$\bar{y} = \frac{2}{c(a + b)}\frac{1}{2} \int_0^c \left(\frac{b - a}{c}x + a\right)^2 dx = \frac{1}{c(a + b)} \int_0^c \left[\left(\frac{b - a}{c}\right)^2 x^2 + \frac{2a(b - a)}{c}x + a^2\right] dx$$

$$= \frac{1}{c(a + b)}\left[\left(\frac{b - a}{c}\right)^2 \frac{x^3}{3} + \frac{2a(b - a)}{c}\frac{x^2}{2} + a^2 x\right]_0^c = \frac{1}{c(a + b)}\left[\frac{(b - a)^2 c}{3} + ac(b - a) + a^2 c\right]$$

$$= \frac{1}{3c(a + b)}[(b^2 - 2ab + a^2)c + 3ac(b - a) + 3a^2 c]$$

$$= \frac{1}{3(a + b)}[b^2 - 2ab + a^2 + 3ab - 3a^2 + 3a^2] = \frac{a^2 + ab + b^2}{3(a + b)}$$

Thus,

$$(\bar{x}, \bar{y}) = \left(\frac{(a + 2b)c}{3(a + b)}, \frac{a^2 + ab + b^2}{3(a + b)}\right).$$

The one line passes through $(0, a/2)$ and $(c, b/2)$. It's equation is

$$y = \frac{b - a}{2c}x + \frac{a}{2}.$$

The other line passes through $(0, -b)$ and $(c, a + b)$. It's equation is

$$y = \frac{a + 2b}{c}x - b.$$

$(\bar{x}, \bar{y})$ is the point of intersection of these two lines.

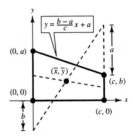

36. $\bar{x} = 0$ by symmetry

$$A = \frac{1}{2}\pi r^2$$

$$\frac{1}{A} = \frac{2}{\pi r^2}$$

$$\bar{y} = \frac{2}{\pi r^2}\frac{1}{2} \int_{-r}^{r} \left(\sqrt{r^2 - x^2}\right)^2 dx$$

$$= \frac{1}{\pi r^2}\left[r^2 x - \frac{x^3}{3}\right]_{-r}^{r} = \frac{1}{\pi r^2}\left[\frac{4r^3}{3}\right] = \frac{4r}{3\pi}$$

$$(\bar{x}, \bar{y}) = \left(0, \frac{4r}{3\pi}\right)$$

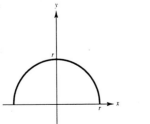

37. $\bar{x} = 0$ by symmetry

$$A = \frac{1}{2}\pi ab$$

$$\frac{1}{A} = \frac{2}{\pi ab}$$

$$\bar{y} = \frac{2}{\pi ab}\frac{1}{2} \int_{-a}^{a} \left(\frac{b}{a}\sqrt{a^2 - x^2}\right)^2 dx$$

$$= \frac{1}{\pi ab}\left(\frac{b^2}{a^2}\right)\left[a^2 x - \frac{x^3}{3}\right]_{-a}^{a} = \frac{b}{\pi a^3}\left[\frac{4a^3}{3}\right] = \frac{4b}{3\pi}$$

$$(\bar{x}, \bar{y}) = \left(0, \frac{4b}{3\pi}\right)$$

38. $A = \int_0^1 [1 - (2x - x^2)]\, dx = \frac{1}{3}$

$\frac{1}{A} = 3$

$\bar{x} = 3 \int_0^1 x[1 - (2x - x^2)]\, dx = 3 \int_0^1 [x - 2x^2 + x^3]\, dx = 3 \left[\frac{x^2}{2} - \frac{2}{3}x^3 + \frac{x^4}{4} \right]_0^1 = \frac{1}{4}$

$\bar{y} = 3 \int_0^1 \frac{[1 + (2x - x^2)]}{2}[1 - (2x - x^2)]\, dx = \frac{3}{2} \int_0^1 [1 - (2x - x^2)^2]\, dx$

$= \frac{3}{2} \int_0^1 [1 - 4x^2 + 4x^3 - x^4]\, dx = \frac{3}{2} \left[x - \frac{4}{3}x^3 + x^4 - \frac{x^5}{5} \right]_0^1 = \frac{7}{10}$

$(\bar{x}, \bar{y}) = \left(\frac{1}{4}, \frac{7}{10} \right)$

39. (a)

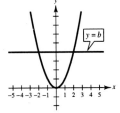

(b) $\bar{x} = 0$ by symmetry

(c) $M_y = \int_{-\sqrt{b}}^{\sqrt{b}} x(b - x^2)\, dx = 0$ because $bx - x^3$ is odd

(d) $\bar{y} > \frac{b}{2}$ since there is more area above $y = \frac{b}{2}$ than below

(e) $M_x = \int_{-\sqrt{b}}^{\sqrt{b}} \frac{(b + x^2)(b - x^2)}{2}\, dx$

$= \int_{-\sqrt{b}}^{\sqrt{b}} \frac{b^2 - x^4}{2}\, dx = \frac{1}{2} \left[b^2 x - \frac{x^5}{5} \right]_{-\sqrt{b}}^{\sqrt{b}}$

$= b^2 \sqrt{b} - \frac{b^2 \sqrt{b}}{5} = \frac{4b^2 \sqrt{b}}{5}$

$A = \int_{-\sqrt{b}}^{\sqrt{b}} (b - x^2)\, dx = \left[bx - \frac{x^3}{3} \right]_{-\sqrt{b}}^{\sqrt{b}}$

$= \left(b\sqrt{b} - \frac{b\sqrt{b}}{3} \right) 2 = 4\frac{b\sqrt{b}}{3}$

$\bar{y} = \frac{M_x}{A} = \frac{4b^2 \sqrt{b}/5}{4b\sqrt{b}/3} = \frac{3}{5}b.$

40. (a) $M_y = 0$ by symmetry

$M_y = \int_{-\sqrt[2n]{b}}^{\sqrt[2n]{b}} x(b - x^{2n})\, dx = 0$

because $bx - x^{2n+1}$ is an odd function.

(c) $M_x = \int_{-\sqrt[2n]{b}}^{\sqrt[2n]{b}} \frac{(b + x^{2n})(b - x^{2n})}{2}\, dx = \int_{-\sqrt[2n]{b}}^{\sqrt[2n]{b}} \frac{1}{2}(b^2 - x^{4n})\, dx$

$= \frac{1}{2} \left(b^2 x - \frac{x^{4n+1}}{4n+1} \right) \Big]_{-\sqrt[2n]{b}}^{\sqrt[2n]{b}}$

$= b^2 b^{1/2n} - \frac{b^{(4n+1)/2n}}{4n+1} = \frac{4n}{4n+1} b^{(4n+1)/2n}$

$A = \int_{-\sqrt[2n]{b}}^{\sqrt[2n]{b}} (b - x^{2n})\, dx = 2 \left[bx - \frac{x^{2n+1}}{2n+1} \right]_0^{\sqrt[2n]{b}}$

$= 2 \left[b \cdot b^{1/2n} - \frac{b^{(2n+1)/2n}}{2n+1} \right] = \frac{4n}{2n+1} b^{(2n+1)/2n}$

$\bar{y} = \frac{M_x}{A} = \frac{4n \, b^{(4n+1)/2n}/(4n+1)}{4n \, b^{(24n+1)/2n}/(2n+1)} = \frac{2n+1}{4n+1}b$

(b) $\bar{y} > \frac{b}{2}$ because there is more area above $y = \frac{b}{2}$ than below.

(d)

n	1	2	3	4
$\bar{y}$	$\frac{3}{5}b$	$\frac{5}{9}b$	$\frac{7}{13}b$	$\frac{9}{17}b$

(e) $\lim\limits_{n \to \infty} \bar{y} = \lim\limits_{n \to \infty} \frac{2n+1}{4n+1}b = \frac{1}{2}b$

(f) As $n \to \infty$, the figure gets narrower.

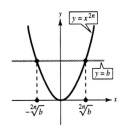

41. (a) $\bar{x} = 0$ by symmetry

$$A = 2 \int_0^{40} f(x)\,dx = \frac{2(40)}{3(4)}[30 + 4(29) + 2(26) + 4(20) + 0] = \frac{20}{3}(278) = \frac{5560}{3}$$

$$M_x = \int_{-40}^{40} \frac{f(x)^2}{2}\,dx = \frac{40}{3(4)}[30^2 + 4(29)^2 + 2(26)^2 + 4(20)^2 + 0] = \frac{10}{3}(7216) = \frac{72160}{3}$$

$$\bar{y} = \frac{M_x}{A} = \frac{72160/3}{5560/3} = \frac{72160}{5560} \approx 12.98$$

$(\bar{x}, \bar{y}) = (0, 12.98)$

(b) $y = (-1.02 \times 10^{-5})x^4 - 0.0019x^2 + 29.28$

(c) $\bar{y} = \frac{M_x}{A} \approx \frac{23697.68}{1843.54} \approx 12.85$

$(\bar{x}, \bar{y}) = (0, 12.85)$

42. Let $f(x)$ be the top curve, given by $l + d$. The bottom curve is $d(x)$.

x	0	0.5	1.0	1.5	2.0
f	2.0	1.93	1.73	1.32	0
d	0.50	0.48	0.43	0.33	0

(a) Area $= 2 \int_0^2 [f(x) - d(x)]\,dx$

$$\approx 2\frac{2}{3(4)}[1.50 + 4(1.45) + 2(1.30) + 4(.99) + 0]$$

$$= \frac{1}{3}[13.86] = 4.62$$

$$M_x = \int_{-2}^2 \frac{f(x) + d(x)}{2}(f(x) - d(x))\,dx$$

$$= \int_0^2 [f(x)^2 - d(x)^2]\,dx$$

$$= \frac{2}{3(4)}[3.75 + 4(3.4945) + 2(2.808) + 4(1.6335) + 0]$$

$$= \frac{1}{6}[29.878] = 4.9797$$

$$\bar{y} = \frac{M_x}{A} = \frac{4.9797}{4.62} = 1.078$$

$(\bar{x}, \bar{y}) = (0, 1.078)$

(b) $f(x) = -0.1061x^4 - 0.06126x^2 + 1.9527$

$d(x) = -0.02648x^4 - 0.01497x^2 + .4862$

(c) $\bar{y} = \frac{M_x}{A} \approx \frac{4.9133}{4.59998} = 1.068$

$(\bar{x}, \bar{y}) = (0, 1.068)$

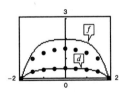

43. Centroids of the given regions: $(1, 0)$ and $(3, 0)$

Area: $A = 4 + \pi$

$$\bar{x} = \frac{4(1) + \pi(3)}{4 + \pi} = \frac{4 + 3\pi}{4 + \pi}$$

$$\bar{y} = \frac{4(0) + \pi(0)}{4 + \pi} = 0$$

$$(\bar{x}, \bar{y}) = \left(\frac{4 + 3\pi}{4 + \pi}, 0\right) \approx (1.88, 0)$$

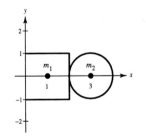

44. Centroids of the given regions: $\left(\frac{1}{2}, \frac{3}{2}\right)$, $\left(2, \frac{1}{2}\right)$, and $\left(\frac{7}{2}, 1\right)$

Area: $A = 3 + 2 + 2 = 7$

$$\bar{x} = \frac{3(1/2) + 2(2) + 2(7/2)}{7} = \frac{25/2}{7} = \frac{25}{14}$$

$$\bar{y} = \frac{3(3/2) + 2(1/2) + 2(1)}{7} = \frac{15/2}{7} = \frac{15}{14}$$

$$(\bar{x}, \bar{y}) = \left(\frac{25}{14}, \frac{15}{14}\right)$$

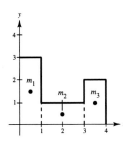

45. Centroids of the given regions: $\left(0, \frac{3}{2}\right)$, $(0, 5)$, and $\left(0, \frac{15}{2}\right)$

Area: $A = 15 + 12 + 7 = 34$

$$\bar{x} = \frac{15(0) + 12(0) + 7(0)}{34} = 0$$

$$\bar{y} = \frac{15(3/2) + 12(5) + 7(15/2)}{34} = \frac{135}{34}$$

$$(\bar{x}, \bar{y}) = \left(0, \frac{135}{34}\right)$$

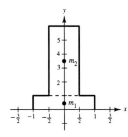

46. $m_1 = \frac{7}{8}(2) = \frac{7}{4}, P_1 = \left(0, \frac{7}{16}\right)$

$$m_2 = \frac{7}{8}\left(6 - \frac{7}{8}\right) = \frac{287}{64}, P_2 = \left(0, \frac{55}{16}\right)$$

By symmetry, $\bar{x} = 0$.

$$\bar{y} = \frac{(7/4)(7/16) + (287/64)(55/16)}{(7/4) + (287/64)} = \frac{16,569}{6384} = \frac{5523}{2128}$$

$$(\bar{x}, \bar{y}) = \left(0, \frac{5523}{2128}\right) \approx (0, 2.595)$$

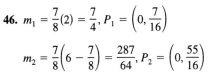

47. Centroids of the given regions: $(1, 0)$ and $(3, 0)$

Mass: $4 + 2\pi$

$$\bar{x} = \frac{4(1) + 2\pi(3)}{4 + 2\pi} = \frac{2 + 3\pi}{2 + \pi}$$

$$\bar{y} = 0$$

$$(\bar{x}, \bar{y}) = \left(\frac{2 + 3\pi}{2 + \pi}, 0\right) \approx (2.22, 0)$$

48. Centroids of the given regions: $(3, 0)$ and $(1, 0)$

Mass: $8 + \pi$

$$\bar{y} = 0$$

$$\bar{x} = \frac{8(1) + \pi(3)}{8 + \pi} = \frac{8 + 3\pi}{8 + \pi}$$

$$(\bar{x}, \bar{y}) = \left(\frac{8 + 3\pi}{8 + \pi}, 0\right) \approx (1.56, 0)$$

49. $V = 2\pi r A = 2\pi(5)(16\pi) = 160\pi^2 \approx 1579.14$

50. $V = 2\pi r A = 2\pi(3)(4\pi) = 24\pi^2$

51. $A = \frac{1}{2}(4)(4) = 8$

$$\bar{y} = \left(\frac{1}{8}\right)\frac{1}{2}\int_0^4 (4 + x)(4 - x)\, dx = \frac{1}{16}\left[16x - \frac{x^3}{3}\right]_0^4 = \frac{8}{3}$$

$$r = \bar{y} = \frac{8}{3}$$

$$V = 2\pi r A = 2\pi\left(\frac{8}{3}\right)(8) = \frac{128\pi}{3} \approx 134.04$$

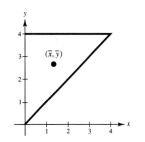

52. $A = \int_1^5 \sqrt{x-1}\, dx = \left[\frac{2}{3}(x-1)^{3/2} \right]_1^5 = \frac{16}{3}$

$\bar{x} = \frac{3}{16} \int_1^5 x\sqrt{x-1}\, dx$ Substitution: $u = x - 1$

$= \frac{3}{16} \left[\frac{2}{5}(x-1)^{5/2} + \frac{2}{3}(x-1)^{3/2} \right]_1^5 = \frac{17}{5}$

$r = \bar{x} = \frac{17}{5}$

$V = 2\pi r A = 2\pi \left(\frac{17}{5} \right) \left(\frac{16}{3} \right) = \frac{544\pi}{15} \approx 113.94$

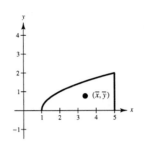

53. The surface area of the sphere is $S = 4\pi r^2$. The arc length of C is $s = \pi r$. The distance traveled by the centroid is

$$d = \frac{S}{s} = \frac{4\pi r^2}{\pi r} = 4r.$$

This distance is also the circumference of the circle of radius y.

$$d = 2\pi y$$

Thus, $2\pi y = 4r$ and we have $y = 2r/\pi$. Therefore, the centroid of the semicircle $y = \sqrt{r^2 - x^2}$ is $(0, 2r/\pi)$.

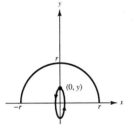

54. The centroid of the circle is $(1, 0)$. The distance traveled by the centroid is 2π. The arc length of the circle is also 2π. Therefore, $S = (2\pi)(2\pi) = 4\pi^2$.

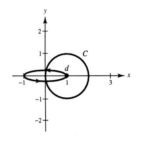

55. $A = \int_0^1 x^n\, dx = \left[\frac{x^{n+1}}{n+1} \right]_0^1 = \frac{1}{n+1}$

$m = \rho A = \frac{\rho}{n+1}$

$M_x = \frac{\rho}{2} \int_0^1 (x^n)^2\, dx = \left[\frac{\rho}{2} \cdot \frac{x^{2n+1}}{2n+1} \right]_0^1 = \frac{\rho}{2(2n+1)}$

$M_y = \rho \int_0^1 x(x^n)\, dx = \left[\rho \cdot \frac{x^{n+2}}{n+2} \right]_0^1 = \frac{\rho}{n+2}$

$\bar{x} = \frac{M_y}{m} = \frac{n+1}{n+2}$

$\bar{y} = \frac{M_x}{m} = \frac{n+1}{2(2n+1)} = \frac{n+1}{4n+2}$

Centroid: $\left(\frac{n+1}{n+2}, \frac{n+1}{4n+2} \right)$

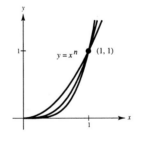

As $n \to \infty$, $(\bar{x}, \bar{y}) \to \left(1, \frac{1}{4} \right)$. The graph approaches the x-axis and the line $x = 1$ as $n \to \infty$.

Section 6.7   Fluid Pressure and Fluid Force

1. $F = PA = [62.4(5)](3) = 936$ lb

2. $F = PA = [62.4(5)](18) = 5616$ lb

3. $F = 62.4(h + 2)(6) - (62.4)(h)(6)$

$= 62.4(2)(6) = 748.8$ lb

4. $F = 62.4(h + 4)(48) - (62.4)(h)(48)$

$= 62.4(4)(48) = 11{,}980.8$ lb

5. $h(y) = 3 - y$

$L(y) = 4$

$F = 62.4 \displaystyle\int_0^3 (3 - y)(4) \, dy$

$= 249.6 \displaystyle\int_0^3 (3 - y) \, dy$

$= 249.6 \left[3y - \dfrac{y^2}{2} \right]_0^3 = 1123.2$ lb

6. $h(y) = 3 - y$

$L(y) = \dfrac{4}{3}y$

$F = 62.4 \displaystyle\int_0^3 (3 - y)\left(\dfrac{4}{3}y \right) dy$

$= \dfrac{4}{3}(62.4) \displaystyle\int_0^3 (3y - y^2) \, dy$

$= \dfrac{4}{3}(62.4) \left[\dfrac{3y^2}{2} - \dfrac{y^3}{3} \right]_0^3 = 374.4$ lb

Force is one-third that of Exercise 5.

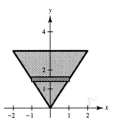

7. $h(y) = 3 - y$

$L(y) = 2\left(\dfrac{y}{3} + 1 \right)$

$F = 2(62.4) \displaystyle\int_0^3 (3 - y)\left(\dfrac{y}{3} + 1 \right) dy$

$= 124.8 \displaystyle\int_0^3 \left(3 - \dfrac{y^2}{3} \right) dy$

$= 124.8 \left[3y - \dfrac{y^3}{9} \right]_0^3 = 748.8$ lb

8. $h(y) = -y$

$L(y) = 2\sqrt{4 - y^2}$

$F = 62.4 \displaystyle\int_{-2}^0 (-y)(2)\sqrt{4 - y^2} \, dy$

$= \left[62.4\left(\dfrac{2}{3} \right)(4 - y^2)^{3/2} \right]_{-2}^0 = 332.8$ lb

9. $h(y) = 4 - y$

$L(y) = 2\sqrt{y}$

$F = 2(62.4) \displaystyle\int_0^4 (4 - y)\sqrt{y}\, dy$

$\quad = 124.8 \displaystyle\int_0^4 (4y^{1/2} - y^{3/2})\, dy$

$\quad = 124.8 \left[\dfrac{8y^{3/2}}{3} - \dfrac{2y^{5/2}}{5}\right]_0^4 = 1064.96 \text{ lb}$

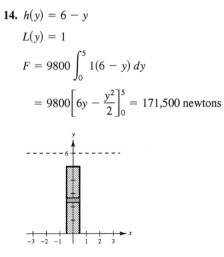

10. $h(y) = -y$

$L(y) = \dfrac{4}{3}\sqrt{9 - y^2}$

$F = 62.4 \displaystyle\int_{-3}^0 (-y)\dfrac{4}{3}\sqrt{9 - y^2}\, dy$

$\quad = 62.4\left(\dfrac{2}{3}\right)\displaystyle\int_{-3}^0 (9 - y^2)^{1/2}(-2y)\, dy$

$\quad = \left[62.4\left(\dfrac{4}{9}\right)(9 - y^2)^{3/2}\right]_{-3}^0 = 748.8 \text{ lb}$

11. $h(y) = 4 - y$

$L(y) = 2$

$F = 9800 \displaystyle\int_0^2 2(4 - y)\, dy$

$\quad = 9800\left[8y - y^2\right]_0^2 = 117{,}600 \text{ newtons}$

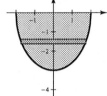

12. $h(y) = \left(1 + 2\sqrt{2}\right) - y$

$L_1(y) = 2y$ [lower part]

$L_2(y) = 2\left(2\sqrt{2} - y\right)$ [upper part]

$F = 19{,}600\left[\displaystyle\int_0^{\sqrt{2}} \left(1 + 2\sqrt{2} - y\right)y\, dy + \displaystyle\int_{\sqrt{2}}^{2\sqrt{2}} \left(1 + 2\sqrt{2} - y\right)\left(2\sqrt{2} - y\right)\, dy\right]$

$\quad = 19{,}600\left(\left[\dfrac{y^2}{2} + \sqrt{2}\,y^2 - \dfrac{y^3}{3}\right]_0^{\sqrt{2}} + \left[2\sqrt{2}\,y + 8y - 2\sqrt{2}\,y^2 - \dfrac{y^2}{2} + \dfrac{y^3}{3}\right]_{\sqrt{2}}^{2\sqrt{2}}\right)$

$\quad = 39{,}200\left(1 + \sqrt{2}\right) \approx 94{,}637.17 \text{ newtons}$

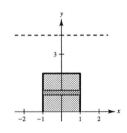

13. $h(y) = 12 - y$

$L(y) = 6 - \dfrac{2y}{3}$

$F = 9800 \displaystyle\int_0^9 (12 - y)\left(6 - \dfrac{2y}{3}\right) dy$

$\quad = 9800\left[72y - 7y^2 + \dfrac{2y^3}{9}\right]_0^9 = 2{,}381{,}400 \text{ newtons}$

14. $h(y) = 6 - y$

$L(y) = 1$

$F = 9800 \displaystyle\int_0^5 1(6 - y)\, dy$

$\quad = 9800\left[6y - \dfrac{y^2}{2}\right]_0^5 = 171{,}500 \text{ newtons}$

15. $h(y) = 2 - y$

$L(y) = 10$

$F = 140.7 \displaystyle\int_0^2 (2 - y)(10)\, dy$

$\quad = 1407 \displaystyle\int_0^2 (2 - y)\, dy$

$\quad = 1407 \left[2y - \dfrac{y^2}{2} \right]_0^2 = 2814 \text{ lb}$

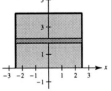

16. $h(y) = -y$

$L(y) = 2\left(\dfrac{4}{3}\sqrt{9 - y^2} \right)$

$F = 140.7 \displaystyle\int_{-3}^0 (-y)(2)\left(\dfrac{4}{3}\sqrt{9 - y^2} \right) dy$

$\quad = \dfrac{(140.7)(4)}{3} \displaystyle\int_{-3}^0 \sqrt{9 - y^2}\,(-2y)\, dy$

$\quad = \left[\dfrac{(140.7)(4)}{3}\left(\dfrac{2}{3} \right)(9 - y^2)^{3/2} \right]_{-3}^0 = 3376.8 \text{ lb}$

17. $h(y) = 4 - y$

$L(y) = 6$

$F = 140.7 \displaystyle\int_0^4 (4 - y)(6)\, dy$

$\quad = 844.2 \displaystyle\int_0^4 (4 - y)\, dy$

$\quad = 844.2 \left[4y - \dfrac{y^2}{2} \right]_0^4 = 6753.6 \text{ lb}$

18. $h(y) = -y$

$L(y) = 6 + \dfrac{3}{2}y$

$F = 140.7 \displaystyle\int_{-4}^0 (-y)\left(6 + \dfrac{3}{2}y \right) dy$

$\quad = \left[-140.7\left(3y^2 + \dfrac{y^3}{2} \right) \right]_{-4}^0 = 2251.2 \text{ lb}$

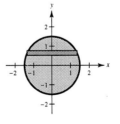

19. $h(y) = -y$

$L(y) = 2\left(\dfrac{1}{2} \right)\sqrt{9 - 4y^2}$

$F = 42 \displaystyle\int_{-3/2}^0 (-y)\sqrt{9 - 4y^2}\, dy$

$\quad = \dfrac{42}{8} \displaystyle\int_{-3/2}^0 (9 - 4y^2)^{1/2}(-8y)\, dy$

$\quad = \left[\left(\dfrac{21}{4} \right)\left(\dfrac{2}{3} \right)(9 - 4y^2)^{3/2} \right]_{-3/2}^0 = 94.5 \text{ lb}$

20. $h(y) = \dfrac{3}{2} - y$

$L(y) = 2\left(\dfrac{1}{2}\right)\sqrt{9 - 4y^2}$

$F = 42 \displaystyle\int_{-3/2}^{3/2} \left(\dfrac{3}{2} - y\right)\sqrt{9 - 4y^2}\, dy = 63 \int_{-3/2}^{3/2} \sqrt{9 - 4y^2}\, dy + \dfrac{21}{4}\int_{-3/2}^{3/2} \sqrt{9 - 4y^2}\,(-8y)\, dy$

The second integral is zero since it is an odd function and the limits of integration are symmetric to the origin. The first integral is twice the area of a semicircle of radius $\dfrac{3}{2}$.

$$\left(\sqrt{9 - 4y^2} = 2\sqrt{(9/4) - y^2}\right)$$

Thus, the force is $63\left(\dfrac{9}{4}\pi\right) = 141.75\pi \approx 445.32$ lb.

21. $h(y) = k - y$

$L(y) = 2\sqrt{r^2 - y^2}$

$F = w \displaystyle\int_{-r}^{r} (k - y)\sqrt{r^2 - y^2}\,(2)\, dy$

$= w\left[2k \displaystyle\int_{-r}^{r} \sqrt{r^2 - y^2}\, dy + \int_{-r}^{r} \sqrt{r^2 - y^2}\,(-2y)\, dy\right]$

The second integral is zero since its integrand is odd and the limits of integration are symmetric to the origin. The first integral is the area of a semicircle with radius r.

$$F = w\left[(2k)\dfrac{\pi r^2}{2} + 0\right] = wk\pi r^2$$

22. $h(y) = k - y$

$L(y) = b$

$F = w \displaystyle\int_{-h/2}^{h/2} (k - y)b\, dy$

$= wb\left[ky - \dfrac{y^2}{2}\right]_{-h/2}^{h/2} = wb(hk) = wkhb$

23. From Exercise 22:

$F = 64(15)(1)(1) = 960$ lb

24. From Exercise 21:

$F = 64(15)\pi\left(\dfrac{1}{2}\right)^2 \approx 753.98$ lb

25. $h(y) = 4 - y$

$F = 62.4 \displaystyle\int_{0}^{4} (4 - y)L(y)\, dy$

Using Simpson's Rule with $n = 8$ we have:

$F \approx 62.4\left(\dfrac{4 - 0}{3(8)}\right)[0 + 4(3.5)(3) + 2(3)(5) + 4(2.5)(8) + 2(2)(9) + 4(1.5)(10) + 2(1)(10.25) + 4(0.5)(10.5) + 0]$

$= 3010.8$ lb

26. $h(y) = 3 - y$

Solving $y = 5x^2/(x^2 + 4)$ for x, you obtain

$$x = \sqrt{4y/(5 - y)}.$$

$$L(y) = 2\sqrt{\frac{4y}{5 - y}}$$

$$F = 62.4(2)\int_0^3 (3 - y)\sqrt{\frac{4y}{5 - y}}\, dy$$

$$= 2(124.8)\int_0^3 (3 - y)\sqrt{\frac{y}{5 - y}}\, dy \approx 546.265 \text{ lb}$$

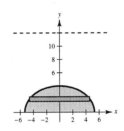

27. $h(y) = 12 - y$

$$L(y) = 2(4^{2/3} - y^{2/3})^{3/2}$$

$$F = 62.4\int_0^4 2(12 - y)(4^{2/3} - y^{2/3})^{3/2}\, dy$$

$$\approx 6448.73 \text{ lb}$$

28. $h(y) = 12 - y$

$$L(y) = 2\frac{\sqrt{7(16 - y^2)}}{2} = \sqrt{7(16 - y^2)}$$

$$F = 62.4\int_0^4 (12 - y)\sqrt{7(16 - y^2)}\, dy$$

$$= 62.4\sqrt{7}\int_0^4 (12 - y)\sqrt{16 - y^2}\, dy \approx 21373.7 \text{ lb}$$

29. (a) If the fluid force is one half of 1123.2 lb, and the height of the water is b, then

$$h(y) = b - y$$

$$L(y) = 4$$

$$F = 62.4\int_0^b (b - y)(4)\, dy = \frac{1}{2}(1123.2)$$

$$\int_0^b (b - y)\, dy = 2.25$$

$$\left[by - \frac{y^2}{2}\right]_0^b = 2.25$$

$$b^2 - \frac{b^2}{2} = 2.25$$

$$b^2 = 4.5 \implies b \approx 2.12 \text{ ft.}$$

(b) The pressure increases with increasing depth.

30. (a) Wall at shallow end

From Exercise 22: $F = 62.4(2)(4)(20) = 9984$ lb

(b) Wall at deep end

From Exercise 22: $F = 62.4(4)(8)(20) = 39{,}936$ lb

(c) Side wall

From Exercise 22: $F_1 = 62.4(2)(4)(40) = 19{,}968$ lb

$$F_2 = 62.4\int_0^4 (8 - y)(10y)\, dy$$

$$= 624\int_0^4 (8y - y^2)\, dy = 624\left[4y^2 - \frac{y^3}{3}\right]_0^4$$

$$= 26{,}624 \text{ lb}$$

Total force: $F_1 + F_2 = 46{,}592$ lb

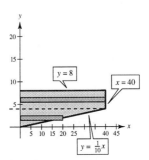

Review Exercises for Chapter 6

1. $A = \int_1^5 \dfrac{1}{x^2}\,dx = \left[-\dfrac{1}{x}\right]_1^5 = \dfrac{4}{5}$

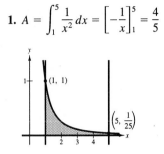

2. $A = \int_{1/2}^5 \left(4 - \dfrac{1}{x^2}\right) dx$

$= \left[4x + \dfrac{1}{x}\right]_{1/2}^5 = \dfrac{81}{5}$

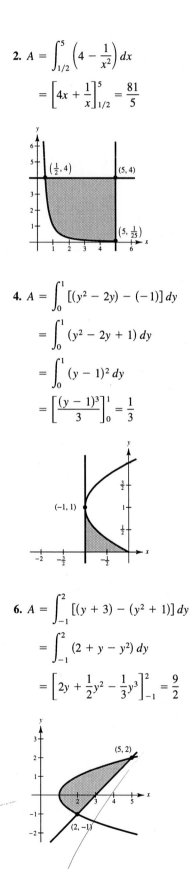

3. $A = \int_{-1}^1 \dfrac{1}{x^2 + 1}\,dx$

$= \left[\arctan x\right]_{-1}^1$

$= \dfrac{\pi}{4} - \left(-\dfrac{\pi}{4}\right) = \dfrac{\pi}{2}$

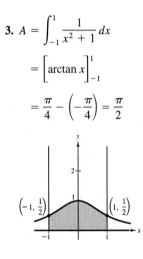

4. $A = \int_0^1 \left[(y^2 - 2y) - (-1)\right] dy$

$= \int_0^1 (y^2 - 2y + 1)\,dy$

$= \int_0^1 (y - 1)^2\,dy$

$= \left[\dfrac{(y - 1)^3}{3}\right]_0^1 = \dfrac{1}{3}$

5. $A = 2\int_0^1 (x - x^3)\,dx$

$= 2\left[\dfrac{1}{2}x^2 - \dfrac{1}{4}x^4\right]_0^1$

$= \dfrac{1}{2}$

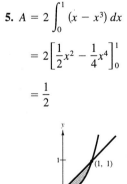

6. $A = \int_{-1}^2 \left[(y + 3) - (y^2 + 1)\right] dy$

$= \int_{-1}^2 (2 + y - y^2)\,dy$

$= \left[2y + \dfrac{1}{2}y^2 - \dfrac{1}{3}y^3\right]_{-1}^2 = \dfrac{9}{2}$

7. $A = \displaystyle\int_0^2 (e^2 - e^x)\, dx$

$\quad = \Big[xe^2 - e^x \Big]_0^2$

$\quad = e^2 + 1$

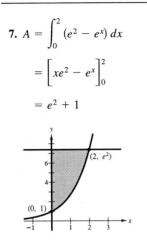

8. $A = 2 \displaystyle\int_{\pi/6}^{\pi/2} (2 - \csc x)\, dx$

$\quad = 2\Big[2x - \ln|\csc x - \cot x| \Big]_{\pi/6}^{\pi/2}$

$\quad = 2\Big([\pi - 0] - \Big[\dfrac{\pi}{3} - \ln(2 - \sqrt{3}) \Big] \Big)$

$\quad = 2\Big[\dfrac{2\pi}{3} + \ln(2 - \sqrt{3}) \Big] \approx 1.555$

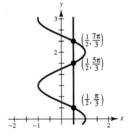

9. $A = \displaystyle\int_{\pi/4}^{5\pi/4} (\sin x - \cos x)\, dx$

$\quad = \Big[-\cos x - \sin x \Big]_{\pi/4}^{5\pi/4}$

$\quad = \Big(\dfrac{1}{\sqrt{2}} + \dfrac{1}{\sqrt{2}} \Big) - \Big(-\dfrac{1}{\sqrt{2}} - \dfrac{1}{\sqrt{2}} \Big)$

$\quad = \dfrac{4}{\sqrt{2}} = 2\sqrt{2}$

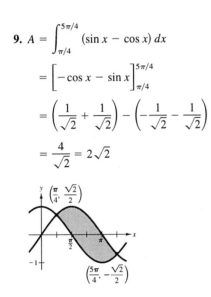

10. $A = \displaystyle\int_{\pi/3}^{5\pi/3} \Big(\dfrac{1}{2} - \cos y \Big)\, dy + \int_{5\pi/3}^{7\pi/3} \Big(\cos y - \dfrac{1}{2} \Big)\, dy$

$\quad = \Big[\dfrac{y}{2} - \sin y \Big]_{\pi/3}^{5\pi/3} + \Big[\sin y - \dfrac{y}{2} \Big]_{5\pi/3}^{7\pi/3}$

$\quad = \dfrac{\pi}{3} + 2\sqrt{3}$

11. $A = \displaystyle\int_0^8 \big[(3 + 8x - x^2) - (x^2 - 8x + 3) \big]\, dx$

$\quad = \displaystyle\int_0^8 (16x - 2x^2)\, dx$

$\quad = \Big[8x^2 - \dfrac{2}{3}x^3 \Big]_0^8 = \dfrac{512}{3} \approx 170.667$

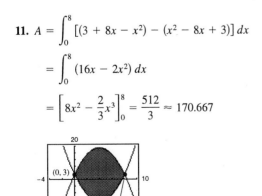

12. Point of intersection is given by:

$\quad x^3 - x^2 + 4x - 3 = 0 \implies x \approx 0.783.$

$\quad A \approx \displaystyle\int_0^{0.783} (3 - 4x + x^2 - x^3)\, dx$

$\quad = \Big[3x - 2x^2 + \dfrac{1}{3}x^3 - \dfrac{1}{4}x^4 \Big]_0^{0.783}$

$\quad \approx 1.189$

13. $y = \left(1 - \sqrt{x}\right)^2$

$$A = \int_0^1 \left(1 - \sqrt{x}\right)^2 dx$$

$$= \int_0^1 \left(1 - 2x^{1/2} + x\right) dx$$

$$= \left[x - \frac{4}{3}x^{3/2} + \frac{1}{2}x^2\right]_0^1 = \frac{1}{6} \approx 0.1667$$

14. $A = 2\int_0^2 \left[2x^2 - (x^4 - 2x^2)\right] dx$

$$= 2\int_0^2 (4x^2 - x^4) dx$$

$$= 2\left[\frac{4}{3}x^3 - \frac{1}{5}x^5\right]_0^2 = \frac{128}{15} \approx 8.5333$$

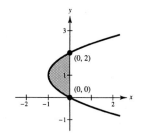

15. $x = y^2 - 2y \Rightarrow x + 1 = (y - 1)^2 \Rightarrow y = 1 \pm \sqrt{x + 1}$

$$A = \int_{-1}^0 \left[\left(1 + \sqrt{x + 1}\right) - \left(1 - \sqrt{x + 1}\right)\right] dx = \int_{-1}^0 2\sqrt{x + 1}\, dx$$

$$A = \int_0^2 \left[0 - (y^2 - 2y)\right] dy = \int_0^2 (2y - y^2)\, dy = \left[y^2 - \frac{1}{3}y^3\right]_0^2 = \frac{4}{3}$$

16. $y = \sqrt{x - 1} \Rightarrow x = y^2 + 1$

$$y = \frac{x - 1}{2} \Rightarrow x = 2y + 1$$

$$A = \int_0^2 \left[(2y + 1) - (y^2 + 1)\right] dy$$

$$= \int_1^5 \left[\sqrt{x - 1} - \frac{x - 1}{2}\right] dx$$

$$= \left[\frac{2}{3}(x - 1)^{3/2} - \frac{1}{4}(x - 1)^2\right]_1^5 = \frac{4}{3}$$

17. $A = \int_0^2 \left[1 - \left(1 - \frac{x}{2}\right)\right] dx + \int_2^3 \left[1 - (x - 2)\right] dx$

$$= \int_0^2 \frac{x}{2}\, dx + \int_2^3 (3 - x)\, dx$$

$$y = 1 - \frac{x}{2} \Rightarrow x = 2 - 2y$$

$$y = x - 2 \Rightarrow x = y + 2,\ y = 1$$

$$A = \int_0^1 \left[(y + 2) - (2 - 2y)\right] dy$$

$$= \int_0^1 3y\, dy = \left[\frac{3}{2}y^2\right]_0^1 = \frac{3}{2}$$

18. $A = \int_0^1 2\, dx + \int_1^5 \left[2 - \sqrt{x - 1}\right] dx$

$$x = y^2 + 1$$

$$A = \int_0^2 (y^2 + 1)\, dy$$

$$= \left[\frac{1}{3}y^3 + y\right]_0^2 = \frac{14}{3}$$

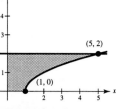

19. Job 1 is better. The salary for Job 1 is greater than the salary for Job 2 for all the years except the first and 10th years.

20. (a) $\displaystyle\int_{3}^{12} [(15.9696t - 6.318) - (8.581t + 6.965)]\, dt = \int_{3}^{12} [7.3886t - 13.283]\, dt$

$$= \left[\frac{7.388t^2}{2} - 13.283t\right]_{3}^{12}$$

$$= 379.1835 \approx 379 \text{ million dollars}$$

(b) $\displaystyle\int_{3}^{12} [(15.9696t - 6.318) - (6.214t + 10.345)]\, dt = \int_{3}^{12} [9.7556t - 16.663]\, dt$

$$= \left[\frac{9.7556t^2}{2} - 16.663\right]_{3}^{12}$$

$$= 508.536 \approx 509 \text{ million dollars}$$

21. (a) **Disc**

$$V = \pi \int_{0}^{4} x^2\, dx = \left[\frac{\pi x^3}{3}\right]_{0}^{4} = \frac{64\pi}{3}$$

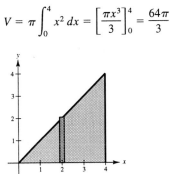

(b) **Shell**

$$V = 2\pi \int_{0}^{4} x^2\, dx = \left[\frac{2\pi}{3}x^3\right]_{0}^{4} = \frac{128\pi}{3}$$

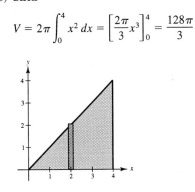

(c) **Shell**

$$V = 2\pi \int_{0}^{4} (4 - x)x\, dx$$

$$= 2\pi \int_{0}^{4} (4x - x^2)\, dx$$

$$= 2\pi \left[2x^2 - \frac{x^3}{3}\right]_{0}^{4} = \frac{64\pi}{3}$$

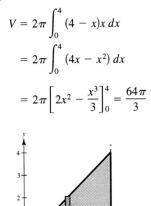

(d) **Shell**

$$V = 2\pi \int_{0}^{4} (6 - x)x\, dx$$

$$= 2\pi \int_{0}^{4} (6x - x^2)\, dx$$

$$= 2\pi \left[3x^2 - \frac{1}{3}x^3\right]_{0}^{4} = \frac{160\pi}{3}$$

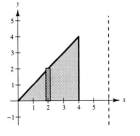

22. (a) Shell

$$V = 2\pi \int_0^2 y^3 \, dy = \left[\frac{\pi}{2}y^4\right]_0^2 = 8\pi$$

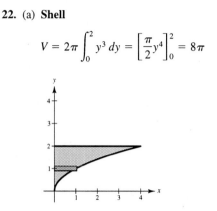

(c) Disc

$$V = \pi \int_0^2 y^4 \, dy = \left[\frac{\pi}{5}y^5\right]_0^2 = \frac{32\pi}{5}$$

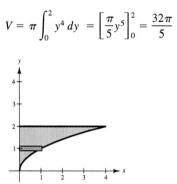

(b) Shell

$$V = 2\pi \int_0^2 (2 - y)y^2 \, dy$$

$$= 2\pi \int_0^2 (2y^2 - y^3) \, dy$$

$$= 2\pi\left[\frac{2}{3}y^3 - \frac{1}{4}y^4\right]_0^2 = \frac{8\pi}{3}$$

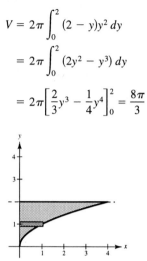

(d) Disc

$$V = \pi \int_0^2 \left[(y^2 + 1)^2 - 1^2\right] dy$$

$$= \pi \int_0^2 (y^4 + 2y^2) \, dy$$

$$= \pi\left[\frac{1}{5}y^5 + \frac{2}{3}y^3\right]_0^2 = \frac{176\pi}{15}$$

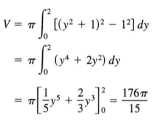

23. (a) Shell

$$V = 4\pi \int_0^4 x\left(\frac{3}{4}\right)\sqrt{16 - x^2} \, dx$$

$$= \left[3\pi\left(-\frac{1}{2}\right)\left(\frac{2}{3}\right)(16 - x^2)^{3/2}\right]_0^4 = 64\pi$$

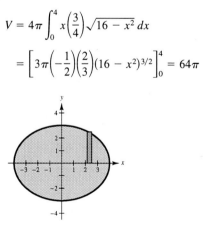

(b) Disc

$$V = 2\pi \int_0^4 \left[\frac{3}{4}\sqrt{16 - x^2}\right]^2 dx$$

$$= \frac{9\pi}{8}\left[16x - \frac{x^3}{3}\right]_0^4 = 48\pi$$

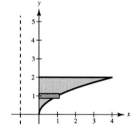

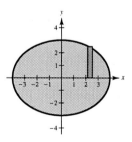

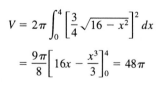

24. (a) Shell

$$V = 4\pi \int_0^a (x) \frac{b}{a} \sqrt{a^2 - x^2}\, dx$$

$$= \frac{-2\pi b}{a} \int_0^a (a^2 - x^2)^{1/2}(-2x)\, dx$$

$$= \left[\frac{-4\pi b}{3a}(a^2 - x^2)^{3/2} \right]_0^a = \frac{4}{3}\pi a^2 b$$

(b) Disc

$$V = 2\pi \int_0^a \frac{b^2}{a^2}(a^2 - x^2)\, dx$$

$$= \frac{2\pi b^2}{a^2} \left[a^2 x - \frac{1}{3}x^3 \right]_0^a$$

$$= \frac{4}{3}\pi a b^2$$

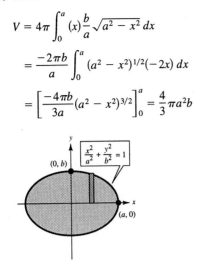

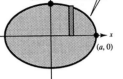

25. Shell

$$V = 2\pi \int_0^1 \frac{x}{x^4 + 1}\, dx$$

$$= \pi \int_0^1 \frac{(2x)}{(x^2)^2 + 1}\, dx$$

$$= \left[\pi \arctan(x^2) \right]_0^1$$

$$= \pi \left[\frac{\pi}{4} - 0 \right] = \frac{\pi^2}{4}$$

26. Disc

$$V = 2\pi \int_0^1 \left[\frac{1}{\sqrt{1 + x^2}} \right]^2 dx$$

$$= \left[2\pi \arctan x \right]_0^1$$

$$= 2\pi \left(\frac{\pi}{4} - 0 \right)$$

$$= \frac{\pi^2}{2}$$

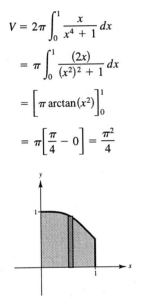

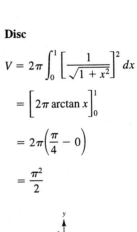

27. Shell

$$u = \sqrt{x - 2}$$

$$x = u^2 + 2$$

$$dx = 2u\, du$$

$$V = 2\pi \int_2^6 \frac{x}{1 + \sqrt{x - 2}}\, dx = 4\pi \int_0^2 \frac{(u^2 + 2)u}{1 + u}\, du$$

$$= 4\pi \int_0^2 \frac{u^3 + 2u}{1 + u}\, du = 4\pi \int_0^2 \left(u^2 - u + 3 - \frac{3}{1 + u} \right) du$$

$$= 4\pi \left[\frac{1}{3}u^3 - \frac{1}{2}u^2 + 3u - 3\ln(1 + u) \right]_0^2 = \frac{4\pi}{3}(20 - 9\ln 3) \approx 42.359$$

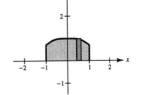

28. Disc

$$V = \pi \int_0^1 (e^{-x})^2 \, dx$$

$$= \pi \int_0^1 e^{-2x} \, dx = \left[-\frac{\pi}{2} e^{-2x} \right]_0^1$$

$$= \left(\frac{-\pi}{2e^2} + \frac{\pi}{2} \right) = \frac{\pi}{2}\left(1 - \frac{1}{e^2} \right)$$

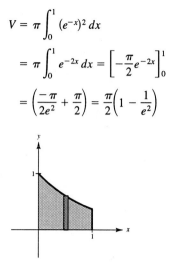

29. Since $y \le 0$, $A = -\int_{-1}^0 x\sqrt{x+1} \, dx$.

$$u = x + 1$$

$$x = u - 1$$

$$dx = du$$

$$A = -\int_0^1 (u-1)\sqrt{u} \, du = -\int_0^1 (u^{3/2} - u^{1/2}) \, du$$

$$= -\left[\frac{2}{5} u^{5/2} - \frac{2}{3} u^{3/2} \right]_0^1 = \frac{4}{15}$$

30. (a) Disc

$$V = \pi \int_{-1}^0 x^2(x+1) \, dx$$

$$= \pi \int_{-1}^0 (x^3 + x^2) \, dx$$

$$= \pi \left[\frac{x^4}{4} + \frac{x^3}{3} \right]_{-1}^0 = \frac{\pi}{12}$$

(b) Shell

$$u = \sqrt{x+1}$$

$$x = u^2 - 1$$

$$dx = 2u \, du$$

$$V = 2\pi \int_{-1}^0 x^2 \sqrt{x+1} \, dx$$

$$= 4\pi \int_0^1 (u^2 - 1)^2 u^2 \, du$$

$$= 4\pi \int_0^1 (u^6 - 2u^4 + u^2) \, du$$

$$= 4\pi \left[\frac{1}{7} u^7 - \frac{2}{5} u^5 + \frac{1}{3} u^3 \right]_0^1 = \frac{32\pi}{105}$$

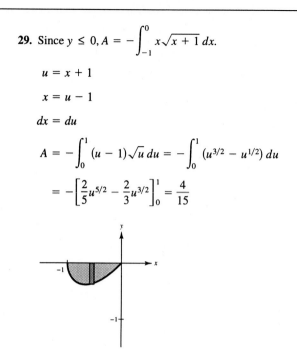

31. From Exercise 23(a) we have: $V = 64\pi$ ft^3

$$\frac{1}{4}V = 16\pi$$

Disc: $\pi \int_{-3}^{y_0} \frac{16}{9}(9 - y^2)\,dy = 16\pi$

$$\frac{1}{9} \int_{-3}^{y_0} (9 - y^2)\,dy = 1$$

$$\left[9y - \frac{1}{3}y^3 \right]_{-3}^{y_0} = 9$$

$$\left(9y_0 - \frac{1}{3}y_0^3 \right) - (-27 + 9) = 9$$

$$y_0^3 - 27y_0 - 27 = 0$$

By Newton's Method, $y_0 \approx -1.042$ and the depth of the gasoline is $3 - 1.042 = 1.958$ ft.

32. $A(x) = \frac{1}{2}bh = \frac{1}{2}\left(2\sqrt{a^2 - x^2}\right)\left(\sqrt{3}\sqrt{a^2 - x^2}\right)$

$$= \sqrt{3}(a^2 - x^2)$$

$$V = \sqrt{3} \int_{-a}^{a} (a^2 - x^2)\,dx = \sqrt{3}\left[a^2x - \frac{x^3}{3} \right]_{-a}^{a}$$

$$= \sqrt{3}\left(\frac{4a^3}{3} \right)$$

Since $(4\sqrt{3}\,a^3)/3 = 10$, we have $a^3 = (5\sqrt{3})/2$. Thus,

$$a = \sqrt[3]{\frac{5\sqrt{3}}{2}} \approx 1.630 \text{ meters.}$$

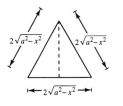

33. $f(x) = \frac{4}{5}x^{5/4}$

$$f'(x) = x^{1/4}$$

$$1 + [f'(x)]^2 = 1 + \sqrt{x}$$

$$u = 1 + \sqrt{x}$$

$$x = (u - 1)^2$$

$$dx = 2(u - 1)\,du$$

$$s = \int_0^4 \sqrt{1 + \sqrt{x}}\,dx = 2\int_1^3 \sqrt{u}(u - 1)\,du$$

$$= 2\int_1^3 (u^{3/2} - u^{1/2})\,du$$

$$= 2\left[\frac{2}{5}u^{5/2} - \frac{2}{3}u^{3/2} \right]_1^3 = \frac{4}{15}\left[u^{3/2}(3u - 5) \right]_1^3$$

$$= \frac{8}{15}\left(1 + 6\sqrt{3} \right) \approx 6.076$$

34. $y = \frac{x^3}{6} + \frac{1}{2x}$

$$y' = \frac{1}{2}x^2 - \frac{1}{2x^2}$$

$$1 + (y')^2 = \left(\frac{1}{2}x^2 + \frac{1}{2x^2} \right)^2$$

$$s = \int_1^3 \left(\frac{1}{2}x^2 + \frac{1}{2x^2} \right)dx = \left[\frac{1}{6}x^3 - \frac{1}{2x} \right]_1^3 = \frac{14}{3}$$

35. $y = 300 \cosh\left(\frac{x}{2000} \right) - 280,\ -2000 \le x \le 2000$

$$y' = \frac{3}{20}\sinh\left(\frac{x}{2000} \right)$$

$$s = \int_{-2000}^{2000} \sqrt{1 + \left[\frac{3}{20}\sinh\left(\frac{x}{2000} \right) \right]^2}\,dx = \frac{1}{20}\int_{-2000}^{2000} \sqrt{400 + 9\sinh^2\left(\frac{x}{2000} \right)}\,dx$$

$$\approx 4018.2 \text{ ft (by Simpson's Rule or graphing utility)}$$

36. Since $f(x) = \tan x$ has $f'(x) = \sec^2 x$, this integral represents the length of the graph of $\tan x$ from $x = 0$ to $x = \pi/4$. This

37. $y = \dfrac{3}{4}x$

$$y' = \frac{3}{4}$$

$$1 + (y')^2 = \frac{25}{16}$$

$$S = 2\pi \int_0^4 \left(\frac{3}{4}x\right)\sqrt{\frac{25}{16}}\, dx = \left[\left(\frac{15\pi}{8}\right)\frac{x^2}{2}\right]_0^4 = 15\pi$$

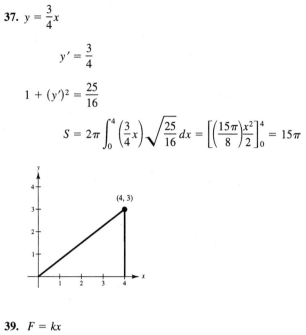

38. $y = 2\sqrt{x}$

$$y' = \frac{1}{\sqrt{x}}$$

$$1 + (y')^2 = 1 + \frac{1}{x} = \frac{x+1}{x}$$

$$S = 2\pi \int_0^3 2\sqrt{x}\sqrt{\frac{x+1}{x}}\, dx = 4\pi \int_0^3 \sqrt{x+1}\, dx$$

$$= 4\pi\left[\left(\frac{2}{3}\right)(x+1)^{3/2}\right]_0^3 = \frac{56\pi}{3}$$

39. $F = kx$

$$4 = k(1)$$

$$F = 4x$$

$$W = \int_0^5 4x\, dx = \left[2x^2\right]_0^5$$

$$= 50 \text{ in} \cdot \text{lb} \approx 4.167 \text{ ft} \cdot \text{lb}$$

40. $F = kx$

$$50 = k(9) \implies k = \frac{50}{9}$$

$$F = \frac{50}{9}x$$

$$W = \int_0^9 \frac{50}{9}x\, dx = \left[\frac{25}{9}x^2\right]_0^9$$

$$= 225 \text{ in} \cdot \text{lb} = 18.75 \text{ ft} \cdot \text{lb}$$

41. Volume of disc: $\pi\left(\dfrac{1}{3}\right)^2 \Delta y$

Weight of disc: $62.4\pi\left(\dfrac{1}{3}\right)^2 \Delta y$

Distance: $175 - y$

$$W = \frac{62.4\pi}{9}\int_0^{150}(175 - y)\, dy = \frac{62.4\pi}{9}\left[175y - \frac{y^2}{2}\right]_0^{150}$$

$$= 104{,}000\pi \text{ ft} \cdot \text{lb} \approx 163.4 \text{ ft} \cdot \text{ton}$$

42. We know that

$$\frac{dV}{dt} = \frac{4 \text{ gal/min} - 12 \text{ gal/min}}{7.481 \text{ gal/ft}^3} = -\frac{8}{7.481} \text{ ft}^3/\text{min}$$

$$V = \pi r^2 h = \pi\left(\frac{1}{9}\right)h$$

$$\frac{dV}{dt} = \frac{\pi}{9}\left(\frac{dh}{dt}\right)$$

$$\frac{dh}{dt} = \frac{9}{\pi}\left(\frac{dV}{dt}\right) = \frac{9}{\pi}\left(-\frac{8}{7.481}\right) \approx -3.064 \text{ ft/min.}$$

Depth of water: $-3.064t + 150$

Time to drain well: $t = \dfrac{150}{3.064} \approx 49$ minutes

$(49)(12) = 588$ gallons pumped

Volume of water pumped in Exercise 41: 391.7 gallons

$$\frac{391.7}{52\pi} = \frac{588}{x\pi}$$

$$x = \frac{588(52)}{391.7} \approx 78$$

Work $\approx 78\pi$ ft $\cdot$ ton

43. Weight of section of chain: $5 \, \Delta x$

Distance moved: $10 - x$

$$W = 5 \int_0^{10} (10 - x) \, dx = \left[-\frac{5}{2}(10 - x)^2 \right]_0^{10} = 250 \text{ ft} \cdot \text{lb}$$

44. (a) Weight of section of cable: $4 \, \Delta x$

Distance: $200 - x$

$$W = 4 \int_0^{200} (200 - x) \, dx = \left[-2(200 - x)^2 \right]_0^{200} = 80,000 \text{ ft} \cdot \text{lb} = 40 \text{ ft} \cdot \text{ton}$$

(b) Work to move 300 pounds 200 feet vertically: $200(300) = 60,000 \text{ ft} \cdot \text{lb} = 30 \text{ ft} \cdot \text{ton}$

Total work = work for drawing up the cable + work of lifting the load

$$= 40 \text{ ft} \cdot \text{ton} + 30 \text{ ft} \cdot \text{ton} = 70 \text{ ft} \cdot \text{ton}$$

45. $W = \int_a^b F(x) \, dx$

$$80 = \int_0^4 ax^2 \, dx = \frac{ax^3}{3} \Big]_0^4 = \frac{64}{3} a$$

$$a = \frac{3(80)}{64} = \frac{15}{4} = 3.75$$

46. $W = \int_a^b F(x) \, dx$

$$F(x) = \begin{cases} -(2/9)x + 6, & 0 \le x \le 9 \\ -(4/3)x + 16, & 9 \le x \le 12 \end{cases}$$

$$W = \int_0^9 \left(-\frac{2}{9}x + 6 \right) dx + \int_9^{12} \left(-\frac{4}{3}x + 16 \right) dx$$

$$= \left[-\frac{1}{9}x^2 + 6x \right]_0^9 + \left[-\frac{2}{3}x^2 + 16x \right]_9^{12}$$

$$= (-9 + 54) + (-96 + 192 + 54 - 144)$$

$$= 51 \text{ ft} \cdot \text{lbs}$$

47. $A = \int_0^a \left(\sqrt{a} - \sqrt{x} \right)^2 dx = \int_0^a \left(a - 2\sqrt{a}\, x^{1/2} + x \right) dx = \left[ax - \frac{4}{3}\sqrt{a}\, x^{3/2} + \frac{1}{2}x^2 \right]_0^a = \frac{a^2}{6}$

$$\frac{1}{A} = \frac{6}{a^2}$$

$$\bar{x} = \frac{6}{a^2} \int_0^a x\left(\sqrt{a} - \sqrt{x} \right)^2 dx = \frac{6}{a^2} \int_0^a \left(ax - 2\sqrt{a}\, x^{3/2} + x^2 \right) dx$$

$$\bar{y} = \left(\frac{6}{a^2} \right) \frac{1}{2} \int_0^a \left(\sqrt{a} - \sqrt{x} \right)^4 dx$$

$$= \frac{3}{a^2} \int_0^a \left(a^2 - 4a^{3/2}x^{1/2} + 6ax - 4a^{1/2}x^{3/2} + x^2 \right) dx$$

$$= \frac{3}{a^2} \left[a^2 x - \frac{8}{3}a^{3/2}x^{3/2} + 3ax^2 - \frac{8}{5}a^{1/2}x^{5/2} + \frac{1}{3}x^3 \right]_0^a = \frac{a}{5}$$

$$(\bar{x}, \bar{y}) = \left(\frac{a}{5}, \frac{a}{5} \right)$$

48. $A = \int_{-1}^{3} [(2x + 3) - x^2]\, dx = \left[x^2 + 3x - \frac{1}{3}x^3 \right]_{-1}^{3} = \frac{32}{3}$

$\dfrac{1}{A} = \dfrac{3}{32}$

$\bar{x} = \dfrac{3}{32} \int_{-1}^{3} x(2x + 3 - x^2)\, dx = \dfrac{3}{32} \int_{-1}^{3} (3x + 2x^2 - x^3)\, dx = \dfrac{3}{32} \left[\dfrac{3}{2}x^2 + \dfrac{2}{3}x^3 - \dfrac{1}{4}x^4 \right]_{-1}^{3} = 1$

$\bar{y} = \left(\dfrac{3}{32} \right) \dfrac{1}{2} \int_{-1}^{3} [(2x + 3)^2 - x^4]\, dx = \dfrac{3}{64} \int_{-1}^{3} (9 + 12x + 4x^2 - x^4)\, dx$

$= \dfrac{3}{64} \left[9x + 6x^2 + \dfrac{4}{3}x^3 - \dfrac{1}{5}x^5 \right]_{-1}^{3} = \dfrac{17}{5}$

$(\bar{x}, \bar{y}) = \left(1, \dfrac{17}{5} \right)$

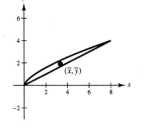

49. By symmetry, $x = 0$.

$A = 2 \int_{0}^{1} (a^2 - x^2)\, dx = 2 \left[a^2 x - \dfrac{x^3}{3} \right]_{0}^{a} = \dfrac{4a^3}{3}$

$\dfrac{1}{A} = \dfrac{3}{4a^3}$

$\bar{y} = \left(\dfrac{3}{4a^3} \right) \dfrac{1}{2} \int_{-a}^{a} (a^2 - x^2)^2\, dx$

$= \dfrac{6}{8a^3} \int_{0}^{a} (a^4 - 2a^2 x^2 + x^4)\, dx$

$= \dfrac{6}{8a^3} \left[a^4 x - \dfrac{2a^2}{3}x^3 + \dfrac{1}{5}x^5 \right]_{0}^{a}$

$= \dfrac{6}{8a^3} \left(a^5 - \dfrac{2}{3}a^5 + \dfrac{1}{5}a^5 \right) = \dfrac{2a^2}{5}$

$(\bar{x}, \bar{y}) = \left(0, \dfrac{2a^2}{5} \right)$

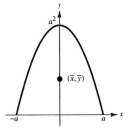

50. $A = \int_{0}^{8} \left(x^{2/3} - \dfrac{1}{2}x \right) dx = \left[\dfrac{3}{5}x^{5/3} - \dfrac{1}{4}x^2 \right]_{0}^{8} = \dfrac{16}{5}$

$\dfrac{1}{A} = \dfrac{5}{16}$

$\bar{x} = \dfrac{5}{16} \int_{0}^{8} x \left(x^{2/3} - \dfrac{1}{2}x \right) dx$

$= = \dfrac{5}{16} \left[\dfrac{3}{8}x^{8/3} - \dfrac{1}{6}x^3 \right]_{0}^{8} = \dfrac{10}{3}$

$\bar{y} = \left(\dfrac{5}{16} \right) \dfrac{1}{2} \int_{0}^{8} \left(x^{4/3} - \dfrac{1}{4}x^2 \right) dx$

$= \dfrac{1}{2} \left(\dfrac{5}{16} \right) \left[\dfrac{3}{7}x^{7/3} - \dfrac{1}{12}x^3 \right]_{0}^{8} = \dfrac{40}{21}$

$(\bar{x}, \bar{y}) = \left(\dfrac{10}{3}, \dfrac{40}{21} \right)$

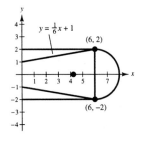

51. $\bar{y} = 0$ by symmetry

For the trapezoid:

$m = [(4)(6) - (1)(6)]\rho = 18\rho$

$M_y = \rho \int_{0}^{6} x \left[\left(\dfrac{1}{6}x + 1 \right) - \left(-\dfrac{1}{6}x - 1 \right) \right] dx$

$= \rho \int_{0}^{6} \left(\dfrac{1}{3}x^2 + 2x \right) dx = \rho \left[\dfrac{x^3}{9} + x^2 \right]_{0}^{6} = 60\rho$

—CONTINUED—

51. —CONTINUED—

For the semicircle:

$$m = \left(\frac{1}{2}\right)(\pi)(2)^2\rho = 2\pi\rho$$

$$M_y = \rho \int_6^8 x\left[\sqrt{4-(x-6)^2} - \left(-\sqrt{4-(x-6)^2}\right)\right] dx = 2\rho \int_6^8 x\sqrt{4-(x-6)^2}\, dx$$

Let $u = x - 6$, then $x = u + 6$ and $dx = du$. When $x = 6$, $u = 0$. When $x = 8$, $u = 2$.

$$M_y = 2\rho \int_0^2 (u+6)\sqrt{4-u^2}\, du = 2\rho \int_0^2 u\sqrt{4-u^2}\, du + 12\rho \int_0^2 \sqrt{4-u^2}\, du$$

$$= 2\rho\left[\left(-\frac{1}{2}\right)\left(\frac{2}{3}\right)(4-u^2)^{3/2}\right]_0^2 + 12\rho\left[\frac{\pi(2)^2}{4}\right] = \frac{16\rho}{3} + 12\pi\rho = \frac{4\rho(4+9\pi)}{3}$$

Thus, we have:

$$\bar{x}(18\rho + 2\pi\rho) = 60\rho + \frac{4\rho(4+9\pi)}{3}$$

$$\bar{x} = \frac{180\rho + 4\rho(4+9\pi)}{3} \cdot \frac{1}{2\rho(9+\pi)} = \frac{2(9\pi+49)}{3(\pi+9)}$$

The centroid of the blade is $\left(\dfrac{2(9\pi+49)}{3(\pi+9)}, 0\right)$.

52. The numerical centroid should agree with the centroid located in part a.

53. Wall at shallow end:

$$F = 62.4 \int_0^5 y(20)\, dy = \left[(1248)\frac{y^2}{2}\right]_0^5 = 15,600 \text{ lb}$$

Wall at deep end:

$$F = 62.4 \int_0^{10} y(20)\, dy = \left[(624)y^2\right]_0^{10} = 62,400 \text{ lb}$$

Side wall:

$$F_1 = 62.4 \int_0^5 y(40)\, dy = \left[(1248)y^2\right]_0^5 = 31,200 \text{ lb}$$

$$F_2 = 62.4 \int_0^5 (10-y)8y\, dy = 62.4 \int_0^5 (80y - 8y^2)\, dy$$

$$F = F_1 + F_2 = 72,800 \text{ lb}$$

54. Let D = surface of liquid; ρ = weight per cubic volume.

$$F = \rho \int_c^d (D-y)[f(y) - g(y)]\, dy$$

$$= \rho\left[\int_c^d D[f(y) - g(y)]\, dy - \int_c^d y[f(y) - g(y)]\, dy\right]$$

$$= \rho\left[\int_c^d [f(y) - g(y)]\, dy\right]\left[D - \frac{\int_c^d y[f(y) - g(y)]\, dy}{\int_c^d [f(y) - g(y)]\, dy}\right]$$

$$= \rho(\text{Area})(D - \bar{y})$$

$$= \rho(\text{Area})(\text{depth of centroid})$$

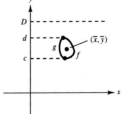

55. $F = 62.4(16\pi)5 = 4992\pi \text{ lb}$

C H A P T E R 7
Integration Techniques, L'Hôpital's Rule, and Improper Integrals

C H A P T E R   7
Integration Techniques, L'Hôpital's Rule, and Improper Integrals

Section 7.1   Basic Integration Formulas

Solutions to Selected Odd-Numbered Exercises

1. (a) $\dfrac{d}{dx}\left[2\sqrt{x^2+1}+C\right]=2\left(\dfrac{1}{2}\right)(x^2+1)^{-1/2}(2x)=\dfrac{2x}{\sqrt{x^2+1}}$

    (b) $\dfrac{d}{dx}\left[\sqrt{x^2+1}+C\right]=\dfrac{1}{2}(x^2+1)^{-1/2}(2x)=\dfrac{x}{\sqrt{x^2+1}}$

    (c) $\dfrac{d}{dx}\left[\dfrac{1}{2}\sqrt{x^2+1}+C\right]=\dfrac{1}{2}\left(\dfrac{1}{2}\right)(x^2+1)^{-1/2}(2x)=\dfrac{x}{2\sqrt{x^2+1}}$

    (d) $\dfrac{d}{dx}\left[\ln(x^2+1)+C\right]=\dfrac{2x}{x^2+1}$

$\displaystyle\int\dfrac{x}{\sqrt{x^2+1}}\,dx$ matches (b).

2. (a) $\dfrac{d}{dx}\left[\ln\sqrt{x^2+1}+C\right]=\dfrac{1}{2}\left(\dfrac{2x}{x^2+1}\right)=\dfrac{x}{x^2+1}$

    (b) $\dfrac{d}{dx}\left[\dfrac{2x}{(x^2+1)^2}+C\right]=\dfrac{(x^2+1)^2(2)-(2x)(2)(x^2+1)(2x)}{(x^2+1)^4}=\dfrac{2(1-3x^2)}{(x^2+1)^3}$

    (c) $\dfrac{d}{dx}[\arctan x+C]=\dfrac{1}{1+x^2}$

    (d) $\dfrac{d}{dx}\left[\ln(x^2+1)+C\right]=\dfrac{2x}{x^2+1}$

$\displaystyle\int\dfrac{x}{x^2+1}\,dx$ matches (a).

3. (a) $\dfrac{d}{dx}\left[\ln\sqrt{x^2+1}+C\right]=\dfrac{1}{2}\left(\dfrac{2x}{x^2+1}\right)=\dfrac{x}{x^2+1}$

    (b) $\dfrac{d}{dx}\left[\dfrac{2x}{(x^2+1)^2}+C\right]=\dfrac{(x^2+1)^2(2)-(2x)(2)(x^2+1)(2x)}{(x^2+1)^4}=\dfrac{2(1-3x^2)}{(x^2+1)^3}$

    (c) $\dfrac{d}{dx}[\arctan x+C]=\dfrac{1}{1+x^2}$

    (d) $\dfrac{d}{dx}\left[\ln(x^2+1)+C\right]=\dfrac{2x}{x^2+1}$

$\displaystyle\int\dfrac{1}{x^2+1}\,dx$ matches (c).

4. (a) $\dfrac{d}{dx}[2x\sin(x^2+1)+C] = 2x[\cos(x^2+1)(2x)]+2\sin(x^2+1) = 2[2x^2\cos(x^2+1)+\sin(x^2+1)]$

(b) $\dfrac{d}{dx}\left[-\dfrac{1}{2}\sin(x^2+1)+C\right] = -\dfrac{1}{2}\cos(x^2+1)(2x) = -x\cos(x^2+1)$

(c) $\dfrac{d}{dx}\left[\dfrac{1}{2}\sin(x^2+1)+C\right] = \dfrac{1}{2}\cos(x^2+1)(2x) = x\cos(x^2+1)$

(d) $\dfrac{d}{dx}[-2x\sin(x^2+1)+C] = -2x[\cos(x^2+1)(2x)]-2\sin(x^2+1) = -2[2x^2\cos(x^2+1)+\sin(x^2+1)]$

$\displaystyle\int x\cos(x^2+1)\,dx$ matches (c).

5. $\displaystyle\int(3x-2)^4\,dx$

$u=3x-2,\ du=3\,dx,\ n=4$

Use $\displaystyle\int u^n\,du.$

6. $\displaystyle\int\dfrac{2t-1}{t^2-t+2}\,dt$

$u=t^2-t+2,\ du=(2t-1)\,dt$

Use $\displaystyle\int\dfrac{du}{u}.$

7. $\displaystyle\int\dfrac{1}{\sqrt{x}\left(1-2\sqrt{x}\right)}\,dx$

$u=1-2\sqrt{x},\ du=-\dfrac{1}{\sqrt{x}}\,dx$

Use $\displaystyle\int\dfrac{du}{u}.$

8. $\displaystyle\int\dfrac{2}{(2t-1)^2+4}\,dt$

$u=2t-1,\ du=2dt,\ a=2$

Use $\displaystyle\int\dfrac{du}{u^2+a^2}.$

9. $\displaystyle\int\dfrac{3}{\sqrt{1-t^2}}\,dt$

$u=t,\ du=dt,\ a=1$

Use $\displaystyle\int\dfrac{du}{\sqrt{a^2-u^2}}$

10. $\displaystyle\int\dfrac{-2x}{\sqrt{x^2-4}}\,dx$

$u=x^2-4,\ du=2x\,dx,\ n=-\dfrac{1}{2}$

Use $\displaystyle\int u^n\,du.$

11. $\displaystyle\int t\sin t^2\,dt$

$u=t^2,\ du=2t\,dt$

Use $\displaystyle\int\sin u\,du.$

12. $\displaystyle\int\sec 3x\tan 3x\,dx$

$u=3x,\ du=3\,dx$

Use $\displaystyle\int\sec u\tan u\,du.$

13. $\displaystyle\int\cos x e^{\sin x}\,dx$

$u=\sin x,\ du=\cos x\,dx$

Use $\displaystyle\int e^u\,du.$

14. $\displaystyle\int\dfrac{1}{x\sqrt{x^2-4}}\,dx$

$u=x,\ du=dx,\ a=2$

Use $\displaystyle\int\dfrac{du}{u\sqrt{u^2-a^2}}.$

15. Let $u=-2x+5,\ du=-2\,dx.$

$\displaystyle\int(-2x+5)^{3/2}\,dx = -\dfrac{1}{2}\int(-2x+5)^{3/2}(-2)\,dx$

$\qquad = -\dfrac{1}{5}(-2x+5)^{5/2}+C$

16. Let $u=t-9,\ du=dt.$

$\displaystyle\int\dfrac{2}{(t-9)^2}\,dt = 2\int(t-9)^{-2}\,dt = \dfrac{-2}{t-9}+C$

17. $\displaystyle\int\left[v+\dfrac{1}{(3v-1)^3}\right]dv = \int v\,dv+\dfrac{1}{3}\int(3v-1)^{-3}(3)dv$

$\qquad = \dfrac{1}{2}v^2 - \dfrac{1}{6(3v-1)^2}+C$

18. Let $u = 4 - 2x^2$, $du = -4x\, dx$.

$$\int x\sqrt{4 - 2x^2}\, dx = -\frac{1}{4}\int (4 - 2x^2)^{1/2}(-4x)dx = -\frac{1}{6}(4 - 2x^2)^{3/2} + C$$

19. Let $u = -t^3 + 9t + 1$, $du = (-3t^2 + 9)\, dt = -3(t^2 - 3)\, dt$.

$$\int \frac{t^2 - 3}{-t^3 + 9t + 1}\, dt = -\frac{1}{3}\int \frac{-3(t^2 - 3)}{-t^3 + 9t + 1}\, dt = -\frac{1}{3}\ln|-t^3 + 9t + 1| + C$$

20. $\displaystyle\int \frac{2x}{x - 4}\, dx = \int 2\, dx + \int \frac{8}{x - 4}\, dx$

$$= 2x + 8\ln|x - 4| + C$$

21. $\displaystyle\int \frac{x^2}{x - 1}\, dx = \int (x + 1)\, dx + \int \frac{1}{x - 1}\, dx$

$$= \frac{1}{2}x^2 + x + \ln|x - 1| + C$$

22. Let $u = x^2 + 2x - 4$, $du = 2(x + 1)\, dx$.

$$\int \frac{x + 1}{\sqrt{x^2 + 2x - 4}}\, dx = \frac{1}{2}\int (x^2 + 2x - 4)^{-1/2}(2)(x + 1)\, dx$$

$$= \sqrt{x^2 + 2x - 4} + C$$

23. Let $u = 1 + e^x$, $du = e^x\, dx$.

$$\int \frac{e^x}{1 + e^x}\, dx = \ln(1 + e^x) + C$$

24. $\displaystyle\int \left(\frac{1}{3x - 1} - \frac{1}{3x + 1}\right) dx = \frac{1}{3}\int \frac{1}{3x - 1}(3)\, dx - \frac{1}{3}\int \frac{1}{3x + 1}(3)\, dx$

$$= \frac{1}{3}\ln|3x - 1| - \frac{1}{3}\ln|3x + 1| + C = \frac{1}{3}\ln\left|\frac{3x - 1}{3x + 1}\right| + C$$

25. $\displaystyle\int (1 + 2x^2)^2\, dx = \int (4x^4 + 4x^2 + 1)dx = \frac{4}{5}x^5 + \frac{4}{3}x^3 + x + C = \frac{x}{15}(12x^4 + 20x^2 + 15) + C$

26. $\displaystyle\int x\left(1 + \frac{1}{x}\right)^3 = \int x\left(1 + \frac{3}{x} + \frac{3}{x^2} + \frac{1}{x^3}\right) dx = \int \left(x + 3 + \frac{3}{x} + \frac{1}{x^2}\right) dx = \frac{1}{2}x^2 + 3x + 3\ln|x| - \frac{1}{x} + C$

27. Let $u = 2\pi x^2$, $du = 4\pi x\, dx$.

$$\int x(\cos 2\pi x^2)\, dx = \frac{1}{4\pi}\int (\cos 2\pi x^2)(4\pi x)\, dx$$

$$= \frac{1}{4\pi}\sin 2\pi x^2 + C$$

28. Let $v = 4u$, $dv = 4\, du$.

$$\int \sec(4u)\, du = \frac{1}{4}\int \sec(4u)(4)\, du$$

$$= \frac{1}{4}\ln|\sec(4u) + \tan(4u)| + C$$

29. Let $u = \pi x$, $du = \pi\, dx$.

$$\int \csc(\pi x)\cot(\pi x)\, dx = \frac{1}{\pi}\int \csc(\pi x)\cot(\pi x)\pi\, dx = -\frac{1}{\pi}\csc(\pi x) + C$$

30. Let $u = \cos x$, $du = -\sin x\, dx$.

$$\int \frac{\sin x}{\sqrt{\cos x}}\, dx = -\int (\cos x)^{-1/2}(-\sin x)\, dx$$

$$= -2\sqrt{\cos x} + C$$

31. Let $u = 5x$, $du = 5\, dx$.

$$\int e^{5x}\, dx = \frac{1}{5}\int e^{5x}(5)\, dx = \frac{1}{5}e^{5x} + C$$

32. Let $u = \cot x$, $du = -\csc^2 x \, dx$.

$$\int \csc^2 x e^{\cot x} \, dx = -\int e^{\cot x}(-\csc^2 x) \, dx = -e^{\cot x} + C$$

33. Let $u = 1 + e^x$, $du = e^x \, dx$.

$$\int \frac{2}{e^{-x} + 1} \, dx = 2 \int \left(\frac{1}{e^{-x} + 1} \right) \left(\frac{e^x}{e^x} \right) dx$$

$$= 2 \int \frac{e^x}{1 + e^x} \, dx = 2 \ln(1 + e^x) + C$$

34. $\displaystyle \int \frac{1}{2e^x - 3} \, dx = \int \frac{1}{2e^x - 3} \left(\frac{e^{-x}}{e^{-x}} \right) dx = \int \frac{e^{-x}}{2 - 3e^{-x}} = \frac{1}{3} \int \frac{3e^{-x}}{2 - 3e^{-x}} \, dx = \frac{1}{3} \ln|2 - 3e^{-x}| + C$

35. $\displaystyle \int \frac{1 + \sin x}{\cos x} \, dx = \int (\sec x + \tan x) \, dx = \ln|\sec x + \tan x| + \ln|\sec x| + C = \ln|\sec x(\sec x + \tan x)| + C$

36. $\displaystyle \int \frac{1}{\sec x - 1} \, dx = \int \frac{1}{\sec x - 1} \left(\frac{\sec x + 1}{\sec x + 1} \right) dx = \int \frac{(\sec x + 1)}{\tan^2 x} = \int \frac{\sec x}{\tan^2 x} \, dx + \int \cot^2 x \, dx$

$$= \int \frac{\cos x}{\sin^2 x} \, dx + \int (\csc^2 x - 1) \, dx$$

$$= -\frac{1}{\sin x} - \cot x - x + C$$

$$= -(\csc x + \cot x + x) + C$$

37. $\displaystyle \int \frac{2t - 1}{t^2 + 4} \, dt = \int \frac{2t}{t^2 + 4} \, dt - \int \frac{1}{4 + t^2} \, dt$

$$= \ln(t^2 + 4) - \frac{1}{2} \arctan \frac{t}{2} + C$$

38. $\displaystyle \int \frac{3}{t^2 + 1} \, dt = 3 \arctan t + C$

39. Let $u = 2t - 1$, $du = 2 \, dt$.

$$\int \frac{-1}{\sqrt{1 - (2t - 1)^2}} \, dt = -\frac{1}{2} \int \frac{2}{\sqrt{1 - (2t - 1)^2}} \, dt$$

$$= -\frac{1}{2} \arcsin(2t - 1) + C$$

40. Let $u = \sqrt{3}x$, $du = \sqrt{3} \, dx$.

$$\int \frac{1}{4 + 3x^2} \, dx = \frac{1}{\sqrt{3}} \int \frac{\sqrt{3}}{4 + (\sqrt{3}x)^2} \, dx$$

$$= \frac{1}{2\sqrt{3}} \arctan \left(\frac{\sqrt{3}x}{2} \right) + C$$

41. Let $u = \cos\left(\frac{2}{t} \right)$, $du = \frac{2 \sin(2/t)}{t^2} \, dt$.

$$\int \frac{\tan(2/t)}{t^2} \, dt = \frac{1}{2} \int \frac{1}{\cos(2/t)} \left[\frac{2 \sin(2/t)}{t^2} \right] dt$$

$$= \frac{1}{2} \ln \left| \cos\left(\frac{2}{t} \right) \right| + C$$

42. Let $u = \frac{1}{t}$, $du = \frac{-1}{t^2} \, dt$.

$$\int \frac{e^{1/t}}{t^2} \, dt = -\int e^{1/t} \left(\frac{-1}{t^2} \right) dt = -e^{1/t} + C$$

43. $\displaystyle \int \frac{3}{\sqrt{6x - x^2}} \, dx = 3 \int \frac{1}{\sqrt{9 - (x - 3)^2}} \, dx = 3 \arcsin\left(\frac{x - 3}{3} \right) + C$

44. $\displaystyle \int \frac{1}{(x - 1)\sqrt{4x^2 - 8x + 3}} \, dx = \int \frac{2}{[2(x - 1)]\sqrt{[2(x - 1)]^2 - 1}} \, dx = \text{arcsec}|2(x - 1)| + C$

45. $\displaystyle \int \frac{4}{4x^2 + 4x + 65} \, dx = \int \frac{1}{[x + (1/2)]^2 + 16} \, dx = \frac{1}{4} \arctan\left[\frac{x + (1/2)}{4} \right] + C = \frac{1}{4} \arctan\left(\frac{2x + 1}{8} \right) + C$

46. $\int \dfrac{1}{\sqrt{2 - 2x - x^2}}\, dx = \int \dfrac{1}{\sqrt{3 - (x + 1)^2}}\, dx = \arcsin\left(\dfrac{x + 1}{\sqrt{3}}\right) + C$

47. $\dfrac{ds}{dt} = \dfrac{t}{\sqrt{1 - t^4}}, \left(0, -\dfrac{1}{2}\right)$

(a)

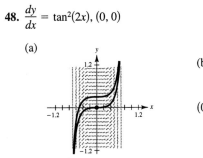

(b) $u = t^2,\ du = 2t\, dt$

$\int \dfrac{t}{\sqrt{1 - t^4}}\, dt = \dfrac{1}{2} \int \dfrac{2t}{\sqrt{1 - (t^2)^2}}\, dt = \dfrac{1}{2} \arcsin t^2 + C$

$\left(0, -\dfrac{1}{2}\right): \ -\dfrac{1}{2} = \dfrac{1}{2} \arcsin 0 + C \Rightarrow C = -\dfrac{1}{2}$

$s = \dfrac{1}{2} \arcsin t^2 - \dfrac{1}{2}$

48. $\dfrac{dy}{dx} = \tan^2(2x), (0, 0)$

(a)

(b) $\int \tan^2(2x)\, dx = \int (\sec^2(2x) - 1)\, dx = \dfrac{1}{2} \tan(2x) - x + C$

$(0, 0): \ 0 = C$

$y = \dfrac{1}{2} \tan(2x) - x$

49. $y = \int (1 + e^x)^2\, dx = \int (e^{2x} + 2e^x + 1)\, dx$

$= \dfrac{1}{2} e^{2x} + 2e^x + x + C$

50. $r = \int \dfrac{(1 + e^t)^2}{e^t}\, dt = \int \dfrac{1 + 2e^t + e^{2t}}{e^t}\, dt$

$= \int (e^{-t} + 2 + e^t)\, dt = -e^{-t} + 2t + e^t + C$

51. $\dfrac{dy}{dx} = \dfrac{\sec^2 x}{4 + \tan^2 x}$

Let $u = \tan x,\ du = \sec^2 x\, dx$.

$y = \int \dfrac{\sec^2 x}{4 + \tan^2 x}\, dx = \dfrac{1}{2} \arctan\left(\dfrac{\tan x}{2}\right) + C$

52. Let $u = 2x,\ du = 2\, dx$.

$y = \int \dfrac{1}{x\sqrt{4x^2 - 1}}\, dx = \int \dfrac{2}{2x\sqrt{(2x)^2 - 1}}\, dx$

$= \operatorname{arcsec}|2x| + C$

53. Let $u = 2x,\ du = 2\, dx$.

$\int_0^{\pi/4} \cos 2x\, dx = \dfrac{1}{2} \int_0^{\pi/4} \cos 2x(2)\, dx$

$= \left[\dfrac{1}{2} \sin 2x\right]_0^{\pi/4} = \dfrac{1}{2}$

54. Let $u = \sin t,\ du = \cos t\, dt$.

$\int_0^{\pi} \sin^2 t \cos t\, dt = \left[\dfrac{1}{3} \sin^3 t\right]_0^{\pi} = 0$

55. Let $u = -x^2,\ du = -2x\, dx$.

$\int_0^1 xe^{-x^2}\, dx = -\dfrac{1}{2} \int_0^1 e^{-x^2}(-2x)\, dx = \left[-\dfrac{1}{2} e^{-x^2}\right]_0^1$

$= \dfrac{1}{2}(1 - e^{-1}) \approx 0.316$

56. Let $u = 1 - \ln x,\ du = \dfrac{-1}{x}\, dx$.

$\int_1^e \dfrac{1 - \ln x}{x}\, dx = -\int_1^e (1 - \ln x)\left(\dfrac{-1}{x}\right) dx$

$= \left[-\dfrac{1}{2}(1 - \ln x)^2\right]_1^e = \dfrac{1}{2}$

57. Let $u = x^2 + 9$, $du = 2x\ dx$.

$$\int_0^4 \frac{2x}{\sqrt{x^2 + 9}}\ dx = \int_0^4 (x^2 + 9)^{-1/2}(2x)\ dx$$

$$= \left[2\sqrt{x^2 + 9} \right]_0^4 = 4$$

58. $\int_1^2 \frac{x - 2}{x}\ dx = \int_1^2 \left(1 - \frac{2}{x} \right) dx$

$$= \left[x - 2 \ln x \right]_1^2 = 1 - \ln 4 \approx -0.386$$

59. Let $u = 3x$, $du = 3\ dx$.

$$\int_0^{2/\sqrt{3}} \frac{1}{4 + 9x^2}\ dx = \frac{1}{3} \int_0^{2/\sqrt{3}} \frac{3}{4 + (3x)^2}\ dx$$

$$= \left[\frac{1}{6} \arctan\left(\frac{3x}{2} \right) \right]_0^{2/\sqrt{3}}$$

$$= \frac{\pi}{18} \approx 0.175$$

60. $\int_0^4 \frac{1}{\sqrt{25 - x^2}}\ dx = \left[\arcsin \frac{x}{5} \right]_0^4 = \arcsin \frac{4}{5} \approx 0.927$

61. $\int \frac{1}{x^2 + 4x + 13}\ dx = \frac{1}{3} \arctan\left(\frac{x + 2}{3} \right) + C$

The antiderivatives are vertical translations of each other.

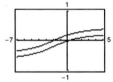

62. $\int \frac{x - 2}{x^2 + 4x + 13}\ dx = \frac{1}{2} \ln(x^2 + 4x + 13) - \frac{4}{3} \arctan\left(\frac{x + 2}{3} \right) + C$

The antiderivatives are vertical translations of each other.

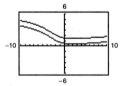

63. $\int \frac{1}{1 + \sin\theta}\ d\theta = \tan\theta - \sec\theta + C \left(\text{or } \frac{-2}{1 + \tan(\theta/2)} \right)$

The antiderivatives are vertical translations of each other.

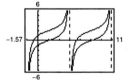

64. $\int \left(\frac{e^x + e^{-x}}{2} \right)^3 dx = \frac{1}{24}[e^{3x} + 9e^x - 9e^{-x} - e^{-3x}] + C$

The antiderivatives are vertical translations of each other.

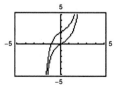

65. $\sin x + \cos x = a \sin(x + b)$

$\sin x + \cos x = a \sin x \cos b + a \cos x \sin b$

$\sin x + \cos x = (a \cos b) \sin x + (a \sin b) \cos x$

Equate coefficients of like terms to obtain the following.

$1 = a \cos b$ and $1 = a \sin b$

Thus, $a = 1/\cos b$. Now, substitute for a in $1 = a \sin b$.

$$1 = \left(\frac{1}{\cos b}\right) \sin b$$

$$1 = \tan b \implies b = \frac{\pi}{4}$$

Since $b = \frac{\pi}{4}, a = \dfrac{1}{\cos(\pi/4)} = \sqrt{2}$. Thus, $\sin x + \cos x = \sqrt{2} \sin\left(x + \frac{\pi}{4}\right)$.

$$\int \frac{dx}{\sin x + \cos x} = \int \frac{dx}{\sqrt{2}\sin(x + (\pi/4))} = \frac{1}{\sqrt{2}}\int \csc\left(x + \frac{\pi}{4}\right) dx = -\frac{1}{\sqrt{2}} \ln\left|\csc\left(x + \frac{\pi}{4}\right) + \cot\left(x + \frac{\pi}{4}\right)\right| + C$$

66. $f(x) = \frac{1}{5}(x^3 - 7x^2 + 10x)$

$\displaystyle\int_0^5 f(x)\, dx < 0$ because

more area is below the x-axis than above.

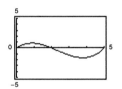

67. $\displaystyle\int_0^2 \frac{4x}{x^2 + 1} dx \approx 3$

Matches (a).

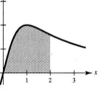

68. $\displaystyle\int_0^2 \frac{4}{x^2 + 1} dx \approx 4$

Matches (d).

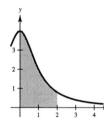

69. Let $u = 1 - x^2, du = -2x\, dx$.

$$A = 4 \int_0^1 x\sqrt{1 - x^2}\, dx$$

$$= -2 \int_0^1 (1 - x^2)^{1/2}(-2x)\, dx$$

$$= \left[-\frac{4}{3}(1 - x^2)^{3/2}\right]_0^1 = \frac{4}{3}$$

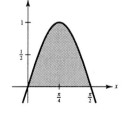

70. $A = \displaystyle\int_0^{\pi/2} \sin 2x\, dx$

$$= \left[-\frac{1}{2}\cos 2x\right]_0^{\pi/2} = 1$$

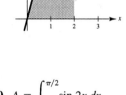

71. $\displaystyle\int_0^{1/a} (x - ax^2)\, dx = \left[\frac{1}{2}x^2 - \frac{a}{3}x^3\right]_0^{1/a}$

$$= \frac{1}{6a^2}$$

Let $\dfrac{1}{6a^2} = \dfrac{2}{3}, 12a^2 = 3, a = \dfrac{1}{2}$.

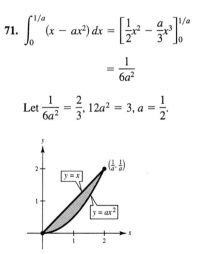

72. $\displaystyle\int_0^2 2\pi x^2 \, dx$

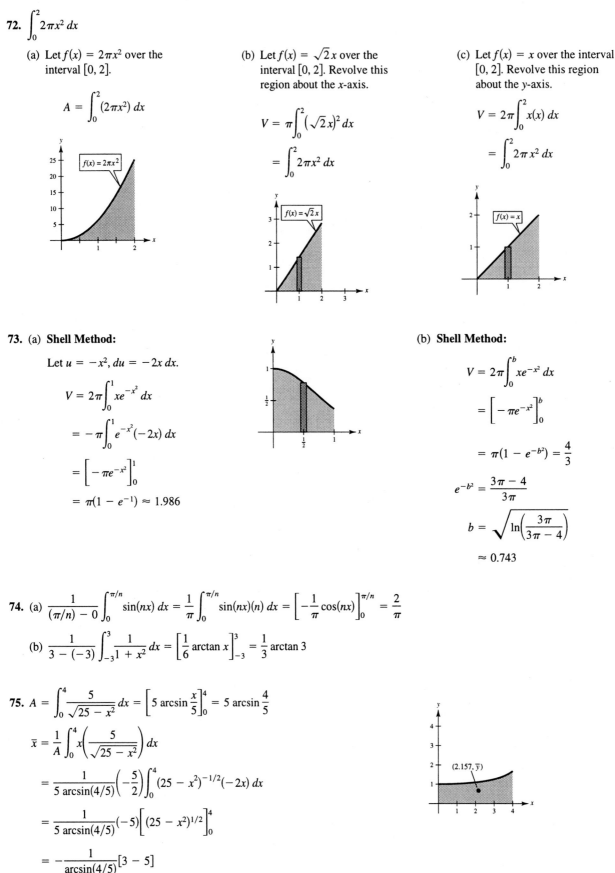

(a) Let $f(x) = 2\pi x^2$ over the interval $[0, 2]$.

$$A = \int_0^2 (2\pi x^2) \, dx$$

(b) Let $f(x) = \sqrt{2}\,x$ over the interval $[0, 2]$. Revolve this region about the x-axis.

$$V = \pi \int_0^2 (\sqrt{2}\,x)^2 \, dx$$

$$= \int_0^2 2\pi x^2 \, dx$$

(c) Let $f(x) = x$ over the interval $[0, 2]$. Revolve this region about the y-axis.

$$V = 2\pi \int_0^2 x(x) \, dx$$

$$= \int_0^2 2\pi x^2 \, dx$$

73. (a) Shell Method:

Let $u = -x^2$, $du = -2x \, dx$.

$$V = 2\pi \int_0^1 x e^{-x^2} \, dx$$

$$= -\pi \int_0^1 e^{-x^2}(-2x) \, dx$$

$$= \left[-\pi e^{-x^2} \right]_0^1$$

$$= \pi(1 - e^{-1}) \approx 1.986$$

(b) Shell Method:

$$V = 2\pi \int_0^b x e^{-x^2} \, dx$$

$$= \left[-\pi e^{-x^2} \right]_0^b$$

$$= \pi(1 - e^{-b^2}) = \frac{4}{3}$$

$$e^{-b^2} = \frac{3\pi - 4}{3\pi}$$

$$b = \sqrt{\ln\left(\frac{3\pi}{3\pi - 4} \right)}$$

$$\approx 0.743$$

74. (a) $\displaystyle \frac{1}{(\pi/n) - 0} \int_0^{\pi/n} \sin(nx) \, dx = \frac{1}{\pi} \int_0^{\pi/n} \sin(nx)(n) \, dx = \left[-\frac{1}{\pi} \cos(nx) \right]_0^{\pi/n} = \frac{2}{\pi}$

(b) $\displaystyle \frac{1}{3 - (-3)} \int_{-3}^3 \frac{1}{1 + x^2} \, dx = \left[\frac{1}{6} \arctan x \right]_{-3}^3 = \frac{1}{3} \arctan 3$

75. $A = \displaystyle\int_0^4 \frac{5}{\sqrt{25 - x^2}} \, dx = \left[5 \arcsin \frac{x}{5} \right]_0^4 = 5 \arcsin \frac{4}{5}$

$$\bar{x} = \frac{1}{A} \int_0^4 x\left(\frac{5}{\sqrt{25 - x^2}} \right) dx$$

$$= \frac{1}{5 \arcsin(4/5)} \left(-\frac{5}{2} \right) \int_0^4 (25 - x^2)^{-1/2}(-2x) \, dx$$

$$= \frac{1}{5 \arcsin(4/5)} (-5) \left[(25 - x^2)^{1/2} \right]_0^4$$

$$= -\frac{1}{\arcsin(4/5)} [3 - 5]$$

$$= \frac{2}{\arcsin(4/5)} \approx 2.157$$

76. $y = 2\sqrt{x}$

$y' = \dfrac{1}{\sqrt{x}}$

$1 + (y')^2 = 1 + \dfrac{1}{x} = \dfrac{x+1}{x}$

$S = 2\pi \displaystyle\int_0^9 2\sqrt{x}\sqrt{\dfrac{x+1}{x}}\,dx$

$\quad = 2\pi \displaystyle\int_0^9 2\sqrt{x+1}\,dx$

$\quad = \left[4\pi\left(\dfrac{2}{3}\right)(x+1)^{3/2}\right]_0^9$

$\quad = \dfrac{8\pi}{3}\left(10\sqrt{10} - 1\right) \approx 256.545$

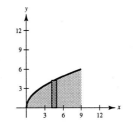

77. $y = \tan(\pi x)$

$y' = \pi\sec^2(\pi x)$

$1 + (y')^2 = 1 + \pi^2\sec^4(\pi x)$

$s = \displaystyle\int_0^{1/4} \sqrt{1 + \pi^2\sec^4(\pi x)}\,dx$

$\quad \approx 1.0320$

78. $y = x^{2/3}$

$y' = \dfrac{2}{3x^{1/3}}$

$1 + (y')^2 = 1 + \dfrac{4}{9x^{2/3}}$

$s = \displaystyle\int_1^8 \sqrt{1 + \dfrac{4}{9x^{2/3}}}\,dx \approx 7.6337$

79. True. If $u = x^4 - 1$, then $du = 4x^3\,dx$.

$\displaystyle\int \dfrac{x^3}{x^4-1}\,dx = \dfrac{1}{4}\int\dfrac{4x^3\,dx}{x^4-1} = \dfrac{1}{4}\int\dfrac{du}{u}$

80. False. If $u = \sin x$, then $du = \cos x\,dx$.

$\displaystyle\int\dfrac{dx}{1+\sin^2 x} \neq \int\dfrac{du}{a^2+u^2}$

Section 7.2   Integration by Parts

1. (a) $\dfrac{d}{dx}[\sin x - x\cos x] = \cos x - (-x\sin x + \cos x) = x\sin x.$ Matches (ii)

(b) $\dfrac{d}{dx}[x^2\sin x + 2x\cos x - 2\sin x] = x^2\cos x + 2x\sin x - 2x\sin x + 2\cos x - 2\cos x = x^2\cos x.$ Matches (iv)

(c) $\dfrac{d}{dx}[x^2 e^x - 2xe^x + 2e^x] = x^2 e^x + 2xe^x - 2xe^x - 2e^x + 2e^x = x^2 e^x.$ Matches (iii)

(d) $\dfrac{d}{dx}[-x + x\ln x] = -1 + x\left(\dfrac{1}{x}\right) + \ln x = \ln x.$ Matches (i)

2. The guidelines help in selecting dv and u:

(a) Let dv be a portion of the integrand that can be integrated.

(b) Let u be a portion of the integrand whose derivative is a simpler function than u.

3. $\displaystyle\int xe^{2x}\,dx$

$u = x,\ dv = e^{2x}\,dx$

4. $\displaystyle\int x^2 e^{2x}\,dx$

$u = x^2,\ dv = e^{2x}\,dx$

5. $\displaystyle\int (\ln x)^2\,dx$

$u = (\ln x)^2,\ dv = dx$

6. $\int \ln 3x\, dx$

$u = \ln 3x, dv = dx$

7. $\int x \sec^2 x\, dx$

$u = x, dv = \sec^2 x\, dx$

8. $\int x^2 \cos x\, dx$

$u = x^2, dv = \cos x\, dx$

9. $dv = e^{-2x}\, dx \implies v = \int e^{-2x}\, dx = -\frac{1}{2}e^{-2x}$

$u = x \qquad \implies du = dx$

$\int xe^{-2x}\, dx = -\frac{1}{2}xe^{-2x} - \int -\frac{1}{2}e^{-2x}\, dx$

$\qquad = -\frac{1}{2}xe^{-2x} - \frac{1}{4}e^{-2x} + C = \frac{-1}{4e^{2x}}(2x + 1) + C$

10. $dv = e^{-x}\, dx \implies v = \int e^{-x}\, dx = -e^{-x}$

$u = x \qquad \implies du = dx$

$\int \frac{x}{e^x}\, dx = \int xe^{-x}\, dx$

$\qquad = -xe^{-x} - \int -e^{-x}\, dx = -xe^{-x} - e^{-x} + C$

11. Use integration by parts three times.

(1) $dv = e^x\, dx \implies v = \int e^x\, dx = e^x$ (2) $dv = e^x\, dx \implies v = \int e^x\, dx = e^x$ (3) $dv = e^x\, dx \implies v = \int e^x\, dx = e^x$

$\quad u = x^3 \implies du = 3x^2\, dx$ $u = x^2 \implies du = 2x\, dx$ $u = x \implies du = dx$

$\int x^3 e^x\, dx = x^3 e^x - 3\int x^2 e^x\, dx = x^3 e^x - 3x^2 e^x + 6\int xe^x\, dx$

$\qquad = x^3 e^x - 3x^2 e^x + 6xe^x - 6e^x + C = e^x(x^3 - 3x^2 + 6x - 6) + C$

12. $\int \frac{e^{1/t}}{t^2}\, dt = -\int e^{1/t}\left(\frac{-1}{t^2}\right) dt = -e^{1/t} + C$

13. $\int x^2 e^{x^3}\, dx = \frac{1}{3}\int e^{x^3}(3x^2)dx = \frac{1}{3}e^{x^3} + C$

14. $dv = x^3\, dx \implies v = \int x^3\, dx = \frac{x^4}{4}$

$u = \ln x \implies du = \frac{1}{x}\, dx$

$\int x^3 \ln x\, dx = \frac{x^4}{4}\ln x - \frac{1}{4}\int x^3\, dx = \frac{x^4}{4}\ln x - \frac{x^4}{16} + C$

$\qquad = \frac{x^4}{16}(4\ln x - 1) + C$

15. $dv = t\, dt \qquad \implies v = \int t\, dt = \frac{t^2}{2}$

$u = \ln(t + 1) \implies du = \frac{1}{t + 1}\, dt$

$\int t \ln(t + 1)\, dt = \frac{t^2}{2}\ln(t + 1) - \frac{1}{2}\int \frac{t^2}{t + 1}\, dt$

$\qquad = \frac{t^2}{2}\ln(t + 1) - \frac{1}{2}\int\left(t - 1 + \frac{1}{t + 1}\right) dt$

$\qquad = \frac{t^2}{2}\ln(t + 1) - \frac{1}{2}\left[\frac{t^2}{2} - t + \ln(t + 1)\right] + C$

$\qquad = \frac{1}{4}[2(t^2 - 1)\ln|t + 1| - t^2 + 2t] + C$

16. Let $u = \ln x, du = \frac{1}{x}\, dx$.

$\int \frac{1}{x(\ln x)^3}\, dx = \int (\ln x)^{-3}\left(\frac{1}{x}\right)dx = \frac{-1}{2(\ln x)^2} + C$

17. Let $u = \ln x, du = \frac{1}{x}\, dx$.

$\int \frac{(\ln x)^2}{x}\, dx = \int (\ln x)^2\left(\frac{1}{x}\right) dx = \frac{(\ln x)^3}{3} + C$

18. $dv = \dfrac{1}{x^2} dx \implies v = \displaystyle\int \dfrac{1}{x^2} dx = -\dfrac{1}{x}$

$u = \ln x \implies du = \dfrac{1}{x} dx$

$\displaystyle\int \dfrac{\ln x}{x^2} dx = -\dfrac{\ln x}{x} + \int \dfrac{1}{x^2} dx = -\dfrac{\ln x}{x} - \dfrac{1}{x} + C$

19. $dv = \dfrac{1}{(2x+1)^2} dx \implies v = \displaystyle\int (2x+1)^{-2} dx$

$\qquad\qquad\qquad\qquad = -\dfrac{1}{2(2x+1)}$

$u = xe^{2x} \qquad\qquad \implies du = (2xe^{2x} + e^{2x}) dx$

$\qquad\qquad\qquad\qquad\qquad = e^{2x}(2x+1) dx$

$\displaystyle\int \dfrac{xe^{2x}}{(2x+1)^2} dx = -\dfrac{xe^{2x}}{2(2x+1)} + \int \dfrac{e^{2x}}{2} dx$

$\qquad\qquad\qquad = \dfrac{-xe^{2x}}{2(2x+1)} + \dfrac{e^{2x}}{4} + C$

$\qquad\qquad\qquad = \dfrac{e^{2x}}{4(2x+1)} + C$

20. $dv = \dfrac{x}{(x^2+1)^2} dx \implies v = \displaystyle\int (x^2+1)^{-2} x\, dx = -\dfrac{1}{2(x^2+1)}$

$u = x^2 e^{x^2} \qquad\qquad \implies du = (2x^3 e^{x^2} + 2xe^{x^2}) dx = 2xe^{x^2}(x^2+1) dx$

$\displaystyle\int \dfrac{x^3 e^{x^2}}{(x^2+1)^2} dx = -\dfrac{x^2 e^{x^2}}{2(x^2+1)} + \int xe^{x^2} dx = -\dfrac{x^2 e^{x^2}}{2(x^2+1)} + \dfrac{e^{x^2}}{2} + C = \dfrac{e^{x^2}}{2(x^2+1)} + C$

21. Use integration by parts twice.

(1) $dv = e^x dx \implies v = \displaystyle\int e^x dx = e^x$

$\quad u = x^2 \implies du = 2x\, dx$

(2) $dv = e^x dx \implies v = \displaystyle\int e^x dx = e^x$

$\quad u = x \implies du = dx$

$\displaystyle\int (x^2 - 1)e^x dx = \int x^2 e^x dx - \int e^x dx = x^2 e^x - 2\int xe^x dx - e^x$

$\qquad\qquad\qquad = x^2 e^x - 2\left[xe^x - \int e^x dx \right] - e^x = x^2 e^x - 2xe^x + e^x + C = (x-1)^2 e^x + C$

22. $dv = \dfrac{1}{x^2} dx \implies v = \displaystyle\int \dfrac{1}{x^2} dx = -\dfrac{1}{x}$

$u = \ln 2x \implies du = \dfrac{1}{x} dx$

$\displaystyle\int \dfrac{\ln(2x)}{x^2} dx = -\dfrac{\ln(2x)}{x} + \int \dfrac{1}{x^2} dx = -\dfrac{\ln(2x)}{x} - \dfrac{1}{x} + C$

$\qquad\qquad = -\dfrac{\ln(2x) + 1}{x} + C$

23. $dv = \sqrt{x-1}\, dx \implies v = \displaystyle\int (x-1)^{1/2} dx = \dfrac{2}{3}(x-1)^{3/2}$

$u = x \qquad\qquad \implies du = dx$

$\displaystyle\int x\sqrt{x-1}\, dx = \dfrac{2}{3}x(x-1)^{3/2} - \dfrac{2}{3}\int (x-1)^{3/2} dx$

$\qquad\qquad = \dfrac{2}{3}x(x-1)^{3/2} - \dfrac{4}{15}(x-1)^{5/2} + C$

$\qquad\qquad = \dfrac{2(x-1)^{3/2}}{15}(3x+2) + C$

24. $dv = \dfrac{1}{\sqrt{2+3x}} dx \implies v = \displaystyle\int (2+3x)^{-1/2} dx = \dfrac{2}{3}\sqrt{2+3x}$

$u = x \qquad\qquad \implies du = dx$

$\displaystyle\int \dfrac{x}{\sqrt{2+3x}} dx = \dfrac{2x\sqrt{2+3x}}{3} - \dfrac{2}{3}\int \sqrt{2+3x}\, dx$

$\qquad\qquad = \dfrac{2x\sqrt{2+3x}}{3} - \dfrac{4}{27}(2+3x)^{3/2} + C = \dfrac{2\sqrt{2+3x}}{27}[9x - 2(2+3x)] + C = \dfrac{2\sqrt{2+3x}}{27}(3x-4) + C$

25. $dv = \cos x\, dx \implies v = \int \cos x\, dx = \sin x$

$u = x \qquad \implies du = dx$

$\int x \cos x\, dx = x \sin x - \int \sin x\, dx = x \sin x + \cos x + C$

26. $dv = \sec\theta \tan\theta\, d\theta \implies v = \int \sec\theta \tan\theta\, d\theta = \sec\theta$

$u = \theta \qquad \implies du = d\theta$

$\int \theta \sec\theta \tan\theta\, d\theta = \theta \sec\theta - \int \sec\theta\, d\theta$

$\qquad\qquad\qquad = \theta \sec\theta - \ln|\sec\theta + \tan\theta| + C$

27. $dv = dx \qquad \implies v = \int dx = x$

$u = \arctan x \implies du = \dfrac{1}{1+x^2}\, dx$

$\int \arctan x\, dx = x \arctan x - \int \dfrac{x}{1+x^2}\, dx$

$\qquad\qquad\quad = x \arctan x - \dfrac{1}{2}\ln(1+x^2) + C$

28. $dv = dx \qquad \implies v = \int dx = x$

$u = \arccos x \implies du = -\dfrac{1}{\sqrt{1-x^2}}\, dx$

$\int \arccos x\, dx = x \arccos x + \int \dfrac{x}{\sqrt{1-x^2}}\, dx$

$\qquad\qquad\quad = x \arccos x - \sqrt{1-x^2} + C$

29. Use integration by parts twice.

(1) $dv = e^{2x}dx \implies v = \int e^{2x}\, dx = \dfrac{1}{2}e^{2x}$

$u = \sin x \implies du = \cos x\, dx$

(2) $dv = e^{2x}\, dx \implies v = \int e^{2x}\, dx = \dfrac{1}{2}e^{2x}$

$u = \cos x \implies du = -\sin x\, dx$

$\int e^{2x} \sin x\, dx = \dfrac{1}{2}e^{2x}\sin x - \dfrac{1}{2}\int e^{2x}\cos x\, dx = \dfrac{1}{2}e^{2x}\sin x - \dfrac{1}{2}\left(\dfrac{1}{2}e^{2x}\cos x + \dfrac{1}{2}\int e^{2x}\sin x\, dx\right)$

$\dfrac{5}{4}\int e^{2x}\sin x\, dx = \dfrac{1}{2}e^{2x}\sin x - \dfrac{1}{4}e^{2x}\cos x$

$\int e^{2x}\sin x\, dx = \dfrac{1}{5}e^{2x}(2\sin x - \cos x) + C$

30. Use integration by parts twice.

(1) $dv = e^x\, dx \implies v = \int e^x\, dx = e^x$

$u = \cos 2x \implies du = -2\sin 2x\, dx$

(2) $dv = e^x\, dx \implies v = \int e^x\, dx = e^x$

$u = \sin 2x \implies du = 2\cos 2x\, dx$

$\int e^x \cos 2x\, dx = e^x \cos 2x + 2\int e^x \sin 2x\, dx = e^x \cos 2x + 2\left(e^x \sin 2x - 2\int e^x \cos 2x\, dx\right)$

$5\int e^x \cos 2x\, dx = e^x \cos 2x + 2e^x \sin 2x$

$\int e^x \cos 2x\, dx = \dfrac{e^x}{5}(\cos 2x + 2\sin 2x) + C$

31. $y' = xe^{x^2}$

$y = \int xe^{x^2}\, dx = \dfrac{1}{2}e^{x^2} + C$

32. $dv = dx \implies v = x$

$u = \ln x \implies du = \dfrac{1}{x}\, dx$

$y' = \ln x$

$y = \int \ln x\, dx = x \ln x - \int x\left(\dfrac{1}{x}\right) dx$

$\quad = x \ln x - x + C = x(-1 + \ln x) + C$

33. Use integration by parts twice.

$\quad$ (1) $dv = \dfrac{1}{\sqrt{2 + 3t}}\, dt \implies v = \displaystyle\int (2 + 3t)^{-1/2}\, dt = \dfrac{2}{3}\sqrt{2 + 3t}$

$\qquad u = t^2 \qquad\qquad \implies du = 2t\, dt$

$\quad$ (2) $dv = \sqrt{2 + 3t}\, dt \implies v = \displaystyle\int (2 + 3t)^{1/2}\, dt = \dfrac{2}{9}(2 + 3t)^{3/2}$

$\qquad u = t \qquad\qquad \implies du = dt$

$y = \displaystyle\int \dfrac{t^2}{\sqrt{2 + 3t}}\, dt = \dfrac{2t^2\sqrt{2 + 3t}}{3} - \dfrac{4}{3}\int t\sqrt{2 + 3t}\, dt$

$\qquad = \dfrac{2t^2\sqrt{2 + 3t}}{3} - \dfrac{4}{3}\left[\dfrac{2t}{9}(2 + 3t)^{3/2} - \dfrac{2}{9}\int (2 + 3t)^{3/2}\, dt \right]$

$\qquad = \dfrac{2t^2\sqrt{2 + 3t}}{3} - \dfrac{8t}{27}(2 + 3t)^{3/2} + \dfrac{16}{405}(2 + 3t)^{5/2} + C$

$\qquad = \dfrac{2\sqrt{2 + 3t}}{405}(27t^2 - 24t + 32) + C$

34. Use integration by parts twice.

$\quad$ (1) $dv = \sqrt{x - 1}\, dx \implies v = \displaystyle\int (x - 1)^{1/2}\, dx = \dfrac{2}{3}(x - 1)^{3/2}$

$\qquad u = x^2 \qquad\qquad \implies du = 2x\, dx$

$\quad$ (2) $dv = (x - 1)^{3/2}dx \implies v = \displaystyle\int (x - 1)^{3/2}\, dx = \dfrac{2}{5}(x - 1)^{5/2}$

$\qquad u = x \qquad\qquad \implies du = dx$

$y = \displaystyle\int x^2\sqrt{x - 1}\, dx$

$\quad = \dfrac{2}{3}x^2(x - 1)^{3/2} - \dfrac{4}{3}\int x(x - 1)^{3/2}\, dx$

$\quad = \dfrac{2}{3}x^2(x - 1)^{3/2} - \dfrac{4}{3}\left[\dfrac{2}{5}x(x - 1)^{5/2} - \dfrac{2}{5}\int (x - 1)^{5/2}\, dx \right]$

$\quad = \dfrac{2}{3}x^2(x - 1)^{3/2} - \dfrac{8}{15}x(x - 1)^{5/2} + \dfrac{16}{105}(x - 1)^{7/2} + C$

$\quad = \dfrac{2(x - 1)^{3/2}}{105}(15x^2 + 12x + 8) + C$

35. $(\cos y)y' = 2x$

$\displaystyle\int \cos y\, dy = \int 2x\, dx$

$\qquad \sin y = x^2 + C$

36. $dv = dx \qquad \implies v = \displaystyle\int dx = x$

$u = \arctan\dfrac{x}{2} \implies du = \dfrac{1}{1 + (x/2)^2}\left(\dfrac{1}{2}\right) dx = \dfrac{2}{4 + x^2}\, dx$

$y = \displaystyle\int \arctan\dfrac{x}{2}\, dx = x\arctan\dfrac{x}{2} - \int \dfrac{2x}{4 + x^2}\, dx$

$\quad = x\arctan\dfrac{x}{2} - \ln(4 + x^2) + C$

37. (a)

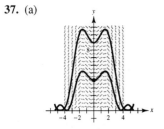

(b) $\dfrac{dy}{dx} = x\sqrt{y}\cos x,\ (0, 4)$

$$\int \frac{dy}{\sqrt{y}} = \int x \cos x\, dx$$

$$\int y^{-1/2}\, dy = \int x \cos x\, dx \qquad (u = x,\, du = dx,\, dv = \cos x\, dx,\, v = \sin x)$$

$$2y^{1/2} = x \sin x - \int \sin x\, dx$$

$$= x \sin x + \cos x + C$$

$(0, 4):\ 2(4)^{1/2} = 0 + 1 + C \implies C = 3$

$2\sqrt{y} = x \sin x + \cos x + 3$

38. (a)

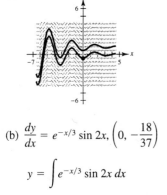

(b) $\dfrac{dy}{dx} = e^{-x/3} \sin 2x,\ \left(0, -\dfrac{18}{37}\right)$

$$y = \int e^{-x/3} \sin 2x\, dx$$

Use integration by parts twice.

(1) $u = \sin 2x,\ du = 2 \cos 2x$

$dv = e^{-x/3}\, dx,\ v = -3e^{-x/3}$

$$\int e^{-x/3} \sin 2x\, dx = -3e^{-x/3} \sin 2x + \int 6e^{-x/3} \cos 2x\, dx$$

(2) $u = \cos 2x,\ du = -2 \sin 2x$

$dv = e^{-x/3}\, dx,\ v = -3e^{-x/3}$

$$\int e^{-x/3} \sin 2x\, dx = -3e^{-x/3} \sin 2x + 6\left[-3e^{-x/3} \cos 2x - \int 6e^{-x/3} \sin 2x\, dx\right] + C$$

$$37 \int e^{-x/3} \sin 2x\, dx = -3e^{-x/3} \sin 2x - 18e^{-x/3} \cos 2x + C$$

$$y = \int e^{-x/3} \sin 2x\, dx = \frac{1}{37}\left[-3e^{-x/3} \sin 2x - 18e^{-x/3} \cos 2x\right] + C$$

$\left(0, \dfrac{-18}{37}\right):\ \dfrac{-18}{37} = \dfrac{1}{37}[0 - 18] + C \implies C = 0$

$$y = \frac{-1}{37}\left[3e^{-x/3} \sin 2x + 18e^{-x/3} \cos 2x\right]$$

39. $dv = \sin 2x \, dx \implies v = \int \sin 2x \, dx = -\frac{1}{2} \cos 2x$

$\qquad u = x \qquad\qquad \implies du = dx$

$\qquad \int x \sin 2x \, dx = \frac{-1}{2} x \cos 2x + \frac{1}{2} \int \cos 2x \, dx$

$\qquad\qquad\qquad = \frac{-1}{2} x \cos 2x + \frac{1}{4} \sin 2x + C$

$\qquad\qquad\qquad = \frac{1}{4}(\sin 2x - 2x \cos 2x) + C$

$\qquad$ Thus, $\int_0^\pi x \sin 2x \, dx = \left[\frac{1}{4}(\sin 2x - 2x \cos 2x)\right]_0^\pi = -\frac{\pi}{2}.$

40. $dv = x \, dx \qquad \implies v = \int x \, dx = \frac{x^2}{2}$

$\qquad u = \arcsin x^2 \implies du = \frac{2x}{\sqrt{1 - x^4}} \, dx$

$\qquad \int x \arcsin x^2 \, dx = \frac{x^2}{2} \arcsin x^2 - \int \frac{x^3}{\sqrt{1 - x^4}} \, dx$

$\qquad\qquad\qquad = \frac{x^2}{2} \arcsin x^2 + \frac{1}{4}(2)(1 - x^4)^{1/2} + C$

$\qquad\qquad\qquad = \frac{1}{2}\left[x^2 \arcsin x^2 + \sqrt{1 - x^4}\right] + C$

$\qquad$ Thus, $\int_0^1 x \arcsin x^2 dx = \frac{1}{2}\left[x^2 \arcsin x^2 + \sqrt{1 - x^4}\right]_0^1$

$\qquad\qquad\qquad\qquad = \frac{1}{4}(\pi - 2).$

41. Use integration by parts twice.

$\quad$ (1) $dv = e^x \, dx \implies v = \int e^x \, dx = e^x$

$\qquad\quad u = \sin x \implies du = \cos x \, dx$

$\quad$ (2) $dv = e^x \, dx \implies v = \int e^x \, dx = e^x$

$\qquad\quad u = \cos x \implies du = -\sin x \, dx$

$\quad \int e^x \sin x \, dx = e^x \sin x - \int e^x \cos x \, dx = e^x \sin x - e^x \cos x - \int e^x \sin x \, dx$

$\quad 2 \int e^x \sin x \, dx = e^x(\sin x - \cos x)$

$\quad \int e^x \sin x \, dx = \frac{e^x}{2}(\sin x - \cos x) + C$

$\quad$ Thus, $\int_0^1 e^x \sin x \, dx = \left[\frac{e^x}{2}(\sin x - \cos x)\right]_0^1 = \frac{e}{2}(\sin 1 - \cos 1) + \frac{1}{2} = \frac{e(\sin 1 - \cos 1) + 1}{2} \approx 0.909.$

42. See Exercise 1.

$\qquad \int_0^1 x^2 e^x \, dx = \left[x^2 e^x - 2xe^x + 2e^x\right]_0^1 = e - 2 \approx 0.718$

43. See Exercise 25.

$\qquad \int_0^{\pi/2} x \cos x \, dx = \left[x \sin x + \cos x\right]_0^{\pi/2} = \frac{\pi}{2} - 1$

44. $dv = dx \qquad\qquad \implies v = \int dx = x$

$\qquad u = \ln(1 + x^2) \implies du = \frac{2x}{1 + x^2} \, dx$

$\qquad \int \ln(1 + x^2) dx = x \ln(1 + x^2) - \int \frac{2x^2}{1 + x^2} \, dx$

$\qquad\qquad\qquad = x \ln(1 + x^2) - 2 \int \left[1 - \frac{1}{1 + x^2}\right] dx = x \ln(1 + x^2) - 2x + 2 \arctan x + C$

$\qquad$ Thus, $\int_0^1 \ln(1 + x^2) \, dx = \left[x \ln(1 + x^2) - 2x + 2 \arctan x\right]_0^1 = \ln 2 - 2 + \frac{\pi}{2}.$

45. $\displaystyle\int x^2 e^{2x}\,dx = x^2\left(\frac{1}{2}e^{2x}\right) - (2x)\left(\frac{1}{4}e^{2x}\right) + 2\left(\frac{1}{8}e^{2x}\right) + C$

$\displaystyle = \frac{1}{2}x^2 e^{2x} - \frac{1}{2}xe^{2x} + \frac{1}{4}e^{2x} + C$

$\displaystyle = \frac{1}{4}e^{2x}(2x^2 - 2x + 1) + C$

Alternate signs	u and its derivatives	v' and its antiderivatives
+	x^2	e^{2x}
−	$2x$	$\frac{1}{2}e^{2x}$
+	2	$\frac{1}{4}e^{2x}$
−	0	$\frac{1}{8}e^{2x}$

46. $\displaystyle\int x^4 e^{-x}\,dx = x^4(-e^{-x}) - 4x^3 e^{-x} + 12x^2(-e^{-x}) - 24xe^{-x} + 24(-e^{-x}) + C$

$\displaystyle = -e^{-x}(x^4 + 4x^3 + 12x^2 + 24x + 24) + C$

Alternate signs	u and its derivatives	v' and its antiderivatives
+	x^4	e^{-x}
−	$4x^3$	$-e^{-x}$
+	$12x^2$	e^{-x}
−	$24x$	$-e^{-x}$
+	24	e^{-x}
−	0	$-e^{-x}$

47. $\displaystyle\int x^3 \sin x\,dx = x^3(-\cos x) - 3x^2(-\sin x) + 6x \cos x - 6 \sin x + C$

$\displaystyle = -x^3 \cos x + 3x^2 \sin x + 6x \cos x - 6 \sin x + C$

$\displaystyle = (3x^2 - 6)\sin x - (x^3 - 6x)\cos x + C$

Alternate signs	u and its derivatives	v' and its antiderivatives
+	x^3	$\sin x$
−	$3x^2$	$-\cos x$
+	$6x$	$-\sin x$
−	6	$\cos x$
+	0	$\sin x$

48. $\displaystyle\int x^3 \cos 2x\,dx = x^3\left(\frac{1}{2}\sin 2x\right) - 3x^2\left(-\frac{1}{4}\cos 2x\right) + 6x\left(-\frac{1}{8}\sin 2x\right) - 6\left(\frac{1}{16}\cos 2x\right) + C$

$\displaystyle = \frac{1}{2}x^3 \sin 2x + \frac{3}{4}x^2 \cos 2x - \frac{3}{4}x \sin 2x - \frac{3}{8}\cos 2x + C$

$\displaystyle = \frac{1}{8}\left[4x^3 \sin 2x + 6x^2 \cos 2x - 6x \sin 2x - 3 \cos 2x\right] + C$

Alternate signs	u and its derivatives	v' and its antiderivatives
+	x^3	$\cos 2x$
−	$3x^2$	$\frac{1}{2}\sin 2x$
+	$6x$	$-\frac{1}{4}\cos 2x$
−	6	$-\frac{1}{8}\sin 2x$
+	0	$\frac{1}{16}\cos 2x$

49. $\displaystyle\int x \sec^2 x\,dx = x \tan x + \ln|\cos x| + C$

Alternate signs	u and its derivatives	v' and its antiderivatives
+	x	$\sec^2 x$
−	1	$\tan x$
+	0	$-\ln|\cos x|$

50. $\int x^2(x-2)^{3/2}dx = \frac{2}{5}x^2(x-2)^{5/2} - \frac{8}{35}x(x-2)^{7/2} + \frac{16}{315}(x-2)^{9/2} + C$

$$= \frac{2}{315}(x-2)^{5/2}(35x^2 + 40x + 32) + C$$

Alternate signs	u and its derivatives	v' and its antiderivatives
$+$	x^2	$(x-2)^{3/2}$
$-$	$2x$	$\frac{2}{5}(x-2)^{5/2}$
$+$	2	$\frac{4}{35}(x-2)^{7/2}$
$-$	0	$\frac{8}{315}(x-2)^{9/2}$

51. $\int t^3 e^{-4t}\, dt = -\frac{e^{-4t}}{128}(32t^3 + 24t^2 + 12t + 3) + C$

52. $\int \alpha^4 \sin \pi\alpha\, d\alpha = \frac{1}{\pi^5}[-(\alpha\pi)^4 \cos \pi\alpha + 4(\alpha\pi)^3 \sin \pi\alpha + 12(\alpha\pi)^2 \cos \pi\alpha - 24(\alpha\pi) \sin \pi\alpha - 24 \cos \pi\alpha] + C$

53. $\int_0^{\pi/2} e^{-2x} \sin 3x\, dx = \left[\frac{e^{-2x}(-2\sin 3x - 3\cos 3x)}{13}\right]_0^{\pi/2} = \frac{1}{13}(2e^{-\pi} + 3) \approx 0.2374$

54. $\int_0^5 x^4(25 - x^2)^{3/2}\, dx = \left[\frac{1{,}171{,}875 \arcsin(x/5)}{128} - \frac{x(2x^2 + 25)(25 - x^2)^{5/2}}{16} + \frac{625x(25 - x^2)^{3/2}}{64} + \frac{46{,}875x\sqrt{25 - x^2}}{128}\right]_0^5$

$$\approx 14{,}381.0699$$

55. (a) $dv = \sqrt{2x - 3}\, dx \implies v = \int (2x - 3)^{1/2}\, dx = \frac{1}{3}(2x - 3)^{3/2}$

$\quad\quad u = 2x \quad\quad\quad \implies du = 2\, dx$

$\quad\quad \int 2x\sqrt{2x - 3}\, dx = \frac{2}{3}x(2x - 3)^{3/2} - \frac{2}{3}\int (2x - 3)^{3/2}\, dx$

$\quad\quad\quad\quad\quad = \frac{2}{3}x(2x - 3)^{3/2} - \frac{2}{15}(2x - 3)^{5/2} + C$

$\quad\quad\quad\quad\quad = \frac{2}{15}(2x - 3)^{3/2}(3x + 3) + C = \frac{2}{5}(2x - 3)^{3/2}(x + 1) + C$

$\quad$ (b) $u = 2x - 3 \implies x = \frac{u + 3}{2}$ and $dx = \frac{1}{2}du$

$\quad\quad \int 2x\sqrt{2x - 3}\, dx = \int 2\left(\frac{u + 3}{2}\right)u^{1/2}\left(\frac{1}{2}\right) du = \frac{1}{2}\int (u^{3/2} + 3u^{1/2})\, du = \frac{1}{2}\left[\frac{2}{5}u^{5/2} + 2u^{3/2}\right] + C$

$\quad\quad\quad\quad\quad = \frac{1}{5}u^{3/2}(u + 5) + C = \frac{1}{5}(2x - 3)^{3/2}[(2x - 3) + 5] + C = \frac{2}{5}(2x - 3)^{3/2}(x + 1) + C$

56. (a) $dv = \sqrt{4 + x}\, dx \implies v = \int (4 + x)^{1/2}\, dx = \frac{2}{3}(4 + x)^{3/2}$

$\quad\quad u = x \quad\quad\quad \implies du = dx$

$\quad\quad \int x\sqrt{4 + x}\, dx = \frac{2}{3}x(4 + x)^{3/2} - \frac{2}{3}\int (4 + x)^{3/2}\, dx$

$\quad\quad\quad\quad\quad = \frac{2}{3}x(4 + x)^{3/2} - \frac{4}{15}(4 + x)^{5/2} + C = \frac{2}{15}(4 + x)^{3/2}(3x - 8) + C$

—CONTINUED—

56. —CONTINUED—

(b) $u = 4 + x \implies x = u - 4$ and $dx = du$

$$\int x\sqrt{4 + x}\, dx = \int (u - 4)u^{1/2}du = \int (u^{3/2} - 4u^{1/2})\, du$$

$$= \frac{2}{5}u^{5/2} - \frac{8}{3}u^{3/2} + C = \frac{2}{15}u^{3/2}(3u - 20) + C$$

$$= \frac{2}{15}(4 + x)^{3/2}[3(4 + x) - 20] + C = \frac{2}{15}(4 + x)^{3/2}(3x - 8) + C$$

57. (a) $dv = \dfrac{x}{\sqrt{4 + x^2}}\, dx \implies v = \int (4 + x^2)^{-1/2}x\, dx = \sqrt{4 + x^2}$

$u = x^2 \implies du = 2x\, dx$

$$\int \frac{x^3}{\sqrt{4 + x^2}}\, dx = x^2\sqrt{4 + x^2} - 2\int x\sqrt{4 + x^2}\, dx$$

$$= x^2\sqrt{4 + x^2} - \frac{2}{3}(4 + x^2)^{3/2} + C = \frac{1}{3}\sqrt{4 + x^2}(x^2 - 8) + C$$

(b) $u = 4 + x^2 \implies x^2 = u - 4$ and $2x\, dx = du \implies x\, dx = \dfrac{1}{2}du$

$$\int \frac{x^3}{\sqrt{4 + x^2}}\, dx = \int \frac{x^2}{\sqrt{4 + x^2}}x\, dx = \int \frac{u - 4}{\sqrt{u}}\frac{1}{2}\, du$$

$$= \frac{1}{2}\int (u^{1/2} - 4u^{-1/2})\, du = \frac{1}{2}\left(\frac{2}{3}u^{3/2} - 8u^{1/2}\right) + C$$

$$= \frac{1}{3}u^{1/2}(u - 12) + C = \frac{1}{3}\sqrt{4 + x^2}[(4 + x^2) - 12] + C = \frac{1}{3}\sqrt{4 + x^2}(x^2 - 8) + C$$

58. (a) $dv = \sqrt{4 - x}\, dx \implies v = \int (4 - x)^{1/2}\, dx$

$$= -\frac{2}{3}(4 - x)^{3/2}$$

$u = x \qquad \implies du = dx$

$$\int x\sqrt{4 - x}\, dx = -\frac{2}{3}x(4 - x)^{3/2} + \frac{2}{3}\int (4 - x)^{3/2}\, dx$$

$$= -\frac{2}{3}x(4 - x)^{3/2} - \frac{4}{15}(4 - x)^{5/2} + C$$

$$= -\frac{2}{15}(4 - x)^{3/2}[5x + 2(4 - x)] + C$$

$$= -\frac{2}{15}(4 - x)^{3/2}(3x + 8) + C$$

(b) $u = 4 - x \implies x = 4 - u$ and $dx = -du$

$$\int x\sqrt{4 - x}\, dx = -\int (4 - u)\sqrt{u}\, du$$

$$= -\int (4u^{1/2} - u^{3/2})\, du$$

$$= -\frac{8}{3}u^{3/2} + \frac{2}{5}u^{5/2} + C$$

$$= -\frac{2}{15}u^{3/2}(20 - 3u) + C$$

$$= -\frac{2}{15}(4 - x)^{3/2}[20 - 3(4 - x)] + C$$

$$= -\frac{2}{15}(4 - x)^{3/2}(3x + 8) + C$$

59. $n = 0$: $\int \ln x \, dx = x(\ln x - 1) + C$

$n = 1$: $\int x \ln x \, dx = \frac{x^2}{4}(2 \ln x - 1) + C$

$n = 2$: $\int x^2 \ln x \, dx = \frac{x^3}{9}(3 \ln x - 1) + C$

$n = 3$: $\int x^3 \ln x \, dx = \frac{x^4}{16}(4 \ln x - 1) + C$

$n = 4$: $\int x^4 \ln x \, dx = \frac{x^5}{25}(5 \ln x - 1) + C$

In general,

$$\int x^n \ln x \, dx = \frac{x^{n+1}}{(n+1)^2}[(n+1)\ln x - 1] + C.$$

(See Exercise 63)

60. $n = 0$: $\int e^x \, dx = e^x + C$

$n = 1$: $\int xe^x \, dx = xe^x - e^x + C = xe^x - \int e^x \, dx$

$n = 2$: $\int x^2 e^x \, dx = x^2 e^x - 2xe^x + 2e^x + C$

$$= x^2 e^x - 2 \int xe^x \, dx$$

$n = 3$: $\int x^3 e^x \, dx = x^3 e^x - 3x^2 e^x + 6xe^x - 6e^x + C$

$$= x^3 e^x - 3 \int x^2 e^x \, dx$$

$n = 4$: $\int x^4 e^x \, dx$

$$= x^4 e^x - 4x^3 e^x + 12x^2 e^x - 24xe^x + 24e^x + C$$

$$= x^4 e^x - 4 \int x^3 e^x \, dx$$

In general, $\int x^n e^x \, dx = x^n e^x - n\int x^{n-1} e^x \, dx$.

61. $dv = \sin x \, dx \implies v = -\cos x$

$u = x^n \qquad \implies du = nx^{n-1} \, dx$

$\int x^n \sin x \, dx = -x^n \cos x + n \int x^{n-1} \cos x \, dx$

62. $dv = \cos x \, dx \implies v = \sin x$

$u = x^n \qquad \implies du = nx^{n-1} \, dx$

$\int x^n \cos x \, dx = x^n \sin x - n \int x^{n-1} \sin x \, dx$

63. $dv = x^n \, dx \implies v = \frac{x^{n+1}}{n+1}$

$u = \ln x \implies du = \frac{1}{x} \, dx$

$\int x^n \ln x \, dx = \frac{x^{n+1}}{n+1} \ln x - \int \frac{x^n}{n+1} \, dx$

$$= \frac{x^{n+1}}{n+1} \ln x - \frac{x^{n+1}}{(n+1)^2} + C$$

$$= \frac{x^{n+1}}{(n+1)^2}[(n+1)\ln x - 1] + C$$

64. $dv = e^{ax} \, dx \implies v = \frac{1}{a} e^{ax}$

$u = x^n \implies du = nx^{n-1} \, dx$

$\int x^n e^{ax} \, dx = \frac{x^n e^{ax}}{a} - \frac{n}{a} \int x^{n-1} e^{ax} \, dx$

65. Use integration by parts twice.

(1) $dv = e^{ax} \, dx \implies v = \frac{1}{a} e^{ax}$

$u = \sin bx \implies du = b \cos bx \, dx$

$\int e^{ax} \sin bx \, dx = \frac{e^{ax} \sin bx}{a} - \frac{b}{a} \int e^{ax} \cos bx \, dx$

(2) $dv = e^{ax} \, dx \implies v = \frac{1}{a} e^{ax}$

$u = \cos bx \implies du = -b \sin bx \, dx$

$$= \frac{e^{ax} \sin bx}{a} - \frac{b}{a}\left[\frac{e^{ax} \cos bx}{a} + \frac{b}{a} \int e^{ax} \sin bx \, dx \right] = \frac{e^{ax} \sin bx}{a} - \frac{b^2}{a^2} \int e^{ax} \sin bx \, dx$$

—CONTINUED—

65. —CONTINUED—

Therefore, $\left(1 + \dfrac{b^2}{a^2}\right)\displaystyle\int e^{ax} \sin bx \, dx = \dfrac{e^{ax}(a \sin bx - b \cos bx)}{a^2}$

$$\int e^{ax} \sin bx \, dx = \frac{e^{ax}(a \sin bx - b \cos bx)}{a^2 + b^2} + C.$$

66. Use integration by parts twice.

(1) $dv = e^{ax} \, dx \implies v = \dfrac{1}{a}e^{ax}$ (2) $dv = e^{ax} \, dx \implies v = \dfrac{1}{a}e^{ax}$

     $u = \cos bx \implies du = -b \sin bx$ $u = \sin bx \implies du = b \cos bx$

$\displaystyle\int e^{ax} \cos bx \, dx = \frac{e^{ax} \cos bx}{a} + \frac{b}{a}\int e^{ax} \sin bx \, dx = \frac{e^{ax} \cos bx}{a} + \frac{b}{a}\left[\frac{e^{ax} \sin bx}{a} - \frac{b}{a}\int e^{ax} \cos bx \, dx\right]$

$\qquad\qquad = \dfrac{e^{ax} \cos bx}{a} + \dfrac{be^{ax} \sin bx}{a^2} - \dfrac{b^2}{a^2}\displaystyle\int e^{ax} \cos bx \, dx$

Therefore, $\left(1 + \dfrac{b^2}{a^2}\right)\displaystyle\int e^{ax} \cos bx \, dx = \dfrac{e^{ax}(a \cos bx + b \sin bx)}{a^2}$

$$\int e^{ax} \cos bx \, dx = \frac{e^{ax}(a \cos bx + b \sin bx)}{a^2 + b^2} + C.$$

67. $n = 3$ (Use formula in Exercise 63.)

$$\int x^3 \ln x \, dx = \frac{x^4}{16}[4 \ln x - 1] + C$$

68. $n = 2$ (Use formula in Exercise 62.)

$$\int x^2 \cos x \, dx = x^2 \sin x - 2\int x \sin x \, dx \text{ (Use formula in Exercise 61.) } (n = 1)$$

$$= x^2 \sin x - 2\left[-x \cos x + \int \cos x \, dx\right] = x^2 \sin x + 2x \cos x - 2 \sin x + C$$

69. $a = 2, b = 3$ (Use formula in Exercise 66.)

$$\int e^{2x} \cos 3x \, dx = \frac{e^{2x}(2 \cos 3x + 3 \sin 3x)}{13} + C$$

70. $n = 3, a = 2$ (Use formula in Exercise 64 three times.)

$$\int x^3 e^{2x} \, dx = \frac{x^3 e^{2x}}{2} - \frac{3}{2}\int x^2 e^{2x} \, dx \quad (n = 3, a = 2)$$

$$= \frac{x^3 e^{2x}}{2} - \frac{3}{2}\left[\frac{x^2 e^{2x}}{2} - \int x e^{2x} \, dx\right] \quad (n = 2, a = 2)$$

$$= \frac{x^3 e^{2x}}{2} - \frac{3x^2 e^{2x}}{4} + \frac{3}{2}\left[\frac{x e^{2x}}{2} - \frac{1}{2}\int e^{2x} \, dx\right] = \frac{x^3 e^{2x}}{2} - \frac{3x^2 e^{2x}}{4} + \frac{3x e^{2x}}{4} - \frac{3e^{2x}}{8} + C \quad (n = 1, a = 2)$$

$$= \frac{e^{2x}}{8}(4x^3 - 6x^2 + 6x - 3) + C$$

71. $dv = e^{-x} dx \implies v = -e^{-x}$

$u = x \implies du = dx$

$$A = \int_0^4 xe^{-x} dx = \left[-xe^{-x} \right]_0^4 + \int_0^4 e^{-x} dx = \frac{-4}{e^4} - \left[e^{-x} \right]_0^4$$

$$= 1 - \frac{5}{e^4} \approx 0.908$$

72. $dv = e^{-x/3} dx \implies v = -3e^{-x/3}$

$u = x \implies du = dx$

$$A = \frac{1}{9} \int_0^3 xe^{-x/3} dx$$

$$= \frac{1}{9} \left(\left[-3xe^{-x/3} \right]_0^3 + 3 \int_0^3 e^{-x/3} dx \right)$$

$$= \frac{1}{9} \left(\frac{-9}{e} - \left[9e^{-x/3} \right]_0^3 \right)$$

$$= -\frac{1}{e} - \frac{1}{e} + 1 = 1 - \frac{2}{e} \approx 0.264$$

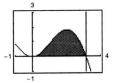

73. $A = \int_0^1 e^{-x} \sin(\pi x) dx$

$$= \left[\frac{e^{-x}(-\sin \pi x - \pi \cos \pi x)}{1 + \pi^2} \right]_0^1$$

$$= \frac{1}{1 + \pi^2} \left(\frac{\pi}{e} + \pi \right) = \frac{\pi}{1 + \pi^2} \left(\frac{1}{e} + 1 \right)$$

$$\approx 0.395 \text{ (See Exercise 65.)}$$

74. $A = \int_0^\pi x \sin x \, dx = \left[-x \cos x + \sin x \right]_0^\pi$

$$= \pi \text{ (See Exercise 61.)}$$

75. (a) $A = \int_1^e \ln x \, dx = \left[-x + x \ln x \right]_1^e = 1 \text{ (See Exercise 1.)}$

(b) $R(x) = \ln x, \, r(x) = 0$

$$V = \pi \int_1^e (\ln x)^2 dx$$

$$= \pi \left[x(\ln x)^2 - 2x \ln x + 2x \right]_1^e \text{ (Use integration by parts twic, see Exercise 5.)}$$

$$= \pi(e - 2) \approx 2.257$$

(c) $p(x) = x, \, h(x) = \ln x$

$$V = 2\pi \int_1^e x \ln x \, dx = 2\pi \left[\frac{x^2}{4}(-1 + 2 \ln x) \right]_1^e$$

$$= \frac{(e^2 + 1)\pi}{2} \approx 13.177 \text{ (See Exercise 63.)}$$

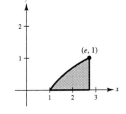

(d) $\bar{x} = \dfrac{\int_1^e x \ln x \, dx}{1} = \dfrac{e^2 + 1}{4} \approx 2.097$

$\bar{y} = \dfrac{\frac{1}{2}\int_1^e (\ln x)^2 dx}{1} = \dfrac{e - 2}{2} \approx 0.359$

$(\bar{x}, \bar{y}) = \left(\dfrac{e^2 + 1}{4}, \dfrac{e - 2}{2} \right) \approx (2.097, 0.359)$

76. In Example 6, we showed that the centroid of an equivalent region was $(1, \pi/8)$. By symmetry, the centroid of this region is $(\pi/8, 1)$.

You can also solve this problem directly.

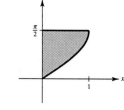

$$A = \int_0^1 \left(\frac{\pi}{2} - \arcsin x\right)dx = \left[\frac{\pi}{2}x - x \arcsin x - \sqrt{1 - x^2}\right]_0^1 \text{ (Example 3)}$$

$$= \left(\frac{\pi}{2} - \frac{\pi}{2} - 0\right) - (-1) = 1$$

$$\bar{x} = \frac{M_y}{A} = \int_0^1 x\left[\frac{\pi}{2} - \arcsin x\right]dx = \frac{\pi}{8}$$

$$\bar{y} = \frac{M_x}{A} = \int_0^1 \frac{(\pi/2) + \arcsin x}{2}\left[\frac{\pi}{2} - \arcsin x\right]dx = 1$$

77. Average value $= \dfrac{1}{\pi}\displaystyle\int_0^\pi e^{-4t}(\cos 2t + 5 \sin 2t)\,dt$

$$= \frac{1}{\pi}\left[e^{-4t}\left(\frac{-4\cos 2t + 2\sin 2t}{20}\right) + 5e^{-4t}\left(\frac{-4\sin 2t - 2\cos 2t}{20}\right)\right]_0^\pi \text{ (From Exercises 65 and 66)}$$

$$= \frac{7}{10\pi}(1 - e^{-4t}) \approx 0.223$$

78. (a) Average $= \displaystyle\int_1^2 (1.6t \ln t + 1)\,dt = \left[0.8t^2 \ln t - 0.4t^2 + t\right]_1^2 = 3.2(\ln 2) - 0.2 \approx 2.018$

(b) Average $= \displaystyle\int_3^4 (1.6t \ln t + 1)\,dt = \left[0.8t^2 \ln t - 0.4t^2 + t\right]_3^4 = 12.8(\ln 4) - 7.2(\ln 3) - 1.8 \approx 8.035$

79. $c(t) = 100{,}000 + 4000t,\ r = 5\%,\ t_1 = 10$

$$P = \int_0^{10} (100{,}000 + 4000t)e^{-0.05t}\,dt = 4000\int_0^{10}(25 + t)e^{-0.05t}\,dt$$

Let $u = 25 + t$, $dv = e^{-0.05t}dt$, $du = dt$, $v = -\dfrac{100}{5}e^{-0.05t}$

$$P = 4000\left\{\left[(25 + t)\left(-\frac{100}{5}e^{-0.05t}\right)\right]_0^{10} + \frac{100}{5}\int_0^{10}e^{-0.05t}\,dt\right\}$$

$$= 4000\left\{\left[(25 + t)\left(-\frac{100}{5}e^{-0.05t}\right)\right]_0^{10} - \left[\frac{10{,}000}{25}e^{-0.05t}\right]_0^{10}\right\} \approx \$931{,}265$$

80. $c(t) = 30{,}000 + 500t,\ r = 7\%,\ t_1 = 5$

$$P\int_0^5 (30{,}000 + 500t)e^{-0.07t}\,dt = 500\int_0^5 (60 + t)e^{-0.07t}\,dt$$

Let $u = 60 + t$, $dv = e^{-0.07t}\,dt$, $du = dt$, $v = -\dfrac{100}{7}e^{-0.07t}$.

$$P = 500\left\{\left[(60 + t)\left(-\frac{100}{7}e^{-0.07t}\right)\right]_0^5 + \frac{100}{7}\int_0^5 e^{-0.07t}\,dt\right\}$$

$$= 500\left\{\left[(60 + t)\left(-\frac{100}{7}e^{-0.07t}\right)\right]_0^5 - \left[\frac{10{,}000}{49}e^{-0.07t}\right]_0^5\right\} \approx \$131{,}528.68$$

81. $\displaystyle\int_{-\pi}^{\pi} x \sin nx\, dx = \left[-\frac{x}{n}\cos nx + \frac{1}{n^2}\sin nx \right]_{-\pi}^{\pi}$

$$= -\frac{\pi}{n}\cos \pi n - \frac{\pi}{n}\cos(-\pi n)$$

$$= -\frac{2\pi}{n}\cos \pi n$$

$$= \begin{cases} -(2\pi/n), & \text{if } n \text{ is even} \\ (2\pi/n), & \text{if } n \text{ is odd} \end{cases}$$

82. $\displaystyle\int_{-\pi}^{\pi} x^2 \cos nx\, dx = \left[\frac{x^2}{n}\sin nx + \frac{2x}{n^2}\cos nx - \frac{2}{n^3}\sin nx \right]_{-\pi}^{\pi}$

$$= \frac{2\pi}{n^2}\cos n\pi + \frac{2\pi}{n^2}\cos(-n\pi)$$

$$= \frac{4\pi}{n^2}\cos n\pi$$

$$= \begin{cases} (4\pi/n^2), & \text{if } n \text{ is even} \\ -(4\pi/n^2), & \text{if } n \text{ is odd} \end{cases}$$

$$= \frac{(-1)^n 4\pi}{n^2}$$

83. Let $u = x,\, dv = \sin\left(\dfrac{n\pi}{2}x\right)dx,\, du = dx,\, v = -\dfrac{2}{n\pi}\cos\left(\dfrac{n\pi}{2}x\right).$

$$I_1 = \int_0^1 x \sin\left(\frac{n\pi}{2}x\right)dx = \left[\frac{-2x}{n\pi}\cos\left(\frac{n\pi}{2}x\right) \right]_0^1 + \frac{2}{n\pi}\int_0^1 \cos\left(\frac{n\pi}{2}x\right)dx$$

$$= -\frac{2}{n\pi}\cos\left(\frac{n\pi}{2}\right) + \left[\left(\frac{2}{n\pi}\right)^2 \sin\left(\frac{n\pi}{2}x\right) \right]_0^1$$

$$= -\frac{2}{n\pi}\cos\left(\frac{n\pi}{2}\right) + \left(\frac{2}{n\pi}\right)^2 \sin\left(\frac{n\pi}{2}\right)$$

Let $u = (-x + 2),\, dv = \sin\left(\dfrac{n\pi}{2}x\right)dx,\, du = -dx,\, v = -\dfrac{2}{n\pi}\cos\left(\dfrac{n\pi}{2}x\right).$

$$I_2 = \int_1^2 (-x + 2)\sin\left(\frac{n\pi}{2}x\right)dx = \left[\frac{-2(-x+2)}{n\pi}\cos\left(\frac{n\pi}{2}x\right) \right]_1^2 - \frac{2}{n\pi}\int_1^2 \cos\left(\frac{n\pi}{2}x\right)dx$$

$$= \frac{2}{n\pi}\cos\left(\frac{n\pi}{2}\right) - \left[\left(\frac{2}{n\pi}\right)^2 \sin\left(\frac{n\pi}{2}x\right) \right]_1^2$$

$$= \frac{2}{n\pi}\cos\left(\frac{n\pi}{2}\right) + \left(\frac{2}{n\pi}\right)^2 \sin\left(\frac{n\pi}{2}\right)$$

$$h(I_1 + I_2) = b_n = h\left[\left(\frac{2}{n\pi}\right)^2 \sin\left(\frac{n\pi}{2}\right) + \left(\frac{2}{n\pi}\right)^2 \sin\left(\frac{n\pi}{2}\right) \right] = \frac{8h}{(n\pi)^2}\sin\left(\frac{n\pi}{2}\right)$$

84. For any integrable function, $\int f(x)\, dx = C + \int f(x)\, dx$, but this cannot be used to imply that $C = 0$.

85. Shell Method:

$$V = 2\pi \int_a^b x f(x)\, dx$$

$$dv = x\, dx \implies v = \frac{x^2}{2}$$

$$u = f(x) \implies du = f'(x)\, dx$$

$$V = 2\pi\left[\frac{x^2}{2}f(x) - \int \frac{x^2}{2}f'(x)\, dx \right]_a^b$$

$$= \pi\left[(b^2 f(b) - a^2 f(a)) - \int_a^b x^2 f'(x)\, dx \right]$$

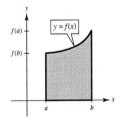

—**CONTINUED**—

85. —CONTINUED—

Disc Method:

$$V = \pi \int_0^{f(a)} (b^2 - a^2)\, dy + \pi \int_{f(a)}^{f(b)} \left[b^2 - [f^{-1}(y)]^2 \right] dy$$

$$= \pi(b^2 - a^2) f(a) + \pi b^2 (f(b) - f(a)) - \pi \int_{f(a)}^{f(b)} [f^{-1}(y)]^2\, dy$$

$$= \pi \left[(b^2 f(b) - a^2 f(a)) - \int_{f(a)}^{f(b)} [f^{-1}(y)]^2\, dy \right]$$

Since $x = f^{-1}(y)$, we have $f(x) = y$ and $f'(x)dx = dy$. When $y = f(a)$, $x = a$. When $y = f(b)$, $x = b$. Thus,

$$\int_{f(a)}^{f(b)} [f^{-1}(y)]^2\, dy = \int_a^b x^2 f'(x)\, dx$$

and the volumes are the same.

86. $f'(x) = xe^{-x}$

(a) $f(x) = \displaystyle\int xe^{-x}\, dx = -xe^{-x} - e^{-x} + C$

   (Parts: $u = x$, $dv = e^{-x}\, dx$)

   $f(0) = 0 = -1 + C \implies C = 1$

   $f(x) = -xe^{-x} - e^{-x} + 1$

(b)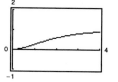

(c) You obtain the points

n	x_n	y_n
0	0	0
1	0.05	0
2	0.10	2.378×10^{-3}
3	0.15	0.0069
4	0.20	0.0134
$\vdots$	$\vdots$	$\vdots$
80	4.0	0.9064

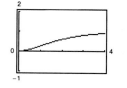

(d) You obtain the points

n	x_n	y_n
0	0	0
1	0.1	0
2	0.2	0.0090484
3	0.3	0.025423
4	0.4	0.047648
$\vdots$	$\vdots$	$\vdots$
40	4.0	0.9039

(e) $f(4) = 0.9084$

The approximations are tangent line approximations. The results in (c) are better because Δx is smaller.

87. $f'(x) = \cos \sqrt{x}, f(0) = 2$

(a) It cannot be solved by integration.

(b) You obtain the points

n	x_n	y_n
0	0	0
1	0.05	2.05
2	0.10	2.098755
3	0.15	2.146276
$\vdots$	$\vdots$	$\vdots$
80	4.0	2.8403565

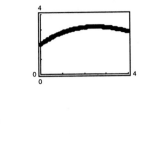

Section 7.3   Trigonometric Integrals

1. $f(x) = \sin^4 x + \cos^4 x$

(a) $\sin^4 x + \cos^4 x = \left(\dfrac{1 - \cos 2x}{2}\right)^2 + \left(\dfrac{1 + \cos 2x}{2}\right)^2$

$$= \frac{1}{4}[1 - 2\cos 2x + \cos^2 2x + 1 + 2\cos 2x + \cos^2 2x]$$

$$= \frac{1}{4}\left[2 + 2\frac{1 + \cos 4x}{2}\right]$$

$$= \frac{1}{4}[3 + \cos 4x]$$

(b) $\sin^4 x + \cos^4 x = (\sin^2 x)^2 + \cos^4 x$

$$= (1 - \cos^2 x)^2 + \cos^4 x$$

$$= 1 - 2\cos^2 x + 2\cos^4 x$$

(c) $\sin^4 x + \cos^4 x = \sin^4 x + 2\sin^2 x \cos^2 x + \cos^4 x - 2\sin^2 x \cos^2 x$

$$= (\sin^2 x + \cos^2 x)^2 - 2\sin^2 x \cos^2 x$$

$$= 1 - 2\sin^2 x \cos^2 x$$

(d) $1 - 2\sin^2 x \cos^2 x = 1 - (2\sin x \cos x)(\sin x \cos x)$

$$= 1 - (\sin 2x)\left(\frac{1}{2}\sin 2x\right)$$

$$= 1 - \frac{1}{2}\sin^2(2x)$$

(e) Four ways. There is often more than one way to rewrite a trigonometric expression.

2. (a) $y = \sec x \Longrightarrow y' = \sec x \tan x = \sin x \sec^2 x.$

Matches (iii)

(b) $y = \cos x + \sec x \Longrightarrow y' = -\sin x + \sec x \tan x$

$$= -\sin x + \sec^2 x \sin x$$

$$= \sin x(-1 + \sec^2 x)$$

$$= \sin x \tan^2 x \quad \text{Matches (i)}$$

(c) $y = x - \tan x + \dfrac{1}{3}\tan^3 x \Longrightarrow y' = 1 - \sec^2 x + \tan^2 x \sec^2 x$

$$= -\tan^2 x + \tan^2 x(1 + \tan^2 x)$$

$$= \tan^4 x \quad \text{Matches (iv)}$$

—CONTINUED—

2. —CONTINUED—

(d) $y = 3x + 2 \sin x \cos^3 x + 3 \sin x \cos x \implies$

$y' = 3 + 2 \cos x(\cos^3 x) + 6 \sin x \cos^2 x(-\sin x) + 3 \cos^2 x - 3 \sin^2 x$

$= 3 + 2 \cos^4 x - 6 \cos^2 x(1 - \cos^2 x) + 3 \cos^2 x - 3(1 - \cos^2 x)$

$= 8 \cos^4 x$ Matches (ii)

3. Let $u = \cos x$, $du = -\sin x \, dx$.

$$\int \cos^3 x \sin x \, dx = -\int \cos^3 x(-\sin x) \, dx$$

$$= -\frac{1}{4} \cos^4 x + C$$

4. Let $u = \sin x$, $du = \cos x \, dx$.

$$\int \cos^3 x \sin^2 x \, dx = \int \cos x(1 - \sin^2 x)\sin^2 x \, dx$$

$$= \int (\sin^2 x - \sin^4 x)\cos x \, dx$$

$$= \frac{1}{3} \sin^3 x - \frac{1}{5} \sin^5 x + C$$

5. Let $u = \sin 2x$, $du = 2 \cos 2x \, dx$.

$$\int \sin^5 2x \cos 2x \, dx = \frac{1}{2} \int \sin^5 2x(2 \cos 2x)dx$$

$$= \frac{1}{12} \sin^6 2x + C$$

6. Let $u = \cos x$, $du = -\sin x \, dx$.

$$\int \sin^3 x \, dx = \int \sin x(1 - \cos^2 x) \, dx$$

$$= \int \cos^2 x(-\sin x) \, dx + \int \sin x \, dx$$

$$= \frac{1}{3} \cos^3 x - \cos x + C$$

7. Let $u = \cos x$, $du = -\sin x \, dx$.

$$\int \sin^5 x \cos^2 x \, dx = \int \sin x(1 - \cos^2 x)^2 \cos^2 x \, dx$$

$$= -\int (\cos^2 x - 2 \cos^4 x + \cos^6 x)(-\sin x) \, dx = \frac{-1}{3} \cos^3 x + \frac{2}{5} \cos^5 x - \frac{1}{7} \cos^7 x + C$$

8. Let $u = \sin \frac{x}{3}$, $du = \frac{1}{3} \cos \frac{x}{3} \, dx$.

$$\int \cos^3 \frac{x}{3} \, dx = \int \left(\cos \frac{x}{3}\right)\left(1 - \sin^2 \frac{x}{3}\right) dx$$

$$= 3 \int \left(1 - \sin^2 \frac{x}{3}\right)\left(\frac{1}{3} \cos \frac{x}{3}\right) dx$$

$$= 3\left(\sin \frac{x}{3} - \frac{1}{3} \sin^3 \frac{x}{3}\right) + C$$

$$= 3 \sin \frac{x}{3} - \sin^3 \frac{x}{3} + C$$

9. $\displaystyle\int \cos^2 3x \, dx = \int \frac{1 + \cos 6x}{2} \, dx$

$$= \frac{1}{2}\left(x + \frac{1}{6} \sin 6x\right) + C$$

$$= \frac{1}{12}(6x + \sin 6x) + C$$

10. $\displaystyle\int \sin^2 2x \, dx = \int \frac{1 - \cos 4x}{2} \, dx = \frac{1}{2}\left(x - \frac{1}{4} \sin 4x\right) + C = \frac{1}{8}(4x - \sin 4x) + C$

11. Integration by parts.

$$dv = \sin^2 x\, dx = \frac{1 - \cos 2x}{2} \implies v = \frac{x}{2} - \frac{\sin 2x}{4} = \frac{1}{4}(2x - \sin 2x)$$

$$u = x \implies du = dx$$

$$\int x \sin^2 x\, dx = \frac{1}{4} x(2x - \sin 2x) - \frac{1}{4} \int (2x - \sin 2x)\, dx$$

$$= \frac{1}{4} x(2x - \sin 2x) - \frac{1}{4}\left(x^2 + \frac{1}{2}\cos 2x\right) + C = \frac{1}{8}(2x^2 - 2x \sin 2x - \cos 2x) + C$$

12. Use integration by parts twice.

$$dv = \sin^2 x\, dx = \frac{1 - \cos 2x}{2} \implies v = \frac{x}{2} - \frac{\sin 2x}{4} = \frac{1}{4}(2x - \sin 2x)$$

$$u = x^2 \implies du = 2x\, dx$$

$$dv = \sin 2x\, dx \implies v = -\frac{1}{2}\cos 2x$$

$$u = x \qquad \implies du = dx$$

$$\int x^2 \sin^2 x\, dx = \frac{1}{4} x^2(2x - \sin 2x) - \frac{1}{2} \int (2x^2 - x \sin 2x)\, dx$$

$$= \frac{1}{2} x^3 - \frac{1}{4} x^2 \sin 2x - \frac{1}{3} x^3 + \frac{1}{2} \int x \sin 2x\, dx$$

$$= \frac{1}{6} x^3 - \frac{1}{4} x^2 \sin 2x + \frac{1}{2}\left[-\frac{1}{2} x \cos 2x + \frac{1}{2} \int \cos 2x\, dx\right]$$

$$= \frac{1}{6} x^3 - \frac{1}{4} x^2 \sin 2x - \frac{1}{4} x \cos 2x + \frac{1}{8} \sin 2x + C$$

$$= \frac{1}{24}(4x^3 - 6x^2 \sin 2x - 6x \cos 2x + 3 \sin 2x) + C$$

13. Let $u = \sin x$, $du = \cos x\, dx$.

$$\int_0^{\pi/2} \cos^3 x\, dx = \int_0^{\pi/2} (1 - \sin^2 x) \cos x\, dx$$

$$= \left[\sin x - \frac{1}{3} \sin^3 x\right]_0^{\pi/2} = \frac{2}{3}$$

14. Let $u = \sin x$, $du = \cos x\, dx$.

$$\int_0^{\pi/2} \cos^5 x\, dx = \int_0^{\pi/2} (1 - \sin^2 x)^2 \cos x\, dx$$

$$= \int_0^{\pi/2} (1 - 2 \sin^2 x + \sin^4 x) \cos x\, dx$$

$$= \left[\sin x - \frac{2}{3} \sin^3 x + \frac{1}{5} \sin^5 x\right]_0^{\pi/2}$$

$$= \frac{8}{15}$$

15. Let $u = \sin x$, $du = \cos x\, dx$.

$$\int_0^{\pi/2} \cos^7 x\, dx = \int_0^{\pi/2} (1 - \sin^2 x)^3 \cos x\, dx = \int_0^{\pi/2} (1 - 3 \sin^2 x + 3 \sin^4 x - \sin^6 x) \cos x\, dx$$

$$= \left[\sin x - \sin^3 x + \frac{3}{5} \sin^5 x - \frac{1}{7} \sin^7 x\right]_0^{\pi/2} = \frac{16}{35}$$

16. $\displaystyle\int_0^{\pi/2} \sin^2 x \, dx = \frac{1}{2}\int_0^{\pi/2}(1-\cos 2x)\, dx$

$$= \frac{1}{2}\left[x - \frac{1}{2}\sin 2x\right]_0^{\pi/2} = \frac{\pi}{4}$$

17. $\displaystyle\int \sec(3x)\, dx = \frac{1}{3}\ln|\sec 3x + \tan 3x| + C$

18. $\displaystyle\int \sec^2(2x-1)\, dx = \frac{1}{2}\tan(2x-1) + C$

19. $\displaystyle\int \sec^4 5x \, dx = \int (1 + \tan^2 5x)\sec^2 5x \, dx$

$$= \frac{1}{5}\left(\tan 5x + \frac{\tan^3 5x}{3}\right) + C$$

$$= \frac{\tan 5x}{15}(3 + \tan^2 5x) + C$$

20. $\displaystyle\int \sec^6 \frac{x}{2}\, dx = \int\left[1 + \tan^2\frac{x}{2}\right]^2 \sec^2\frac{x}{2}\, dx = \int\left(1 + 2\tan^2\frac{x}{2} + \tan^4\frac{x}{2}\right)\sec^2\frac{x}{2}\, dx$

$$= 2\tan\frac{x}{2} + \frac{4\tan^3(x/2)}{3} + \frac{2\tan^5(x/2)}{5} + C$$

$$= \frac{2}{15}\tan\frac{x}{2}\left(15 + 10\tan^2\frac{x}{2} + 3\tan^4\frac{x}{2}\right) + C$$

21. $dv = \sec^2 \pi x \, dx \implies v = \frac{1}{\pi}\tan \pi x$

$u = \sec \pi x \implies du = \pi \sec \pi x \tan \pi x \, dx$

$\displaystyle\int \sec^3 \pi x \, dx = \frac{1}{\pi}\sec \pi x \tan \pi x - \int \sec \pi x \tan^2 \pi x \, dx = \frac{1}{\pi}\sec \pi x \tan \pi x - \int \sec \pi x(\sec^2 \pi x - 1)\, dx$

$2\displaystyle\int \sec^3 \pi x \, dx = \frac{1}{\pi}(\sec \pi x \tan \pi x + \ln|\sec \pi x + \tan \pi x|) + C_1$

$\displaystyle\int \sec^3 \pi x \, dx = \frac{1}{2\pi}(\sec \pi x \tan \pi x + \ln|\sec \pi x + \tan \pi x|) + C$

22. $\displaystyle\int \tan^2 x \, dx = \int (\sec^2 x - 1)\, dx = \tan x - x + C$

23. $\displaystyle\int \tan^5 \frac{x}{4}\, dx = \int\left(\sec^2\frac{x}{4} - 1\right)\tan^3\frac{x}{4}\, dx$

$$= \int \tan^3\frac{x}{4}\sec^2\frac{x}{4}\, dx - \int \tan^3\frac{x}{4}\, dx$$

$$= \tan^4\frac{x}{4} - \int\left(\sec^2\frac{x}{4} - 1\right)\tan\frac{x}{4}\, dx$$

$$= \tan^4\frac{x}{4} - 2\tan^2\frac{x}{4} - 4\ln\left|\cos\frac{x}{4}\right| + C$$

24. $\displaystyle\int \tan^3 \frac{\pi x}{2}\sec^2 \frac{\pi x}{2}\, dx = \frac{1}{2\pi}\tan^4 \frac{\pi x}{2} + C$

25. $u = \tan x, \, du = \sec^2 x \, dx$

$$\int \sec^2 x \tan x \, dx = \frac{1}{2}\tan^2 x + C$$

26. Let $u = \sec t, \, du = \sec t \tan t \, dt.$

$$\int \tan^3 t \sec^3 t \, dt = \int (\sec^2 t - 1)\sec^3 t \tan t \, dt$$

$$= \int \sec^4 t(\sec t \tan t)\, dt - \int \sec^2 t(\sec t \tan t)\, dt = \frac{1}{5}\sec^5 t - \frac{1}{3}\sec^3 t + C$$

27. $\displaystyle\int \tan^2 x \sec^2 x \, dx = \frac{\tan^3 x}{3} + C$

28. $\displaystyle\int \tan^5 2x \sec^2 2x \, dx = \frac{1}{12} \tan^6 2x + C$

29. $\displaystyle\int \sec^6 4x \tan 4x \, dx = \frac{1}{4} \int \sec^5 4x (4 \sec 4x \tan 4x) \, dx$

$$= \frac{\sec^6 4x}{24} + C$$

30. $\displaystyle\int \sec^2 \frac{x}{2} \tan \frac{x}{2} \, dx = 2 \int \sec \frac{x}{2} \left(\frac{1}{2} \sec \frac{x}{2} \tan \frac{x}{2} \right) dx$

$$= \sec^2 \frac{x}{2} + C$$

or $\displaystyle\int \sec^2 \frac{x}{2} \tan \frac{x}{2} \, dx = 2 \int \tan \frac{x}{2} \left(\frac{1}{2} \sec^2 \frac{x}{2} \right) dx$

$$= \tan^2 \frac{x}{2} + C$$

31. Let $u = \sec x$, $du = \sec x \tan x \, dx$.

$$\int \sec^3 x \tan x \, dx = \int \sec^2 x (\sec x \tan x) \, dx$$

$$= \frac{1}{3} \sec^3 x + C$$

32. $\displaystyle\int \tan^3 3x \, dx = \int (\sec^2 3x - 1) \tan 3x \, dx$

$$= \frac{1}{3} \int \tan 3x (3 \sec^2 3x) \, dx + \frac{1}{3} \int \frac{-3 \sin 3x}{\cos 3x} \, dx$$

$$= \frac{1}{6} \tan^2 3x + \frac{1}{3} \ln|\cos 3x| + C$$

33. $\displaystyle r = \int \sin^4(\pi\theta) \, d\theta = \frac{1}{4} \int [1 - \cos(2\pi\theta)]^2 \, d\theta$

$$= \frac{1}{4} \int [1 - 2 \cos(2\pi\theta) + \cos^2(2\pi\theta)] \, d\theta$$

$$= \frac{1}{4} \int \left[1 - 2 \cos(2\pi\theta) + \frac{1 + \cos(4\pi\theta)}{2} \right] d\theta$$

$$= \frac{1}{4} \left[\theta - \frac{1}{\pi} \sin(2\pi\theta) + \frac{\theta}{2} + \frac{1}{8\pi} \sin(4\pi\theta) \right] + C$$

$$= \frac{1}{32\pi} [12\pi\theta - 8 \sin(2\pi\theta) + \sin(4\pi\theta)] + C$$

34. $\displaystyle s = \int \sin^2 \frac{x}{\alpha} \cos^2 \frac{\alpha}{2} \, d\alpha$

$$= \int \left(\frac{1 - \cos \alpha}{2} \right) \left(\frac{1 + \cos \alpha}{2} \right) d\alpha = \int \frac{1 - \cos^2 \alpha}{4} \, d\alpha$$

$$= \frac{1}{4} \int \sin^2 \alpha \, d\alpha = \frac{1}{8} \int (1 - \cos 2\alpha) \, d\alpha$$

$$= \frac{1}{8} \left[\theta - \frac{\sin 2\alpha}{2} \right] + C$$

$$= \frac{1}{16} (2\alpha - \sin 2\alpha) + C$$

35. $\displaystyle y = \int \tan^3 3x \sec 3x \, dx$

$$= \int (\sec^2 3x - 1) \sec 3x \tan 3x \, dx$$

$$= \frac{1}{3} \int \sec^2 3x (3 \sec 3x \tan 3x) \, dx - \frac{1}{3} \int 3 \sec 3x \tan 3x \, dx$$

$$= \frac{1}{9} \sec^3 3x - \frac{1}{3} \sec 3x + C$$

36. $\displaystyle y = \int \sqrt{\tan x} \sec^4 x \, dx$

$$= \int \tan^{1/2} x (\tan^2 x + 1) \sec^2 x \, dx$$

$$= \int (\tan^{5/2} x + \tan^{1/2} x) \sec^2 x \, dx$$

$$= \frac{2}{7} \tan^{7/2} x + \frac{2}{3} \tan^{3/2} x + C$$

37. (a)

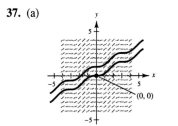

(b) $\dfrac{dy}{dx} = \sin^2 x$, $(0, 0)$

$$y = \int \sin^2 x \, dx = \int \frac{1 - \cos 2x}{2} \, dx$$

$$= \frac{1}{2} x - \frac{\sin 2x}{4} + C$$

$(0, 0)$: $0 = C$, $y = \dfrac{1}{2} x - \dfrac{\sin 2x}{4}$

38. (a)

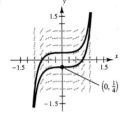

(b) $\dfrac{dy}{dx} = \sec^2 x \tan^2 x, \left(0, -\dfrac{1}{4}\right)$

$y = \displaystyle\int \sec^2 x \tan^2 x \, dx \quad u = \tan x, \, du = \sec^2 x \, dx$

$y = \dfrac{\tan^3 x}{3} + C$

$\left(0, -\dfrac{1}{4}\right): \; -\dfrac{1}{4} = C \implies y = \dfrac{1}{3}\tan^3 x - \dfrac{1}{4}$

39. $\displaystyle\int \sin 3x \cos 2x \, dx = \dfrac{1}{2}\int (\sin 5x + \sin x) \, dx$

$\qquad = \dfrac{-1}{2}\left(\dfrac{1}{5}\cos 5x + \cos x\right) + C$

$\qquad = \dfrac{-1}{10}(\cos 5x + 5\cos x) + C$

40. $\displaystyle\int \cos 3\theta \cos(-2\theta) \, d\theta = \int \cos 3\theta \cos 2\theta \, d\theta$

$\qquad = \dfrac{1}{2}\int (\cos 5\theta + \cos \theta) \, d\theta$

$\qquad = \dfrac{1}{2}\sin \theta + \dfrac{1}{10}\sin 5\theta + C$

41. $\displaystyle\int \sin \theta \sin 3\theta \, d\theta = \dfrac{1}{2}\int (\cos 2\theta - \cos 4\theta) \, d\theta$

$\qquad = \dfrac{1}{2}\left(\dfrac{1}{2}\sin 2\theta - \dfrac{1}{4}\sin 4\theta\right) + C$

$\qquad = \dfrac{1}{8}(2\sin 2\theta - \sin 4\theta) + C$

42. $\displaystyle\int \sin(-4x)\cos 3x \, dx = -\int \sin 4x \cos 3x \, dx$

$\qquad = -\dfrac{1}{2}\int (\sin x + \sin 7x) \, dx$

$\qquad = -\dfrac{1}{2}\left[-\cos x - \dfrac{1}{7}\cos 7x\right] + C$

$\qquad = \dfrac{1}{14}[7\cos x + \cos 7x] + C$

43. $\displaystyle\int \cot^3 2x \, dx = \int (\csc^2 2x - 1)\cot 2x \, dx$

$\qquad = -\dfrac{1}{2}\int \cot 2x(-2\csc^2 2x) \, dx - \dfrac{1}{2}\int \dfrac{2\cos 2x}{\sin 2x} \, dx$

$\qquad = -\dfrac{1}{4}\cot^2 2x - \dfrac{1}{2}\ln|\sin 2x| + C$

$\qquad = \dfrac{1}{4}(\ln|\csc^2 2x| - \cot^2 2x) + C$

44. Let $u = \tan \dfrac{x}{2}, \, du = \dfrac{1}{2}\sec^2 \dfrac{x}{2} \, dx.$

$\displaystyle\int \tan^4 \dfrac{x}{2}\sec^4 \dfrac{x}{2} \, dx = \int \tan^4 \dfrac{x}{2}\left(\tan^2 \dfrac{x}{2} + 1\right)\sec^2 \dfrac{x}{2} \, dx$

$\qquad = 2\displaystyle\int \left(\tan^6 \dfrac{x}{2} + \tan^4 \dfrac{x}{2}\right)\left(\dfrac{1}{2}\sec^2 \dfrac{x}{2}\right) dx$

$\qquad = \dfrac{2}{7}\tan^7 \dfrac{x}{2} + \dfrac{2}{5}\tan^5 \dfrac{x}{2} + C$

45. Let $u = \cot \theta, \, du = -\csc^2 \theta \, d\theta.$

$\displaystyle\int \csc^4 \theta \, d\theta = \int \csc^2 \theta(1 + \cot^2 \theta) \, d\theta$

$\qquad = \displaystyle\int \csc^2 \theta \, d\theta + \int \csc^2 \theta \cot^2 \theta \, d\theta$

$\qquad = -\cot \theta - \dfrac{1}{3}\cot^3 \theta + C$

46. $u = \cot 3x, \, du = -3\csc^2 3x \, dx$

$\displaystyle\int \csc^2 3x \cot 3x \, dx = -\dfrac{1}{3}\int \cot 3x(-3\csc^2 3x) \, dx$

$\qquad = -\dfrac{1}{6}\cot^2 3x + C$

47. $\displaystyle\int \dfrac{\cot^2 t}{\csc t} \, dt = \int \dfrac{\csc^2 t - 1}{\csc t} \, dt = \int (\csc t - \sin t) \, dt = \ln|\csc t - \cot t| + \cos t + C$

48. $\displaystyle \int \frac{\cot^3 t}{\csc t}\, dt = \int \frac{\cos^3 t}{\sin^2 t}\, dt = \int \frac{(1 - \sin^2 t)\cos t}{\sin^2 t}\, dt$

$\displaystyle \qquad = \int \frac{\cos t}{\sin^2 t}\, dt - \int \cos t\, dt$

$\displaystyle \qquad = \frac{-1}{\sin t} - \sin t + C$

$\displaystyle \qquad = -\csc t - \sin t + C$

49. $\displaystyle \int \frac{1}{\sec x \tan x}\, dx = \int \frac{\cos^2 x}{\sin x}\, dx = \int \frac{1 - \sin^2 x}{\sin x}\, dx$

$\displaystyle \qquad = \int (\csc x - \sin x)\, dx$

$\displaystyle \qquad = \ln|\csc x - \cot x| + \cos x + C$

50. $\displaystyle \int \frac{\sin^2 x - \cos^2 x}{\cos x}\, dx = \int \frac{1 - 2\cos^2 x}{\cos x}\, dx = \int (\sec x - 2\cos x)\, dx = \ln|\sec x + \tan x| - 2\sin x + C$

51. $\displaystyle \int (\tan^4 t - \sec^4 t)\, dt = \int (\tan^2 t + \sec^2 t)(\tan^2 t - \sec^2 t)\, dt \qquad (\tan^2 t - \sec^2 t = -1)$

$\displaystyle \qquad\qquad = -\int (\tan^2 t + \sec^2 t)\, dt = -\int (2\sec^2 t - 1)\, dt = -2\tan t + t + C$

52. $\displaystyle \int \frac{1 - \sec t}{\cos t - 1}\, dt = \int \frac{\cos t - 1}{(\cos t - 1)\cos t}\, dt$

$\displaystyle \qquad = \int \sec t\, dt = \ln|\sec t + \tan t| + C$

53. $\displaystyle \int_{-\pi}^{\pi} \sin^2 x\, dx = 2\int_0^{\pi} \frac{1 - \cos 2x}{2}\, dx$

$\displaystyle \qquad = \left[x - \frac{1}{2}\sin 2x \right]_0^{\pi} = \pi$

54. $\displaystyle \int_0^{\pi/4} \tan^2 x\, dx = \int_0^{\pi/4} (\sec^2 x - 1)\, dx$

$\displaystyle \qquad = \left[\tan x - x \right]_0^{\pi/4} = 1 - \frac{\pi}{4}$

55. $\displaystyle \int_0^{\pi/4} \tan^3 x\, dx = \int_0^{\pi/4} (\sec^2 x - 1)\tan x\, dx$

$\displaystyle \qquad = \int_0^{\pi/4} \sec^2 x \tan x\, dx - \int_0^{\pi/4} \frac{\sin x}{\cos x}\, dx$

$\displaystyle \qquad = \left[\frac{1}{2}\tan^2 x + \ln|\cos x| \right]_0^{\pi/4}$

$\displaystyle \qquad = \frac{1}{2}(1 - \ln 2)$

56. Let $u = \tan t$, $du = \sec^2 t\, dt$.

$\displaystyle \int_0^{\pi/4} \sec^2 t \sqrt{\tan t}\, dt = \left[\frac{2}{3}\tan^{3/2} t \right]_0^{\pi/4} = \frac{2}{3}$

57. Let $u = 1 + \sin t$, $du = \cos t\, dt$.

$\displaystyle \int_0^{\pi/2} \frac{\cos t}{1 + \sin t}\, dt = \left[\ln|1 + \sin t| \right]_0^{\pi/2} = \ln 2$

58. $\displaystyle \int_{-\pi}^{\pi} \sin 3\theta \cos \theta\, d\theta = \frac{1}{2}\int_{-\pi}^{\pi} (\sin 4\theta + \sin 2\theta)\, d\theta$

$\displaystyle \qquad = -\frac{1}{2}\left[\frac{1}{4}\cos 4\theta + \frac{1}{2}\cos 2\theta \right]_{-\pi}^{\pi} = 0$

59. Let $u = \sin x$, $du = \cos x\, dx$.

$\displaystyle \int_{-\pi/2}^{\pi/2} \cos^3 x\, dx = 2\int_0^{\pi/2} (1 - \sin^2 x)\cos x\, dx$

$\displaystyle \qquad = 2\left[\sin x - \frac{1}{3}\sin^3 x \right]_0^{\pi/2} = \frac{4}{3}$

60. $\displaystyle \int_{-\pi/2}^{\pi/2} (\sin^2 x + 1)\, dx = \int_{-\pi/2}^{\pi/2} \left(\frac{1 - \cos 2x}{2} + 1 \right)\, dx$

$\displaystyle \qquad = \int_{-\pi/2}^{\pi/2} \left(\frac{3}{2} - \frac{1}{2}\cos 2x \right)\, dx = \left[\frac{3}{2}x - \frac{1}{4}\sin 2x \right]_{-\pi/2}^{\pi/2} = \frac{3\pi}{2}$

61. $\displaystyle\int \cos^4 \frac{x}{2}\,dx = \frac{1}{16}[6x + 8\sin x + \sin 2x] + C$

62. $\displaystyle\int \sin^2 x \cos^2 x\,dx = \frac{1}{32}[4x - \sin 4x] + C$

63. $\displaystyle\int \sec^5 \pi x\,dx = \frac{1}{4\pi}\left\{\sec^3 \pi x \tan \pi x + \frac{3}{2}[\sec \pi x \tan \pi x + \ln|\sec \pi x + \tan \pi x|]\right\} + C$

64. $\displaystyle\int \tan^3(1 - x)\,dx = -\frac{\tan^2(1 - x)}{2} - \ln|\cos(1 - x)| + C$

65. $\displaystyle\int \sec^5 \pi x \tan \pi x\,dx = \frac{1}{5\pi}\sec^5 \pi x + C$

66. $\displaystyle\int \sec^4(1 - x)\tan(1 - x)\,dx = -\frac{\sec^4(1 - x)}{4} + C$

67. $\displaystyle\int_0^{\pi/4} \sin 2\theta \sin 3\theta\,d\theta = \frac{1}{2}\left[\sin \theta - \frac{1}{5}\sin 5\theta\right]_0^{\pi/4} = \frac{3\sqrt{2}}{10}$

68. $\displaystyle\int_0^{\pi/2}(1 - \cos \theta)^2\,d\theta = \left[\frac{3}{2}\theta - 2\sin\theta + \frac{1}{4}\sin 2\theta\right]_0^{\pi/2}$

$$= \frac{3\pi}{4} - 2$$

69. $\displaystyle\int_0^{\pi/2}\sin^4 x\,dx = \frac{1}{4}\left[\frac{3x}{2} - \sin 2x + \frac{1}{8}\sin 4x\right]_0^{\pi/2}$

$$= \frac{3\pi}{16}$$

70. $\displaystyle\int_0^{\pi/2}\sin^6 x\,dx = \frac{1}{8}\left[\frac{5x}{2} - 2\sin 2x + \frac{3}{8}\sin 4x + \frac{1}{6}\sin^3 2x\right]_0^{\pi/2} = \frac{5\pi}{32}$

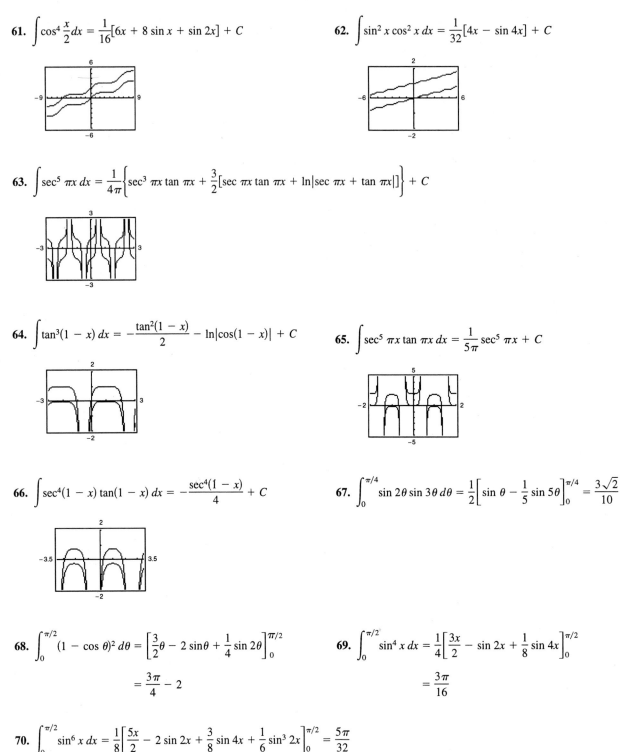

71. (a) Let $u = \tan 3x$, $du = 3 \sec^2 3x \, dx$.

$$\int \sec^4 3x \tan^3 3x \, dx = \int \sec^2 3x \tan^3 3x \sec^2 3x \, dx$$

$$= \frac{1}{3} \int (\tan^2 3x + 1) \tan^3 3x (3 \sec^2 3x) \, dx$$

$$= \frac{1}{3} \int (\tan^5 3x + \tan^3 3x)(3 \sec^2 3x) \, dx$$

$$= \frac{\tan^6 3x}{18} + \frac{\tan^4 3x}{12} + C_1$$

Or let $u = \sec 3x$, $du = 3 \sec 3x \tan 3x \, dx$.

$$\int \sec^4 3x \tan^3 3x \, dx = \int \sec^3 3x \tan^2 3x \sec 3x \tan 3x \, dx$$

$$= \frac{1}{3} \int \sec^3 3x (\sec^2 3x - 1)(3 \sec 3x \tan 3x) \, dx$$

$$= \frac{\sec^6 3x}{18} - \frac{\sec^4 3x}{12} + C$$

(c) $\dfrac{\sec^6 3x}{18} - \dfrac{\sec^4 3x}{12} + C = \dfrac{(1 + \tan^2 3x)^3}{18} - \dfrac{(1 + \tan^2 3x)^2}{12} + C$

$$= \frac{1}{18} \tan^6 3x + \frac{1}{6} \tan^4 3x + \frac{1}{6} \tan^2 3x + \frac{1}{18} - \frac{1}{12} \tan^4 3x - \frac{1}{6} \tan^2 3x - \frac{1}{12} + C$$

$$= \frac{\tan^6 3x}{18} + \frac{\tan^4 3x}{12} + \left(\frac{1}{18} - \frac{1}{12}\right) + C$$

$$= \frac{\tan^6 3x}{18} + \frac{\tan^4 3x}{12} + C_2$$

(b)

72. (a) Let $u = \tan x$, $du = \sec^2 x \, dx$.

$$\int \sec^2 x \tan x \, dx = \frac{1}{2} \tan^2 x + C_1$$

Or let $u = \sec x$, $du = \sec x \tan x \, dx$.

$$\int \sec x (\sec x \tan x) \, dx = \frac{1}{2} \sec^2 x + C$$

(c) $\dfrac{1}{2} \sec^2 x + C = \dfrac{1}{2}(\tan^2 x + 1) + C = \dfrac{1}{2} \tan^2 x + \left(\dfrac{1}{2} + C\right) = \dfrac{1}{2} \tan^2 x + C_2$

(b)

73. $A = \displaystyle\int_0^1 \sin^2(\pi x) \, dx$

$$= \int_0^1 \frac{1 - \cos(2\pi x)}{2} \, dx$$

$$= \left[\frac{x}{2} - \frac{1}{4\pi} \sin(2\pi x)\right]_0^1$$

$$= \frac{1}{2}$$

74. Discs

$R(x) = \tan x$

$r(x) = 0$

$$V = 2\pi \int_0^{\pi/4} \tan^2 x \, dx$$

$$= 2\pi \int_0^{\pi/4} (\sec^2 x - 1) \, dx$$

$$= 2\pi \left[\tan x - x\right]_0^{\pi/4}$$

$$= 2\pi \left(1 - \frac{\pi}{4}\right) \approx 1.348$$

75. (a) $V = \pi \int_0^\pi \sin^2 x \, dx = \frac{\pi}{2} \int_0^\pi (1 - \cos 2x) \, dx = \frac{\pi}{2}\left[x - \frac{1}{2}\sin 2x \right]_0^\pi = \frac{\pi^2}{2}$

(b) $A = \int_0^\pi \sin x \, dx = \left[-\cos x \right]_0^\pi = 1 + 1 = 2$

Let $u = x$, $dv = \sin x \, dx$, $du = dx$, $v = -\cos x$.

$\bar{x} = \frac{1}{A}\int_0^\pi x \sin x \, dx = \frac{1}{2}\left[\left[-x \cos x \right]_0^\pi + \int_0^\pi \cos x \, dx \right] = \frac{1}{2}\left[-x\cos x + \sin x \right]_0^\pi = \frac{\pi}{2}$

$\bar{y} = \frac{1}{2A}\int_0^\pi \sin^2 x \, dx$

$\quad = \frac{1}{8}\int_0^\pi (1 - \cos 2x)\, dx$

$\quad = \frac{1}{8}\left[x - \frac{1}{2}\sin 2x \right]_0^\pi = \frac{\pi}{8}$

$(\bar{x}, \bar{y}) = \left(\frac{\pi}{2}, \frac{\pi}{8} \right)$

76. (a) $V = \pi \int_0^{\pi/2} \cos^2 x \, dx = \frac{\pi}{2} \int_0^{\pi/2} (1 + \cos 2x) \, dx = \frac{\pi}{2}\left[x + \frac{1}{2}\sin 2x \right]_0^{\pi/2} = \frac{\pi^2}{4}$

(b) $A = \int_0^{\pi/2} \cos x \, dx = \left[\sin x \right]_0^{\pi/2} = 1$

Let $u = x$, $dv = \cos x \, dx$, $du = dx$, $v = \sin x$.

$\bar{x} = \int_0^{\pi/2} x \cos x \, dx = \left[x \sin x \right]_0^{\pi/2} - \int_0^{\pi/2} \sin x \, dx = \left[x \sin x + \cos x \right]_0^{\pi/2} = \frac{\pi}{2} - 1 = \frac{\pi - 2}{2}$

$\bar{y} = \frac{1}{2}\int_0^{\pi/2} \cos^2 x \, dx$

$\quad = \frac{1}{4}\int_0^{\pi/2} (1 + \cos 2x)\, dx$

$\quad = \frac{1}{4}\left[x + \frac{1}{2}\sin 2x \right]_0^{\pi/2} = \frac{\pi}{8}$

$(\bar{x}, \bar{y}) = \left(\frac{\pi - 2}{2}, \frac{\pi}{8} \right)$

77. $dv = \sin x \, dx \implies v = -\cos x$

$u = \sin^{n-1} x \implies du = (n - 1)\sin^{n-2} x \cos x \, dx$

$\int \sin^n x \, dx = -\sin^{n-1} x \cos x + (n - 1)\int \sin^{n-2} x \cos^2 x \, dx$

$\quad = -\sin^{n-1} x \cos x + (n - 1)\int \sin^{n-2} x(1 - \sin^2 x)\, dx$

$\quad = -\sin^{n-1} x \cos x + (n - 1)\int \sin^{n-2} x \, dx - (n - 1)\int \sin^n x \, dx$

Therefore, $n\int \sin^n x \, dx = -\sin^{n-1} x \cos x + (n - 1)\int \sin^{n-2} x \, dx$

$\int \sin^n x \, dx = \frac{-\sin^{n-1} x \cos x}{n} + \frac{n - 1}{n}\int \sin^{n-2} x \, dx.$

78. $dv = \cos x \, dx \implies v = \sin x$

$u = \cos^{n-1} x \implies du = -(n-1)\cos^{n-2} x \sin x \, dx$

$$\int \cos^n x \, dx = \cos^{n-1} x \sin x + (n-1) \int \cos^{n-2} x \sin^2 x \, dx$$

$$= \cos^{n-1} x \sin x + (n-1) \int \cos^{n-2} x (1 - \cos^2 x) \, dx$$

$$= \cos^{n-1} x \sin x + (n-1) \int \cos^{n-2} x \, dx - (n-1) \int \cos^n x \, dx$$

Therefore, $n \int \cos^n x \, dx = \cos^{n-1} x \sin x + (n-1) \int \cos^{n-2} x \, dx$

$$\int \cos^n x \, dx = \frac{\cos^{n-1} x \sin x}{n} + \frac{n-1}{n} \int \cos^{n-2} x \, dx.$$

79. Let $u = \sin^{n-1} x$, $du = (n-1)\sin^{n-2} x \cos x \, dx$, $dv = \cos^m x \sin x \, dx$, $v = \dfrac{-\cos^{m+1} x}{m+1}$.

$$\int \cos^m x \sin^n x \, dx = \frac{-\sin^{n-1} x \cos^{m+1} x}{m+1} + \frac{n-1}{m+1} \int \sin^{n-2} x \cos^{m+2} x \, dx$$

$$= \frac{-\sin^{n-1} x \cos^{m+1} x}{m+1} + \frac{n-1}{m+1} \int \sin^{n-2} x \cos^m x (1 - \sin^2 x) \, dx$$

$$= \frac{-\sin^{n-1} x \cos^{m+1} x}{m+1} + \frac{n-1}{m+1} \int \sin^{n-2} x \cos^m x \, dx - \frac{n-1}{m+1} \int \sin^n x \cos^m x \, dx$$

$$\frac{m+n}{m+1} \int \cos^m x \sin^n x \, dx = \frac{-\sin^{n-1} x \cos^{m+1} x}{m+1} + \frac{n-1}{m+1} \int \sin^{n-2} x \cos^m x \, dx$$

$$\int \cos^m x \sin^n x \, dx = \frac{-\cos^{m+1} x \sin^{n-1}}{m+n} + \frac{n-1}{m+n} \int \cos^m x \sin^{n-2} x \, dx$$

80. Let $u = \sec^{n-2} x$, $du = (n-2)\sec^{n-2} x \tan x \, dx$, $dv = \sec^2 x \, dx$, $v = \tan x$.

$$\int \sec^n x \, dx = \sec^{n-2} x \tan x - \int (n-2) \sec^{n-2} x \tan^2 x \, dx$$

$$= \sec^{n-2} x \tan x - (n-2) \int \sec^{n-2} x (\sec^2 x - 1) \, dx$$

$$= \sec^{n-2} x \tan x - (n-2) \left[\int \sec^n x \, dx - \int \sec^{n-2} x \, dx \right]$$

$$(n-1) \int \sec^n x \, dx = \sec^{n-2} x \tan x + (n-2) \int \sec^{n-2} x \, dx$$

$$\int \sec^n x \, dx = \frac{1}{n-1} \sec^{n-2} x \tan x + \frac{n-2}{n-1} \int \sec^{n-2} x \, dx$$

81. $\displaystyle \int \sin^5 x \, dx = -\frac{\sin^4 x \cos x}{5} + \frac{4}{5} \int \sin^3 x \, dx$

$$= -\frac{\sin^4 x \cos x}{5} + \frac{4}{5} \left[-\frac{\sin^2 x \cos x}{3} + \frac{2}{3} \int \sin x \, dx \right]$$

$$= -\frac{1}{5} \sin^4 x \cos x - \frac{4}{15} \sin^2 x \cos x - \frac{8}{15} \cos x + C$$

$$= -\frac{\cos x}{15} [3 \sin^4 x + 4 \sin^2 x + 8] + C$$

82. $\displaystyle\int \cos^4 x \, dx = \frac{\cos^3 x \sin x}{4} + \frac{3}{4}\int \cos^2 x \, dx$

$\displaystyle = \frac{\cos^3 x \sin x}{4} + \frac{3}{4}\left[\frac{\cos x \sin x}{2} + \frac{1}{2}\int dx\right]$

$\displaystyle = \frac{1}{4}\cos^3 x \sin x + \frac{3}{8}\cos x \sin x + \frac{3}{8}x + C$

$\displaystyle = \frac{1}{8}[2\cos^3 x \sin x + 3\cos x \sin x + 3x] + C$

83. $\displaystyle\int \sec^4\left(\frac{2\pi x}{5}\right) dx = \frac{5}{2\pi}\int \sec^4\left(\frac{2\pi x}{5}\right)\frac{2\pi}{5} \, dx$

$\displaystyle = \frac{5}{2\pi}\left[\frac{1}{3}\sec^2\left(\frac{2\pi x}{5}\right)\tan\left(\frac{2\pi x}{5}\right) + \frac{2}{3}\int \sec^2\left(\frac{2\pi x}{5}\right)\frac{2\pi}{5} \, dx\right]$

$\displaystyle = \frac{5}{6\pi}\left[\sec^2\left(\frac{2\pi x}{5}\right)\tan\left(\frac{2\pi x}{5}\right) + 2\tan\left(\frac{2\pi x}{5}\right)\right] + C$

$\displaystyle = \frac{5}{6\pi}\tan\left(\frac{2\pi x}{5}\right)\left[\sec^2\left(\frac{2\pi x}{5}\right) + 2\right] + C$

84. $\displaystyle\int \sin^4 x \cos^2 x \, dx = -\frac{\cos^3 x \sin^3 x}{6} + \frac{1}{2}\int \cos^2 x \sin^2 x \, dx$

$\displaystyle = -\frac{\cos^3 x \sin^3 x}{6} + \frac{1}{2}\left[-\frac{\cos^3 x \sin x}{4} + \frac{1}{4}\int \cos^2 x \, dx\right]$

$\displaystyle = -\frac{1}{6}\cos^3 x \sin^3 x - \frac{1}{8}\cos^3 x \sin x + \frac{1}{8}\left[\frac{\cos x \sin x}{2} + \frac{x}{2}\right] + C$

$\displaystyle = -\frac{1}{48}[8\cos^3 x \sin^3 x + 6\cos^3 x \sin x - 3\cos x \sin x - 3x] + C$

85. (a) n is odd and $n \geq 3$.

$\displaystyle\int_0^{\pi/2} \cos^n x \, dx = \left[\frac{\cos^{n-1} x \sin x}{n}\right]_0^{\pi/2} + \frac{n-1}{n}\int_0^{\pi/2} \cos^{n-2} x \, dx$

$\displaystyle = \frac{n-1}{n}\left[\left[\frac{\cos^{n-3} x \sin x}{n-2}\right]_0^{\pi/2} + \frac{n-3}{n-2}\int_0^{\pi/2} \cos^{n-4} x \, dx\right]$

$\displaystyle = \frac{n-1}{n}\cdot\frac{n-3}{n-2}\left[\left[\frac{\cos^{n-5} x \sin x}{n-4}\right]_0^{\pi/2} + \frac{n-5}{n-4}\int_0^{\pi/2} \cos^{n-6} x \, dx\right]$

$\displaystyle = \frac{n-1}{n}\cdot\frac{n-3}{n-2}\cdot\frac{n-5}{n-4}\int_0^{\pi/2} \cos^{n-6} x \, dx$

$\displaystyle = \frac{n-1}{n}\cdot\frac{n-3}{n-2}\cdot\frac{n-5}{n-4}\cdots\int_0^{\pi/2} \cos x \, dx$

$\displaystyle = \left[\frac{n-1}{n}\cdot\frac{n-3}{n-2}\cdot\frac{n-5}{n-4}\cdots(\sin x)\right]_0^{\pi/2}$

$\displaystyle = \frac{n-1}{n}\cdot\frac{n-3}{n-2}\cdot\frac{n-5}{n-4}\cdots 1 \quad \text{(Reverse the order)}$

$\displaystyle = (1)\left(\frac{2}{3}\right)\left(\frac{4}{5}\right)\left(\frac{6}{7}\right)\cdots\left(\frac{n-1}{n}\right)$

$\displaystyle = \left(\frac{2}{3}\right)\left(\frac{4}{5}\right)\left(\frac{6}{7}\right)\cdots\left(\frac{n-1}{n}\right)$

—CONTINUED—

85. —CONTINUED—

(b) n is even and $n \geq 2$.

$$\int_0^{\pi/2} \cos^n x \, dx = \frac{n-1}{n} \cdot \frac{n-3}{n-2} \cdot \frac{n-5}{n-4} \cdots \int_0^{\pi/2} \cos^2 x \, dx \quad \text{(From part (a).)}$$

$$= \left[\frac{n-1}{n} \cdot \frac{n-3}{n-2} \cdot \frac{n-5}{n-4} \cdots \left(\frac{x}{2} + \frac{1}{4} \sin 2x \right) \right]_0^{\pi/2}$$

$$= \frac{n-1}{n} \cdot \frac{n-3}{n-2} \cdot \frac{n-5}{n-4} \cdots \frac{\pi}{4} \quad \text{(Reverse the order)}$$

$$= \left(\frac{\pi}{2} \cdot \frac{1}{2} \right) \left(\frac{3}{4} \right) \left(\frac{5}{6} \right) \cdots \left(\frac{n-1}{n} \right)$$

$$= \left(\frac{1}{2} \right) \left(\frac{3}{4} \right) \left(\frac{5}{6} \right) \cdots \left(\frac{n-1}{n} \right) \left(\frac{\pi}{2} \right)$$

86. $\displaystyle\int_{-\pi}^{\pi} \cos(mx) \cos(nx) \, dx = \frac{1}{2} \left[\frac{\sin(m+n)x}{m+n} + \frac{\sin(m-n)x}{m-n} \right]_{-\pi}^{\pi} = 0, \quad (m \neq n)$

$$\int_{-\pi}^{\pi} \sin(mx) \sin(nx) \, dx = \frac{1}{2} \int_{-\pi}^{\pi} \left[\cos(m-n)x - \cos(m+n)x \right] dx$$

$$= \frac{1}{2} \left[\frac{\sin(m-n)x}{m-n} - \frac{\sin(m+n)x}{m+n} \right]_{-\pi}^{\pi} = 0, \quad (m \neq n)$$

$$\int_{-\pi}^{\pi} \sin(mx) \cos(nx) \, dx = \frac{1}{2} \int_{-\pi}^{\pi} \left[\sin(m+n)x + \sin(m-n)x \right] dx$$

$$= -\frac{1}{2} \left[\frac{\cos(m+n)x}{m+n} + \frac{\cos(m-n)x}{m-n} \right]_{-\pi}^{\pi}, \quad (m \neq n)$$

$$= -\frac{1}{2} \left[\left(\frac{\cos(m+n)\pi}{m+n} + \frac{\cos(m-n)\pi}{m-n} \right) - \left(\frac{\cos(m+n)(-\pi)}{m+n} + \frac{\cos(m-n)(-\pi)}{m-n} \right) \right]$$

$$= 0, \text{ since } \cos(-\theta) = \cos\theta.$$

$$\int_{-\pi}^{\pi} \sin(mx) \cos(mx) \, dx = \frac{1}{m} \frac{\sin^2(mx)}{2} \bigg]_{-\pi}^{\pi} = 0$$

87. (a) $f(t) = a_0 + a_1 \cos \dfrac{\pi t}{6} + b_1 \sin \dfrac{\pi t}{6}$ where:

$$a_0 = \frac{1}{12} \int_0^{12} f(t) \, dt$$

$$a_1 = \frac{1}{6} \int_0^{12} f(t) \cos \frac{\pi t}{6} \, dt$$

$$b_1 = \frac{1}{6} \int_0^{12} f(t) \sin \frac{\pi t}{6} \, dt$$

—CONTINUED—

87. —CONTINUED—

$$a_0 \approx \frac{12-0}{3(12)^2}[30.9 + 4(32.2) + 2(41.1) + 4(53.7) + 2(64.6) + 4(74.0) + 2(78.2) + 4(77.0) + 2(71.0) +$$

$$4(60.1) + 2(47.1) + 4(35.7) + 30.9] \approx 55.46$$

$$a_1 \approx \frac{12-0}{6(3)(12)}\left[30.9 \cos 0 + 4\left(32.2 \cos \frac{\pi}{6}\right) + 2\left(41.1 \cos \frac{\pi}{3}\right) + 4\left(53.7 \cos \frac{\pi}{2}\right) + 2\left(64.6 \cos \frac{2\pi}{3}\right) + \right.$$

$$4\left(74.0 \cos \frac{5\pi}{6}\right) + 2(78.2 \cos \pi) + 4\left(77.0 \cos \frac{7\pi}{6}\right) + 2\left(71.0 \cos \frac{4\pi}{3}\right) +$$

$$\left. 4\left(60.1 \cos \frac{3\pi}{2}\right) + 2\left(47.1 \cos \frac{5\pi}{3}\right) + 4\left(35.7 \cos \frac{11\pi}{6}\right) + 30.9 \cos 2\pi\right] \approx -23.88$$

$$b_1 \approx \frac{12-0}{6(3)(12)}\left[30.9 \sin 0 + 4\left(32.2 \sin \frac{\pi}{6}\right) + 2\left(41.1 \sin \frac{\pi}{3}\right) + 4\left(53.7 \sin \frac{\pi}{2}\right) + 2\left(64.6 \sin \frac{2\pi}{3}\right) + \right.$$

$$4\left(74.0 \sin \frac{5\pi}{6}\right) + 2(78.2 \sin \pi) + 4\left(77.0 \sin \frac{7\pi}{6}\right) + 2\left(71.0 \sin \frac{4\pi}{3}\right) +$$

$$\left. 4\left(60.1 \sin \frac{3\pi}{2}\right) + 2\left(47.1 \sin \frac{5\pi}{3}\right) + 4\left(35.7 \sin \frac{11\pi}{6}\right) + 30.9 \sin 2\pi\right] \approx -3.34$$

$$H(t) \approx 55.46 - 23.88 \cos \frac{\pi t}{6} - 3.34 \sin \frac{\pi t}{6}$$

(b) $$a_0 \approx \frac{12-0}{3(12)^2}[18.0 + 4(17.7) + 2(25.8) + 4(36.1) + 2(45.4) + 4(55.2) + 2(59.9) + 4(59.4) + 2(53.1) +$$

$$4(43.2) + 2(34.3) + 4(24.2) + 18.0] \approx 39.34$$

$$a_1 \approx \frac{12-0}{6(3)(12)}\left[18.0 \cos 0 + 4\left(17.7 \cos \frac{\pi}{6}\right) + 2\left(25.8 \cos \frac{\pi}{3}\right) + 4\left(36.1 \cos \frac{\pi}{2}\right) + 2\left(45.4 \cos \frac{2\pi}{3}\right) + \right.$$

$$4\left(55.2 \cos \frac{5\pi}{6}\right) + 2(59.9 \cos \pi) + 4\left(59.4 \cos \frac{7\pi}{6}\right) + 2\left(53.1 \cos \frac{4\pi}{3}\right) +$$

$$\left. 4\left(43.2 \cos \frac{3\pi}{2}\right) + 2\left(34.3 \cos \frac{5\pi}{3}\right) + 4\left(24.2 \cos \frac{11\pi}{6}\right) + 18 \cos 2\pi\right] \approx -20.78$$

$$b_1 \approx \frac{12-0}{6(3)(12)}\left[18.0 \sin 0 + 4\left(17.7 \sin \frac{\pi}{6}\right) + 2\left(25.8 \sin \frac{\pi}{3}\right) + 4\left(36.1 \sin \frac{\pi}{2}\right) + 2\left(45.4 \sin \frac{2\pi}{3}\right) + \right.$$

$$4\left(55.2 \sin \frac{5\pi}{6}\right) + 2(59.9 \sin \pi) + 4\left(59.4 \sin \frac{7\pi}{6}\right) + 2\left(53.1 \sin \frac{4\pi}{3}\right) +$$

$$\left. 4\left(43.2 \sin \frac{3\pi}{2}\right) + 2\left(34.3 \sin \frac{5\pi}{3}\right) + 4\left(24.2 \sin \frac{11\pi}{6}\right) + 18 \sin 2\pi\right] \approx -4.33$$

$$L(t) \approx 39.34 - 20.78 \cos \frac{\pi t}{6} - 4.33 \sin \frac{\pi t}{6}$$

(c) The difference between the maximum and minimum temperatures is greatest in the summer.

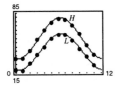

88. Let $x = \sin\theta$, then $dx = \cos\theta\,d\theta$, $1 - x^2 = \cos^2\theta$.

$$\int_{-1}^{1} (1 - x^2)^n\,dx = \int_{-\pi/2}^{\pi/2} (\cos^2\theta)^n \cos\theta\,d\theta = \int_{-\pi/2}^{\pi/2} \cos^{2n+1}\theta\,d\theta = 2\int_{0}^{\pi/2} \cos^{2n+1}\theta\,d\theta$$

$$= 2\left(\frac{2}{3}\right)\left(\frac{4}{5}\right)\left(\frac{6}{7}\right)\cdots\left(\frac{2n}{2n+1}\right) = \frac{2 \cdot 2^n \cdot n!}{3 \cdot 5 \cdots (2n+1)} \cdot \frac{(2n)!}{(2n)!}$$

$$= \frac{2 \cdot 2^n \cdot n!\, 2^n\, n!}{(2n+1)!} = \frac{2^{2n+1}(n!)^2}{(2n+1)!}$$

Section 7.4   Trigonometric Substitution

1. $\dfrac{d}{dx}\left[4\ln\left|\dfrac{\sqrt{x^2+16}-4}{x}\right| + \sqrt{x^2+16} + C \right] = \dfrac{d}{dx}\left[4\ln\left|\sqrt{x^2+16}-4\right| - 4\ln|x| + \sqrt{x^2+16} + C \right]$

$$= 4\left[\frac{x/\sqrt{x^2+16}}{\sqrt{x^2+16}-4}\right] - \frac{4}{x} + \frac{x}{\sqrt{x^2+16}}$$

$$= \frac{4x}{\sqrt{x^2+16}\left(\sqrt{x^2+16}-4\right)} - \frac{4}{x} + \frac{x}{\sqrt{x^2+16}}$$

$$= \frac{4x^2 - 4\sqrt{x^2+16}\left(\sqrt{x^2+16}-4\right) + x^2\left(\sqrt{x^2+16}-4\right)}{x\sqrt{x^2+16}\left(\sqrt{x^2+16}-4\right)}$$

$$= \frac{4x^2 - 4(x^2+16) + 16\sqrt{x^2+16} + x^2\sqrt{x^2+16} - 4x^2}{x\sqrt{x^2+16}\left(\sqrt{x^2+16}-4\right)}$$

$$= \frac{\sqrt{x^2+16}(x^2+16) - 4(x^2+16)}{x\sqrt{x^2+16}\left(\sqrt{x^2+16}-4\right)}$$

$$= \frac{(x^2+16)\left(\sqrt{x^2+16}-4\right)}{x\sqrt{x^2+16}\left(\sqrt{x^2+16}-4\right)} = \frac{\sqrt{x^2+16}}{x}$$

Indefinite integral: $\displaystyle\int \frac{\sqrt{x^2+16}}{x}\,dx$ Matches (b)

2. $\dfrac{d}{dx}\left[8\ln\left|\sqrt{x^2-16}+x\right| + \dfrac{1}{2}x\sqrt{x^2-16} + C \right] = 8\left[\dfrac{\left(x/\sqrt{x^2-16}\right)+1}{\sqrt{x^2-16}+x}\right] + \dfrac{1}{2}x\left(\dfrac{x}{\sqrt{x^2-16}}\right) + \dfrac{1}{2}\sqrt{x^2-16}$

$$= \frac{8\left(x + \sqrt{x^2-16}\right)}{\sqrt{x^2-16}\left(\sqrt{x^2-16}+x\right)} + \frac{x^2}{2\sqrt{x^2+16}} + \frac{\sqrt{x^2-16}}{2}$$

$$= \frac{16 + x^2 + x^2 - 16}{2\sqrt{x^2-16}}$$

$$= \frac{x^2}{\sqrt{x^2-16}}$$

Indefinite integral: $\displaystyle\int \frac{x^2}{\sqrt{x^2-16}}$ Matches (d)

3. $\dfrac{d}{dx}\left[8\arcsin\dfrac{x}{4} - \dfrac{x\sqrt{16-x^2}}{2} + C \right] = 8\dfrac{1/4}{\sqrt{1-(x/4)^2}} - \dfrac{x(1/2)(16-x^2)^{-1/2}(-2x) + \sqrt{16-x^2}}{2}$

$$= \frac{8}{\sqrt{16-x^2}} + \frac{x^2}{2\sqrt{16-x^2}} - \frac{\sqrt{16-x^2}}{2}$$

$$= \frac{16}{2\sqrt{16-x^2}} + \frac{x^2}{2\sqrt{16-x^2}} - \frac{(16-x^2)}{2\sqrt{16-x^2}} = \frac{x^2}{\sqrt{16-x^2}}$$

Matches (a)

4. $\dfrac{d}{dx}\left[8\arcsin\dfrac{x-3}{4}+\dfrac{(x-3)\sqrt{7+6x-x^2}}{2}+C\right]=8\left[\dfrac{1}{\sqrt{1-[(x-3)/4]^2}}\cdot\dfrac{1}{4}\right]+\dfrac{1}{2}(x-3)\dfrac{3-x}{\sqrt{7+6x-x^2}}+\dfrac{1}{2}\sqrt{7+6x-x^2}$

$$=\dfrac{8}{\sqrt{16-(x-3)^2}}-\dfrac{(x-3)^2}{2\sqrt{16-(x-3)^2}}+\dfrac{\sqrt{16-(x-3)^2}}{2}$$

$$=\dfrac{16-(x^2-6x+9)+16-(x^2-6x+9)}{2\sqrt{16-(x-3)^2}}$$

$$=\dfrac{2[16-(x-3)^2]}{2\sqrt{16-(x-3)^2}}$$

$$=\sqrt{16-(x-3)^2}$$

$$=\sqrt{7+6x-x^2}$$

Indefinite integral: $\displaystyle\int\sqrt{7+6x-x^2}\,dx$ Matches (c)

5. Let $x=5\sin\theta$, $dx=5\cos\theta\,d\theta$, $\sqrt{25-x^2}=5\cos\theta$.

$$\int\dfrac{1}{(25-x^2)^{3/2}}\,dx=\int\dfrac{5\cos\theta}{(5\cos\theta)^3}\,d\theta$$

$$=\dfrac{1}{25}\int\sec^2\theta\,d\theta$$

$$=\dfrac{1}{25}\tan\theta+C$$

$$=\dfrac{x}{25\sqrt{25-x^2}}+C$$

6. Same substitution as in Exercise 5

$$\int\dfrac{1}{x^2\sqrt{25-x^2}}\,dx=\int\dfrac{5\cos\theta\,d\theta}{(25\sin^2\theta)(5\cos\theta)}=\dfrac{1}{25}\int\csc^2\theta\,d\theta=-\dfrac{1}{25}\cot\theta+C=\dfrac{-\sqrt{25-x^2}}{25x}+C$$

7. Same substitution as in Exercise 5

$$\int\dfrac{\sqrt{25-x^2}}{x}\,dx=\int\dfrac{25\cos^2\theta\,d\theta}{5\sin\theta}=5\int\dfrac{1-\sin^2\theta}{\sin\theta}\,d\theta=5\int(\csc\theta-\sin\theta)\,d\theta$$

$$=5[\ln|\csc\theta-\cot\theta|+\cos\theta]+C=5\ln\left|\dfrac{5-\sqrt{25-x^2}}{x}\right|+\sqrt{25-x^2}+C$$

8. Same substitution as in Exercise 5

$$\int\dfrac{x^2}{\sqrt{25-x^2}}\,dx=\int\dfrac{25\sin^2\theta}{5\cos\theta}(5\cos\theta)\,d\theta=\dfrac{25}{2}\int(1-\cos 2\theta)\,d\theta$$

$$=\dfrac{25}{2}\left(\theta-\dfrac{1}{2}\sin 2\theta\right)+C=\dfrac{25}{2}(\theta-\sin\theta\cos\theta)+C$$

$$=\dfrac{25}{2}\left[\arcsin\left(\dfrac{x}{5}\right)-\left(\dfrac{x}{5}\right)\left(\dfrac{\sqrt{25-x^2}}{5}\right)\right]+C=\dfrac{1}{2}\left[25\arcsin\left(\dfrac{x}{5}\right)-x\sqrt{25-x^2}\right]+C$$

9. Let $x = 2 \sec\theta$, $dx = 2 \sec\theta \tan\theta \, d\theta$, $\sqrt{x^2 - 4} = 2\tan\theta$.

$$\int \frac{1}{\sqrt{x^2 - 4}}\, dx = \int \frac{2 \sec\theta \tan\theta \, d\theta}{2 \tan\theta} = \int \sec\theta \, d\theta = \ln|\sec\theta + \tan\theta| + C_1$$

$$= \ln\left| \frac{x}{2} + \frac{\sqrt{x^2 - 4}}{2} \right| + C_1$$

$$= \ln\left| x + \sqrt{x^2 - 4} \right| - \ln 2 + C_1 = \ln\left| x + \sqrt{x^2 - 4} \right| + C$$

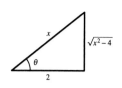

10. Same substitution as in Exercise 9

$$\int \frac{\sqrt{x^2 - 4}}{x}\, dx = \int \frac{2\tan\theta}{2\sec\theta}(2\sec\theta\tan\theta)\, d\theta = 2\int \tan^2\theta \, d\theta = 2\int (\sec^2\theta - 1)\, d\theta$$

$$= 2(\tan\theta - \theta) + C = 2\left[\frac{\sqrt{x^2 - 4}}{2} - \operatorname{arcsec}\left(\frac{x}{2}\right) \right] + C = \sqrt{x^2 - 4} - 2\operatorname{arcsec}\left(\frac{x}{2}\right) + C$$

11. Same substitution as in Exercise 9

$$\int x^3\sqrt{x^2 - 4}\, dx = \int (8\sec^3\theta)(2\tan\theta)(2\sec\theta\tan\theta)\, d\theta = 32\int \tan^2\theta \sec^4\theta \, d\theta$$

$$= 32\int \tan^2\theta(1 + \tan^2\theta)\sec^2\theta \, d\theta = 32\left(\frac{\tan^3\theta}{3} + \frac{\tan^5\theta}{5} \right) + C$$

$$= \frac{32}{15}\tan^3\theta[5 + 3\tan^2\theta] + C = \frac{32}{15}\frac{(x^2 - 4)^{3/2}}{8}\left[5 + 3\frac{(x^2 - 4)}{4} \right] + C$$

$$= \frac{1}{15}(x^2 - 4)^{3/2}[20 + 3(x^2 - 4)] + C = \frac{1}{15}(x^2 - 4)^{3/2}(3x^2 + 8) + C$$

12. Same substitution as in Exercise 9

$$\int \frac{x^3}{\sqrt{x^2 - 4}}\, dx = \int \frac{8\sec^3\theta}{2\tan\theta}(2\sec\theta\tan\theta)\, d\theta = 8\int \sec^4\theta \, d\theta$$

$$= 8\int (1 + \tan^2\theta)\sec^2\theta \, d\theta = 8\left(\tan\theta + \frac{\tan^3\theta}{3} \right) + C = \frac{8}{3}\tan\theta(3 + \tan^2\theta) + C$$

$$= \frac{8}{3}\left(\frac{\sqrt{x^2 - 4}}{2} \right)\left(3 + \frac{x^2 - 4}{4} \right) + C = \frac{1}{3}\sqrt{x^2 - 4}\,(12 + x^2 - 4) + C = \frac{1}{3}\sqrt{x^2 - 4}\,(x^2 + 8) + C$$

13. Let $x = \tan\theta$, $dx = \sec^2\theta \, d\theta$, $\sqrt{1 + x^2} = \sec\theta$.

$$\int x\sqrt{1 + x^2}\, dx = \int \tan\theta(\sec\theta)\sec^2\theta \, d\theta = \frac{\sec^3\theta}{3} + C = \frac{1}{3}(1 + x^2)^{3/2} + C$$

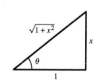

Note: This integral could have been evaluated with the Power Rule.

14. Same substitution as in Exercise 13

$$\int \frac{x^3}{\sqrt{1 + x^2}}\, dx = \int \frac{\tan^3\theta}{\sec\theta}\sec^2\theta \, d\theta = \int (\sec^2\theta - 1)\ \sec\theta\tan\theta \, d\theta = \frac{\sec^3\theta}{3} - \sec\theta + C$$

$$= \frac{1}{3}\sec\theta(\sec^2\theta - 3) + C = \frac{1}{3}\sqrt{1 + x^2}[(1 + x^2) - 3] + C = \frac{1}{3}\sqrt{1 + x^2}(x^2 - 2) + C$$

15. Same substitution as in Exercise 13

$$\int \frac{1}{(1 + x^2)^2}\, dx = \int \frac{1}{\left(\sqrt{1 + x^2}\right)^4}\, dx$$

$$= \int \frac{\sec^2 \theta\, d\theta}{\sec^4 \theta}$$

$$= \int \cos^2 \theta\, d\theta = \frac{1}{2}\int (1 + \cos 2\theta)\, d\theta$$

$$= \frac{1}{2}\left[\theta + \frac{\sin 2\theta}{2}\right]$$

$$= \frac{1}{2}\left[\theta + \sin \theta \cos \theta\right] + C$$

$$= \frac{1}{2}\left[\arctan x + \left(\frac{x}{\sqrt{1 + x^2}}\right)\left(\frac{1}{\sqrt{1 + x^2}}\right)\right] + C$$

$$= \frac{1}{2}\left[\arctan x + \frac{x}{1 + x^2}\right] + C$$

16. Same substitution as in Exercise 13

$$\int \frac{x^2}{(1 + x^2)^2}\, dx = \int \frac{x^2}{\left(\sqrt{1 + x^2}\right)^4}\, dx = \int \frac{\tan^2 \theta \sec^2 \theta\, d\theta}{\sec^4 \theta} = \int \sin^2 \theta\, d\theta$$

$$= \frac{1}{2}\int (1 - \cos 2\theta)\, d\theta = \frac{1}{2}\left[\theta - \frac{\sin 2\theta}{2}\right] = \frac{1}{2}\left[\theta - \sin \theta \cos \theta\right] + C$$

$$= \frac{1}{2}\left[\arctan x - \left(\frac{x}{\sqrt{1 + x^2}}\right)\left(\frac{1}{\sqrt{1 + x^2}}\right)\right] + C = \frac{1}{2}\left[\arctan x - \frac{x}{1 + x^2}\right] + C$$

17. Let $u = 3x$, $a = 2$, and $du = 3\, dx$.

$$\int \sqrt{4 + 9x^2}\, dx = \frac{1}{3}\int \sqrt{(2)^2 + (3x)^2}\, 3\, dx$$

$$= \frac{1}{3}\left(\frac{1}{2}\right)\left(3x\sqrt{4 + 9x^2} + 4\ln\left|3x + \sqrt{4 + 9x^2}\right|\right) + C$$

$$= \frac{1}{2}x\sqrt{4 + 9x^2} + \frac{2}{3}\ln\left|3x + \sqrt{4 + 9x^2}\right| + C$$

18. Let $u = x$, $a = 1$, and $du = dx$.

$$\int \sqrt{1 + x^2}\, dx = \frac{1}{2}\left(x\sqrt{1 + x^2} + \ln\left|x + \sqrt{1 + x^2}\right|\right) + C$$

19. $\displaystyle\int \frac{x}{\sqrt{x^2 + 9}}\, dx = \frac{1}{2}\int (x^2 + 9)^{-1/2}(2x)\, dx$

$$= \sqrt{x^2 + 9} + C$$

(Power Rule)

20. $\displaystyle\int \frac{1}{\sqrt{25 - x^2}}\, dx = \arcsin \frac{x}{5} + C$

21. Let $x = 2 \sin \theta$, $dx = 2 \cos \theta \, d\theta$, $\sqrt{4 - x^2} = 2 \cos \theta$.

$$
\begin{aligned}
\int_0^2 \sqrt{16 - 4x^2} \, dx &= 2 \int_0^2 \sqrt{4 - x^2} \, dx \\
&= 2 \int_0^{\pi/2} 2 \cos \theta (2 \cos \theta \, d\theta) \\
&= 8 \int_0^{\pi/2} \cos^2 \theta \, d\theta \\
&= 4 \int_0^{\pi/2} (1 + \cos 2\theta) \, d\theta \\
&= 4 \left[\theta + \frac{1}{2} \sin 2\theta \right]_0^{\pi/2} = 2\pi
\end{aligned}
$$

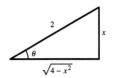

22. Let $u = 16 - 4x^2$, $du = -8x \, dx$.

$$
\int_0^2 x \sqrt{16 - 4x^2} \, dx = -\frac{1}{8} \int_0^2 (16 - 4x^2)^{1/2} (-8x) \, dx = \left[-\frac{1}{12} (16 - 4x^2)^{3/2} \right]_0^2 = \frac{16}{3}
$$

23. Let $x = 3 \sec \theta$, $dx = 3 \sec \theta \tan \theta \, d\theta$, $\sqrt{x^2 - 9} = 3 \tan \theta$.

$$
\begin{aligned}
\int \frac{1}{\sqrt{x^2 - 9}} \, dx &= \int \frac{3 \sec \theta \tan \theta \, d\theta}{3 \tan \theta} \\
&= \int \sec \theta \, d\theta \\
&= \ln |\sec \theta + \tan \theta| + C_1 \\
&= \ln \left| \frac{x}{3} + \frac{\sqrt{x^2 - 9}}{3} \right| + C_1 \\
&= \ln \left| x + \sqrt{x^2 - 9} \right| + C
\end{aligned}
$$

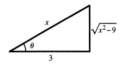

24. Let $u = 1 - t^2$, $du = -2t \, dt$.

$$
\int \frac{t}{(1 - t^2)^{3/2}} \, dt = -\frac{1}{2} \int (1 - t^2)^{-3/2} (-2t) \, dt = \frac{1}{\sqrt{1 - t^2}} + C
$$

25. Let $x = \sin \theta$, $dx = \cos \theta \, d\theta$, $\sqrt{1 - x^2} = \cos \theta$.

$$
\begin{aligned}
\int \frac{\sqrt{1 - x^2}}{x^4} \, dx &= \int \frac{\cos \theta (\cos \theta \, d\theta)}{\sin^4 \theta} \\
&= \int \cot^2 \theta \csc^2 \theta \, d\theta \\
&= -\frac{1}{3} \cot^3 \theta + C \\
&= \frac{-(1 - x^2)^{3/2}}{3x^3} + C
\end{aligned}
$$

26. Let $2x = 3 \tan \theta$, $dx = \frac{3}{2} \sec^2 \theta \, d\theta$, $\sqrt{4x^2 + 9} = 3 \sec \theta$.

$$
\begin{aligned}
\int \frac{\sqrt{4x^2 + 9}}{x^4} \, dx &= \int \frac{3 \sec \theta [(3/2) \sec^2 \theta \, d\theta]}{(3/2)^4 \tan^4 \theta} \\
&= \frac{8}{9} \int \frac{\cos \theta}{\sin^4 \theta} \, d\theta \\
&= \frac{-8}{27 \sin^3 \theta} + C \\
&= -\frac{8}{27} \csc^3 \theta + C \\
&= \frac{-(4x^2 + 9)^{3/2}}{27x^3} + C
\end{aligned}
$$

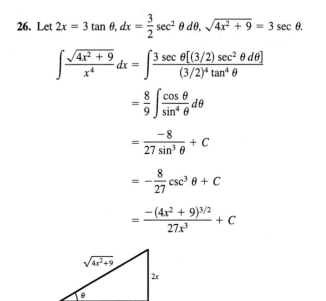

27. Same substitutions as in Exercise 26

$$\int \frac{1}{x\sqrt{4x^2 + 9}} \, dx = \int \frac{(3/2) \sec^2 \theta \, d\theta}{(3/2) \tan \theta \, 3\sec \theta}$$

$$= \frac{1}{3} \int \csc \theta \, d\theta = -\frac{1}{3} \ln|\csc \theta + \cot \theta| + C = -\frac{1}{3} \ln \left| \frac{\sqrt{4x^2 + 9} + 3}{2x} \right| + C$$

28. Let $x = \sqrt{3} \tan \theta$, $dx = \sqrt{3} \sec^2 \theta \, d\theta$, $x^2 + 3 = 3 \sec^2 \theta$.

$$\int \frac{1}{(x^2 + 3)^{3/2}} \, dx = \int \frac{\sqrt{3} \sec^2 \theta \, d\theta}{3\sqrt{3} \sec^3 \theta}$$

$$= \frac{1}{3} \int \cos \theta \, d\theta$$

$$= \frac{1}{3} \sin \theta + C$$

$$= \frac{x}{3\sqrt{x^2 + 3}} + C$$

29. Same substitutions as in Exercise 28

$$\int \frac{x}{(x^2 + 3)^{3/2}} \, dx = \int \frac{\sqrt{3} \tan \theta \left(\sqrt{3} \sec^2 \theta \, d\theta \right)}{3\sqrt{3} \sec^3 \theta} = \frac{\sqrt{3}}{3} \int \sin \theta \, d\theta = -\frac{\sqrt{3}}{3} \cos \theta + C = \frac{-1}{\sqrt{x^2 + 3}} + C$$

Note: This integral could have been evaluated with the Power Rule: $u = x^2 + 3$, $du = 2x \, dx$

30. Let $2x = 4 \tan \theta$, $dx = 2 \sec^2 \theta \, d\theta$, $\sqrt{4x^2 + 16} = 4 \sec \theta$.

$$\int \frac{1}{x\sqrt{4x^2 + 16}} \, dx = \int \frac{2 \sec^2 \theta \, d\theta}{2 \tan \theta (4 \sec \theta)}$$

$$= \frac{1}{4} \int \frac{\sec \theta}{\tan \theta} \, d\theta = \frac{1}{4} \int \csc \theta \, d\theta$$

$$= -\frac{1}{4} \ln|\csc \theta + \cot \theta| + C = -\frac{1}{4} \ln \left| \frac{\sqrt{x^2 + 4} + 2}{x} \right| + C$$

31. Let $u = 1 + e^{2x}$, $du = 2e^{2x} \, dx$.

$$\int e^{2x} \sqrt{1 + e^{2x}} \, dx = \frac{1}{2} \int (1 + e^{2x})^{1/2} (2e^{2x}) dx = \frac{1}{3} (1 + e^{2x})^{3/2} + C$$

32. Let $u = x^2 + 2x + 2$, $du = (2x + 2) \, dx$.

$$\int (x + 1) \sqrt{x^2 + 2x + 2} \, dx = \frac{1}{2} \int (x^2 + 2x + 2)^{1/2} (2x + 2) \, dx = \frac{1}{3} (x^2 + 2x + 2)^{3/2} + C$$

33. Let $e^x = \sin \theta$, $e^x \, dx = \cos \theta \, d\theta$, $\sqrt{1 - e^{2x}} = \cos \theta$.

$$\int e^x \sqrt{1 - e^{2x}} \, dx = \int \cos^2 \theta \, d\theta$$

$$= \frac{1}{2} \int (1 + \cos 2\theta) d\theta$$

$$= \frac{1}{2} \left[\theta + \frac{\sin 2\theta}{2} \right]$$

$$= \frac{1}{2} (\theta + \sin \theta \cos \theta) + C = \frac{1}{2} \left(\arcsin e^x + e^x \sqrt{1 - e^{2x}} \right) + C$$

34. Let $\sqrt{x} = \sin\theta$, $x = \sin^2\theta$, $dx = 2\sin\theta\cos\theta\,d\theta$, $\sqrt{1-x} = \cos\theta$.

$$\int \frac{\sqrt{1-x}}{\sqrt{x}}\,dx = \int \frac{\cos\theta(2\sin\theta\cos\theta\,d\theta)}{\sin\theta}$$

$$= 2\int \cos^2\theta\,d\theta$$

$$= \int (1 + \cos 2\theta)\,d\theta$$

$$= (\theta + \sin\theta\cos\theta) + C$$

$$= \arcsin\sqrt{x} + \sqrt{x}\sqrt{1-x} + C$$

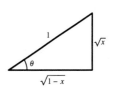

35. Let $x = \sqrt{2}\tan\theta$, $dx = \sqrt{2}\sec^2\theta\,d\theta$, $x^2 + 2 = 2\sec^2\theta$.

$$\int \frac{1}{4 + 4x^2 + x^4}\,dx = \int \frac{1}{(x^2 + 2)^2}\,dx$$

$$= \int \frac{\sqrt{2}\sec^2\theta\,d\theta}{4\sec^4\theta}$$

$$= \frac{\sqrt{2}}{4}\int \cos^2\theta\,d\theta$$

$$= \frac{\sqrt{2}}{4}\left(\frac{1}{2}\right)\int (1 + \cos 2\theta)\,d\theta$$

$$= \frac{\sqrt{2}}{8}\left(\theta + \frac{1}{2}\sin 2\theta\right) + C$$

$$= \frac{\sqrt{2}}{8}(\theta + \sin\theta\cos\theta) + C$$

$$= \frac{1}{4}\left[\frac{x}{x^2 + 2} + \frac{1}{\sqrt{2}}\arctan\frac{x}{\sqrt{2}}\right] + C$$

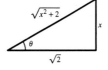

36. Let $x = \tan\theta$, $dx = \sec^2\theta\,d\theta$, $x^2 + 1 = \sec^2\theta$.

$$\int \frac{x^3 + x + 1}{x^4 + 2x^2 + 1}\,dx = \frac{1}{4}\int \frac{4x^3 + 4x}{x^4 + 2x^2 + 1}\,dx + \int \frac{1}{(x^2 + 1)^2}\,dx$$

$$= \frac{1}{4}\ln(x^4 + 2x^2 + 1) + \int \frac{\sec^2\theta\,d\theta}{\sec^4\theta}$$

$$= \frac{1}{2}\ln(x^2 + 1) + \frac{1}{2}\int (1 + \cos 2\theta)\,d\theta$$

$$= \frac{1}{2}\ln(x^2 + 1) + \frac{1}{2}(\theta + \sin\theta\cos\theta) + C$$

$$= \frac{1}{2}\left[\ln(x^2 + 1) + \arctan x + \frac{x}{x^2 + 1}\right] + C$$

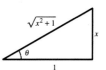

37. Assume $x > 0$.

$$u = \text{arcsec } 2x, \implies du = \frac{1}{x\sqrt{4x^2 - 1}} dx, dv = dx \implies v = x$$

$$\int \text{arcsec } 2x\, dx = x \text{ arcsec } 2x - \int \frac{1}{\sqrt{4x^2 - 1}} dx$$

$$2x = \sec\theta, dx = \frac{1}{2}\sec\theta\tan\theta\, d\theta, \sqrt{4x^2 - 1} = \tan\theta$$

$$\int \text{arcsec } 2x\, dx = x\text{ arcsec } 2x - \int \frac{(1/2)\sec\theta\tan\theta\, d\theta}{\tan\theta} = x\text{ arcsec } 2x - \frac{1}{2}\int \sec\theta\, d\theta$$

$$= x\text{ arcsec } 2x - \frac{1}{2}\ln|\sec\theta + \tan\theta| + C = x\text{ arcsec } 2x - \frac{1}{2}\ln\left|2x + \sqrt{4x^2 - 1}\right| + C$$

A similar argument applies if $x < 0$.

38. $u = \arcsin x, \implies du = \frac{1}{\sqrt{1 - x^2}} dx, dv = x\, dx \implies v = \frac{x^2}{2}$

$$\int x \arcsin x\, dx = \frac{x^2}{2}\arcsin x - \frac{1}{2}\int \frac{x^2}{\sqrt{1 - x^2}} dx$$

$$x = \sin\theta, dx = \cos\theta\, d\theta, \sqrt{1 - x^2} = \cos\theta$$

$$\int x \arcsin x\, dx = \frac{x^2}{2}\arcsin x = \frac{1}{2}\int \frac{\sin^2\theta}{\cos\theta}\cos\theta\, d\theta = \frac{x^2}{2}\arcsin x - \frac{1}{4}\int (1 - \cos 2\theta)\, d\theta$$

$$= \frac{x^2}{2}\arcsin x - \frac{1}{4}\left[\theta - \frac{1}{2}\sin 2\theta\right] + C = \frac{x^2}{2}\arcsin x - \frac{1}{4}[\theta - \sin\theta\cos\theta] + C$$

$$= \frac{x^2}{2}\arcsin x - \frac{1}{4}\left[\arcsin x - x\sqrt{1 - x^2}\right] + C = \frac{1}{4}\left[(2x^2 - 1)\arcsin x + x\sqrt{1 - x^2}\right] + C$$

39. $\displaystyle\int \frac{1}{\sqrt{4x - x^2}} dx = \int \frac{1}{\sqrt{4 - (x - 2)^2}} dx = \arcsin\left(\frac{x - 2}{2}\right) + C$

40. Let $x - 1 = \sin\theta, dx = \cos\theta\, d\theta, \sqrt{1 - (x - 1)^2} = \sqrt{2x - x^2} = \cos\theta$.

$$\int \frac{x^2}{\sqrt{2x - x^2}} dx = \int \frac{x^2}{\sqrt{1 - (x - 1)^2}} dx$$

$$= \int \frac{(1 + \sin\theta)^2(\cos\theta\, d\theta)}{\cos\theta}$$

$$= \int (1 + 2\sin\theta + \sin^2\theta)\, d\theta$$

$$= \int \left(\frac{3}{2} + 2\sin\theta - \frac{1}{2}\cos 2\theta\right) d\theta$$

$$= \frac{3}{2}\theta - 2\cos\theta - \frac{1}{4}\sin 2\theta + C$$

$$= \frac{3}{2}\theta - 2\cos\theta - \frac{1}{2}\sin\theta\cos\theta + C$$

$$= \frac{3}{2}\arcsin(x - 1) - 2\sqrt{2x - x^2} - \frac{1}{2}(x - 1)\sqrt{2x - x^2} + C$$

$$= \frac{3}{2}\arcsin(x - 1) - \frac{1}{2}\sqrt{2x - x^2}(x + 3) + C$$

41. Let $x + 2 = 2 \tan \theta$, $dx = 2 \sec^2 \theta \, d\theta$, $\sqrt{(x + 2)^2 + 4} = 2 \sec \theta$.

$$\int \frac{x}{\sqrt{x^2 + 4x + 8}} \, dx = \int \frac{x}{\sqrt{(x + 2)^2 + 4}} \, dx = \int \frac{(2 \tan \theta - 2)(2 \sec^2 \theta) \, d\theta}{2 \sec \theta}$$

$$= 2 \int (\tan \theta - 1)(\sec \theta) \, d\theta$$

$$= 2[\sec \theta - \ln|\sec \theta + \tan \theta|] + C_1$$

$$= 2\left[\frac{\sqrt{(x + 2)^2 + 4}}{2} - \ln\left|\frac{\sqrt{(x + 2)^2 + 4}}{2} + \frac{x + 2}{2}\right|\right] + C_1$$

$$= \sqrt{x^2 + 4x + 8} - 2\left[\ln\left|\sqrt{x^2 + 4x + 8} + (x + 2)\right| - \ln 2\right] + C_1$$

$$= \sqrt{x^2 + 4x + 8} - 2 \ln\left|\sqrt{x^2 + 4x + 8} + (x + 2)\right| + C$$

42. Let $x - 3 = 2 \sec \theta$, $dx = 2 \sec \theta \tan \theta \, d\theta$, $\sqrt{(x - 3)^2 - 4} = 2 \tan \theta$.

$$\int \frac{x}{\sqrt{x^2 - 6x + 5}} \, dx = \int \frac{x}{\sqrt{(x - 3)^2 - 4}} \, dx = \int \frac{(2 \sec \theta + 3)}{2 \tan \theta}(2 \sec \theta \tan \theta) \, d\theta$$

$$= \int (2 \sec^2 \theta + 3 \sec \theta) \, d\theta$$

$$= 2 \tan \theta + 3 \ln|\sec \theta + \tan \theta| + C_1$$

$$= 2\left(\frac{\sqrt{(x - 3)^3 - 4}}{2}\right) + 3 \ln\left|\frac{x - 3}{2} + \frac{\sqrt{(x - 3)^2 - 4}}{2}\right| + C_1$$

$$= \sqrt{x^2 - 6x + 5} + 3 \ln\left|(x - 3) + \sqrt{x^2 - 6x + 5}\right| + C$$

43. Let $t = \sin \theta$, $dt = \cos \theta \, d\theta$, $1 - t^2 = \cos^2 \theta$.

(a) $\int \dfrac{t^2}{(1 - t^2)^{3/2}} \, dt = \int \dfrac{\sin^2 \theta \cos \theta \, d\theta}{\cos^3 \theta}$

$$= \int \tan^2 \theta \, d\theta$$

$$= \int (\sec^2 \theta - 1) \, d\theta$$

$$= \tan \theta - \theta + C$$

$$= \frac{t}{\sqrt{1 - t^2}} - \arcsin t + C$$

Thus, $\displaystyle\int_0^{\sqrt{3}/2} \frac{t^2}{(1 - t^2)^{3/2}} \, dt = \left[\frac{t}{\sqrt{1 - t^2}} - \arcsin t\right]_0^{\sqrt{3}/2} = \frac{\sqrt{3}/2}{\sqrt{1/4}} - \arcsin \frac{\sqrt{3}}{2} = \sqrt{3} - \frac{\pi}{3} \approx 0.685.$

(b) When $t = 0$, $\theta = 0$. When $t = \sqrt{3}/2$, $\theta = \pi/3$. Thus,

$$\int_0^{\sqrt{3}/2} \frac{t^2}{(1 - t^2)^{3/2}} \, dt = \left[\tan \theta - \theta\right]_0^{\pi/3} = \sqrt{3} - \frac{\pi}{3} \approx 0.685.$$

44. Same substitution as in Exercise 43

(a) $\displaystyle\int \frac{1}{(1-t^2)^{5/2}}\,dt = \int \frac{\cos\theta\,d\theta}{\cos^5\theta} = \int \sec^4\theta\,d\theta = \int (\tan^2\theta + 1)\sec^2\theta\,d\theta$

$\displaystyle = \frac{1}{3}\tan^3\theta + \tan\theta + C = \frac{1}{3}\left(\frac{t}{\sqrt{1-t^2}}\right)^3 + \frac{t}{\sqrt{1-t^2}} + C$

Thus, $\displaystyle\int_0^{\sqrt{3}/2} \frac{1}{(1-t^2)^{5/2}}\,dt = \left[\frac{t^3}{3(1-t^2)^{3/2}} + \frac{t}{\sqrt{1-t^2}}\right]_0^{\sqrt{3}/2}$

$\displaystyle = \frac{3\sqrt{3}/8}{3(1/4)^{3/2}} + \frac{\sqrt{3}/2}{\sqrt{1/4}} = \sqrt{3} + \sqrt{3} = 2\sqrt{3} \approx 3.464.$

(b) When $t = 0$, $\theta = 0$. When $t = \sqrt{3}/2$, $\theta = \pi/3$. Thus,

$\displaystyle\int_0^{\sqrt{3}/2} \frac{1}{(1-t^2)^{5/2}}\,dt = \left[\frac{1}{3}\tan^3\theta + \tan\theta\right]_0^{\pi/3} = \frac{1}{3}(\sqrt{3})^3 + \sqrt{3} = 2\sqrt{3} \approx 3.464.$

45. (a) Let $x = 3\tan\theta$, $dx = 3\sec^2\theta\,d\theta$, $\sqrt{x^2 + 9} = 3\sec\theta$.

$\displaystyle\int \frac{x^3}{\sqrt{x^2 + 9}}\,dx = \int \frac{(27\tan^3\theta)(3\sec^2\theta\,d\theta)}{3\sec\theta}$

$\displaystyle = 27\int (\sec^2\theta - 1)\sec\theta\tan\theta\,d\theta$

$\displaystyle = 27\left[\frac{1}{3}\sec^3\theta - \sec\theta\right] + C = 9[\sec^3\theta - 3\sec\theta] + C$

$\displaystyle = 9\left[\left(\frac{\sqrt{x^2+9}}{3}\right)^3 - 3\left(\frac{\sqrt{x^2+9}}{3}\right)\right] + C = \frac{1}{3}(x^2+9)^{3/2} - 9\sqrt{x^2+9} + C$

Thus, $\displaystyle\int_0^3 \frac{x^3}{\sqrt{x^2+9}}\,dx = \left[\frac{1}{3}(x^2+9)^{3/2} - 9\sqrt{x^2+9}\right]_0^3$

$\displaystyle = \left(\frac{1}{3}(54\sqrt{2}) - 27\sqrt{2}\right) - (9 - 27)$

$\displaystyle = 18 - 9\sqrt{2} = 9(2 - \sqrt{2}) \approx 5.272.$

(b) When $x = 0$, $\theta = 0$. When $x = 3$, $\theta = \pi/4$. Thus,

$\displaystyle\int_0^3 \frac{x^3}{\sqrt{x^2+9}}\,dx = 9\left[\sec^3\theta - 3\sec\theta\right]_0^{\pi/4} = 9(2\sqrt{2} - 3\sqrt{2}) - 9(1 - 3) = 9(2 - \sqrt{2}) \approx 5.272.$

46. (a) Let $3x = 5\sin\theta$, $dx = (5/3)\cos\theta\,d\theta$, $\sqrt{25 - 9x^2} = 5\cos\theta$.

$\displaystyle\int \sqrt{25 - 9x^2}\,dx = \frac{25}{3}\int \cos^2\theta\,d\theta$

$\displaystyle = \frac{25}{6}\int (1 + \cos 2\theta)\,d\theta$

$\displaystyle = \frac{25}{6}\left[\theta + \frac{1}{2}\sin 2\theta\right] + C$

$\displaystyle = \frac{25}{6}[\theta + \sin\theta\cos\theta] + C$

$\displaystyle = \frac{25}{6}\left[\arcsin\frac{3x}{5} + \left(\frac{3x}{5}\right)\left(\frac{\sqrt{25-9x^2}}{5}\right)\right] + C$

Thus, $\displaystyle\int_0^{5/3} \sqrt{25 - 9x^2}\,dx = \frac{25}{6}\left[\arcsin\frac{3x}{5} + \frac{3x\sqrt{25-9x^2}}{25}\right]_0^{5/3} = \frac{25}{6}\left[\frac{\pi}{2}\right] = \frac{25\pi}{12} \approx 6.545.$

—CONTINUED—

46. —CONTINUED—

(b) When $x = 0$, $\theta = 0$. When $x = 5/3$, $\theta = \pi/2$. Thus,

$$\int_0^{5/3} \sqrt{25 - 9x^2}\, dx = \frac{25}{6}\left[\theta + \frac{1}{2}\sin 2\theta\right]_0^{\pi/2} = \frac{25}{6}\left[\frac{\pi}{2}\right] = \frac{25\pi}{12} \approx 6.545.$$

47. $\displaystyle\int \frac{x^2}{\sqrt{x^2 + 10x + 9}}\, dx = \frac{1}{2}\sqrt{x^2 + 10x + 9}\,(x - 15) + 33\ln\left|(x + 5) + \sqrt{x^2 + 10x + 9}\right| + C$

48. $\displaystyle\int (x^2 + 2x + 11)^{3/2}\, dx = \frac{1}{4}(x + 1)(x^2 + 2x + 26)\sqrt{x^2 + 2x + 11} + \frac{75}{2}\ln\left|\sqrt{x^2 + 2x + 11} + (x + 1)\right| + C$

49. $\displaystyle\int \frac{x^2}{\sqrt{x^2 - 1}}\, dx = \frac{1}{2}\left(x\sqrt{x^2 - 1} + \ln\left|x + \sqrt{x^2 - 1}\right|\right) + C$

50. $\displaystyle\int x^2\sqrt{x^2 - 4}\, dx = \frac{1}{4}x^3\sqrt{x^2 - 4} - \frac{1}{2}x\sqrt{x^2 - 4} - 2\ln\left|x + \sqrt{x^2 - 4}\right| + C$

51. $\displaystyle A = 4\int_0^a \frac{b}{a}\sqrt{a^2 - x^2}\, dx$

$$= \frac{4b}{a}\int_0^a \sqrt{a^2 - x^2}\, dx$$

$$= \left[\frac{4b}{a}\left(\frac{1}{2}\right)\left(a^2\arcsin\frac{x}{a} + x\sqrt{a^2 - x^2}\right)\right]_0^a$$

$$= \frac{2b}{a}\left(a^2\left(\frac{\pi}{2}\right)\right)$$

$$= \pi ab$$

Note: See Theorem 7.2 for $\int \sqrt{a^2 - x^2}\, dx$.

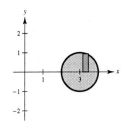

52. (a) $x^2 + (y - k)^2 = 25$

Radius of circle $= 5$

$k^2 = 5^2 + 5^2 = 50$

$k = 5\sqrt{2}$

(b) Area $=$ square $- \dfrac{1}{4}$(circle)

$$= 25 - \frac{1}{4}\pi(5)^2 = 25\left(1 - \frac{\pi}{4}\right)$$

(c) Area $= r^2 - \dfrac{1}{4}\pi r^2 = r^2\left(1 - \dfrac{\pi}{4}\right)$

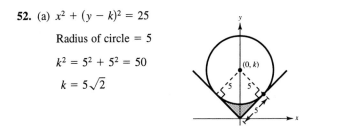

53. Let $x - 3 = \sin\theta$, $dx = \cos\theta\, d\theta$, $\sqrt{1 - (x - 3)^2} = \cos\theta$.

Shell Method:

$$V = 4\pi\int_2^4 x\sqrt{1 - (x - 3)^2}\, dx$$

$$= 4\pi\int_{-\pi/2}^{\pi/2} (3 + \sin\theta)\cos^2\theta\, d\theta$$

$$= 4\pi\left[\frac{3}{2}\int_{-\pi/2}^{\pi/2}(1 + \cos 2\theta)\, d\theta + \int_{-\pi/2}^{\pi/2}\cos^2\theta\sin\theta\, d\theta\right]$$

$$= 4\pi\left[\frac{3}{2}\left(\theta + \frac{1}{2}\sin 2\theta\right) - \frac{1}{3}\cos^3\theta\right]_{-\pi/2}^{\pi/2} = 6\pi^2$$

54. Let $x - h = r \sin \theta$, $dx = r \cos \theta \, d\theta$, $\sqrt{r^2 - (x - h)^2} = r \cos \theta$.

Shell Method:

$$V = 4\pi \int_{h-r}^{h+r} x \sqrt{r^2 - (x - h)^2} \, dx$$

$$= 4\pi \int_{-\pi/2}^{\pi/2} (h + r \sin \theta) r \cos \theta (r \cos \theta) \, d\theta$$

$$= 4\pi r^2 \int_{-\pi/2}^{\pi/2} (h + r \sin \theta) \cos^2 \theta \, d\theta$$

$$= 4\pi r^2 \left[\frac{h}{2} \int_{-\pi/2}^{\pi/2} (1 + \cos 2\theta) \, d\theta + r \int_{-\pi/2}^{\pi/2} \sin \theta \cos^2 \theta \, d\theta \right]$$

$$= 2\pi r^2 h \left[\theta + \frac{1}{2} \sin 2\theta \right]_{-\pi/2}^{\pi/2} - \left[4\pi r^3 \left(\frac{\cos^3 \theta}{3} \right) \right]_{-\pi/2}^{\pi/2}$$

$$= 2\pi^2 r^2 h$$

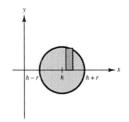

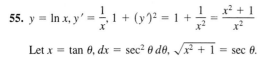

55. $y = \ln x$, $y' = \dfrac{1}{x}$, $1 + (y')^2 = 1 + \dfrac{1}{x^2} = \dfrac{x^2 + 1}{x^2}$

Let $x = \tan \theta$, $dx = \sec^2 \theta \, d\theta$, $\sqrt{x^2 + 1} = \sec \theta$.

$$s = \int_1^5 \sqrt{\frac{x^2 + 1}{x^2}} \, dx = \int_1^5 \frac{\sqrt{x^2 + 1}}{x} \, dx$$

$$= \int_a^b \frac{\sec \theta}{\tan \theta} \sec^2 \theta \, d\theta = \int_a^b \frac{\sec \theta}{\tan \theta} (1 + \tan^2 \theta) \, d\theta$$

$$= \int_a^b (\csc \theta + \sec \theta \tan \theta) \, d\theta$$

$$= \left[-\ln|\csc \theta + \cot \theta| + \sec \theta \right]_a^b$$

$$= \left[-\ln \left| \frac{\sqrt{x^2 + 1}}{x} + \frac{1}{x} \right| + \sqrt{x^2 + 1} \right]_1^5$$

$$= \left[-\ln \left(\frac{\sqrt{26} + 1}{5} \right) + \sqrt{26} \right] - \left[-\ln(\sqrt{2} + 1) + \sqrt{2} \right]$$

$$= \ln \left[\frac{5(\sqrt{2} + 1)}{\sqrt{26} + 1} \right] + \sqrt{26} - \sqrt{2} \approx 4.367 \quad \text{or} \quad \ln \left[\frac{\sqrt{26} - 1}{5(\sqrt{2} - 1)} \right] + \sqrt{26} - \sqrt{2}$$

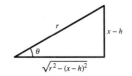

56. $y = x^2$, $y' = 2x$, $1 + (y')^2 = 1 + 4x^2$

Let $u = 2x$, $du = 2 \, dx$, $a = 1$. (See Theorem 7.2)

$$s = \int_0^3 \sqrt{1 + 4x^2} \, dx = \frac{1}{2} \int_0^3 \sqrt{4x^2 + 1} \, (2) \, dx$$

$$= \frac{1}{4} \left[2x \sqrt{4x^2 + 1} + \ln \left| 2x + \sqrt{4x^2 + 1} \right| \right]_0^3$$

$$= \frac{1}{4} \left[6\sqrt{37} + \ln \left(6 + \sqrt{37} \right) \right] \approx 9.747$$

57. Length of one arch of sine curve: $y = \sin x,\ y' = \cos x$

$$L_1 = \int_0^\pi \sqrt{1 + \cos^2 x}\, dx$$

Length of one arch of cosine curve: $y = \cos x,\ y' = -\sin x$

$$
\begin{aligned}
L_2 &= \int_{-\pi/2}^{\pi/2} \sqrt{1 + \sin^2 x}\, dx \\
&= \int_{-\pi/2}^{\pi/2} \sqrt{1 + \cos^2\!\left(x - \frac{\pi}{2}\right)}\, dx \qquad u = x - \frac{\pi}{2},\ du = dx \\
&= \int_{-\pi}^{0} \sqrt{1 + \cos^2 u}\, du \\
&= \int_0^\pi \sqrt{1 + \cos^2 u}\, du = L_1
\end{aligned}
$$

58. (a) Place the center of the circle at $(0, 1)$; $x^2 + (y - 1)^2 = 1$. The depth d satisfies $0 \le d \le 2$. The volume is

$$
\begin{aligned}
V &= 3 \cdot 2 \int_0^d \sqrt{1 - (y - 1)^2}\, dy \\
&= 6 \cdot \frac{1}{2}\left[\arcsin(y - 1) + (y - 1)\sqrt{1 - (y - 1)^2} \right]_0^d \quad \text{(Theorem 7.2 (1))} \\
&= 3\left[\arcsin(d - 1) + (d - 1)\sqrt{1 - (d - 1)^2} - \arcsin(-1) \right] \\
&= \frac{3\pi}{2} + 3\arcsin(d - 1) + 3(d - 1)\sqrt{2d - d^2}.
\end{aligned}
$$

(b)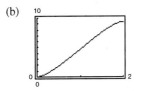

(c) The full tank holds $3\pi \approx 9.4248$ cubic meters. The horizontal lines

$$y = \frac{3\pi}{4},\ y = \frac{3\pi}{2},\ y = \frac{9\pi}{4}$$

intersect the curve at $d = 0.596,\ 1.0,\ 1.404$. The dipstick would have these markings on it.

(d) $\displaystyle V = 6 \int_0^d \sqrt{1 - (y - 1)^2}\, dy$

$$\frac{dV}{dt} = \frac{dV}{dd} \cdot \frac{dd}{dt} = 6\sqrt{1 - (d - 1)^2} \cdot d'(t) = \frac{1}{4}$$

$$\Rightarrow d'(t) = \frac{1}{24\sqrt{1 - (d - 1)^2}}$$

(e)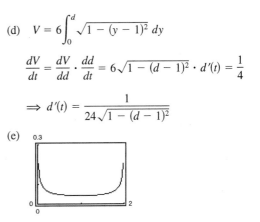

The minimum occurs at $d = 1$, which is the widest part of the tank.

59. (a)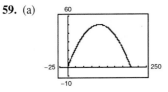

(b) $y = 0$ for $x = 200$ (range)

—CONTINUED—

59. —CONTINUED—

(c) $y = x - 0.005x^2$, $y' = 1 - 0.01x$, $1 + (y')^2 = 1 + (1 - 0.01x)^2$

Let $u = 1 - 0.01x$, $du = -0.01\, dx$, $a = 1$. (See Theorem 7.2.)

$$s = \int_0^{200} \sqrt{1 + (1 - 0.01x)^2}\, dx = -100 \int_0^{200} \sqrt{(1 - 0.01x)^2 + 1}\,(-0.01)\, dx$$

$$= -50\left[(1 - 0.01x)\sqrt{(1 - 0.01x)^2 + 1} + \ln\left|(1 - 0.01x) + \sqrt{(1 - 0.01x)^2 + 1}\right| \right]_0^{200}$$

$$= -50\left[\left(-\sqrt{2} + \ln\left|-1 + \sqrt{2}\right|\right) - \left(\sqrt{2} + \ln\left|1 + \sqrt{2}\right|\right) \right]$$

$$= 100\sqrt{2} + 50 \ln\left(\frac{\sqrt{2} + 1}{\sqrt{2} - 1}\right) \approx 229.559$$

60. (a)

(b) $y = 0$ for $x = 72$

(c) $y = x - \dfrac{x^2}{72}$, $y' = 1 - \dfrac{x}{36}$, $1 + (y')^2 = 1 + \left(1 - \dfrac{x}{36}\right)^2$

$$s = \int_0^{72} \sqrt{1 + \left(1 - \frac{x}{36}\right)^2}\, dx$$

$$= -36 \int_0^{72} \sqrt{1 + \left(1 - \frac{x}{36}\right)^2}\left(-\frac{1}{36}\right) dx$$

$$= -\frac{36}{2}\left[\left(1 - \frac{x}{36}\right)\sqrt{1 + \left(1 - \frac{x}{36}\right)^2} + \ln\left|\left(1 - \frac{x}{36}\right) + \sqrt{1 + \left(1 - \frac{x}{36}\right)^2}\right| \right]_0^{72}$$

$$= -18\left[\left(-\sqrt{2} + \ln\left|-1 + \sqrt{2}\right|\right) - \left(\sqrt{2} + \ln\left|1 + \sqrt{2}\right|\right) \right]$$

$$= 36\sqrt{2} + 18 \ln\left(\frac{\sqrt{2} + 1}{\sqrt{2} - 1}\right) \approx 82.641$$

61. $y = x^2$, $\quad y' = 2x$, $\qquad 1 + (y') = 1 + 4x^2$

$2x = \tan\theta$, $dx = \dfrac{1}{2}\sec^2\theta\, d\theta$, $\sqrt{1 + 4x^2} = \sec\theta$

(For $\int\sec^5\theta\, d\theta$ and $\int\sec^3\theta\, d\theta$, see Exercise 80 in Section 7.3)

$$S = 2\pi \int_0^{\sqrt{2}} x^2\sqrt{1 + 4x^2}\, dx$$

$$= 2\pi \int_a^b \left(\frac{\tan\theta}{2}\right)^2 (\sec\theta)\left(\frac{1}{2}\sec^2\theta\right) d\theta$$

$$= \frac{\pi}{4}\int_a^b \sec^3\theta \tan^2\theta\, d\theta = \frac{\pi}{4}\left[\int_a^b \sec^5\theta\, d\theta - \int_a^b \sec^3\theta\, d\theta \right]$$

$$= \frac{\pi}{4}\left\{ \frac{1}{4}\left[\sec^3\theta \tan\theta + \frac{3}{2}(\sec\theta \tan\theta + \ln|\sec\theta + \tan\theta|) \right] - \frac{1}{2}(\sec\theta \tan\theta + \ln|\sec\theta + \tan\theta|) \right\}_a^b$$

$$= \frac{\pi}{4}\left[\frac{1}{4}[(1 + 4x^2)^{3/2}(2x)] - \frac{1}{8}[(1 + 4x^2)^{1/2}(2x) + \ln|\sqrt{1 + 4x^2} + 2x|] \right]_0^{\sqrt{2}}$$

$$= \frac{\pi}{4}\left[\frac{54\sqrt{2}}{4} - \frac{6\sqrt{2}}{6} = \frac{1}{8}\ln(3 + 2\sqrt{2}) \right]$$

$$= \frac{\pi}{4}\left(\frac{51\sqrt{2}}{4} - \frac{\ln(3 + 2\sqrt{2})}{8} \right) = \frac{\pi}{32}\left[102\sqrt{2} - \ln(3 + 2\sqrt{2}) \right] \approx 13.989$$

62. Let $x = 3 \tan \theta$, $dx = 3 \sec^2 \theta \, d\theta$, $\sqrt{x^2 + 9} = 3 \sec \theta$.

$$A = 2 \int_0^4 \frac{3}{\sqrt{x^2 + 9}} \, dx = 6 \int_0^4 \frac{dx}{\sqrt{x^2 + 9}} = 6 \int_a^b \frac{3 \sec^2 \theta \, d\theta}{3 \sec \theta}$$

$$= 6 \int_a^b \sec \theta \, d\theta = \left[6 \ln |\sec \theta + \tan \theta| \right]_a^b = \left[6 \ln \left| \frac{\sqrt{x^2 + 9} + x}{3} \right| \right]_0^4 = 6 \ln 3$$

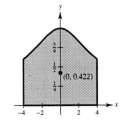

$\bar{x} = 0$ (by symmetry)

$$\bar{y} = \frac{1}{2}\left(\frac{1}{A}\right) \int_{-4}^4 \left(\frac{3}{\sqrt{x^2 + 9}}\right)^2 dx$$

$$= \frac{9}{12 \ln 3} \int_{-4}^4 \frac{1}{x^2 + 9} \, dx$$

$$= \frac{3}{4 \ln 3} \left[\frac{1}{3} \arctan \frac{x}{3}\right]_{-4}^4$$

$$= \frac{2}{4 \ln 3} \arctan \frac{4}{3} \approx 0.422$$

$$(\bar{x}, \bar{y}) = \left(0, \frac{1}{2 \ln 3} \arctan \frac{4}{3}\right) \approx (0, 0.422)$$

63. First find where the curves intersect.

$$y^2 = 16 - (x - 4)^2 = \frac{1}{16}x^4$$

$$16^2 - 16(x - 4)^2 = x^4$$

$$16^2 - 16x^2 + 128x - 16^2 = x^4$$

$$x^4 + 16x^2 - 128x = 0$$

$$x(x - 4)(x^2 + 4x + 32) \implies x = 0, 4$$

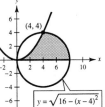

$$A = \int_0^4 \frac{1}{4}x^2 dx + \frac{1}{4}\pi(4)^2 = \frac{1}{12}x^3 \Big]_0^4 + 4\pi = \frac{16}{3} + 4\pi$$

$$M_y = \int_0^4 x\left[\frac{1}{4}x^2\right] dx + \int_4^8 x\sqrt{16 - (x - 4)^2} \, dx$$

$$= \frac{x^4}{16}\Big]_0^4 + \int_4^8 (x - 4)\sqrt{16 - (x - 4)^2} \, dx + \int_4^8 4\sqrt{16 - (x - 4)^2} \, dx$$

$$= 16 + \left[\frac{-1}{3}(16 - (x - 4)^2)^{3/2}\right]_4^8 + 2\left[16 \arcsin \frac{x - 4}{4} + (x - 4)\sqrt{16 - (x - 4)^2}\right]_4^8$$

$$= 16 + \frac{1}{3}16^{3/2} + 2\left[16\left(\frac{\pi}{2}\right)\right] = 16 + \frac{64}{3} + 16\pi = \frac{112}{3} + 16\pi$$

$$M_x = \int_0^4 \frac{1}{2}\left(\frac{1}{4}x^2\right)^2 dx + \int_4^8 \frac{1}{2}(16 - (x - 4)^2) \, dx$$

$$= \left[\frac{1}{32} \cdot \frac{x^5}{5}\right]_0^4 + \left[8x - \frac{(x - 4)^3}{6}\right]_4^8$$

$$= \frac{32}{5} + \left(64 - \frac{64}{6}\right) - 32 = \frac{416}{15}$$

$$\bar{x} = \frac{M_y}{A} = \frac{112/3 + 16\pi}{16/3 + 4\pi} = \frac{112 + 48\pi}{16 + 12\pi} = \frac{28 + 12\pi}{4 + 3\pi} \approx 4.89$$

$$\bar{y} = \frac{M_x}{A} = \frac{416/15}{(16/3) + 4\pi} = \frac{104}{5(4 + 3\pi)} \approx 1.55$$

$$(\bar{x}, \bar{y}) \approx (4.89, 1.55)$$

64. Let $r = L \tan \theta$, $dr = L \sec^2 \theta \, d\theta$, $r^2 + L^2 = L^2 \sec^2 \theta$.

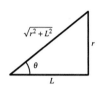

$$\frac{1}{R} \int_0^R \frac{2mL}{(r^2 + L^2)^{3/2}} \, dr = \frac{2mL}{R} \int_a^b \frac{L \sec^2 \theta \, d\theta}{L^3 \sec^3 \theta}$$

$$= \frac{2m}{RL} \int_a^b \cos \theta \, d\theta$$

$$= \left[\frac{2m}{RL} \sin \theta \right]_a^b$$

$$= \left[\frac{2m}{RL} \frac{r}{\sqrt{r^2 + L^2}} \right]_0^R$$

$$= \frac{2m}{L\sqrt{R^2 + L^2}}$$

65. (a) Area of representative rectangle: $2\sqrt{1 - y^2} \, \Delta y$

Pressure: $2(62.4)(3 - y)\sqrt{1 - y^2} \, \Delta y$

$$F = 124.8 \int_{-1}^1 (3 - y)\sqrt{1 - y^2} \, dy$$

$$= 124.8 \left[3 + \int_{-1}^1 \sqrt{1 - y^2} \, dy - \int_{-1}^1 y\sqrt{1 - y^2} \, dy \right]$$

$$= 124.8 \left[\frac{3}{2} \left(\arcsin y + y\sqrt{1 - y^2} \right) + \frac{1}{2} \left(\frac{2}{3} \right)(1 - y^2)^{3/2} \right]_{-1}^1$$

$$= (62.4)3[\arcsin 1 - \arcsin(-1)] = 187.2\pi \text{ lb}$$

(b) $F = 124.8 \int_{-1}^1 (d - y)\sqrt{1 - y^2} \, dy = 124.8d \int_{-1}^1 \sqrt{1 - y^2} \, dy - 124.8 \int_{-1}^1 y\sqrt{1 - y^2} \, dy$

$$= 124.8 \left(\frac{d}{2} \right) \left[\arcsin y + y\sqrt{1 - y^2} \right]_{-1}^1 - 124.8(0) = 62.4\pi d \text{ lb}$$

66. (a) $F_{\text{inside}} = 48 \int_{-1}^{0.8} (0.8 - y)(2)\sqrt{1 - y^2} \, dy$

$$= 96 \left[0.8 \int_{-1}^{0.8} \sqrt{1 - y^2} \, dy - \int_{-1}^{0.8} y\sqrt{1 - y^2} \, dy \right]$$

$$= 96 \left[\frac{0.8}{2} \left(\arcsin y + y\sqrt{1 - y^2} \right) + \frac{1}{3}(1 - y^2)^{3/2} \right]_{-1}^{0.8} \approx 96(1.263) \approx 121.3 \text{ lbs}$$

(b) $F_{\text{outside}} = 64 \int_{-1}^{0.4} (0.4 - y)(2)\sqrt{1 - y^2} \, dy$

$$= 128 \left[0.4 \int_{-1}^{0.4} \sqrt{1 - y^2} \, dy - \int_{-1}^{0.4} y\sqrt{1 - y^2} \, dy \right]$$

$$= 128 \left[\frac{0.4}{2} \left(\arcsin y + y\sqrt{1 - y^2} \right) + \frac{1}{3}(1 - y^2)^{3/2} \right]_{-1}^{0.4} \approx 92.98$$

67. (a) $m = \dfrac{dy}{dx} = \dfrac{y - \left(y + \sqrt{144 - x^2}\right)}{x - 0}$

$\qquad\qquad = -\dfrac{\sqrt{144 - x^2}}{x}$

(b) $y = -\displaystyle\int \dfrac{\sqrt{144 - x^2}}{x}\, dx$

Let $x = 12 \sin \theta$, $dx = 12 \cos \theta\, d\theta$, $\sqrt{144 - x^2} = 12 \cos \theta$.

$\quad y = -\displaystyle\int \dfrac{12 \cos \theta}{12 \sin \theta} 12 \cos \theta\, d\theta = -12 \int \dfrac{1 - \sin^2 \theta}{\sin \theta}\, d\theta$

$\qquad = -12 \displaystyle\int (\csc \theta - \sin \theta)\, d\theta = -12 \ln|\csc \theta - \cot \theta| - 12 \cos \theta + C$

$\qquad = -12 \ln\left|\dfrac{12}{x} - \dfrac{\sqrt{144 - x^2}}{x}\right| - 12\left(\dfrac{\sqrt{144 - x^2}}{12}\right) + C$

$\qquad = -12 \ln\left|\dfrac{12 - \sqrt{144 - x^2}}{x}\right| - \sqrt{144 - x^2} + C$

When $x = 12$, $y = 0 \implies C = 0$. Thus, $y = -12 \ln\left(\dfrac{12 - \sqrt{144 - x^2}}{x}\right) - \sqrt{144 - x^2}$.

Note: $\dfrac{12 - \sqrt{144 - x^2}}{x} > 0$ for $0 < x \le 12$

(c) Vertical asymptote: $x = 0$

(d) $y + \sqrt{144 - x^2} = 12 \implies y = 12 - \sqrt{144 - x^2}$

Thus,

$$12 - \sqrt{144 - x^2} = -12 \ln\left(\dfrac{12 - \sqrt{144 - x^2}}{x}\right) - \sqrt{144 - x^2}$$

$$-1 = \ln\left(\dfrac{12 - \sqrt{144 - x^2}}{x}\right)$$

$$xe^{-1} = 12 - \sqrt{144 - x^2}$$

$$(xe^{-1} - 12)^2 = \left(-\sqrt{144 - x^2}\right)^2$$

$$x^2 e^{-2} - 24xe^{-1} + 144 = 144 - x^2$$

$$x^2(e^{-2} + 1) - 24xe^{-1} = 0$$

$$x[x(e^{-2} + 1) - 24e^{-1}] = 0$$

$$x = 0 \text{ or } x = \dfrac{24e^{-1}}{e^{-2} + 1} \approx 7.77665.$$

Therefore,

$$s = \int_{7.77665}^{12} \sqrt{1 + \left(-\dfrac{\sqrt{144 - x^2}}{x}\right)^2}\, dx = \int_{7.77665}^{12} \sqrt{\dfrac{x^2 + (144 - x^2)}{x^2}}\, dx$$

68. $S = \sqrt{1445.6 + 228.5t - 4.4t^2}$

(a)

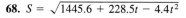
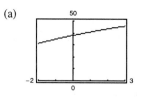

(b) $S'(t) = \frac{1}{2}(1445.6 + 228.5t - 4.4t^2)^{-1/2}(228.5 - 8.8t)$

$S'(1) \approx 2.69$

(c) average value $= \frac{1}{2}\int_8^{10} S(t)\,dt \approx 56.07$

69. True

$$\int \frac{dx}{\sqrt{1 - x^2}} = \int \frac{\cos\theta\,d\theta}{\cos\theta} = \int d\theta$$

70. False

$$\int \frac{\sqrt{x^2 - 1}}{x}\,dx = \int \frac{\tan\theta}{\sec\theta}(\sec\theta\tan\theta\,d\theta)$$

$$= \int \tan^2\theta\,d\theta$$

71. False

$$\int_0^{\sqrt{3}} \frac{dx}{\left(\sqrt{1 + x^2}\right)^3} = \int_0^{\pi/3} \frac{\sec^2\theta\,d\theta}{\sec^3\theta} = \int_0^{\pi/3} \cos\theta\,d\theta$$

72. True

$$\int_{-1}^{1} x^2\sqrt{1 - x^2}\,dx = 2\int_0^1 x^2\sqrt{1 - x^2}\,dx = 2\int_0^{\pi/2} (\sin^2\theta)(\cos\theta)(\cos\theta\,d\theta) = 2\int_0^{\pi/2} \sin^2\theta\cos^2\theta\,d\theta$$

73. Let $u = a\sin\theta,\ du = a\cos\theta\,d\theta,\ \sqrt{a^2 - u^2} = a\cos\theta$.

$$\int \sqrt{a^2 - u^2}\,du = \int a^2\cos^2\theta\,d\theta = a^2\int \frac{1 + \cos 2\theta}{2}\,d\theta$$

$$= \frac{a^2}{2}\left(\theta + \frac{1}{2}\sin 2\theta\right) + C = \frac{a^2}{2}(\theta + \sin\theta\cos\theta) + C$$

$$= \frac{a^2}{2}\left[\arcsin\frac{u}{a} + \left(\frac{u}{a}\right)\left(\frac{\sqrt{a^2 + u^2}}{a}\right)\right] + C = \frac{1}{2}\left[a^2\arcsin\frac{u}{a} + u\sqrt{a^2 - u^2}\right] + C$$

Let $u = a\sec\theta,\ du = a\sec\theta\tan\theta\,d\theta,\ \sqrt{u^2 - a^2} = a\tan\theta$.

$$\int \sqrt{u^2 - a^2}\,du = \int a\tan\theta(a\sec\theta\tan\theta)\,d\theta = a^2\int \tan^2\theta\sec\theta\,d\theta$$

$$= a^2\int (\sec^2\theta - 1)\sec\theta\,d\theta = a^2\int (\sec^3\theta - \sec\theta)\,d\theta$$

$$= a^2\left[\frac{1}{2}\sec\theta\tan\theta + \frac{1}{2}\int \sec\theta\,d\theta\right] - a^2\int \sec\theta\,d\theta = a^2\left[\frac{1}{2}\sec\theta\tan\theta - \frac{1}{2}\ln|\sec\theta + \tan\theta|\right]$$

$$= \frac{a^2}{2}\left[\frac{u}{a}\cdot\frac{\sqrt{u^2 - a^2}}{a} - \ln\left|\frac{u}{a} + \frac{\sqrt{u^2 - a^2}}{a}\right|\right] + C_1$$

$$= \frac{1}{2}\left[u\sqrt{u^2 - a^2} - a^2\ln|u + \sqrt{u^2 - a^2}|\right] + C$$

Let $u = a\tan\theta,\ du = a\sec^2\theta\,d\theta,\ \sqrt{u^2 + a^2} = a\sec\theta\,d\theta$.

$$\int \sqrt{u^2 + a^2}\,du = \int (a\sec\theta)(a\sec^2\theta)\,d\theta$$

$$= a^2\int \sec^3\theta\,d\theta = a^2\left[\frac{1}{2}\sec\theta\tan\theta + \frac{1}{2}\ln|\sec\theta + \tan\theta|\right] + C_1$$

$$= \frac{a^2}{2}\left[\frac{\sqrt{u^2 + a^2}}{a}\cdot\frac{u}{a} + \ln\left|\frac{\sqrt{u^2 + a^2}}{a} + \frac{u}{a}\right|\right] + C_1 = \frac{1}{2}\left[u\sqrt{u^2 + a^2} + a^2\ln|u + \sqrt{u^2 + a^2}|\right] + C$$

Section 7.5   Partial Fractions

1. $\dfrac{5}{x^2 - 10x} = \dfrac{5}{x(x - 10)} = \dfrac{A}{x} + \dfrac{B}{x - 10}$

2. $\dfrac{4x^2 + 3}{(x - 5)^3} = \dfrac{A}{x - 5} + \dfrac{B}{(x - 5)^2} + \dfrac{C}{(x - 5)^3}$

3. $\dfrac{2x - 3}{x^3 + 10x} = \dfrac{2x - 3}{x(x^2 + 10)} = \dfrac{A}{x} + \dfrac{Bx + C}{x^2 + 10}$

4. $\dfrac{x - 2}{x^2 + 4x + 3} = \dfrac{x - 2}{(x + 1)(x + 3)} = \dfrac{A}{x + 1} + \dfrac{B}{x + 3}$

5. $\dfrac{16x}{x^3 - 10x^2} = \dfrac{16x}{x^2(x - 10)} = \dfrac{A}{x} + \dfrac{B}{x^2} + \dfrac{C}{x - 10}$

6. $\dfrac{2x - 1}{x(x^2 + 1)^2} = \dfrac{A}{x} + \dfrac{Bx + C}{x^2 + 1} + \dfrac{Dx + E}{(x^2 + 1)^2}$

7. $\dfrac{1}{x^2 - 1} = \dfrac{1}{(x + 1)(x - 1)} = \dfrac{A}{x + 1} + \dfrac{B}{x - 1}$

$1 = A(x - 1) + B(x + 1)$

When $x = -1, 1 = -2A, A = -\frac{1}{2}$.

When $x = 1, 1 = 2B, B = \frac{1}{2}$.

$\displaystyle\int \dfrac{1}{x^2 - 1}\, dx = -\dfrac{1}{2}\int \dfrac{1}{x + 1}\, dx + \dfrac{1}{2}\int \dfrac{1}{x - 1}\, dx$

$\qquad = -\dfrac{1}{2}\ln|x + 1| + \dfrac{1}{2}\ln|x - 1| + C$

$\qquad = \dfrac{1}{2}\ln\left|\dfrac{x - 1}{x + 1}\right| + C$

8. $\dfrac{1}{4x^2 - 9} = \dfrac{1}{(2x - 3)(2x + 3)} = \dfrac{A}{2x - 3} + \dfrac{B}{2x + 3}$

$1 = A(2x + 3) + B(2x - 3)$

When $x = \frac{3}{2}, 1 = 6A, A = \frac{1}{6}$.

When $x = -\frac{3}{2}, 1 = -6B, B = -\frac{1}{6}$.

$\displaystyle\int \dfrac{1}{4x^2 - 9}\, dx = \dfrac{1}{6}\left[\int \dfrac{1}{2x - 3}\, dx - \int \dfrac{1}{2x + 3}\, dx\right]$

$\qquad = \dfrac{1}{12}[\ln|2x - 3| - \ln|2x + 3|] + C$

$\qquad = \dfrac{1}{12}\ln\left|\dfrac{2x - 3}{2x + 3}\right| + C$

9. $\dfrac{3}{x^2 + x - 2} = \dfrac{3}{(x - 1)(x + 2)} = \dfrac{A}{x - 1} + \dfrac{B}{x + 2}$

$3 = (x + 2) + B(x - 1)$

When $x = 1, 3 = 3A, A = 1$.

When $x = -2, 3 = -3B, B = -1$.

$\displaystyle\int \dfrac{3}{x^2 + x - 2}\, dx = \int \dfrac{1}{x - 1}\, dx - \int \dfrac{1}{x + 2}\, dx$

$\qquad = \ln|x - 1| - \ln|x + 2| + C$

$\qquad = \ln\left|\dfrac{x - 1}{x + 2}\right| + C$

10. $\displaystyle\int \dfrac{x + 1}{x^2 + 4x + 3}\, dx = \int \dfrac{1}{x + 3}\, dx = \ln|x + 3| + C$

11. $\dfrac{5 - x}{2x^2 + x - 1} = \dfrac{5 - x}{(2x - 1)(x + 1)} = \dfrac{A}{2x - 1} + \dfrac{B}{x + 1}$

$5 - x = A(x + 1) + B(2x - 1)$

When $x = \frac{1}{2}, \frac{9}{2} = \frac{3}{2}A, A = 3$. When $x = -1, 6 = -3B, B = -2$.

$\displaystyle\int \dfrac{5 - x}{2x^2 + x - 1}\, dx = 3\int \dfrac{1}{2x - 1}\, dx - 2\int \dfrac{1}{x + 1}\, dx$

$\qquad = \dfrac{3}{2}\ln|2x - 1| - 2\ln|x + 1| + C$

12. $\dfrac{3x^2 - 7x - 2}{x(x - 1)(x + 1)} = \dfrac{A}{x} + \dfrac{B}{x - 1} + \dfrac{C}{x + 1}$

$\quad\quad 3x^2 - 7x - 2 = A(x^2 - 1) + Bx(x + 1) + Cx(x - 1)$

When $x = 0$, $-2 = -A$, $A = 2$. When $x = 1$, $-6 = 2B$, $B = -3$. When $x = -1$, $8 = 2C$, $C = 4$.

$$\int \dfrac{3x^2 - 7x - 2}{x^3 - x} \, dx = 2\int \dfrac{1}{x}\, dx - 3\int \dfrac{1}{x - 1}\, dx + 4\int \dfrac{1}{x + 1}\, dx$$

$$= 2\ln|x| - 3\ln|x - 1| + 4\ln|x + 1| + C$$

13. $\dfrac{x^2 + 12x + 12}{x(x + 2)(x - 2)} = \dfrac{A}{x} + \dfrac{B}{x + 2} + \dfrac{C}{x - 2}$

$\quad\quad x^2 + 12x + 12 = A(x + 2)(x - 2) + Bx(x - 2) + Cx(x + 2)$

When $x = 0$, $12 = -4A$, $A = -3$. When $x = -2$, $-8 = 8B$, $B = -1$. When $x = 2$, $40 = 8C$, $C = 5$.

$$\int \dfrac{x^2 + 12x + 12}{x^3 - 4x} \, dx = 5\int \dfrac{1}{x - 2}\, dx - \int \dfrac{1}{x + 2}\, dx - 3\int \dfrac{1}{x}\, dx$$

$$= 5\ln|x - 2| - \ln|x + 2| - 3\ln|x| + C$$

14. $\dfrac{x^3 - x + 3}{x^2 + x - 2} = x - 1 + \dfrac{2x + 1}{(x + 2)(x - 1)} = x - 1 + \dfrac{A}{x + 2} + \dfrac{B}{x - 1}$

$\quad\quad 2x + 1 = A(x - 1) + B(x + 2)$

When $x = -2$, $-3 = -3A$, $A = 1$. When $x = 1$, $3 = 3B$, $B = 1$.

$$\int \dfrac{x^3 - x + 3}{x^2 + x - 2} \, dx = \int \left[x - 1 + \dfrac{1}{x + 2} + \dfrac{1}{x - 1} \right] dx$$

$$= \dfrac{x^2}{2} - x + \ln|x + 2| + \ln|x - 1| + C = \dfrac{x^2}{2} - x + \ln|x^2 + x - 2| + C$$

15. $\dfrac{2x^3 - 4x^2 - 15x + 5}{x^2 - 2x - 8} = 2x + \dfrac{x + 5}{(x - 4)(x + 2)} = 2x + \dfrac{A}{x - 4} + \dfrac{B}{x + 2}$

$\quad\quad x + 5 = A(x + 2) + B(x - 4)$

When $x = 4$, $9 = 6A$, $A = \frac{3}{2}$. When $x = -2$, $3 = -6B$, $B = -\frac{1}{2}$.

$$\int \dfrac{2x^3 - 4x^2 - 15x + 5}{x^2 - 2x - 8} \, dx = \int \left[2x + \dfrac{3/2}{x - 4} - \dfrac{1/2}{x + 2} \right] dx$$

$$= x^2 + \dfrac{3}{2}\ln|x - 4| - \dfrac{1}{2}\ln|x + 2| + C$$

16. $\dfrac{x + 2}{x(x - 4)} = \dfrac{A}{x - 4} + \dfrac{B}{x}$

$\quad\quad x + 2 = Ax + B(x - 4)$

When $x = 4$, $6 = 4A$, $A = \frac{3}{2}$.

When $x = 0$, $2 = -4B$, $B = -\frac{1}{2}$.

$$\int \dfrac{x + 2}{x^2 - 4x} \, dx = \int \left[\dfrac{3/2}{x - 4} - \dfrac{1/2}{x} \right] dx$$

$$= \dfrac{3}{2}\ln|x - 4| - \dfrac{1}{2}\ln|x| + C$$

17. $\dfrac{4x^2 + 2x - 1}{x^2(x + 1)} = \dfrac{A}{x} + \dfrac{B}{x^2} + \dfrac{C}{x + 1}$

$\quad\quad 4x^2 + 2x - 1 = Ax(x + 1) + B(x + 1) + Cx^2$

When $x = 0$, $B = -1$. When $x = -1$, $C = 1$.

When $x = 1$, $A = 3$.

$$\int \dfrac{4x^2 + 2x - 1}{x^3 + x^2} \, dx = \int \left[\dfrac{3}{x} - \dfrac{1}{x^2} + \dfrac{1}{x + 1} \right] dx$$

$$= 3\ln|x| + \dfrac{1}{x} + \ln|x + 1| + C$$

$$= \dfrac{1}{x} + \ln|x^4 + x^3| + C$$

18. $\dfrac{2x-3}{(x-1)^2} = \dfrac{A}{x-1} + \dfrac{B}{(x-1)^2}$

$\quad 2x - 3 = A(x-1) + B$

When $x = 1, B = -1$. When $x = 0, A = 2$.

$$\int \frac{2x-3}{(x-1)^2}\,dx = \int \left[\frac{2}{x-1} - \frac{1}{(x-1)^2} \right]\,dx$$

$$= 2\ln|x-1| + \frac{1}{x-1} + C$$

19. $\dfrac{x^2-1}{x(x^2+1)} = \dfrac{A}{x} + \dfrac{Bx+C}{x^2+1}$

$\quad x^2 - 1 = A(x^2+1) + (Bx+C)x$

When $x = 0, A = -1$. When $x = 1, 0 = -2 + B + C$.
When $x = -1, 0 = -2 + B + C$. Solving these
equations we have $A = -1, B = 2, C = 0$.

$$\int \frac{x^2-1}{x^3+x}\,dx = -\int \frac{1}{x}\,dx + \int \frac{2x}{x^2+1}\,dx$$

$$= \ln|x^2+1| - \ln|x| + C$$

$$= \ln\left| \frac{x^2+1}{x} \right| + C$$

20. $\dfrac{x}{(x-1)(x^2+x+1)} = \dfrac{A}{x-1} + \dfrac{Bx+C}{x^2+x+1}$

$$x = A(x^2+x+1) + (Bx+C)(x-1)$$

When $x = 1, 1 = 3A$. When $x = 0, 0 = A - C$. When $x = -1, -1 = A + 2B - 2C$. Solving these equations we have
$A = \frac{1}{3}, B = -\frac{1}{3}, C = \frac{1}{3}$.

$$\int \frac{x}{x^3-1}\,dx = \frac{1}{3} \left[\int \frac{1}{x-1}\,dx + \int \frac{-x+1}{x^2+x+1}\,dx \right]$$

$$= \frac{1}{3} \left[\int \frac{1}{x-1}\,dx - \frac{1}{2}\int \frac{2x+1}{x^2+x+1}\,dx + \frac{3}{2}\int \frac{1}{[x+(1/2)]^2+(3/4)}\,dx \right]$$

$$= \frac{1}{3} \left[\ln|x-1| - \frac{1}{2}\ln|x^2+x+1| + \sqrt{3}\arctan\left(\frac{2x+1}{\sqrt{3}}\right) \right] + C$$

21. $\dfrac{x^2}{x^4-2x^2-8} = \dfrac{A}{x-2} + \dfrac{B}{x+2} + \dfrac{Cx+D}{x^2+2}$

$$x^2 = A(x+2)(x^2+2) + B(x-2)(x^2+2) + (Cx+D)(x+2)(x-2)$$

When $x = 2, 4 = 24A$. When $x = -2, 4 = -24B$. When $x = 0, 0 = 4A - 4B - 4D$, and when $x = 1$,
$1 = 9A - 3B - 3C - 3D$. Solving these equations we have $A = \frac{1}{6}, B = -\frac{1}{6}, C = 0, D = \frac{1}{3}$.

$$\int \frac{x^2}{x^4-2x^2-8}\,dx = \frac{1}{6} \left[\int \frac{1}{x-2}\,dx - \int \frac{1}{x+2}\,dx + 2\int \frac{1}{x^2+2}\,dx \right]$$

$$= \frac{1}{6} \left[\ln\left| \frac{x-2}{x+2} \right| + \sqrt{2}\arctan\frac{x}{\sqrt{2}} \right] + C$$

22. $\dfrac{2x^2+x+8}{(x^2+4)^2} = \dfrac{Ax+B}{x^2+4} + \dfrac{Cx+D}{(x^2+4)^2}$

$\quad 2x^2 + x + 8 = (Ax+B)(x^2+4) + Cx + D$

$$= Ax^3 + Bx^2 + (4A+C)x + (4B+D)$$

By equating the coefficients of like terms we have $A = 0, B = 2, 4A + C = 1, 4B + D = 8$. Solving these equations we have
$A = 0, B = 2, C = 1, D = 0$.

$$\int \frac{2x^2+x+8}{(x^2+4)^2}\,dx = 2\int \frac{1}{x^2+4}\,dx + \frac{1}{2}\int \frac{2x}{(x^2+4)^2}\,dx$$

$$= \arctan\frac{x}{2} - \frac{1}{2(x^2+4)} + C$$

23. $\dfrac{x}{(2x-1)(2x+1)(4x^2+1)} = \dfrac{A}{2x-1} + \dfrac{B}{2x+1} + \dfrac{Cx+D}{4x^2+1}$

$$x = A(2x+1)(4x^2+1) + B(2x-1)(4x^2+1) + (Cx+D)(2x-1)(2x+1)$$

When $x = \frac{1}{2}, \frac{1}{2} = 4A$. When $x = -\frac{1}{2}, -\frac{1}{2} = -4B$. When $x = 0, 0 = A - B - D$, and when $x = 1$, $1 = 15A + 5B + 3C + 3D$. Solving these equations we have $A = \frac{1}{8}, B = \frac{1}{8}, C = -\frac{1}{2}, D = 0$.

$$\int \frac{x}{16x^4 - 1}\,dx = \frac{1}{8}\left[\int \frac{1}{2x-1}\,dx + \int \frac{1}{2x+1}\,dx - 4\int \frac{x}{4x^2+1}\,dx\right]$$

$$= \frac{1}{16}\ln\left|\frac{4x^2-1}{4x^2+1}\right| + C$$

24. $\dfrac{x^2 - 4x + 7}{(x+1)(x^2-2x+3)} = \dfrac{A}{x+1} + \dfrac{Bx+C}{x^2-2x+3}$

$$x^2 - 4x + 7 = A(x^2 - 2x + 3) + (Bx + C)(x + 1)$$

When $x = -1, 12 = 6A$. When $x = 0, 7 = 3A + C$. When $x = 1, 4 = 2A + 2B + 2C$. Solving these equations we have $A = 2, B = -1, C = 1$.

$$\int \frac{x^2 - 4x + 7}{x^3 - x^2 + x + 3}\,dx = 2\int \frac{1}{x+1}\,dx + \int \frac{-x+1}{x^2-2x+3}\,dx$$

$$= 2\ln|x+1| - \frac{1}{2}\ln|x^2 - 2x + 3| + C$$

25. $\dfrac{x^2 + 5}{(x+1)(x^2-2x+3)} = \dfrac{A}{x+1} + \dfrac{Bx+C}{x^2-2x+3}$

$$x^2 + 5 = A(x^2 - 2x + 3) + (Bx + C)(x + 1)$$

$$= (A + B)x^2 + (-2A + B + C)x + (3A + C)$$

When $x = -1, A = 1$. By equating coefficients of like terms, we have $A + B = 1, -2A + B + C = 0, 3A + C = 5$. Solving these equations we have $A = 1, B = 0, C = 2$.

$$\int \frac{x^2 + 5}{x^3 - x^2 + x + 3}\,dx = \int \frac{1}{x+1}\,dx + 2\int \frac{1}{(x-1)^2 + 2}\,dx$$

$$= \ln|x+1| + \sqrt{2}\arctan\left(\frac{x-1}{\sqrt{2}}\right) + C$$

26. $\dfrac{x^2 + x + 3}{(x^2+3)^2} = \dfrac{Ax+B}{x^2+3} + \dfrac{Cx+D}{(x^2+3)^2}$

$$x^2 + x + 3 = (Ax + B)(x^2 + 3) + Cx + D$$

$$= Ax^3 + Bx^2 + (3A + C)x + (3B + D)$$

By equating coefficients of like terms, we have $A = 0$, $B = 1, 3A + C = 1, 3B + D = 3$. Solving these equations we have $A = 0, B = 1, C = 1, D = 0$.

$$\int \frac{x^2 + x + 3}{x^4 + 6x^2 + 9}\,dx = \int \left[\frac{1}{x^2+3} + \frac{x}{(x^2+3)^2}\right]dx$$

$$= \frac{1}{\sqrt{3}}\arctan\frac{x}{\sqrt{3}} - \frac{1}{2(x^2+3)} + C$$

27. $\dfrac{3}{(2x+1)(x+2)} = \dfrac{A}{2x+1} + \dfrac{B}{x+2}$

$$3 = A(x + 2) + B(2x + 1)$$

When $x = -\frac{1}{2}, A = 2$. When $x = -2, B = -1$.

$$\int_0^1 \frac{3}{2x^2 + 5x + 2}\,dx = \int_0^1 \frac{2}{2x+1}\,dx - \int_0^1 \frac{1}{x+2}\,dx$$

$$= \left[\ln|2x-1| - \ln|x+2|\right]_0^1$$

$$= \ln 2$$

28. $\dfrac{x-1}{x^2(x+1)} = \dfrac{A}{x} + \dfrac{B}{x^2} + \dfrac{C}{x+1}$

$x - 1 = Ax(x+1) + B(x+1) + Cx^2$

When $x = 0$, $B = -1$. When $x = -1$, $C = -2$. When $x = 1$, $0 = 2A + 2B + C$. Solving these equations we have $A = 2$, $B = -1$, $C = -2$.

$$\int_1^5 \frac{x-1}{x^2(x+1)}\,dx = 2\int_1^5 \frac{1}{x}\,dx - \int_1^5 \frac{1}{x^2}\,dx - 2\int_1^5 \frac{1}{x+1}\,dx$$

$$= \left[2\ln|x| + \frac{1}{x} - 2\ln|x+1|\right]_1^5$$

$$= \left[2\ln\left|\frac{x}{x+1}\right| + \frac{1}{x}\right]_1^5$$

$$= 2\ln\frac{5}{3} - \frac{4}{5}$$

29. $\dfrac{x+1}{x(x^2+1)} = \dfrac{A}{x} + \dfrac{Bx+C}{x^2+1}$

$x + 1 = A(x^2+1) + (Bx+C)x$

When $x = 0$, $A = 1$. When $x = 1$, $2 = 2A + B + C$. When $x = -1$, $0 = 2A + B - C$. Solving these equations we have $A = 1$, $B = -1$, $C = 1$.

$$\int_1^2 \frac{x+1}{x(x^2+1)}\,dx = \int_1^2 \frac{1}{x}\,dx - \int_1^2 \frac{x}{x^2+1}\,dx + \int_1^2 \frac{1}{x^2+1}\,dx$$

$$= \left[\ln|x| - \frac{1}{2}\ln(x^2+1) + \arctan x\right]_1^2$$

$$= \frac{1}{2}\ln\frac{8}{5} - \frac{\pi}{4} + \arctan 2$$

$$\approx 0.557$$

30. $\displaystyle\int_0^1 \frac{x^2-x}{x^2+x+1}\,dx = \int_0^1 dx - \int_0^1 \frac{2x+1}{x^2+x+1}\,dx$

$$= \left[x - \ln|x^2+x+1|\right]_0^1$$

$$= 1 - \ln 3$$

31. $\displaystyle\int \frac{3x\,dx}{x^2-6x+9} = 3\ln|x-3| - \frac{9}{x-3} + C$

$(4, 0)$: $3\ln|4-3| - \dfrac{9}{4-3} + C = 0 \Rightarrow C = 9$

32. $\displaystyle\int \frac{6x^2+1}{x^2(x-1)^3}\,dx = 3\ln\left|\frac{x-1}{x}\right| + \frac{1}{x} + \frac{2}{x-1} - \frac{7}{2(x-1)^2} + C$

$(2, 1)$: $3\ln\left|\dfrac{1}{2}\right| + \dfrac{1}{2} + \dfrac{2}{1} - \dfrac{7}{2} + C = 1 \Rightarrow C = 2 - 3\ln\dfrac{1}{2}$

33. $\int \dfrac{x^2 + x + 2}{(x^2 + 2)^2}\,dx = \dfrac{\sqrt{2}}{2}\arctan\dfrac{x}{\sqrt{2}} - \dfrac{1}{2(x^2 + 2)} + C$

$(0, 1): \ 0 - \dfrac{1}{4} + C = 1 \Rightarrow C = \dfrac{5}{4}$

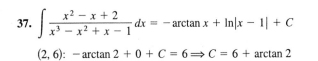

34. $\int \dfrac{x^3}{(x^2 - 4)^2}\,dx = \dfrac{1}{2}\ln|x^2 - 4| - \dfrac{2}{x^2 - 4} + C$

$(3, 4): \ \dfrac{1}{2}\ln 5 - \dfrac{2}{5} + C = 4 \Rightarrow C = \dfrac{22}{5} - \dfrac{1}{2}\ln 5$

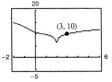

35. $\int \dfrac{2x^2 - 2x + 3}{x^3 - x^2 - x - 2}\,dx = \ln|x - 2| + \dfrac{1}{2}\ln|x^2 + x + 1| - \sqrt{3}\arctan\left(\dfrac{2x + 1}{\sqrt{3}}\right) + C$

$(3, 10): \ 0 + \dfrac{1}{2}\ln 13 - \sqrt{3}\arctan\dfrac{7}{\sqrt{3}} + C = 10 \Rightarrow C = 10 - \dfrac{1}{2}\ln 13 + \sqrt{3}\arctan\dfrac{7}{\sqrt{3}}$

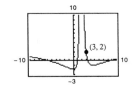

36. $\int \dfrac{x(2x - 9)}{x^3 - 6x^2 + 12x - 8}\,dx = 2\ln|x - 2| + \dfrac{1}{x - 2} + \dfrac{5}{(x - 2)^2} + C$

$(3, 2): \ 0 + 1 + 5 + C = 2 \Rightarrow C = -4$

37. $\int \dfrac{x^2 - x + 2}{x^3 - x^2 + x - 1}\,dx = -\arctan x + \ln|x - 1| + C$

$(2, 6): \ -\arctan 2 + 0 + C = 6 \Rightarrow C = 6 + \arctan 2$

38. $\int \dfrac{1}{x^2 - 4}\,dx = \dfrac{1}{4}\ln\left|\dfrac{x - 2}{x + 2}\right| + C$

$(6, 4): \ \dfrac{1}{4}\ln\left|\dfrac{4}{8}\right| + C = 4 \Rightarrow C = 4 - \dfrac{1}{4}\ln\dfrac{1}{2} = 4 + \dfrac{1}{4}\ln 2$

39. Let $u = \cos x$ $du = -\sin x\,dx$.

$$\dfrac{1}{u(u - 1)} = \dfrac{A}{u} + \dfrac{B}{u - 1}$$

$$1 = A(u - 1) + Bu$$

When $u = 0, A = -1$. When $u = 1, B = 1, u = \cos x$, $du = -\sin x\,dx$.

$$\int \dfrac{\sin x}{\cos x(\cos x - 1)}\,dx = -\int \dfrac{1}{u(u - 1)}\,du$$

$$= \int \dfrac{1}{u}\,du - \int \dfrac{1}{u - 1}\,du$$

$$= \ln|u| - \ln|u - 1| + C$$

$$= \ln\left|\dfrac{u}{u - 1}\right| + C$$

$$= \ln\left|\dfrac{\cos x}{\cos x - 1}\right| + C$$

40. Let $u = \cos x$, $du = \sin x\,dx$.

$$\dfrac{1}{u(u + 1)} = \dfrac{A}{u} + \dfrac{B}{u + 1}$$

$$1 = A(u + 1) + Bu$$

When $u = 0, A = 1$. When $u = -1, B = -1, u = \cos x$. $du = -\sin dx$.

$$\int \dfrac{\sin x}{\cos x + \cos^2 x}\,dx = -\int \dfrac{1}{u(u + 1)}\,du$$

$$= \int \dfrac{1}{u + 1}\,du - \int \dfrac{1}{u}\,du$$

$$= \ln|u + 1| - \ln|u| + C$$

$$= \ln\left|\dfrac{u + 1}{u}\right| + C$$

$$= \ln\left|\dfrac{\cos x + 1}{\cos x}\right| + C$$

$$= \ln|1 + \sec x| + C$$

41. $\displaystyle\int \frac{3\cos x}{\sin^2 x + \sin x - 2}\,dx = 3\int \frac{1}{u^2 + u - 2}\,du$

$$= \ln\left|\frac{u-1}{u+2}\right| + C$$

$$= \ln\left|\frac{-1 + \sin x}{2 + \sin x}\right| + C$$

(From Exercise 9 with $u = \sin x$, $du = \cos x\,dx$)

42. $\displaystyle\frac{1}{u(u+1)} = \frac{A}{u} + \frac{B}{u+1}$, $u = \tan x$, $du = \sec^2 x\,dx$

$$1 = A(u + 1) + Bu$$

When $u = 0$, $A = 1$.
When $u = -1$, $1 = -B \Rightarrow B = -1$.

$$\int \frac{\sec^2 x\,dx}{\tan x(\tan x + 1)} = \int \frac{1}{u(u+1)}\,du$$

$$= \int \left(\frac{1}{u} - \frac{1}{u+1}\right) du$$

$$= \ln|u| - \ln|u+1| + C$$

$$= \ln\left|\frac{u}{u+1}\right| + C$$

$$= \ln\left|\frac{\tan x}{\tan x + 1}\right| + C$$

43. Let $u = e^x$, $du = e^x\,dx$.

$$\frac{1}{(u-1)(u+4)} = \frac{A}{u-1} + \frac{B}{u+4}$$

$$1 = A(u+4) + B(u-1)$$

When $u = 1$, $A = \frac{1}{5}$.
When $u = -4$, $B = -\frac{1}{5}$, $u = e^x$, $du = e^x\,dx$.

$$\int \frac{e^x}{(e^x - 1)(e^x + 4)}\,dx = \int \frac{1}{(u-1)(u+4)}\,du$$

$$= \frac{1}{5}\left(\int \frac{1}{u-1}\,du - \int \frac{1}{u+4}\,du\right)$$

$$= \frac{1}{5}\ln\left|\frac{u-1}{u+4}\right| + C$$

$$= \frac{1}{5}\ln\left|\frac{e^x - 1}{e^x + 4}\right| + C$$

44. Let $u = e^x$, $du = e^x\,dx$.

$$\frac{1}{(u^2+1)(u-1)} = \frac{A}{u-1} + \frac{Bu + C}{u^2 + 1}$$

$$1 = A(u^2 + 1) + (Bu + C)(u - 1)$$

When $u = 1$, $A = \frac{1}{2}$.
When $u = 0$, $1 = A - C$.
When $u = -1$, $1 = 2A + 2B - 2C$. Solving these equations we have $A = \frac{1}{2}$, $B = -\frac{1}{2}$, $C = -\frac{1}{2}$, $u = e^x$, $du = e^x\,dx$.

$$\int \frac{e^x}{(e^{2x} + 1)(e^x - 1)}\,dx = \int \frac{1}{(u^2 + 1)(u - 1)}\,du$$

$$= \frac{1}{2}\left(\int \frac{1}{u-1}\,du - \int \frac{u+1}{u^2+1}\,du\right)$$

$$= \frac{1}{2}\left(\ln|u-1| - \frac{1}{2}\ln|u^2 + 1| - \arctan u\right) + C$$

$$= \frac{1}{4}\left(2\ln|e^x - 1| - \ln|e^{2x} + 1| - 2\arctan e^x\right) + C$$

45. $\dfrac{1}{x(a + bx)} = \dfrac{A}{x} + \dfrac{B}{a + bx}$

$$1 = A(a + bx) + Bx$$

When $x = 0, 1 = aA \Rightarrow A = 1/a$.
When $x = -a/b, 1 = -(a/b)B \Rightarrow B = -b/a$.

$$\int \frac{1}{x(a + bx)}\, dx = \frac{1}{a}\int \left(\frac{1}{x} - \frac{b}{a + bx}\right) dx$$

$$= \frac{1}{a}\left(\ln|x| - \ln|a + bx|\right) + C$$

$$= \frac{1}{a}\ln\left|\frac{x}{a + bx}\right| + C$$

46. $\dfrac{1}{a^2 - x^2} = \dfrac{A}{a - x} + \dfrac{B}{a + x}$

$$1 = A(a + x) + B(a - x)$$

When $x = a, A = 1/2a$.
When $x = -a, B = 1/2a$.

$$\int \frac{1}{a^2 - x^2}\, dx = \frac{1}{2a}\int \left(\frac{1}{a - x} + \frac{1}{a + x}\right) dx$$

$$= \frac{1}{2a}\left(-\ln|a - x| + \ln|a + x|\right) + C$$

$$= \frac{1}{2a}\ln\left|\frac{a + x}{a - x}\right| + C$$

47. $\dfrac{x}{(a + bx)^2} = \dfrac{A}{a + bx} + \dfrac{B}{(a + bx)^2}$

$$x = A(a + bx) + B$$

When $x = -a/b, B = -a/b$.
When $x = 0, 0 = aA + B \Rightarrow A = 1/b$.

$$\int \frac{x}{(a + bx)^2}\, dx = \int \left(\frac{1/b}{a + bx} + \frac{-a/b}{(a + bx)^2}\right) dx$$

$$= \frac{1}{b}\int \frac{1}{a + bx}\, dx - \frac{a}{b}\int \frac{1}{(a + bx)^2}\, dx$$

$$= \frac{1}{b^2}\ln|a + bx| + \frac{a}{b^2}\left(\frac{1}{a + bx}\right) + C$$

$$= \frac{1}{b^2}\left(\frac{a}{a + bx} + \ln|a + bx|\right) + C$$

48. $\dfrac{1}{x^2(a + bx)} = \dfrac{A}{x} + \dfrac{B}{x^2} + \dfrac{C}{a + bx}$

$$1 = Ax(a + bx) + B(a + bx) + Cx^2$$

When $x = 0, 1 = Ba \Rightarrow B = 1/a$.
When $x = -a/b, 1 = C(a^2/b^2) \Rightarrow C = b^2/a^2$.
When $x = 1, 1 = (a + b)A + (a + b)B + C \Rightarrow$
$\qquad A = -b/a^2$.

$$\int \frac{1}{x^2(a + bx)}\, dx = \int \left(\frac{-b/a^2}{x} + \frac{1/a}{x^2} + \frac{b^2/a^2}{a + bx}\right) dx$$

$$= -\frac{b}{a^2}\ln|x| - \frac{1}{ax} + \frac{b}{a^2}\ln|a + bx| + C$$

$$= -\frac{1}{ax} + \frac{b}{a^2}\ln\left|\frac{a + bx}{x}\right| + C$$

$$= -\frac{1}{ax} - \frac{b}{a^2}\ln\left|\frac{x}{a + bx}\right| + C$$

49. $A = \displaystyle\int_1^3 \frac{10}{x(x^2 + 1)}\, dx \approx 3$

Matches (c)

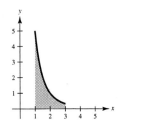

50. $A = 2\displaystyle\int_0^3 \left(1 - \frac{7}{16 - x^2}\right) dx = 2\int_0^3 dx - 14\int_0^3 \frac{1}{16 - x^2}\, dx$

$$= \left[2x - \frac{14}{8}\ln\left|\frac{4 + x}{4 - x}\right|\right]_0^3 \quad \text{(From Exercise 46)}$$

$$= 6 - \frac{7}{4}\ln 7 \approx 2.595$$

51. (a) $V = \pi\displaystyle\int_0^3 \left(\frac{2x}{x^2 + 1}\right)^2 dx = 4\pi\int_0^3 \frac{x^2}{(x^2 + 1)^2}\, dx$

$$= 4\pi\int_0^3 \left(\frac{1}{x^2 + 1} - \frac{1}{(x^2 + 1)^2}\right) dx \qquad \text{(partial fractions)}$$

$$= 4\pi\left[\arctan x - \frac{1}{2}\left(\arctan x + \frac{x}{x^2 + 1}\right)\right]_0^3 \quad \text{(trigonometric substitution)}$$

$$= 2\pi\left[\arctan x - \frac{x}{x^2 + 1}\right]_0^3 = 2\pi\left[\arctan 3 - \frac{3}{10}\right] \approx 5.963$$

—CONTINUED—

51. —CONTINUED—

(b) $A = \int_0^3 \frac{2x}{x^2 + 1}\, dx = \left[\ln(x^2 + 1)\right]_0^3 = \ln 10$

$\bar{x} = \frac{1}{A}\int_0^3 \frac{2x^2}{x^2 + 1}\, dx = \frac{1}{\ln 10}\int_0^3 \left(2 - \frac{2}{x^2 + 1}\right) dx$

$= \frac{1}{\ln 10}\left[2x - 2\arctan x\right]_0^3 = \frac{2}{\ln 10}[3 - \arctan 3] \approx 1.521$

$\bar{y} = \frac{1}{A}\left(\frac{1}{2}\right)\int_0^3 \left(\frac{2x}{x^2 + 1}\right)^2 dx = \frac{2}{\ln 10}\int_0^3 \frac{x^2}{(x^2 + 1)^2}\, dx$

$= \frac{2}{\ln 10}\int_0^3 \left(\frac{1}{x^2 + 1} - \frac{1}{(x^2 + 1)^2}\right) dx$ (partial fractions)

$= \frac{2}{\ln 10}\left[\arctan x - \frac{1}{2}\left(\arctan x + \frac{x}{x^2 + 1}\right)\right]_0^3$ (trigonometric substitution)

$= \frac{2}{\ln 10}\left[\frac{1}{2}\arctan x - \frac{x}{2(x^2 + 1)}\right]_0^3 = \frac{1}{\ln 10}\left[\arctan x - \frac{x}{x^2 + 1}\right]_0^3 = \frac{1}{\ln 10}\left[\arctan 3 - \frac{3}{10}\right] \approx 0.412$

$(\bar{x}, \bar{y}) \approx (1.521, 0.412)$

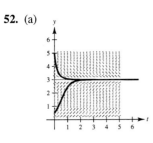

(1.521, 0.412)

52. (a)

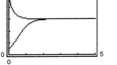

(b) The slope is negative because the function is decreasing.

(c) For $y > 0$, $\lim_{t \to \infty} y(t) = 3$.

(e) $k = 1$, $L = 3$

    (i) $y(0) = 5$: $y = \dfrac{15}{5 - 2e^{-3t}}$

    (ii) $y(0) = \dfrac{1}{2}$: $y = \dfrac{3/2}{(1/2) + (5/2)e^{-3t}} = \dfrac{3}{1 + 5e^{-3t}}$

(f) $\dfrac{dy}{dt} = ky(L - y)$

$\dfrac{d^2y}{dt^2} = k\left[y\left(\dfrac{-dy}{dt}\right) + (L - y)\dfrac{dy}{dt}\right] = 0$

$\Rightarrow y\dfrac{dy}{dt} = (L - y)\dfrac{dy}{dt}$

$\Rightarrow y = \dfrac{L}{2}$

From the first derivative test, this is a maximum.

(d) $\dfrac{dy}{y(L - y)} = \dfrac{A}{y} + \dfrac{B}{L - y}$

$1 = A(L - y) + By \Rightarrow A = \dfrac{1}{L},\ B = \dfrac{1}{L}$

$\displaystyle\int \dfrac{dy}{y(L - y)} = \int k\, dt$

$\dfrac{1}{L}\left[\displaystyle\int \dfrac{1}{y}dy + \int \dfrac{1}{L - y}dy\right] = \int k\, dt$

$\dfrac{1}{L}\left[\ln|y| - \ln|L - y|\right] = kt + C_1$

$\ln\left|\dfrac{y}{L - y}\right| = kLt + LC_1$

$C_2 e^{kLt} = \dfrac{y}{L - y}$

When $t = 0$, $\dfrac{y_0}{L - y_0} = C_2 \Rightarrow \dfrac{y}{L - y} = \dfrac{y_0}{L - y_0}e^{kLt}$.

Solving for y, you obtain $y = \dfrac{y_0 L}{y_0 + (L - y_0)e^{-kLt}}$.

53.
$$\frac{1}{(x+1)(n-x)} = \frac{A}{x+1} + \frac{B}{n-x}, A = B = \frac{1}{n+1}$$

$$\frac{1}{n+1}\int\left(\frac{1}{x+1} + \frac{1}{n-x}\right)dx = kt + C$$

$$\frac{1}{n+1}\ln\left|\frac{x+1}{n-x}\right| = kt + C$$

When $t = 0, x = 0, C = \frac{1}{n+1}\ln\frac{1}{n}$.

$$\frac{1}{n+1}\ln\left|\frac{x+1}{n-x}\right| = kt + \frac{1}{n+1}\ln\frac{1}{n}$$

$$\frac{1}{n+1}\left[\ln\left|\frac{x+1}{n-x}\right| - \ln\frac{1}{n}\right] = kt$$

$$\ln\frac{nx+n}{n-x} = (n+1)kt$$

$$\frac{nx+n}{n-x} = e^{(n+1)kt}$$

$$x = \frac{n\left[e^{(n+1)kt} - 1\right]}{n + e^{(n+1)kt}} \qquad \textbf{Note: } \lim_{t\to\infty} x = n$$

54. (a)
$$\frac{1}{(y_0 - x)(z_0 - x)} = \frac{A}{y_0 - x} + \frac{B}{z_0 - x}, A = \frac{1}{z_0 - y_0}, B = -\frac{1}{z_0 - y_0} \qquad \text{(Assume } y_0 \neq z_0\text{)}$$

$$\frac{1}{z_0 - y_0}\int\left(\frac{1}{y_0 - x} - \frac{1}{z_0 - x}\right)dx = kt + C$$

$$\frac{1}{z_0 - y_0}\ln\left|\frac{z_0 - x}{y_0 - x}\right| = kt + C, \text{ when } t = 0, x = 0$$

$$C = \frac{1}{z_0 - y_0}\ln\frac{z_0}{y_0}$$

$$\frac{1}{z_0 - y_0}\left[\ln\left|\frac{z_0 - x}{y_0 - x}\right| - \ln\left(\frac{z_0}{y_0}\right)\right] = kt$$

$$\ln\left[\frac{y_0(z_0 - x)}{z_0(y_0 - x)}\right] = (z_0 - y_0)kt$$

$$\frac{y_0(z_0 - x)}{z_0(y_0 - x)} = e^{(z_0 - y_0)kt}$$

$$x = \frac{y_0 z_0\left[e^{(z_0 - y_0)kt} - 1\right]}{z_0 e^{(z_0 - y_0)kt} - y_0}$$

(b) (1) If $y_0 < z_0$, $\lim_{t\to\infty} x = y_0$.

(2) If $y_0 > z_0$, $\lim_{t\to\infty} x = z_0$.

(3) If $y_0 = z_0$, then the original equation is

$$\int\frac{1}{(y_0 - x)^2}dx = \int k\,dt$$

$$(y_0 - x)^{-1} = kt + C_1$$

$$x = 0 \text{ when } t = 0 \implies \frac{1}{y_0} = C_1$$

$$\frac{1}{y_0 - x} = kt + \frac{1}{y_0} = \frac{kt\,y_0 + 1}{y_0}$$

$$y_0 - x = \frac{y_0}{kt\,y_0 + 1}$$

$$x = y_0 - \frac{y_0}{kt\,y_0 + 1}$$

As $t\to\infty$, $x\to y_0 = x_0$.

55. $\dfrac{x}{1 + x^4} = \dfrac{Ax + B}{x^2 + \sqrt{2}\,x + 1} + \dfrac{Cx + D}{x^2 - \sqrt{2}\,x + 1}$

$$x = (Ax + B)(x^2 - \sqrt{2}\,x + 1) + (Cx + D)(x^2 + \sqrt{2}\,x + 1)$$

$$= (A + C)x^3 + (B + D - \sqrt{2}\,A + \sqrt{2}\,C)x^2 + (A + C - \sqrt{2}\,B + \sqrt{2}\,D)x + (B + D)$$

$0 = A + C \implies C = -A$

$\left. \begin{array}{l} 0 = B + D - \sqrt{2}\,A + \sqrt{2}\,C \\[2mm] 1 = A + C - \sqrt{2}\,B + \sqrt{2}\,D \end{array} \right\}$ $\begin{array}{l} -2\sqrt{2}\,A = 0 \implies A = 0 \text{ and } C = 0 \\[2mm] -2\sqrt{2}\,B = 1 \implies B = -\dfrac{\sqrt{2}}{4} \text{ and } D = \dfrac{\sqrt{2}}{4} \end{array}$

$0 = B + D \implies D = -B$

Thus,

$$\int_0^1 \frac{x}{1 + x^4}\, dx = \int_0^1 \left[\frac{-\sqrt{2}/4}{x^2 + \sqrt{2}\,x + 1} + \frac{\sqrt{2}/4}{x^2 - \sqrt{2}\,x + 1} \right] dx$$

$$= \frac{\sqrt{2}}{4} \int_0^1 \left[\frac{-1}{[x + (\sqrt{2}/2)]^2 + (1/2)} + \frac{1}{[x - (\sqrt{2}/2)]^2 + (1/2)} \right] dx$$

$$= \frac{\sqrt{2}}{4} \cdot \frac{1}{1/\sqrt{2}} \left[-\arctan\left(\frac{x + (\sqrt{2}/2)}{1/\sqrt{2}} \right) + \arctan\left(\frac{x - (\sqrt{2}/2)}{1/\sqrt{2}} \right) \right]_0^1$$

$$= \frac{1}{2} \left[-\arctan\left(\sqrt{2}\,x + 1 \right) + \arctan\left(\sqrt{2}\,x - 1 \right) \right]_0^1$$

$$= \frac{1}{2} \left[\left(-\arctan\left(\sqrt{2} + 1 \right) + \arctan\left(\sqrt{2} - 1 \right) \right) - \left(-\arctan 1 + \arctan(-1) \right) \right]$$

$$= \frac{1}{2} \left[\arctan\left(\sqrt{2} - 1 \right) - \arctan\left(\sqrt{2} + 1 \right) + \frac{\pi}{4} + \frac{\pi}{4} \right].$$

Since $\arctan x - \arctan y = \arctan[(x - y)/(1 + xy)]$, we have:

$$\int_0^1 \frac{x}{1 + x^4}\, dx = \frac{1}{2} \left[\arctan\left(\frac{(\sqrt{2} - 1) - (\sqrt{2} + 1)}{1 + (\sqrt{2} - 1)(\sqrt{2} + 1)} \right) + \frac{\pi}{2} \right] = \frac{1}{2} \left[\arctan\left(\frac{-2}{2} \right) + \frac{\pi}{2} \right] = \frac{1}{2} \left[-\frac{\pi}{4} + \frac{\pi}{2} \right] = \frac{\pi}{8}$$

56. $\dfrac{N(x)}{D(x)} = \dfrac{P_1}{x - c_1} + \dfrac{P_2}{x - c_2} + \cdots + \dfrac{P_n}{x - c_n}$

$$N(x) = P_1(x - c_2)(x - c_3)\ldots(x - c_n) + P_2(x - c_1)(x - c_3)\ldots(x - c_n) + \cdots + P_n(x - c_1)(x - c_2)\ldots(x - c_{n-1})$$

Let $x = c_1$: $N(c_1) = P_1(c_1 - c_2)(c_1 - c_3)\ldots(c_1 - c_n)$

$$P_1 = \frac{N(c_1)}{(c_1 - c_2)(c_1 - c_3)\ldots(c_1 - c_n)}$$

Let $x = c_2$: $N(c_2) = P_2(c_2 - c_1)(c_2 - c_3)\ldots(c_2 - c_n)$

$$P_2 = \frac{N(c_2)}{(c_2 - c_1)(c_2 - c_3)\ldots(c_2 - c_n)}$$

$\vdots$ $\vdots$

Let $x = c_n$: $N(c_n) = P_n(c_n - c_1)(c_n - c_2)\ldots(c_n - c_{n-1})$

$$P_n = \frac{N(c_n)}{(c_n - c_1)(c_n - c_2)\ldots(c_n - c_{n-1})}$$

—CONTINUED—

56. —CONTINUED—

If $D(x) = (x - c_1)(x - c_2)(x - c_3)...(x - c_n)$, then by the Product Rule

$$D'(x) = (x - c_2)(x - c_3)...(x - c_n) + (x - c_1)(x - c_3)...(x - c_n) + \cdots + (x - c_1)(x - c_2)(x - c_3)...(x - c_{n-1})$$

and

$$D'(c_1) = (c_1 - c_2)(c_1 - c_3)...(c_1 - c_n)$$

$$D'(c_2) = (c_2 - c_1)(c_2 - c_3)...(c_2 - c_n)$$

$$\vdots$$

$$D'(c_n) = (c_n - c_1)(c_n - c_2)...(c_n - c_{n-1}).$$

Thus, $P_k = N(c_k)/D'(c_k)$ for $k = 1, 2, ..., n$.

57. $\dfrac{x^3 - 3x^2 + 1}{x^4 - 13x^2 + 12x} = \dfrac{P_1}{x} + \dfrac{P_2}{x - 1} + \dfrac{P_3}{x + 4} + \dfrac{P_4}{x - 3} \Rightarrow c_1 = 0,\ c_2 = 1,\ c_3 = -4,\ c_4 = 3$

$N(x) = x^3 - 3x^2 + 1$

$D'(x) = 4x^3 - 26x + 12$

$$P_1 = \frac{N(0)}{D'(0)} = \frac{1}{12}$$

$$P_2 = \frac{N(1)}{D'(1)} = \frac{-1}{-10} = \frac{1}{10}$$

$$P_3 = \frac{N(-4)}{D'(-4)} = \frac{-111}{-140} = \frac{111}{140}$$

$$P_4 = \frac{N(3)}{D'(3)} = \frac{1}{42}$$

Thus, $\dfrac{x^3 - 3x^2 + 1}{x^4 - 13x^2 + 12x} = \dfrac{1/12}{x} + \dfrac{1/10}{x - 1} + \dfrac{111/140}{x + 4} + \dfrac{1/42}{x - 3}.$

Section 7.6   Integration by Tables and Other Integration Techniques

1. By Formula 6: $\displaystyle\int \frac{x^2}{1 + x}\, dx = -\frac{x}{2}(2 - x) + \ln|1 + x| + C$

2. By Formula 13: $\displaystyle\int \frac{1}{2x^2(2x - 1)^2}\, dx = \frac{1}{2}\int \frac{1}{x^2(2x - 1)^2}\, dx$

$$= -\frac{1}{2}\left[\frac{4x - 1}{x(2x - 1)} + \frac{4}{-1}\ln\left| \frac{x}{2x - 1} \right| \right] + C$$

$$= \frac{1}{2}\left[4\ln\left| \frac{x}{2x - 1} \right| - \frac{4x - 1}{x(2x - 1)} \right] + C$$

3. By Formula 26: $\displaystyle\int e^x\sqrt{1 + e^{2x}}\, dx = \frac{1}{2}\left[e^x\sqrt{e^{2x} + 1} + \ln\left| e^x + \sqrt{e^{2x} + 1} \right| \right] + C$

$u = e^x,\ du = e^x\, dx$

4. By Formula 29: $\displaystyle\int \frac{\sqrt{x^2 - 4}}{x}\, dx = \sqrt{x^2 - 4} - 2\,\text{arcsec}\,\frac{|x|}{2} + C$

5. By Formula 44: $\int \dfrac{1}{x^2 \sqrt{1 - x^2}}\,dx = -\dfrac{\sqrt{1 - x^2}}{x} + C$

6. By Formula 41: $\int \dfrac{x}{\sqrt{9 - x^4}}\,dx = \dfrac{1}{2} \int \dfrac{2x}{\sqrt{3^2 - (x^2)^2}}\,dx = \dfrac{1}{2} \arcsin \dfrac{x^2}{3} + C$

7. By Formulas 50 and 48: $\int \sin^4(2x)\,dx = \dfrac{1}{2} \int \sin^4(2x)(2)\,dx$

$$= \dfrac{1}{2}\left[\dfrac{-\sin^3(2x)\cos(2x)}{4} + \dfrac{3}{4} \int \sin^2(2x)(2)\,dx \right]$$

$$= \dfrac{1}{2}\left[\dfrac{-\sin^3(2x)\cos(2x)}{4} + \dfrac{3}{8}(2x - \sin 2x \cos 2x) \right] + C$$

$$= \dfrac{1}{16}(6x - 3\sin 2x \cos 2x - 2\sin^3 2x \cos 2x) + C$$

8. By Formulas 51 and 47: $\int \dfrac{\cos^3 \sqrt{x}}{\sqrt{x}}\,dx = 2 \int \cos^3 \sqrt{x}\left(\dfrac{1}{2\sqrt{x}}\right) dx$

$$= 2\left[\dfrac{\cos^2 \sqrt{x} \sin \sqrt{x}}{3} + \dfrac{2}{3} \int \cos \sqrt{x}\left(\dfrac{1}{2\sqrt{x}}\right) dx \right] = \dfrac{2}{3} \sin \sqrt{x}\left(\cos^2 \sqrt{x} + 2\right) + C$$

$$u = \sqrt{x},\ du = \dfrac{1}{2\sqrt{x}}\,dx$$

9. By Formula 57: $\int \dfrac{1}{\sqrt{x}\left(1 - \cos \sqrt{x}\right)}\,dx = 2 \int \dfrac{1}{1 - \cos \sqrt{x}}\left(\dfrac{1}{2\sqrt{x}}\right) dx$

$$= -2\left(\cot \sqrt{x} + \csc \sqrt{x}\right) + C$$

$$u = \sqrt{x},\ du = \dfrac{1}{2\sqrt{x}}\,dx$$

10. By Formula 71:

$$\int \dfrac{1}{1 - \tan 5x}\,dx = \dfrac{1}{5} \int \dfrac{1}{1 - \tan 5x}(5)\,dx$$

$$= \dfrac{1}{5}\left(\dfrac{1}{2}\right)(u - \ln|\cos u - \sin u|) + C$$

$$= \dfrac{1}{10}(5x - \ln|\cos 5x - \sin 5x|) + C$$

$$u = 5x,\ du = 5\,dx$$

11. By Formula 84:

$$\int \dfrac{1}{1 + e^{2x}}\,dx = x - \dfrac{1}{2} \ln(1 + e^{2x}) + C$$

12. By Formula 86:

$$\int e^{-2x} \cos 3x\,dx = \dfrac{e^{-2x}}{13}(-2 \cos 3x + 3 \sin 3x) + C$$

$$u = x,\ du = dx,\ a = -2,\ b = 3$$

13. By Formula 89:

$$\int x^3 \ln x\,dx = \dfrac{x^4}{16}\left(4 \ln|x| - 1\right) + C$$

14. By Formulas 90 and 91: $\int (\ln x)^3\,dx = x(\ln x)^3 - 3 \int (\ln x)^2\,dx$

$$= x(\ln x)^3 - 3x[2 - 2 \ln x + (\ln x)^2] + C$$

$$= x[(\ln x)^3 - 3(\ln x)^2 + 6 \ln x - 6] + C$$

15. (a) By Formulas 83 and 82: $\int x^2 e^x \, dx = x^2 e^x - 2\int x e^x \, dx$

$$= x^2 e^x - 2[(x-1)e^x + C_1]$$

$$= x^2 e^x - 2xe^x + 2e^x + C$$

(b) Integration by parts: $u = x^2, du = 2x \, dx, dv = e^x \, dx, v = e^x$

$$\int x^2 e^x \, dx = x^2 e^x - \int 2x e^x \, dx$$

Parts again: $u = 2x, du = 2 \, dx, dv = e^x \, dx, v = e^x$

$$\int x^2 e^x \, dx = x^2 e^x - \left[2x e^x - \int 2e^x \, dx\right] = x^2 e^x - 2xe^x + 2e^x + C$$

16. (a) By Formula 89: $\int x^4 \ln x \, dx = \dfrac{x^5}{5^2}[-1 + (4+1)\ln x] + C = \dfrac{-x^5}{25} + \dfrac{1}{5}x^5 \ln x + C$

(b) Integration by parts: $u = \ln x, du = \dfrac{1}{x} \, dx, dv = x^4 \, dx, v = \dfrac{x^5}{5}$

$$\int x^4 \ln x \, dx = \frac{x^5}{5} \ln x - \int \frac{x^5}{5}\frac{1}{x} \, dx = \frac{x^5}{5} \ln x - \frac{x^5}{25} + C$$

17. (a) By Formula: 12, $a = b = 1, u = x$, and

$$\int \frac{1}{x^2(x+1)} \, dx = \frac{-1}{1}\left(\frac{1}{x} + \frac{1}{1}\ln\left|\frac{x}{1+x}\right|\right) + C$$

$$= \frac{-1}{x} - \ln\left|\frac{x}{1+x}\right| + C$$

$$= \frac{-1}{x} + \ln\left|\frac{x+1}{x}\right| + C$$

(b) Partial fractions:

$$\frac{1}{x^2(x+1)} = \frac{A}{x} + \frac{B}{x^2} + \frac{C}{x+1}$$

$$1 = Ax(x+1) + B(x+1) + Cx^2$$

$x = 0$: $1 = B$

$x = -1$: $1 = C$

$x = 1$: $1 = 2A + 2 + 1 \Rightarrow A = -1$

$$\int \frac{1}{x^2(x+1)} \, dx = \int \left[\frac{-1}{x} + \frac{1}{x^2} + \frac{1}{x+1}\right] dx$$

$$= -\ln|x| - \frac{1}{x} + \ln|x+1| + C$$

$$= -\frac{1}{x} - \ln\left|\frac{x}{x+1}\right| + C$$

18. (a) By Formula 24: $a = \sqrt{75}, x = u$, and

$$\int \frac{1}{x^2 - 75} \, dx = \frac{1}{2\sqrt{75}} \ln\left|\frac{x - \sqrt{75}}{x + \sqrt{75}}\right| + C$$

$$= \frac{\sqrt{3}}{30} \ln\left|\frac{x - \sqrt{75}}{x + \sqrt{75}}\right| + C$$

(b) Partial fractions:

$$\frac{1}{x^2 - 75} = \frac{A}{x - \sqrt{75}} + \frac{B}{x + \sqrt{75}}$$

$$1 = A(x + \sqrt{75}) + B(x - \sqrt{75})$$

$x = \sqrt{75}$: $1 = 2A\sqrt{75} \Rightarrow A = \dfrac{1}{2\sqrt{75}} = \dfrac{1}{10\sqrt{3}} = \dfrac{\sqrt{3}}{30}$

$x = -\sqrt{75}$: $1 = -2B\sqrt{75} \Rightarrow B = -\dfrac{\sqrt{3}}{30}$

$$\int \frac{1}{x^2 - 75} \, dx = \int \left[\frac{\sqrt{3}/30}{x - \sqrt{75}} - \frac{\sqrt{3}/30}{x + \sqrt{75}}\right] dx$$

$$= \frac{\sqrt{3}}{30} \ln\left|\frac{x - \sqrt{75}}{x + \sqrt{75}}\right| + C$$

19. By Formula 81: $\displaystyle\int xe^{x^2} = \frac{1}{2}e^{x^2} + C$

20. By Formula 21:

$$\int \frac{x}{\sqrt{1+x}}\,dx = -\frac{2}{3}(2-x)\sqrt{1+x} + C$$

21. By Formula 79: $\displaystyle\int x\,\text{arcsec}(x^2+1)\,dx = \frac{1}{2}\int \text{arcsec}(x^2+1)(2x)\,dx$

$$= \frac{1}{2}\Big[(x^2+1)\,\text{arcsec}(x^2+1) - \ln\big((x^2+1) + \sqrt{x^4+2x^2}\big)\Big] + C$$

$u = x^2 + 1,\ du = 2x\,dx$

22. By Formula 79: $\displaystyle\int \text{arcsec}\,2x\,dx = \frac{1}{2}\Big[2x\,\text{arcsec}\,2x - \ln|2x + \sqrt{4x^2-1}|\Big] + C$

$u = 2x,\ du = 2\,dx$

23. By Formula 89:

$$\int x^2\ln x\,dx = \frac{x^3}{9}\big(-1 + 3\ln|x|\big) + C$$

24. By Formula 52:

$$\int x\sin x\,dx = \sin x - x\cos x + C$$

25. By Formula 35: $\displaystyle\int \frac{1}{x^2\sqrt{x^2-4}}\,dx = \frac{\sqrt{x^2-4}}{4x} + C$

26. By Formula 7: $\displaystyle\int \frac{x^2}{(3x-5)^2}\,dx = \frac{1}{27}\Big(3x - \frac{25}{3x-5} + 10\ln|3x-5|\Big) + C$

27. By Formula 4: $\displaystyle\int \frac{2x}{(1-3x)^2}\,dx = 2\int \frac{x}{(1-3x)^2}\,dx = \frac{2}{9}\Big(\ln|1-3x| + \frac{1}{1-3x}\Big) + C$

28. By Formula 14: $\displaystyle\int \frac{1}{x^2+2x+2}\,dx = \frac{2}{\sqrt{4}}\arctan\Big(\frac{2x+2}{2}\Big) + C = \arctan(x+1) + C$

29. By Formula 76:

$$\int e^x \arccos e^x\,dx = e^x \arccos e^x - \sqrt{1-e^{2x}} + C$$

$u = e^x,\ du = e^x\,dx$

30. By Formula 56:

$$\int \frac{\theta^2}{1-\sin\theta^3}\,d\theta = \frac{1}{3}\int \frac{1}{1-\sin\theta^3}3\theta^2\,d\theta$$

$$= \frac{1}{3}(\tan\theta^3 + \sec\theta^3) + C$$

31. By Formula 73:

$$\int \frac{x}{1-\sec x^2}\,dx = \frac{1}{2}\int \frac{2x}{1-\sec x^2}\,dx$$

$$= \frac{1}{2}(x^2 + \cot x^2 + \csc x^2) + C$$

32. By Formula 71:

$$\int \frac{e^x}{1-\tan e^x}\,dx = \frac{1}{2}\big(e^x - \ln|\cos e^x - \sin e^x|\big) + C$$

$u = e^x,\ du = e^x\,dx$

33. By Formula 23:

$$\int \frac{\cos x}{1+\sin^2 x}\,dx = \arctan(\sin x) + C$$

$u = \sin x,\ du = \cos x\,dx$

34. By Formula 23:

$$\int \frac{1}{t[1+(\ln t)^2]}\,dt = \int \frac{1}{1+(\ln t)^2}\Big(\frac{1}{t}\Big)\,dt$$

$$= \arctan(\ln t) + C$$

$u = \ln t,\ du = \dfrac{1}{t}\,dt$

35. By Formula 14: $\int \dfrac{\cos\theta}{3 + 2\sin\theta + \sin^2\theta}\, d\theta = \dfrac{\sqrt{2}}{2}\arctan\left(\dfrac{1 + \sin\theta}{\sqrt{2}}\right) + C$

$u = \sin\theta,\, du = \cos\theta\, d\theta$

36. By Formula 26: $\int \sqrt{3 + x^2}\, dx = \dfrac{1}{2}\left(x\sqrt{x^2 + 3} + 3\ln\left|x + \sqrt{x^2 + 3}\right|\right) + C$

37. By Formula 35: $\displaystyle\int \dfrac{1}{x^2\sqrt{2 + 9x^2}}\, dx = 3\int \dfrac{3}{(3x)^2\sqrt{(\sqrt{2})^2 + (3x)^2}}\, dx$

$= -\dfrac{3\sqrt{2 + 9x^2}}{6x} + C$

$= -\dfrac{\sqrt{2 + 9x^2}}{2x} + C$

38. By Formula 27: $\displaystyle\int x^2\sqrt{2 + (3x)^2}\, dx = \dfrac{1}{27}\int (3x)^2\sqrt{(\sqrt{2})^2 + (3x)^2}\,3\, dx$

$= \dfrac{1}{8(27)}\left[3x(18x^2 + 2)\sqrt{2 + 9x^2} - 4\ln\left|3x + \sqrt{2 + 9x^2}\right|\right] + C$

39. By Formulas 55 and 54: $\displaystyle\int t^4\cos t\, dt = t^4\sin t - 4\int t^3\sin t\, dt$

$= t^4\sin t - 4\left[-t^3\cos t + 3\int t^2\cos t\, dt\right]$

$= t^4\sin t + 4t^3\cos t - 12\left[t^2\sin t - 2\int t\sin t\, dt\right]$

$= t^4\sin t + 4t^3\cos t - 12t^2\sin t + 24(-t\cos t + \sin t) + C$

$= (t^4 - 12t^2 + 24)\sin t + (4t^3 - 24t)\cos t + C$

40. By Formula 77: $\displaystyle\int \sqrt{x}\arctan(x^{3/2})\, dx = \dfrac{2}{3}\int \arctan(x^{3/2})\left(\dfrac{3}{2}\sqrt{x}\right)\, dx$

$= \dfrac{2}{3}\left[x^{3/2}\arctan(x^{3/2}) - \ln\sqrt{1 + x^3}\right] + C$

41. By Formula 3:

$\displaystyle\int \dfrac{\ln x}{x(3 + 2\ln x)}\, dx = \dfrac{1}{4}(2\ln|x| - 3\ln|3 + 2\ln|x||) + C$

$u = \ln x,\, du = \dfrac{1}{x}\, dx$

42. By Formula 45:

$\displaystyle\int \dfrac{e^x}{(1 - e^{2x})^{3/2}}\, dx = \dfrac{e^x}{\sqrt{1 - e^{2x}}} + C$

$u = e^x,\, du = e^x\, dx$

43. By Formulas 1, 25, and 33: $\displaystyle\int \dfrac{x}{(x^2 - 6x + 10)^2}\, dx = \dfrac{1}{2}\int \dfrac{2x - 6 + 6}{(x^2 - 6x + 10)^2}\, dx$

$= \dfrac{1}{2}\int (x^2 - 6x + 10)^{-2}(2x - 6)\, dx + 3\int \dfrac{1}{[(x - 3)^2 + 1]^2}\, dx$

$= -\dfrac{1}{2(x^2 - 6x + 10)} + \dfrac{3}{2}\left[\dfrac{x - 3}{x^2 - 6x + 10} + \arctan(x - 3)\right] + C$

$= \dfrac{3x - 10}{2(x^2 - 6x + 10)} + \dfrac{3}{2}\arctan(x - 3) + C$

44. By Formula 27:

$$\int (2x - 3)^2 \sqrt{(2x - 3)^2 + 4} \, dx = \frac{1}{2} \int (2x - 3)^2 \sqrt{(2x - 3)^2 + 4} (2) \, dx$$

$$= \frac{1}{8}(2x - 3)[(2x - 3)^2 + 2]\sqrt{(2x - 3)^2 + 4} - \ln|2x - 3 + \sqrt{(2x - 3)^2 + 4}| + C$$

$u = 2x - 3, \, du = 2 \, dx$

45. By Formula 31: $\displaystyle\int \frac{x}{\sqrt{x^4 - 6x^2 + 5}} \, dx = \frac{1}{2} \int \frac{2x}{\sqrt{(x^2 - 3)^2 - 4}} \, dx$

$$= \frac{1}{2} \ln|x^2 - 3 + \sqrt{x^4 - 6x^2 + 5}| + C$$

$u = x^2 - 3, \, du = 2x \, dx$

46. By Formula 31: $\displaystyle\int \frac{\cos x}{\sqrt{\sin^2 x + 1}} \, dx = \ln|\sin x + \sqrt{\sin^2 x + 1}| + C$

$u = \sin x, \, du = \cos x \, dx$

47. $\displaystyle\int \frac{x^3}{\sqrt{4 - x^2}} \, dx = \int \frac{8 \sin^3 \theta (2 \cos \theta \, d\theta)}{2 \cos \theta}$

$$= 8 \int (1 - \cos^2 \theta) \sin \theta \, d\theta$$

$$= 8 \int [\sin \theta - \cos^2 \theta (\sin \theta)] \, d\theta$$

$$= -8 \cos \theta + \frac{8 \cos^3 \theta}{3} + C$$

$$= \frac{-\sqrt{4 - x^2}}{3}(x^2 + 8) + C$$

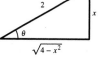

$x = 2 \sin \theta, \, dx = 2 \cos \theta \, d\theta, \, \sqrt{4 - x^2} = 2 \cos \theta$

48. $\displaystyle\int \sqrt{\frac{3 - x}{3 + x}} \, dx = \int \frac{3 - x}{\sqrt{9 - x^2}} \, dx$

$$= 3 \int \frac{1}{\sqrt{9 - x^2}} \, dx + \int \frac{-x}{\sqrt{9 - x^2}} \, dx$$

$$= 3 \arcsin \frac{x}{3} + \sqrt{9 - x^2} + C$$

49. By Formula 8:

$$\int \frac{e^{3x}}{(1 + e^x)^3} \, dx = \int \frac{(e^x)^2}{(1 + e^x)^3}(e^x) \, dx$$

$$= \frac{2}{1 + e^x} - \frac{1}{2(1 + e^x)^2} + \ln|1 + e^x| + C$$

$u = e^x, \, du = e^x \, dx$

50. By Formulas 69 and 61: $\displaystyle\int \sec^5 \theta \, d\theta = \frac{1}{4}(\sec^3 \theta \tan \theta) + \frac{3}{4} \int \sec^3 \theta \, d\theta$

$$= \frac{1}{4}(\sec^3 \theta \tan \theta) + \frac{3}{8}(\sec \theta \tan \theta + \ln|\sec \theta + \tan \theta|) + C$$

$$= \frac{1}{8}[2 \sec^3 \theta \tan \theta + 3 \sec \theta \tan \theta + 3 \ln|\sec \theta + \tan \theta|] + C$$

51. $\dfrac{u^2}{(a + bu)^2} = \dfrac{1}{b^2} - \dfrac{(2a/b)u + (a^2/b^2)}{(a + bu)^2} = \dfrac{1}{b^2} + \dfrac{A}{a + bu} + \dfrac{B}{(a + bu)^2}$

$-\dfrac{2a}{b}u - \dfrac{a^2}{b^2} = A(a + bu) + B = (aA + B) + bAu$

Equating the coefficients of like terms we have $aA + B = -a^2/b^2$ and $bA = -2a/b$. Solving these equations we have $A = -2a/b^2$ and $B = a^2/b^2$.

$\displaystyle \int \dfrac{u^2}{(a + bu)^2} \, du = \dfrac{1}{b^2} \int du - \dfrac{2a}{b^2}\left(\dfrac{1}{b}\right) \int \dfrac{1}{a + bu} b \, du + \dfrac{a^2}{b^2}\left(\dfrac{1}{b}\right) \int \dfrac{1}{(a + bu)^2} b \, du$

$\qquad = \dfrac{1}{b^2}u - \dfrac{2a}{b^3} \ln|a + bu| - \dfrac{a^2}{b^3}\left(\dfrac{1}{a + bu}\right) + C$

$\qquad = \dfrac{1}{b^3}\left(bu - \dfrac{a^2}{a + bu} - 2a \ln|a + bu|\right) + C$

52. Integration by parts: $w = u^n, \, dw = nu^{n-1} \, du, \, dv = \dfrac{du}{\sqrt{a + bu}}, \, v = \dfrac{2}{b}\sqrt{a + bu}$

$\displaystyle \int \dfrac{u^n}{\sqrt{a + bu}} \, du = \dfrac{2u^n}{b}\sqrt{a + bu} - \dfrac{2n}{b} \int u^{n-1}\sqrt{a + bu} \, du$

$\qquad = \dfrac{2u^n}{b}\sqrt{a + bu} - \dfrac{2n}{b} \int u^{n-1}\sqrt{a + bu} \cdot \dfrac{\sqrt{a + bu}}{\sqrt{a + bu}} \, du$

$\qquad = \dfrac{2u^n}{b}\sqrt{a + bu} - \dfrac{2n}{b} \int \dfrac{au^{n-1} + bu^n}{\sqrt{a + bu}} \, du$

$\qquad = \dfrac{2u^n}{b}\sqrt{a + bu} - \dfrac{2na}{b} \int \dfrac{u^{n-1}}{\sqrt{a + bu}} \, du - 2n \int \dfrac{u^n}{\sqrt{a + bu}} \, du$

Therefore, $(2n + 1) \displaystyle \int \dfrac{u^n}{\sqrt{a + bu}} \, du = \dfrac{2}{b}\left[u^n\sqrt{a + bu} - na \int \dfrac{u^{n-1}}{\sqrt{a + bu}} \, du\right]$ and

$\displaystyle \int \dfrac{u^n}{\sqrt{a + bu}} = \dfrac{2}{(2n + 1)b}\left[u^n\sqrt{a + bu} - na \int \dfrac{u^{n-1}}{\sqrt{a + bu}} \, du\right].$

53. When we have $u^2 + a^2$:

$u = a \tan \theta$

$du = a \sec^2 \theta \, d\theta$

$u^2 + a^2 = a^2 \sec^2 \theta$

$\displaystyle \int \dfrac{1}{(u^2 + a^2)^{3/2}} \, du = \int \dfrac{a \sec^2 \theta \, d\theta}{a^3 \sec^3 \theta}$

$\qquad = \dfrac{1}{a^2} \int \cos \theta \, d\theta$

$\qquad = \dfrac{1}{a^2} \sin \theta + C$

$\qquad = \dfrac{u}{a^2\sqrt{u^2 + a^2}} + C$

When we have $u^2 - a^2$:

$u = a \sec \theta$

$du = a \sec \theta \tan \theta \, d\theta$

$u^2 - a^2 = a^2 \tan^2 \theta$

$\displaystyle \int \dfrac{1}{(u^2 - a^2)^{3/2}} \, du = \int \dfrac{a \sec \theta \tan \theta \, d\theta}{a^3 \tan^3 \theta}$

$\qquad = \dfrac{1}{a^2} \int \dfrac{\cos \theta}{\sin^2 \theta} \, d\theta$

$\qquad = -\dfrac{1}{a^2} \csc \theta + C$

$\qquad = \dfrac{-u}{a^2\sqrt{u^2 - a^2}} + C$

54. $\int u^n(\cos u)\, du = u^n \sin u - n\int u^{n-1}(\sin u)\, du$

$w = u^n,\ dv = \cos u\, du,\ dw = nu^{n-1}\, du,\ v = \sin u$

55. $\int (\arctan u)\, du = u \arctan u - \dfrac{1}{2}\int \dfrac{2u}{1+u^2}\, du$

$= u \arctan u - \dfrac{1}{2}\ln(1+u^2) + C$

$= u \arctan u - \ln\sqrt{1+u^2} + C$

$w = \arctan u,\ dv = du,\ dw = \dfrac{du}{1+u^2},\ v = u$

56. $\int (\ln u)^n\, du = u(\ln u)^n - \int n(\ln u)^{n-1}\left(\dfrac{1}{u}\right)u\, du$

$= u(\ln u)^n - n\int (\ln u)^{n-1}\, du$

$w = (\ln u)^n,\ dv = du,\ dw = n(\ln u)^{n-1}\left(\dfrac{1}{u}\right)du,\ v = u$

57. $\int \dfrac{1}{x^{3/2}\sqrt{1-x}}\, dx = \dfrac{-2\sqrt{1-x}}{\sqrt{x}} + C$

$\left(\dfrac{1}{2}, 5\right):\ \dfrac{-2\sqrt{1/2}}{\sqrt{1/2}} + C = 5 \Rightarrow C = 7$

$y = \dfrac{-2\sqrt{1-x}}{\sqrt{x}} + 7$

58. $\int x\sqrt{x^2+2x}\, dx = \dfrac{1}{6}\Big[2(x^2+2x)^{3/2} - 3(x+1)\sqrt{x^2+2x} + 3\ln\big|x+1+\sqrt{x^2+2x}\big|\Big] + C$

$(0,0):\ \dfrac{1}{6}[3\ln|1|] + C = 0 \Rightarrow C = 0$

59. $\int \dfrac{\sqrt{2-2x-x^2}}{x+1}\, dx = \sqrt{2-2x-x^2} - \sqrt{3}\ln\left|\dfrac{\sqrt{3}+\sqrt{2-2x-x^2}}{x+1}\right| + C$

$\left(0, \sqrt{2}\right):\ \sqrt{2} - \sqrt{3}\ln\left(\sqrt{3}+\sqrt{2}\right) + C = \sqrt{2} \Rightarrow C = \sqrt{3}\ln\left(\sqrt{3}+\sqrt{2}\right)$

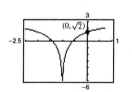

60. $\int \dfrac{1}{(x^2-6x+10)^2}\, dx = \dfrac{1}{2}\left[\dfrac{x-3}{x^2-6x+10} + \arctan(x-3)\right] + C$

$(3,0):\ 0 = C \Rightarrow y = \dfrac{1}{2}\left[\dfrac{x-3}{x^2-6x+10} + \arctan(x-3)\right]$

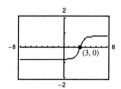

61. $\int \dfrac{1}{\sin\theta \tan\theta}\, d\theta = -\csc\theta + C$

$\left(\dfrac{\pi}{4}, 2\right):\ -\dfrac{2}{\sqrt{2}} + C = 2 \Rightarrow C = 2 + \sqrt{2}$

$y = -\csc\theta + 2 + \sqrt{2}$

62. $\displaystyle\int \frac{\sin\theta}{(\cos\theta)(1+\sin\theta)}\,d\theta = \frac{1}{2}\left[\frac{-\sin\theta}{1+\sin\theta} + \ln\left|\frac{1+\sin\theta}{\cos\theta}\right|\right] + C$

$(0, 1)$: $\;C = 1 \Longrightarrow y = \dfrac{1}{2}\left[\dfrac{-\sin\theta}{1+\sin\theta} + \ln\left|\dfrac{1+\sin\theta}{\cos\theta}\right|\right] + 1$

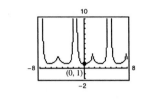

63. $\displaystyle\int \frac{1}{2 - 3\sin\theta}\,d\theta = \int\left[\frac{\dfrac{2\,du}{1+u^2}}{2 - 3\left(\dfrac{2u}{1+u^2}\right)}\right]$

$\displaystyle = \int 2\frac{2}{(1+u^2) - 6u}\,du$

$\displaystyle = \int \frac{1}{u^2 - 3u + 1}\,du$

$\displaystyle = \int \frac{1}{\left(u - \dfrac{3}{2}\right)^2 - \dfrac{5}{4}}\,du$

$\displaystyle = \frac{1}{\sqrt{5}}\ln\left|\frac{\left(u - \dfrac{3}{2}\right) - \dfrac{\sqrt{5}}{2}}{\left(u - \dfrac{3}{2}\right) + \dfrac{\sqrt{5}}{2}}\right| + C$

$\displaystyle = \frac{1}{\sqrt{5}}\ln\left|\frac{2u - 3 - \sqrt{5}}{2u - 3 + \sqrt{5}}\right| + C$

$\displaystyle = \frac{1}{\sqrt{5}}\ln\left|\frac{2\tan\left(\dfrac{\theta}{2}\right) - 3 - \sqrt{5}}{2\tan\left(\dfrac{\theta}{2}\right) - 3 + \sqrt{5}}\right| + C$

$u = \tan\dfrac{\theta}{2}$

64. $\displaystyle\int \frac{\sin\theta}{1 + \cos^2\theta}\,d\theta = -\int \frac{-\sin\theta}{1 + (\cos\theta)^2}\,d\theta$

$\displaystyle = -\arctan(\cos\theta) + C$

65. $\displaystyle\int_0^{\pi/2} \frac{1}{1 + \sin\theta + \cos\theta}\,d\theta = \int_0^1\left[\frac{\dfrac{2\,du}{1+u^2}}{1 + \dfrac{2u}{1+u^2} + \dfrac{1-u^2}{1+u^2}}\right]$

$\displaystyle = \int_0^1 \frac{1}{1+u}\,du$

$\displaystyle = \Big[\ln|1+u|\Big]_0^1$

$\displaystyle = \ln 2$

$u = \tan\dfrac{\theta}{2}$

66. $\displaystyle\int_0^{\pi/2} \frac{1}{3 - 2\cos\theta}\,d\theta = \int_0^1\left[\frac{\dfrac{2u}{1+u^2}}{3 - \dfrac{2(1-u^2)}{1+u^2}}\right]$

$\displaystyle = 2\int_0^1 \frac{1}{5u^2 + 1}\,du$

$\displaystyle = \left[\frac{2}{\sqrt{5}}\arctan\left(\sqrt{5}\,u\right)\right]_0^1$

$\displaystyle = \frac{2}{\sqrt{5}}\arctan\sqrt{5}$

$u = \tan\dfrac{\theta}{2}$

67. $\displaystyle\int \frac{\sin\theta}{3 - 2\cos\theta}\,d\theta = \frac{1}{2}\int \frac{2\sin\theta}{3 - 2\cos\theta}\,d\theta$

$\displaystyle = \frac{1}{2}\ln|u| + C$

$\displaystyle = \frac{1}{2}\ln(3 - 2\cos\theta) + C$

$u = 3 - 2\cos\theta,\, du = 2\sin\theta\,d\theta$

68. $\displaystyle\int \frac{\sin\theta}{1 + \sin\theta}\,d\theta = \int \frac{\sin\theta - \sin^2\theta}{\cos^2\theta}\,d\theta$

$\displaystyle = \int (\sec\theta\tan\theta - \sec^2\theta + 1)\,d\theta$

$\displaystyle = \sec\theta - \tan\theta + C$

69. $\int \dfrac{\cos \sqrt{\theta}}{\sqrt{\theta}} \, d\theta = 2 \int \cos \sqrt{\theta} \left(\dfrac{1}{2\sqrt{\theta}} \right) d\theta$

$\qquad = 2 \sin \sqrt{\theta} + C$

$u = \sqrt{\theta}, \, du = \dfrac{1}{2\sqrt{\theta}} \, d\theta$

70. $\int \dfrac{1}{\sec \theta - \tan \theta} \, d\theta = \int \dfrac{1}{(1/\cos \theta) - (\sin \theta / \cos \theta)} \, d\theta$

$\qquad = -\int \dfrac{-\cos \theta}{1 - \sin \theta} \, d\theta$

$\qquad = -\ln|1 - \sin \theta| + C$

$u = 1 - \sin \theta, \, du = -\cos \theta \, d\theta$

71. $A = \displaystyle\int_0^8 \dfrac{x}{\sqrt{x+1}} \, dx$

$\quad = \left[\dfrac{-2(2-x)}{3} \sqrt{x+1} \right]_0^8$

$\quad = 12 - \left(-\dfrac{4}{3} \right)$

$\quad = \dfrac{40}{3} \approx 13.333$ square units

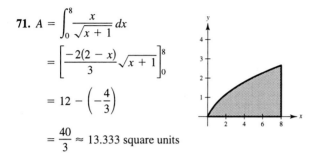

72. $A = \displaystyle\int_0^2 \dfrac{x}{1 + e^{x^2}} \, dx$

$\quad = \dfrac{1}{2} \displaystyle\int_0^2 \dfrac{2x \, dx}{1 + e^{x^2}}$

$\quad = \dfrac{1}{2} \left[x^2 - \ln(1 + e^{x^2}) \right]_0^2$

$\quad = \dfrac{1}{2} \left[4 - \ln(1 + e^4) \right] + \dfrac{1}{2} \ln 2$

$\quad \approx 0.337$ square units

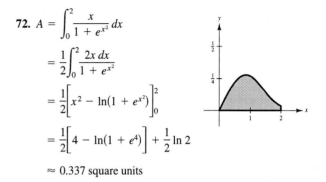

73. $W = \displaystyle\int_0^5 2000xe^{-x} \, dx$

$\quad = -2000 \displaystyle\int_0^5 -xe^{-x} \, dx$

$\quad = 2000 \displaystyle\int_0^5 (-x)e^{-x}(-1) \, dx$

$\quad = 2000 \left[(-x)e^{-x} - e^{-x} \right]_0^5$

$\quad = 2000 \left(-\dfrac{6}{e^5} + 1 \right)$

$\quad \approx 1919.145$ ft · lbs

74. $W = \displaystyle\int_0^5 \dfrac{500x}{\sqrt{26 - x^2}} \, dx$

$\quad = -250 \displaystyle\int_0^5 (26 - x^2)^{-1/2}(-2x) \, dx$

$\quad = \left[-500\sqrt{26 - x^2} \right]_0^5$

$\quad = 500 \left(\sqrt{26} - 1 \right)$

$\quad \approx 2049.51$ ft · lbs

75. (a) $V = 20(2) \displaystyle\int_0^3 \dfrac{2}{\sqrt{1 + y^2}} \, dy$

$\qquad = \left[80 \ln \left| y + \sqrt{1 + y^2} \right| \right]_0^3$

$\qquad = 80 \ln \left(3 + \sqrt{10} \right)$

$\qquad \approx 145.5$ cubic feet

$\quad W = 148 \left(80 \ln \left(3 + \sqrt{10} \right) \right)$

$\qquad = 11{,}840 \ln \left(3 + \sqrt{10} \right)$

$\qquad \approx 21{,}530.4$ lb

(b) By symmetry, $\bar{x} = 0$.

$\quad M = \rho(2) \displaystyle\int_0^3 \dfrac{2}{\sqrt{1 + y^2}} \, dy = \left[4\rho \ln \left| y + \sqrt{1 + y^2} \right| \right]_0^3 = 4\rho \ln \left(3 + \sqrt{10} \right)$

$\quad M_x = 2\rho \displaystyle\int_0^3 \dfrac{2y}{\sqrt{1 + y^2}} \, dy = \left[4\rho \sqrt{1 + y^2} \right]_0^3 = 4\rho \left(\sqrt{10} - 1 \right)$

$\quad \bar{y} = \dfrac{M_x}{M} = \dfrac{4\rho \left(\sqrt{10} - 1 \right)}{4\rho \ln \left(3 + \sqrt{10} \right)} \approx 1.19$

$\quad$ Centroid: $\left(\bar{x}, \bar{y} \right) \approx (0, 1.19)$

76. $\dfrac{1}{2-0}\displaystyle\int_0^2 \dfrac{5000}{1+e^{4.8-1.9t}}\,dt = \dfrac{2500}{-1.9}\displaystyle\int_0^2 \dfrac{-1.9\,dt}{1+e^{4.8-1.9t}}$

$\qquad\qquad = -\dfrac{2500}{1.9}\Big[(4.8-1.9t)-\ln(1+e^{4.8-1.9t})\Big]_0^2$

$\qquad\qquad = -\dfrac{2500}{1.9}\big[(1-\ln(1+e))-(4.8-\ln(1+e^{4.8}))\big]$

$\qquad\qquad = \dfrac{2500}{1.9}\Big[3.8+\ln\Big(\dfrac{1+e}{1+e^{4.8}}\Big)\Big]\approx 401.4$

77. (a) $\displaystyle\int_0^4 \dfrac{k}{2+3x}\,dx = 10$

$\qquad\qquad k = \dfrac{10}{\displaystyle\int_0^4 \dfrac{1}{2+3x}\,dx}\approx \dfrac{10}{0.6486}$

$\qquad\qquad = 15.417 \ \left(= \dfrac{30}{\ln 7}\right)$

$\quad$ (b) $\displaystyle\int_0^4 \dfrac{15.417}{2+3x}\,dx$

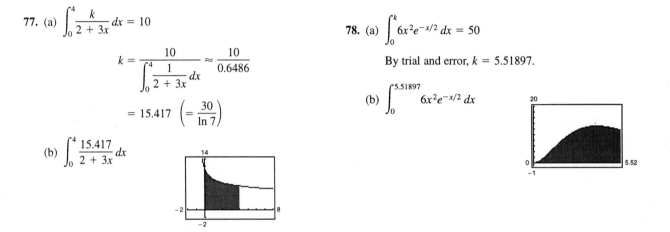

78. (a) $\displaystyle\int_0^k 6x^2 e^{-x/2}\,dx = 50$

$\qquad$ By trial and error, $k = 5.51897.$

$\quad$ (b) $\displaystyle\int_0^{5.51897} 6x^2 e^{-x/2}\,dx$

Section 7.7   Indeterminate Forms and L'Hôpital's Rule

1. $\displaystyle\lim_{x\to 0}\dfrac{\sin 5x}{\sin 2x}\approx 2.5\left(\text{exact: }\dfrac{5}{2}\right)$

x	-0.1	-0.01	-0.001	0.001	0.01	0.1
$f(x)$	2.4132	2.4991	2.500	2.500	2.4991	2.4132

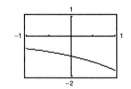

2. $\displaystyle\lim_{x\to 0}\dfrac{1-e^x}{x}\approx -1$

x	-0.1	-0.01	-0.001	0.001	0.01	0.1
$f(x)$	-0.9516	-0.9950	-0.9995	-1.00005	-1.005	-1.0517

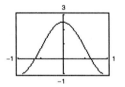

3. $\displaystyle\lim_{x\to\infty} x^5 e^{-x/100}\approx 0$

x	1	10	10^2	10^3	10^4	10^5
$f(x)$	0.9901	90,484	3.7×10^9	4.5×10^{10}	0	0

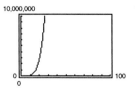

4. $\lim\limits_{x \to \infty} \dfrac{6x}{\sqrt{3x^2 - 2x}} \approx 3.4641 \left(\text{exact: } \dfrac{6}{\sqrt{3}}\right)$

x	1	10	10^2	10^3	10^4	10^5
$f(x)$	6	3.5857	3.4757	3.4653	3.4642	3.4641

5. (a) $\lim\limits_{x \to 3} \dfrac{2(x - 3)}{x^2 - 9} = \lim\limits_{x \to 3} \dfrac{2(x - 3)}{(x + 3)(x - 3)} = \lim\limits_{x \to 3} \dfrac{2}{x + 3} = \dfrac{1}{3}$

(b) $\lim\limits_{x \to 3} \dfrac{2(x - 3)}{x^2 - 9} = \lim\limits_{x \to 3} \dfrac{(d/dx)[2(x - 3)]}{(d/dx)[x^2 - 9]} = \lim\limits_{x \to 3} \dfrac{2}{2x} = \dfrac{2}{6} = \dfrac{1}{3}$

6. (a) $\lim\limits_{x \to -1} \dfrac{2x^2 - x - 3}{x + 1} = \lim\limits_{x \to -1} \dfrac{(2x - 3)(x + 1)}{x + 1} = \lim\limits_{x \to -1} (2x - 3) = -5$

(b) $\lim\limits_{x \to -1} \dfrac{2x^2 - x - 3}{x + 1} = \lim\limits_{x \to -1} \dfrac{(d/dx)[2x^2 - x - 3]}{(d/dx)[x + 1]} = \lim\limits_{x \to -1} \dfrac{4x - 1}{1} = -5$

7. (a) $\lim\limits_{x \to 3} \dfrac{\sqrt{x + 1} - 2}{x - 3} = \lim\limits_{x \to 3} \dfrac{\sqrt{x + 1} - 2}{x - 3} \cdot \dfrac{\sqrt{x + 1} + 2}{\sqrt{x + 1} + 2} = \lim\limits_{x \to 3} \dfrac{(x + 1) - 4}{(x - 3)\left[\sqrt{x + 1} + 2\right]} = \lim\limits_{x \to 3} \dfrac{1}{\sqrt{x + 1} + 2} = \dfrac{1}{4}$

(b) $\lim\limits_{x \to 3} \dfrac{\sqrt{x + 1} - 2}{x - 3} = \lim\limits_{x \to 3} \dfrac{(d/dx)\left[\sqrt{x + 1} - 2\right]}{(d/dx)[x - 3]} = \lim\limits_{x \to 3} \dfrac{1/(2\sqrt{x + 1})}{1} = \dfrac{1}{4}$

8. (a) $\lim\limits_{x \to 0} \dfrac{\sin 4x}{2x} = \lim\limits_{x \to 0} 2\left(\dfrac{\sin 4x}{4x}\right) = 2(1) = 2$ (b) $\lim\limits_{x \to 0} \dfrac{\sin 4x}{2x} = \lim\limits_{x \to 0} \dfrac{(d/dx)[\sin 4x]}{(d/dx)[2x]} = \lim\limits_{x \to 0} \dfrac{4 \cos 4x}{2} = 2$

9. (a) $\lim\limits_{x \to \infty} \dfrac{5x^2 - 3x + 1}{3x^2 - 5} = \lim\limits_{x \to \infty} \dfrac{5 - (3/x) + (1/x^2)}{3 - (5/x^2)} = \dfrac{5}{3}$

(b) $\lim\limits_{x \to \infty} \dfrac{5x^2 - 3x + 1}{3x^2 - 5} = \lim\limits_{x \to \infty} \dfrac{(d/dx)[5x^2 - 3x + 1]}{(d/dx)[3x^2 - 5]} = \lim\limits_{x \to \infty} \dfrac{10x - 3}{6x} = \lim\limits_{x \to \infty} \dfrac{(d/dx)[10x - 3]}{(d/dx)[6x]} = \lim\limits_{x \to \infty} \dfrac{10}{6} = \dfrac{5}{3}$

10. (a) $\lim\limits_{x \to \infty} \dfrac{2x + 1}{4x^2 + x} = \lim\limits_{x \to \infty} \dfrac{(2/x) + (1/x^2)}{4 + (1/x)} = \dfrac{0}{4} = 0$

(b) $\lim\limits_{x \to \infty} \dfrac{2x + 1}{4x^2 + x} = \lim\limits_{x \to \infty} \dfrac{(d/dx)[2x + 1]}{(d/dx)[4x^2 + x]} = \lim\limits_{x \to \infty} \dfrac{2}{8x + 1} = 0$

11. $\lim\limits_{x \to 2} \dfrac{x^2 - x - 2}{x - 2} = \lim\limits_{x \to 2} \dfrac{2x - 1}{1} = 3$ **12.** $\lim\limits_{x \to -1} \dfrac{x^2 - x - 2}{x + 1} = \lim\limits_{x \to -1} \dfrac{2x - 1}{1} = -3$

13. $\lim\limits_{x \to 0} \dfrac{\sqrt{4 - x^2} - 2}{x} = \lim\limits_{x \to 0} \dfrac{-x/\sqrt{4 - x^2}}{1} = 0$ **14.** $\lim\limits_{x \to 2^-} \dfrac{\sqrt{4 - x^2}}{x - 2} = \lim\limits_{x \to 2^-} \dfrac{-x/\sqrt{4 - x^2}}{1}$

$= \lim\limits_{x \to 2^-} \dfrac{-x}{\sqrt{4 - x^2}} = -\infty$

15. $\lim\limits_{x \to 0} \dfrac{e^x - (1 - x)}{x} = \lim\limits_{x \to 0} \dfrac{e^x + 1}{1} = 2$ **16.** $\lim\limits_{x \to 0^+} \dfrac{e^x - (1 + x)}{x^3} = \lim\limits_{x \to 0^+} \dfrac{e^x - 1}{3x^2}$

$= \lim\limits_{x \to 0^+} \dfrac{e^x}{6x} = \infty$

17. Case 1: $n = 1$

$$\lim_{x \to 0^+} \frac{e^x - (1 + x)}{x} = \lim_{x \to 0^+} \frac{e^x - 1}{1} = 0$$

Case 2: $n = 2$

$$\lim_{x \to 0^+} \frac{e^x - (1 + x)}{x^2} = \lim_{x \to 0^+} \frac{e^x - 1}{2x} = \lim_{x \to 0^+} \frac{e^x}{2} = \frac{1}{2}$$

Case 3: $n \geq 3$

$$\lim_{x \to 0^+} \frac{e^x - (1 + x)}{x^n} = \lim_{x \to 0^+} \frac{e^x - 1}{nx^{n-1}} = \lim_{x \to 0^+} \frac{e^x}{n(n - 1)x^{n-2}} = \infty$$

18. $\displaystyle \lim_{x \to 1} \frac{\ln x}{x^2 - 1} = \lim_{x \to 1} \frac{(1/x)}{2x}$

$\displaystyle = \lim_{x \to 1} \frac{1}{2x^2} = \frac{1}{2}$

19. $\displaystyle \lim_{x \to 0} \frac{\sin 2x}{\sin 3x} = \lim_{x \to 0} \frac{2 \cos 2x}{3 \cos 3x} = \frac{2}{3}$

20. $\displaystyle \lim_{x \to 0} \frac{\sin ax}{\sin bx} = \lim_{x \to 0} \frac{a \cos ax}{b \cos bx} = \frac{a}{b}$

21. $\displaystyle \lim_{x \to 0} \frac{\arcsin x}{x} = \lim_{x \to 0} \frac{1/\sqrt{1 - x^2}}{1} = 1$

22. $\displaystyle \lim_{x \to 1} \frac{\arctan x - (\pi/4)}{x - 1} = \lim_{x \to 1} \frac{1/(1 + x^2)}{1} = \frac{1}{2}$

23. $\displaystyle \lim_{x \to \infty} \frac{3x^2 - 2x + 1}{2x^2 + 3} = \lim_{x \to \infty} \frac{6x - 2}{4x}$

$\displaystyle = \lim_{x \to \infty} \frac{6}{4} = \frac{3}{2}$

24. $\displaystyle \lim_{x \to \infty} \frac{x - 1}{x^2 + 2x + 3} = \lim_{x \to \infty} \frac{1}{2x + 2} = 0$

25. $\displaystyle \lim_{x \to \infty} \frac{x^2 + 2x + 3}{x - 1} = \lim_{x \to \infty} \frac{2x + 2}{1} = \infty$

26. $\displaystyle \lim_{x \to \infty} \frac{x^2}{e^x} = \lim_{x \to \infty} \frac{2x}{e^x} = \lim_{x \to \infty} \frac{2}{e^x} = 0$

27. $\displaystyle \lim_{x \to \infty} \frac{x}{\sqrt{x^2 + 1}} = \lim_{x \to \infty} \frac{1}{\sqrt{1 + (1/x^2)}} = 1$

Note: L'Hôpital's Rule does not work on this limit. See Exercise 67.

28. $\displaystyle \lim_{x \to \infty} \frac{\sin x}{x - \pi} = 0$

Note: Use the Squeeze Theorem for $x > \pi$.

$$-\frac{1}{x - \pi} \leq \frac{\sin x}{x - \pi} \leq \frac{1}{x - \pi}$$

29. $\displaystyle \lim_{x \to \infty} \frac{\ln x}{x} = \lim_{x \to \infty} \frac{1/x}{1} = 0$

30. $\displaystyle \lim_{x \to \infty} \frac{e^x}{x} = \lim_{x \to \infty} \frac{e^x}{1} = \infty$

31. (a) $\displaystyle \lim_{x \to 0^+} (-x \ln x) = (-0)(-\infty) = (0)(\infty)$

(b) $\displaystyle \lim_{x \to 0^+} (-x \ln x) = \lim_{x \to 0^+} \frac{\ln x}{-1/x}$

$\displaystyle = \lim_{x \to 0^+} \frac{1/x}{1/x^2}$

$\displaystyle = \lim_{x \to 0^+} x = 0$

(c)

32. (a) $\displaystyle \lim_{x \to 0^+} x^2 \cot x = (0)(\infty)$

(b) $\displaystyle \lim_{x \to 0^+} x^2 \cot x = \lim_{x \to 0^+} \frac{x^2}{\tan x} = \lim_{x \to 0^+} \frac{2x}{\sec^2 x} = 0$

(c)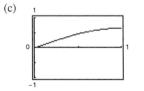

33. (a) $\displaystyle\lim_{x\to\infty} \left(x \sin \frac{1}{x}\right) = (\infty)(0)$

(b) $\displaystyle\lim_{x\to\infty} x \sin \frac{1}{x} = \lim_{x\to\infty} \frac{\sin(1/x)}{1/x}$

$\qquad = \displaystyle\lim_{x\to\infty} \frac{(-1/x^2)\cos(1/x)}{-1/x^2}$

$\qquad = \displaystyle\lim_{x\to\infty} \cos\left(\frac{1}{x}\right) = 1$

(c)

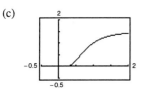

34. (a) $\displaystyle\lim_{x\to\infty} \left(x \tan \frac{1}{x}\right) = (\infty)(0)$

(b) $\displaystyle\lim_{x\to\infty} x \tan \frac{1}{x} = \lim_{x\to\infty} \frac{\tan(1/x)}{1/x}$

$\qquad = \displaystyle\lim_{x\to\infty} \frac{-(1/x^2)\sec^2(1/x)}{-(1/x^2)}$

$\qquad = \displaystyle\lim_{x\to\infty} \sec^2\left(\frac{1}{x}\right) = 1$

(c)

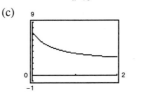

35. (a) $\displaystyle\lim_{x\to 0^+} x^{1/x} = 0^\infty = 0$, not indeterminant
(See Exercise 83)

(b) Let $y = x^{1/x}$

$\ln y = \ln x^{1/x} = \dfrac{1}{x} \ln x.$

Since $x \to 0^+,\ \dfrac{1}{x} \ln x \to (\infty)(-\infty) = -\infty$. Hence,

$\ln y \to -\infty \implies y \to 0^+.$

Therefore, $\displaystyle\lim_{x\to 0^+} x^{1/x} = 0.$

(c)

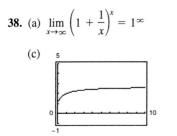

36. (a) $\displaystyle\lim_{x\to 0^+} (e^x + x)^{1/x} = 1^\infty$

(b) Let $y = \displaystyle\lim_{x\to 0^+} (e^x + x)^{1/x}.$

$\ln y = \displaystyle\lim_{x\to 0^+} \frac{\ln(e^x + x)}{x}$

$\qquad = \displaystyle\lim_{x\to 0^+} \left(\frac{(e^x + 1)/(e^x + x)}{1}\right) = 2$

Thus, $\ln y = 2 \implies y = e^2.$
Therefore, $\displaystyle\lim_{x\to 0^+} (e^x + x)^{1/x} = e^2.$

(c)

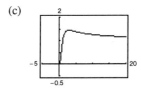

37. (a) $\displaystyle\lim_{x\to\infty} x^{1/x} = \infty^0$
(b) Let $y = \displaystyle\lim_{x\to\infty} x^{1/x}.$

$\ln y = \displaystyle\lim_{x\to\infty} \frac{\ln x}{x} = \lim_{x\to\infty} \left(\frac{1/x}{1}\right) = 0$

Thus, $\ln y = 0 \implies y = e^0 = 1$. Therefore,

$\displaystyle\lim_{x\to\infty} x^{1/x} = 1.$

(c)

(graph)

38. (a) $\displaystyle\lim_{x\to\infty} \left(1 + \frac{1}{x}\right)^x = 1^\infty$

(c)

(graph)

(b) Let $y = \displaystyle\lim_{x\to\infty} \left(1 + \frac{1}{x}\right)^x.$

$\ln y = \displaystyle\lim_{x\to\infty} \left[x \ln\left(1 + \frac{1}{x}\right)\right] = \lim_{x\to\infty} \frac{\ln[1 + (1/x)]}{1/x}$

$\qquad = \displaystyle\lim_{x\to\infty} \frac{\left[\dfrac{(-1/x^2)}{1 + (1/x)}\right]}{(-1/x^2)} = \lim_{x\to\infty} \frac{1}{1 + (1/x)} = 1$

Thus, $\ln y = 1 \implies y = e^1 = e$. Therefore,

$\displaystyle\lim_{x\to\infty} \left(1 + \frac{1}{x}\right)^x = e.$

39. (a) $\lim\limits_{x \to 0^+} (1 + x)^{1/x} = 1^\infty$

(b) Let $y = \lim\limits_{x \to 0^+} (1 + x)^{1/x}$.

$$\ln y = \lim\limits_{x \to 0^+} \frac{\ln(1 + x)}{x}$$

$$= \lim\limits_{x \to 0^+} \left(\frac{1/(1 + x)}{1} \right) = 1$$

Thus, $\ln y = 1 \implies y = e^1 = e$.
Therefore, $\lim\limits_{x \to 0^+} (1 + x)^{1/x} = e$.

(c)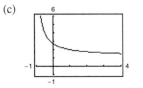

40. (a) $\lim\limits_{x \to \infty} (1 + x)^{1/x} = \infty^0$

(b) Let $y = \lim\limits_{x \to \infty} (1 + x)^{1/x}$.

$$\ln y = \lim\limits_{x \to \infty} \frac{\ln(1 + x)}{x}$$

$$= \lim\limits_{x \to \infty} \left(\frac{1/(1 + x)}{1} \right) = 0$$

Thus, $\ln y = 0 \implies y = e^0 = 1$.
Therefore, $\lim\limits_{x \to \infty} (1 + x)^{1/x} = 1$.

(c)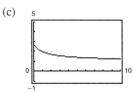

41. (a) $\lim\limits_{x \to 2^+} \left(\frac{8}{x^2 - 4} - \frac{x}{x - 2} \right) = \infty - \infty$

(b) $\lim\limits_{x \to 2^+} \left(\frac{8}{x^2 - 4} - \frac{x}{x - 2} \right) = \lim\limits_{x \to 2^+} \frac{8 - x(x + 2)}{x^2 - 4}$

$$= \lim\limits_{x \to 2^+} \frac{(2 - x)(4 + x)}{(x + 2)(x - 2)}$$

$$= \lim\limits_{x \to 2^+} \frac{-(x + 4)}{x + 2} = \frac{-3}{2}$$

(c)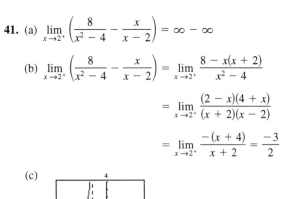

42. (a) $\lim\limits_{x \to 2^+} \left(\frac{1}{x^2 - 4} - \frac{\sqrt{x - 1}}{x^2 - 4} \right) = \infty - \infty$

(b) $\lim\limits_{x \to 2^+} \left(\frac{1}{x^2 - 4} - \frac{\sqrt{x - 1}}{x^2 - 4} \right) = \lim\limits_{x \to 2^+} \frac{1 - \sqrt{x - 1}}{x^2 - 4}$

$$= \lim\limits_{x \to 2^+} \frac{-1/(2\sqrt{x - 1})}{2x}$$

$$= \lim\limits_{x \to 2^+} \frac{-1}{4x\sqrt{x - 1}} = \frac{-1}{8}$$

(c)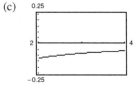

43. (a) $\lim\limits_{x \to 1^+} \left(\frac{3}{\ln x} - \frac{2}{x - 1} \right) = \infty - \infty$

(b) $\lim\limits_{x \to 1^+} \left(\frac{3}{\ln x} - \frac{2}{x - 1} \right) = \lim\limits_{x \to 1^+} \frac{3x - 3 - 2\ln x}{(x - 1)\ln x}$

$$= \lim\limits_{x \to 1^+} \frac{3 - (2/x)}{[(x - 1)/x] + \ln x} = \infty$$

(c)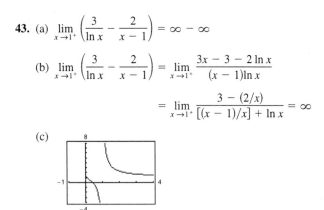

44. (a) $\lim\limits_{x \to 0^+} \left(\frac{1}{x} - \frac{1}{x^2} \right) = \infty - \infty$

(b) $\lim\limits_{x \to 0^+} \left(\frac{1}{x} - \frac{1}{x^2} \right) = \lim\limits_{x \to 0^+} \left(\frac{x - 1}{x^2} \right) = -\infty$

(c)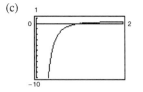

45. (a)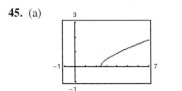

(b) $\lim\limits_{x \to 3} \frac{x - 3}{\ln(2x - 5)} = \lim\limits_{x \to 3} \frac{1}{2/(2x - 5)}$

$$= \lim\limits_{x \to 3} \frac{2x - 5}{2} = \frac{1}{2}$$

46. (a)

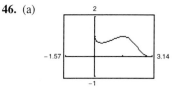

(b) Let $y = (\sin x)^x$, then $\ln y = x \ln(\sin x)$.

$$\lim_{x \to 0^+} \frac{\ln(\sin x)}{1/x} = \lim_{x \to 0^+} \frac{\cos x / \sin x}{-1/x^2} = \lim_{x \to 0^+} \frac{-x^2}{\tan x} = \lim_{x \to 0^+} \frac{-2x}{\sec^2 x} = 0$$

Therefore, since $\ln y = 0$, $y = 1$ and $\lim_{x \to 0^+} (\sin x)^x = 1$.

47. (a)

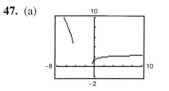

(b) $\displaystyle \lim_{x \to \infty} \left(\sqrt{x^2 + 5x + 2} - x \right) = \lim_{x \to \infty} \left(\sqrt{x^2 + 5x + 2} - x \right) \frac{\left(\sqrt{x^2 + 5x + 2} + x \right)}{\left(\sqrt{x^2 + 5x + 2} + x \right)}$

$$= \lim_{x \to \infty} \frac{(x^2 + 5x + 2) - x^2}{\sqrt{x^2 + 5x + 2} + x}$$

$$= \lim_{x \to \infty} \frac{5x + 2}{\sqrt{x^2 + 5x + 2} + x}$$

$$= \lim_{x \to \infty} \frac{5 + (2/x)}{\sqrt{1 + (5/x) + (2/x^2)} + 1} = \frac{5}{2}$$

48. (a)

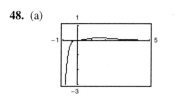

(b) $\displaystyle \lim_{x \to \infty} \frac{x^3}{e^{2x}} = \lim_{x \to \infty} \frac{3x^2}{2e^{2x}} = \lim_{x \to \infty} \frac{6x}{4e^{2x}} = \lim_{x \to \infty} \frac{6}{8e^{2x}} = 0$

49. (a) Let $f(x) = x^2 - 25$ and $g(x) = x - 5$.

    (b) Let $f(x) = (x - 5)^2$ and $g(x) = x^2 - 25$.

    (c) Let $f(x) = x^2 - 25$ and $g(x) = (x - 5)^3$.

50. Let $f(x) = x + 25$ and $g(x) = x$.

51. $\displaystyle \lim_{x \to \infty} \frac{x^2}{e^{5x}} = \lim_{x \to \infty} \frac{2x}{5e^{5x}} = \lim_{x \to \infty} \frac{2}{25e^{5x}} = 0$

52. $\displaystyle \lim_{x \to \infty} \frac{x^3}{e^{2x}} = \lim_{x \to \infty} \frac{3x^2}{2e^{2x}} = \lim_{x \to \infty} \frac{6x}{4e^{2x}} = \lim_{x \to \infty} \frac{6}{8e^{2x}} = 0$

53. $\displaystyle \lim_{x \to \infty} \frac{(\ln x)^3}{x} = \lim_{x \to \infty} \frac{3(\ln x)^2(1/x)}{1}$

$$= \lim_{x \to \infty} \frac{3(\ln x)^2}{x}$$

$$= \lim_{x \to \infty} \frac{6(\ln x)(1/x)}{1}$$

$$= \lim_{x \to \infty} \frac{6(\ln x)}{x} = \lim_{x \to \infty} \frac{6}{x} = 0$$

54. $\displaystyle \lim_{x \to \infty} \frac{(\ln x)^2}{x^3} = \lim_{x \to \infty} \frac{(2 \ln x)/x}{3x^2}$

$$= \lim_{x \to \infty} \frac{2 \ln x}{3x^3}$$

$$= \lim_{x \to \infty} \frac{2/x}{9x^2} = \lim_{x \to \infty} \frac{2}{9x^3} = 0$$

55. $\displaystyle \lim_{x \to \infty} \frac{(\ln x)^n}{x^m} = \lim_{x \to \infty} \frac{n(\ln x)^{n-1}/x}{mx^{m-1}}$

$$= \lim_{x \to \infty} \frac{n(\ln x)^{n-1}}{mx^m}$$

$$= \lim_{x \to \infty} \frac{n(n-1)(\ln x)^{n-2}}{m^2 x^m}$$

$$= \cdots = \lim_{x \to \infty} \frac{n!}{m^n x^m} = 0$$

56. $\displaystyle \lim_{x \to \infty} \frac{x^m}{e^{nx}} = \lim_{x \to \infty} \frac{mx^{m-1}}{ne^{nx}}$

$$= \lim_{x \to \infty} \frac{m(m-1)x^{m-2}}{n^2 e^{nx}}$$

$$= \cdots = \lim_{x \to \infty} \frac{m!}{n^m e^{nx}} = 0$$

57.

x	10	10^2	10^4	10^6	10^8	10^{10}
$\dfrac{(\ln x)^4}{x}$	2.811	4.498	0.720	0.036	0.001	0.000

58.

x	1	5	10	20	30	40	50	100
$\dfrac{e^x}{x^5}$	2.718	0.047	0.220	151.614	4.40×10^5	2.30×10^9	1.66×10^{13}	2.69×10^{33}

59. $y = x^{1/x},\ x > 0$

Horizontal asymptote: $y = 1$ (See Exercise 37)

$$\ln y = \frac{1}{x}\ln x$$

$$\frac{1}{y}\frac{dy}{dx} = \frac{1}{x}\left(\frac{1}{x}\right) + (\ln x)\left(-\frac{1}{x^2}\right)$$

$$\frac{dy}{dx} = x^{1/x}\left(\frac{1}{x^2}\right)(1 - \ln x) = x^{(1/x)-2}(1 - \ln x) = 0$$

Critical number: $x = e$

Intervals: $(0, e)$ (e, ∞)

Sign of dy/dx: $+$ $-$

$y = f(x)$: Increasing   Decreasing

Relative maximum: $(e, e^{1/e})$

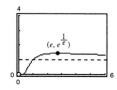

60. $y = x^x,\ x > 0$

$$\lim_{x\to\infty} x^x = \infty \text{ and } \lim_{x\to 0^+} x^x = 1$$

No horizontal asymptotes

$$\ln y = x \ln x$$

$$\frac{1}{y}\frac{dy}{dx} = x\left(\frac{1}{x}\right) + \ln x$$

$$\frac{dy}{dx} = x^x(1 + \ln x) = 0$$

Critical number: $x = e^{-1}$

Intervals: $(0, e^{-1})$ $(e^{-1}, 0)$

Sign of dy/dx: $-$ $+$

$y = f(x)$: Decreasing   Increasing

Relative minimum: $\left(e^{-1}, (e^{-1})^{e^{-1}}\right) = \left(\dfrac{1}{e}, \left(\dfrac{1}{e}\right)^{1/e}\right)$

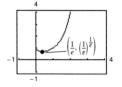

61. $y = 2xe^{-x}$

$$\lim_{x\to\infty}\frac{2x}{e^x} = \lim_{x\to\infty}\frac{2}{e^x} = 0$$

Horizontal asymptote: $y = 0$

$$\frac{dy}{dx} = 2x(-e^{-x}) + 2e^{-x}$$

$$= 2e^{-x}(1 - x) = 0$$

Critical number: $x = 1$

Intervals: $(-\infty, 1)$ $(1, \infty)$

Sign of dy/dx: $+$ $-$

$y = f(x)$: Increasing   Decreasing

Relative maximum: $\left(1, \dfrac{2}{e}\right)$

62. $y = \dfrac{\ln x}{x}$

Horizontal asymptote: $y = 0$ (See Exercise 29)

$\dfrac{dy}{dx} = \dfrac{x(1/x) - (\ln x)(1)}{x^2} = \dfrac{1 - \ln x}{x^2} = 0$

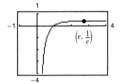

Critical number: $x = e$

Intervals:	$(0, e)$	(e, ∞)
Sign of dy/dx:	$+$	$-$
$y = f(x)$:	Increasing	Decreasing

Relative maximum: $\left(e, \dfrac{1}{e}\right)$

63. $\displaystyle\lim_{x \to 0} \dfrac{e^{2x} - 1}{e^x} = \dfrac{0}{1} = 0$

Limit is not of the form $0/0$ or ∞/∞.
L'Hôpital's Rule does not apply.

64. $\displaystyle\lim_{x \to \infty} \dfrac{\sin \pi x - 1}{x} = 0$ (Numerator is bounded)

Limit is not of the form $0/0$ or ∞/∞.
L'Hôpital's Rule does not apply.

65. $\displaystyle\lim_{x \to \infty} x \cos \dfrac{1}{x} = \infty(1) = \infty$

Limit is not of the form $0/0$ or ∞/∞.
L'Hôpital's Rule does not apply.

66. $\displaystyle\lim_{x \to \infty} \dfrac{e^{-x}}{1 + e^{-x}} = \dfrac{0}{1 + 0} = 0$

Limit is not of the form $0/0$ or ∞/∞.
L'Hôpital's Rule does not apply.

67. (a) $\displaystyle\lim_{x \to \infty} \dfrac{x}{\sqrt{x^2 + 1}} = \lim_{x \to \infty} \dfrac{x/x}{\sqrt{x^2 + 1}/x}$

$= \displaystyle\lim_{x \to \infty} \dfrac{1}{\sqrt{x^2 + 1}/\sqrt{x^2}}$

$= \displaystyle\lim_{x \to \infty} \dfrac{1}{\sqrt{1 + (1/x^2)}}$

$= \dfrac{1}{\sqrt{1 + 0}} = 1$

(b) $\displaystyle\lim_{x \to \infty} \dfrac{x}{\sqrt{x^2 + 1}} = \lim_{x \to \infty} \dfrac{1}{x/\sqrt{x^2 + 1}}$

$= \displaystyle\lim_{x \to \infty} \dfrac{\sqrt{x^2 + 1}}{x} = \lim_{x \to \infty} \dfrac{x/\sqrt{x^2 + 1}}{1}$

$= \displaystyle\lim_{x \to \infty} \dfrac{x}{\sqrt{x^2 + 1}}$

Applying L'Hôpital's rule twice results in the original limit, so L'Hôpital's rule fails.

(c)

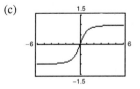

68. $A = P\left(1 + \dfrac{r}{n}\right)^{nt}$

$\ln A = \ln P + nt \ln\left(1 + \dfrac{r}{n}\right) = \ln P + \dfrac{\ln\left(1 + \dfrac{r}{n}\right)}{\dfrac{1}{nt}}$

$\displaystyle\lim_{n \to \infty} \left[\dfrac{\ln\left(1 + \dfrac{r}{n}\right)}{\dfrac{1}{nt}}\right] = \lim_{n \to \infty} \left[\dfrac{-\dfrac{r}{n^2}\left(\dfrac{1}{1 + (r/n)}\right)}{-\left(\dfrac{1}{n^2 t}\right)}\right] = \lim_{n \to \infty} \left[rt\left(\dfrac{1}{1 + \dfrac{r}{n}}\right)\right] = rt$

Since $\displaystyle\lim_{n \to \infty} \ln A = \ln P + rt$, we have $\displaystyle\lim_{n \to \infty} A = e^{(\ln P + rt)} = e^{\ln P}e^{rt} = Pe^{rt}$. Alternatively,

$\displaystyle\lim_{n \to \infty} A = \lim_{n \to \infty} P\left(1 + \dfrac{r}{n}\right)^{nt} = \lim_{n \to \infty} P\left[\left(1 + \dfrac{r}{n}\right)^{n/r}\right]^{rt} = Pe^{rt}$.

69. $\lim\limits_{k \to 0} \dfrac{32\left(1 - e^{-kt} + \dfrac{v_0 k e^{-kt}}{32}\right)}{k} = \lim\limits_{k \to 0} \dfrac{32(1 - e^{-kt})}{k} + \lim\limits_{k \to 0} \left(v_0 e^{-kt}\right)$

$\qquad\qquad\qquad\qquad\qquad = \lim\limits_{k \to 0} \dfrac{32(0 + te^{-kt})}{1} + \lim\limits_{k \to 0} \left(\dfrac{v_0}{e^{kt}}\right) = 32t + v_0$

70. Let N be a fixed value for n. Then

$$\lim_{x \to \infty} \frac{x^{N-1}}{e^x} = \lim_{x \to \infty} \frac{(N-1)x^{N-2}}{e^x} = \lim_{x \to \infty} \frac{(N-1)(N-2)x^{N-3}}{e^x} = \ldots = \lim_{x \to \infty} \left[\frac{(N-1)!}{e^x}\right] = 0. \quad \text{(See Exercise 56)}$$

71. Area of triangle: $\dfrac{1}{2}(2x)(1 - \cos x) = x - x \cos x$

Shaded area: Area of rectangle − Area under curve

$$2x(1 - \cos x) - 2\int_0^x (1 - \cos t)\, dt = 2x(1 - \cos x) - 2\Big[t - \sin t\Big]_0^x$$

$$= 2x(1 - \cos x) - 2(x - \sin x) = 2 \sin x - 2x \cos x$$

Ratio: $\lim\limits_{x \to 0} \dfrac{x - x \cos x}{2 \sin x - 2x \cos x} = \lim\limits_{x \to 0} \dfrac{1 + x \sin x - \cos x}{2 \cos x + 2x \sin x - 2 \cos x}$

$\qquad\qquad\qquad\qquad = \lim\limits_{x \to 0} \dfrac{1 + x \sin x - \cos x}{2x \sin x}$

$\qquad\qquad\qquad\qquad = \lim\limits_{x \to 0} \dfrac{x \cos x + \sin x + \sin x}{2x \cos x + 2 \sin x}$

$\qquad\qquad\qquad\qquad = \lim\limits_{x \to 0} \dfrac{x \cos x + 2 \sin x}{2x \cos x + 2 \sin x} \cdot \dfrac{1/\cos x}{1/\cos x}$

$\qquad\qquad\qquad\qquad = \lim\limits_{x \to 0} \dfrac{x + 2 \tan x}{2x + 2 \tan x}$

$\qquad\qquad\qquad\qquad = \lim\limits_{x \to 0} \dfrac{1 + 2 \sec^2 x}{2 + 2 \sec^2 x} = \dfrac{3}{4}$

72. $f(x) = \dfrac{x^k - 1}{k}$

$k = 1, \qquad f(x) = x - 1$

$k = 0.1, \qquad f(x) = \dfrac{x^{0.1} - 1}{0.1} = 10(x^{0.1} - 1)$

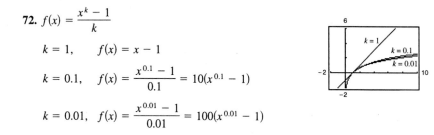

$k = 0.01, \quad f(x) = \dfrac{x^{0.01} - 1}{0.01} = 100(x^{0.01} - 1)$

$\lim\limits_{k \to 0^+} \dfrac{x^k - 1}{k} = \lim\limits_{k \to 0^+} \dfrac{x^k(\ln x)}{1} = \ln x$

73. $f(x) = x^3,\ g(x) = x^2 + 1,\ [0, 1]$

$\dfrac{f(b) - f(a)}{g(b) - g(a)} = \dfrac{f'(c)}{g'(c)}$

$\dfrac{f(1) - f(0)}{g(1) - g(0)} = \dfrac{3c^2}{2c}$

$\dfrac{1}{1} = \dfrac{3c}{2}$

$c = \dfrac{2}{3}$

74. $f(x) = \dfrac{1}{x},\ g(x) = x^2 - 4,\ [1, 2]$

$\dfrac{f(2) - f(1)}{g(2) - g(1)} = \dfrac{f'(c)}{g'(c)}$

$\dfrac{-1/2}{3} = \dfrac{-1/c^2}{2c}$

$-\dfrac{1}{6} = -\dfrac{1}{2c^3}$

$2c^3 = 6$

$c = \sqrt[3]{3}$

75. $f(x) = \sin x, g(x) = \cos x, \left[0, \dfrac{\pi}{2}\right]$

$$\frac{f(\pi/2) - f(0)}{g(\pi/2) - g(0)} = \frac{f'(c)}{g'(c)}$$

$$\frac{1}{-1} = \frac{\cos c}{-\sin c}$$

$$-1 = -\cot c$$

$$c = \frac{\pi}{4}$$

76. $f(x) = \ln x, g(x) = x^3, [1, 4]$

$$\frac{f(4) - f(1)}{g(4) - g(1)} = \frac{f'(c)}{g'(c)}$$

$$\frac{\ln 4}{63} = \frac{1/c}{3c^2} = \frac{1}{3c^3}$$

$$3c^3 \ln 4 = 63$$

$$c^3 = \frac{21}{\ln 4}$$

$$c = \sqrt[3]{\frac{21}{\ln 4}} \approx 2.474$$

77. False. L'Hôpital's Rule does not apply since

$$\lim_{x \to 0} (x^2 + x + 1) \neq 0.$$

$$\lim_{x \to 0} \frac{x^2 + x + 1}{x} = \lim_{x \to 0} \left(x + 1 + \frac{1}{x}\right) = 1 + \infty = \infty$$

78. False. If $y = e^x/x^2$, then

$$y' = \frac{x^2 e^x - 2xe^x}{x^4} = \frac{xe^x(x - 2)}{x^4} = \frac{e^x(x - 2)}{x^3}.$$

79. True

80. False. Let $f(x) = x$ and $g(x) = x + 1$. Then

$$\lim_{x \to \infty} \frac{x}{x + 1} = 1, \text{ but } \lim_{x \to \infty} [x - (x + 1)] = -1.$$

81. (a) $\sin \theta = BD$

$\cos \theta = DO \implies AD = 1 - \cos \theta$

Area $\triangle ABD = \dfrac{1}{2} bh = \dfrac{1}{2}(1 - \cos \theta) \sin \theta = \dfrac{1}{2} \sin \theta - \dfrac{1}{2} \sin \theta \cos \theta$

(b) Area of sector: $\dfrac{1}{2}\theta$

Shaded area: $\dfrac{1}{2}\theta - $ Area $\triangle OBD = \dfrac{1}{2}\theta - \dfrac{1}{2}(\cos \theta)(\sin \theta) = \dfrac{1}{2}\theta - \dfrac{1}{2}\sin \theta \cos \theta$

(c) $R = \dfrac{(1/2)\sin \theta - (1/2)\sin \theta \cos \theta}{(1/2)\theta - (1/2)\sin \theta \cos \theta} = \dfrac{\sin \theta - \sin \theta \cos \theta}{\theta - \sin \theta \cos \theta}$

(d) $\displaystyle\lim_{\theta \to 0} R = \lim_{\theta \to 0} \dfrac{\sin \theta - (1/2)\sin 2\theta}{\theta - (1/2)\sin 2\theta}$

$= \displaystyle\lim_{\theta \to 0} \dfrac{\cos \theta - \cos 2\theta}{1 - \cos 2\theta} = \lim_{\theta \to 0} \dfrac{-\sin \theta + 2\sin 2\theta}{2\sin 2\theta} = \lim_{\theta \to 0} \dfrac{-\cos \theta + 4\cos 2\theta}{4\cos 2\theta} = \dfrac{3}{4}$

82. $g(x) = \begin{cases} e^{-1/x^2}, & x \neq 0 \\ 0, & x = 0 \end{cases}$

$g'(0) = \displaystyle\lim_{x \to 0} \dfrac{g(x) - g(0)}{x - 0} = \lim_{x \to 0} \dfrac{e^{-1/x^2}}{x}$

Let $y = \dfrac{e^{-1/x^2}}{x}$, then $\ln y = \ln\left(\dfrac{e^{-1/x^2}}{x}\right) = -\dfrac{1}{x^2} - \ln x = \dfrac{-1 - x^2 \ln x}{x^2}$. Since

$$\lim_{x \to 0} x^2 \ln x = \lim_{x \to 0} \frac{\ln x}{1/x^2} = \lim_{x \to 0} \frac{1/x}{-2/x^3} = \lim_{x \to 0} \left(-\frac{x^2}{2}\right) = 0$$

we have $\displaystyle\lim_{x \to 0} \left(\dfrac{-1 - x^2 \ln x}{x^2}\right) = -\infty$. Thus, $\displaystyle\lim_{x \to 0} y = e^{-\infty} = 0 \implies g'(0) = 0$.

Note: The graph appears to support this conclusion—the tangent line is horizontal at $(0, 0)$.

83. $\lim\limits_{x \to a} f(x)^{g(x)}$

$$y = f(x)^{g(x)}$$

$$\ln y = g(x) \ln f(x)$$

$$\lim\limits_{x \to a} g(x) \ln f(x) = (\infty)(-\infty) = -\infty$$

As $x \to a$, $\ln y \implies -\infty$, and hence $y = 0$. Thus,

$$\lim\limits_{x \to a} f(x)^{g(x)} = 0.$$

84. $\lim\limits_{x \to a} f(x)^{g(x)}$

$$y = f(x)^{g(x)}$$

$$\ln y = g(x) \ln f(x)$$

$$\lim\limits_{x \to a} g(x) \ln f(x) = (-\infty)(-\infty) = \infty$$

As $x \to a$, $\ln y \implies \infty$, and hence $y = \infty$. Thus,

$$\lim\limits_{x \to a} f(x)^{g(x)} = \infty.$$

85. $f'(a)(b - a) - \displaystyle\int_a^b f''(t)(t - b)\, dt = f'(a)(b - a) - \left\{ \left[f'(t)(t - b) \right]_a^b - \int_a^b f'(t)\, dt \right\}$

$$= f'(a)(b - a) + f'(a)(a - b) + \left[f(t) \right]_a^b = f(b) - f(a)$$

$$dv = f''(t)dt \implies v = f'(t)$$

$$u = t - b \implies du = dt$$

86. $\lim\limits_{x \to 0^+} x^{\ln 2/(1 + \ln x)}$

Let $y = x^{\ln 2/(1 + \ln x)}$, then:

$$\ln y = \frac{\ln 2}{1 + \ln x} \cdot \ln x = \frac{(\ln 2)(\ln x)}{1 + \ln x}$$

$$\lim\limits_{x \to 0^+} \ln y = \lim\limits_{x \to 0^+} \frac{(\ln 2)(\ln x)}{1 + \ln x} = \lim\limits_{x \to 0^+} \frac{(\ln 2)/x}{1/x} = \lim\limits_{x \to 0^+} (\ln 2) = \ln 2$$

Thus, $\lim\limits_{x \to \infty} y = e^{\ln 2} = 2.$

Section 7.8   Improper Integrals

1. Infinite discontinuity at $x = 0$.

$$\int_0^4 \frac{1}{\sqrt{x}}\, dx = \lim\limits_{b \to 0^+} \int_b^4 \frac{1}{\sqrt{x}}\, dx$$

$$= \lim\limits_{b \to 0^+} \left[2\sqrt{x} \right]_b^4$$

$$= \lim\limits_{b \to 0^+} \left(4 - 2\sqrt{b} \right) = 4$$

Converges

2. Infinite discontinuity at $x = 3$.

$$\int_3^4 \frac{1}{(x - 3)^{3/2}}\, dx = \lim\limits_{b \to 3^+} \int_b^4 (x - 3)^{-3/2}\, dx$$

$$= \lim\limits_{b \to 3^+} \left[-2(x - 3)^{-1/2} \right]_b^4$$

$$= \lim\limits_{b \to 3^+} \left[-2 + \frac{2}{\sqrt{b - 3}} \right] = \infty$$

Diverges

3. Infinite discontinuity at $x = 1$.

$$\int_0^2 \frac{1}{(x - 1)^2}\, dx = \int_0^1 \frac{1}{(x - 1)^2}\, dx + \int_1^2 \frac{1}{(x - 1)^2}\, dx$$

$$= \lim\limits_{b \to 1^-} \int_0^b \frac{1}{(x - 1)^2}\, dx + \lim\limits_{c \to 1^+} \int_c^2 \frac{1}{(x - 1)^2}\, dx$$

$$= \lim\limits_{b \to 1^-} \left[-\frac{1}{x - 1} \right]_0^b + \lim\limits_{c \to 1^+} \left[-\frac{1}{x - 1} \right]_c^2 = (\infty - 1) + (-1 + \infty)$$

Diverges

4. Infinite discontinuity at $x = 1$.

$$\int_0^2 \frac{1}{(x-1)^{2/3}}\, dx = \int_0^1 \frac{1}{(x-1)^{2/3}}\, dx + \int_1^2 \frac{1}{(x-1)^{2/3}}\, dx$$

$$= \lim_{b \to 1^-} \int_0^b \frac{1}{(x-1)^{2/3}}\, dx + \lim_{c \to 1^+} \int_c^2 \frac{1}{(x-1)^{2/3}}\, dx$$

$$= \lim_{b \to 1^-} \left[3\sqrt[3]{x-1} \right]_0^b + \lim_{c \to 1^+} \left[3\sqrt[3]{x-1} \right]_c^2 = (0+3) + (3-0) = 6$$

Converges

5. Infinite limit of integration.

$$\int_0^\infty e^{-x}\, dx = \lim_{b \to \infty} \int_0^b e^{-x}\, dx$$

$$= \lim_{b \to \infty} \left[-e^{-x} \right]_0^b = 0 + 1 = 1$$

Converges

6. Infinite limit of integration.

$$\int_{-\infty}^0 e^{2x}\, dx = \lim_{b \to -\infty} \int_b^0 e^{2x}\, dx$$

$$= \lim_{b \to -\infty} \left[\frac{1}{2} e^{2x} \right]_b^0 = \frac{1}{2} - 0 = \frac{1}{2}$$

Converges

7. $\int_{-1}^1 \frac{1}{x^2}\, dx \neq -2$

because the integrand is not defined at $x = 0$.
Diverges

8. $\int_0^\infty e^{-x}\, dx \neq 0$. You need to evaluate the limit.

$$\lim_{b \to \infty} \int_0^b e^{-x}\, dx = \lim_{b \to \infty} \left[-e^{-x} \right]_0^b$$

$$= \lim_{b \to \infty} \left[-e^{-b} + 1 \right] = 1$$

9. $\int_{-\infty}^0 x e^{-2x}\, dx = \lim_{b \to -\infty} \int_b^0 x e^{-2x}\, dx = \lim_{b \to -\infty} \frac{1}{4}\left[(-2x-1)e^{-2x} \right]_b^0 = \lim_{b \to -\infty} \frac{1}{4}\left[-1 + (2b+1)e^{-2b} \right] = -\infty$ (Integration by parts)

Diverges

10. $\int_0^\infty x e^{-x}\, dx = \lim_{b \to \infty} \int_0^b x e^{-x}\, dx = \lim_{b \to \infty} \left[-e^{-x}(x+1) \right]_0^b = \lim_{b \to \infty} \left[1 - e^{-b}(b+1) \right] = 1$

Since $\lim_{b \to \infty} \left(\dfrac{b+1}{e^b} \right) = 0$ by L'Hôpital's Rule.

11. $\int_0^\infty x^2 e^{-x}\, dx = \lim_{b \to \infty} \int_0^b x^2 e^{-x}\, dx = \lim_{b \to \infty} \left[-e^{-x}(x^2 + 2x + 2) \right]_0^b = \lim_{b \to \infty} \left(-\dfrac{b^2 + 2b + 2}{e^b} + 2 \right) = 2$

Since $\lim_{b \to \infty} \left(-\dfrac{b^2 + 2b + 2}{e^b} \right) = 0$ by L'Hôpital's Rule.

12. $\int_0^\infty (x-1)e^{-x}\, dx = \lim_{b \to \infty} \int_0^b (x-1)e^{-x}\, dx = \lim_{b \to \infty} \left[-x e^{-x} \right]_0^b = \lim_{b \to \infty} \left(\dfrac{-b}{e^b} + 0 \right) = 0$ by L'Hôpital's Rule.

13. $\int_1^\infty \frac{1}{x^2}\, dx = \lim_{b \to \infty} \int_1^b \frac{1}{x^2}\, dx$

$$= \lim_{b \to \infty} \left[-\frac{1}{x} \right]_1^b = 1$$

14. $\int_1^\infty \frac{1}{\sqrt{x}}\, dx = \lim_{b \to \infty} \int_1^b \frac{1}{\sqrt{x}}\, dx$

$$= \lim_{b \to \infty} \left[2\sqrt{x} \right]_1^b = \infty - 2$$

Diverges

15. $\displaystyle\int_0^\infty e^{-x}\cos x\,dx = \lim_{b\to\infty}\frac{1}{2}\Big[e^{-x}(-\cos x+\sin x)\Big]_0^b$

$$= \frac{1}{2}[0-(-1)] = \frac{1}{2}$$

16. $\displaystyle\int_0^\infty e^{-ax}\sin bx\,dx = \lim_{c\to\infty}\left[\frac{e^{-ax}(-a\sin bx - b\cos bx)}{a^2+b^2}\right]_0^c$

$$= 0 - \frac{-b}{a^2+b^2} = \frac{b}{a^2+b^2}$$

17. $\displaystyle\int_{-\infty}^\infty \frac{1}{1+x^2}\,dx = \int_{-\infty}^0 \frac{1}{1+x^2}\,dx + \int_0^\infty \frac{1}{1+x^2}\,dx$

$$= \lim_{b\to-\infty}\int_b^0 \frac{1}{1+x^2}\,dx + \lim_{c\to\infty}\int_0^c \frac{1}{1+x^2}\,dx$$

$$= \lim_{b\to-\infty}\Big[\arctan x\Big]_b^0 + \lim_{c\to\infty}\Big[\arctan x\Big]_0^c$$

$$= \frac{\pi}{2} + \frac{\pi}{2} = \pi$$

18. $\displaystyle\int_0^\infty \frac{x^3}{(x^2+1)^2}\,dx = \lim_{b\to\infty}\int_0^b \frac{x}{x^2+1}\,dx - \lim_{b\to\infty}\int_0^b \frac{x}{(x^2+1)^2}\,dx$

$$= \lim_{b\to\infty}\left[\frac{1}{2}\ln(x^2+1) + \frac{1}{2(x^2+1)}\right]_0^b$$

$$= \infty - \frac{1}{2}$$

Diverges

19. $\displaystyle\int_0^\infty \frac{1}{e^x+e^{-x}}\,dx = \lim_{b\to\infty}\int_0^b \frac{e^x}{1+e^{2x}}\,dx$

$$= \lim_{b\to\infty}\Big[\arctan(e^x)\Big]_0^b$$

$$= \frac{\pi}{2} - \frac{\pi}{4} = \frac{\pi}{4}$$

20. $\displaystyle\int_0^\infty \frac{e^x}{1+e^x}\,dx = \lim_{b\to\infty}\Big[\ln(1+e^x)\Big]_0^b = \infty - \ln 2$

Diverges

21. $\displaystyle\int_0^\infty \cos \pi x\,dx = \lim_{b\to\infty}\left[\frac{1}{\pi}\sin \pi x\right]_0^b$

Diverges since $\sin \pi x$ does not approach a limit as $x\to\infty$.

22. $\displaystyle\int_0^\infty \sin\frac{x}{2}\,dx = \lim_{b\to\infty}\left[-2\cos\frac{x}{2}\right]_0^b$

Diverges since $\cos\dfrac{x}{2}$ does not approach a limit as $x\to\infty$.

23. $\displaystyle\int_0^1 \frac{1}{x^2}\,dx = \lim_{b\to0^+}\left[\frac{-1}{x}\right]_b^1 = -1 + \infty$

Diverges

24. $\displaystyle\int_0^1 \frac{1}{x}\,dx = \lim_{b\to0^+}\Big[\ln|x|\Big]_b^1 = 0 - (-\infty)$

Diverges

25. $\displaystyle\int_0^8 \frac{1}{\sqrt[3]{8-x}}\,dx = \lim_{b\to8^-}\int_0^b \frac{1}{\sqrt[3]{8-x}}\,dx = \lim_{b\to8^-}\left[\frac{-3}{2}(8-x)^{2/3}\right]_0^b = 6$

26. $\displaystyle\int_0^e \ln x\,dx = \lim_{b\to0^+}\Big[x\ln x - x\Big]_b^e = (e-e)-(0-0) = 0$ since $\displaystyle\lim_{b\to0^+}b(\ln b) = \lim_{b\to0^+}\frac{\ln b}{1/b} = 0$ by L'Hôpital's Rule.

27. $\displaystyle\int_0^1 x\ln x\,dx = \lim_{b\to0^+}\left[\frac{x^2}{2}\ln|x| - \frac{x^2}{4}\right]_b^1 = \lim_{b\to0^+}\left[\frac{-1}{4} - \frac{b^2\ln b}{2} + \frac{b^2}{4}\right] = \frac{-1}{4}$ since $\displaystyle\lim_{b\to0^+}(b^2\ln b) = 0$ by L'Hopital's Rule.

28. $\displaystyle\int_0^{\pi/2} \sec\theta\,d\theta = \lim_{b\to(\pi/2)}\Big[\ln|\sec\theta + \tan\theta|\Big]_0^b = \infty,$

Diverges

29. $\displaystyle\int_0^{\pi/2} \tan\theta\,d\theta = \lim_{b\to(\pi/2)^-}\Big[\ln|\sec\theta|\Big]_0^b = \infty,$

Diverges

30. $\displaystyle\int_0^2 \frac{1}{\sqrt{4-x^2}}\,dx = \lim_{b\to2^-}\left[\arcsin\left(\frac{x}{2}\right)\right]_0^b = \frac{\pi}{2}$

31. $\displaystyle\int_2^4 \frac{1}{\sqrt{x^2-4}}\,dx = \lim_{b\to2^+}\left[\ln\left|x + \sqrt{x^2-4}\right|\right]_b^4$

$$= \ln\left(4 + 2\sqrt{3}\right) - \ln 2$$

$$= \ln\left(2 + \sqrt{3}\right) \approx 1.317$$

32. $\int_0^2 \dfrac{1}{4 - x^2}\, dx = \lim_{b \to 2^-} \int_0^b \dfrac{1}{4}\left(\dfrac{1}{2 + x} + \dfrac{1}{2 - x}\right) dx = \lim_{b \to 2^-} \left[\dfrac{1}{4} \ln\left|\dfrac{2 + x}{2 - x}\right|\right]_0^b = \infty - 0$

Diverges

33. $\int_0^2 \dfrac{1}{\sqrt[3]{x - 1}}\, dx = \int_0^1 \dfrac{1}{\sqrt[3]{x - 1}}\, dx + \int_1^2 \dfrac{1}{\sqrt[3]{x - 1}}\, dx$

$= \lim_{b \to 1^-} \left[\dfrac{3}{2}(x - 1)^{2/3}\right]_0^b + \lim_{c \to 1^+} \left[\dfrac{3}{2}(x - 1)^{2/3}\right]_c^2 = \dfrac{-3}{2} + \dfrac{3}{2} = 0$

34. $\int_0^2 \dfrac{1}{(x - 1)^{4/3}}\, dx = \int_0^1 \dfrac{1}{(x - 1)^{4/3}}\, dx + \int_1^2 \dfrac{1}{(x - 1)^{4/3}}\, dx$

$= \lim_{b \to 1^-} \left[\dfrac{-3}{(x - 1)^{1/3}}\right]_0^b + \lim_{c \to 1^+} \left[\dfrac{-3}{(x - 1)^{1/3}}\right]_c^2 = \dfrac{-3}{0^-} - 3 + 3 + \dfrac{3}{0^+} = \infty + \infty$

Diverges

35. If $p = 1$, $\int_1^\infty \dfrac{1}{x}\, dx = \lim_{b \to \infty} \int_1^b \dfrac{1}{x}\, dx = \lim_{b \to \infty} \ln x \Big]_1^b$.

Diverges. For $p \neq 1$,

$\int_1^\infty \dfrac{1}{x^p}\, dx = \lim_{b \to \infty} \left[\dfrac{x^{1-p}}{1 - p}\right]_1^b = \lim_{b \to \infty} \left[\dfrac{b^{1-p}}{1 - p} - \dfrac{1}{1 - p}\right]$.

This converges to $\dfrac{1}{p - 1}$ if $1 - p < 0$ or $p > 1$.

36. If $p = 1$, $\int_0^1 \dfrac{1}{x}\, dx = \lim_{a \to 0^+} \ln x \Big]_a^1 = \lim_{a \to 0^+} -\ln a = \infty$.

Diverges. If $p \neq 1$,

$\int_0^1 \dfrac{1}{x^p}\, dx = \lim_{a \to 0^+} \left[\dfrac{x^{1-p}}{1 - p}\right]_a^1 = \lim_{a \to 0^+} \left[\dfrac{1}{1 - p} - \dfrac{a^{1-p}}{1 - p}\right]$.

This converges to $\dfrac{1}{1 - p}$ if $1 - p > 0$ or $p < 1$.

37. For $n = 1$ we have $\int_0^\infty xe^{-x}\, dx = 1$ (see Exercise 10). Assume $\int_0^\infty x^n e^{-x}\, dx$ converges, then for $n + 1$ we have $\int_0^\infty x^{n+1} e^{-x}\, dx$. Using integration by parts, $u = x^{n+1}$, $du = (n + 1)x^n\, dx$, $dv = e^{-x}\, dx$, $v = -e^{-x}$,

$\int_0^\infty x^{n+1} e^{-x}\, dx = \lim_{b \to \infty} \left[-x^{n+1} e^{-x}\right]_0^b + (n + 1) \int_0^\infty x^n e^{-x}\, dx = (n + 1) \int_0^\infty x^n e^{-x}\, dx$ which converges.

38. (a) Assume $\int_a^\infty g(x)\, dx = L$.

$0 \leq \int_a^\infty f(x)\, dx \leq \int_a^\infty g(x)\, dx = L$

Therefore, $\int_a^\infty f(x)\, dx$ converges.

(b) Assume $\int_a^\infty f(x)\, dx = \infty$.

$\int_a^\infty g(x)\, dx \geq \int_a^\infty f(x)\, dx = \infty$

Therefore, $\int_a^\infty g(x)\, dx$ diverges.

39. $\int_0^1 \dfrac{1}{x^3}\, dx$ diverges.

(See Exercise 36, $p = 3 \nless 1$.)

40. $\int_0^1 \dfrac{1}{\sqrt[3]{x}}\, dx = \dfrac{1}{1 - (1/3)} = \dfrac{3}{2}$ converges.

$\left(\text{See Exercise 36, } p = \dfrac{1}{3}.\right)$

41. $\int_1^\infty \dfrac{1}{x^3}\, dx = \dfrac{1}{3 - 1} = \dfrac{1}{2}$ converges.

(See Exercise 35, $p = 3$.)

42. $\int_0^\infty x^4 e^{-x}\, dx$ converges.

(See Exercise 37.)

43. Since $\dfrac{1}{x^2 + 5} \leq \dfrac{1}{x^2}$ on $[1, \infty)$ and $\int_1^\infty \dfrac{1}{x^2}\, dx$ converges by Exercise 35, $\int_1^\infty \dfrac{1}{x^2 + 5}\, dx$ converges.

44. Since $\dfrac{1}{\sqrt{x-1}} \geq \dfrac{1}{x}$ on $[2, \infty)$ and $\displaystyle\int_2^\infty \dfrac{1}{x}\,dx$ diverges by Exercise 35, $\displaystyle\int_2^\infty \dfrac{1}{\sqrt{x-1}}\,dx$ diverges.

45. Since $\dfrac{1}{\sqrt[3]{x(x-1)}} \geq \dfrac{1}{\sqrt[3]{x^2}}$ on $[2, \infty)$ and $\displaystyle\int_2^\infty \dfrac{1}{\sqrt[3]{x^2}}\,dx$ diverges by Exercise 35, $\displaystyle\int_2^\infty \dfrac{1}{\sqrt[3]{x(x-1)}}\,dx$ diverges.

46. Since $\dfrac{1}{\sqrt{x}(1+x)} \leq \dfrac{1}{x^{3/2}}$ on $[1, \infty)$ and $\displaystyle\int_1^\infty \dfrac{1}{x^{3/2}}\,dx$ converges by Exercise 35, $\displaystyle\int_1^\infty \dfrac{1}{\sqrt{x}(1+x)}\,dx$ converges.

47. Since $e^{-x^2} \leq e^{-x}$ on $[1, \infty)$ and $\displaystyle\int_0^\infty e^{-x}\,dx$ converges (see Exercise 5), $\displaystyle\int_0^\infty e^{-x^2}\,dx$ converges.

48. $\dfrac{1}{\sqrt{x}\ln x} \geq \dfrac{1}{x}$ since $\sqrt{x}\ln x < x$ on $[2, \infty)$. Since $\displaystyle\int_2^\infty \dfrac{1}{x}\,dx$ diverges by Exercise 35, $\displaystyle\int_2^\infty \dfrac{1}{\sqrt{x}\ln x}\,dx$ diverges.

49. $f(t) = 1$

$$F(s) = \int_0^\infty e^{-st}\,dt = \lim_{b\to\infty}\left[-\frac{1}{s}e^{-st}\right]_0^b = \frac{1}{s},\ s > 0$$

50. $f(t) = t$

$$F(s) = \int_0^\infty te^{-st}\,dt = \lim_{b\to\infty}\left[\frac{1}{s^2}(-st-1)e^{-st}\right]_0^b$$

$$= \frac{1}{s^2},\ s > 0$$

51. $f(t) = t^2$

$$F(s) = \int_0^\infty t^2 e^{-st}\,dt = \lim_{b\to\infty}\left[\frac{1}{s^3}(-s^2 t^2 - 2st - 2)e^{-st}\right]_0^b$$

$$= \frac{2}{s^3},\ s > 0$$

52. $f(t) = e^{at}$

$$F(s) = \int_0^\infty e^{at}e^{-st}\,dt = \int_0^\infty e^{t(a-s)}\,dt$$

$$= \lim_{b\to\infty}\left[\frac{1}{a-s}e^{t(a-s)}\right]_0^b$$

$$= 0 - \frac{1}{a-s} = \frac{1}{s-a},\ s > a$$

53. $f(t) = \cos at$

$$F(s) = \int_0^\infty e^{-st}\cos at\,dt$$

$$= \lim_{b\to\infty}\left[\frac{e^{-st}}{s^2 + a^2}(-s\cos at + a\sin at)\right]_0^b$$

$$= 0 + \frac{s}{s^2 + a^2} = \frac{s}{s^2 + a^2},\ s > 0$$

54. $f(t) = \sin at$

$$F(s) = \int_0^\infty e^{-st}\sin at\,dt$$

$$= \lim_{b\to\infty}\left[\frac{e^{-st}}{s^2 + a^2}(-s\sin at - a\cos at)\right]_0^b$$

$$= 0 + \frac{a}{s^2 + a^2} = \frac{a}{s^2 + a^2},\ s > 0$$

55. $f(t) = \cosh at$

$$F(s) = \int_0^\infty e^{-st}\cosh at\,dt = \int_0^\infty e^{-st}\left(\frac{e^{at} + e^{-at}}{2}\right)dt = \frac{1}{2}\int_0^\infty\left[e^{t(-s+a)} + e^{t(-s-a)}\right]dt$$

$$= \lim_{b\to\infty}\frac{1}{2}\left[\frac{1}{(-s+a)}e^{t(-s+a)} + \frac{1}{(-s-a)}e^{t(-s-a)}\right]_0^b = 0 - \frac{1}{2}\left[\frac{1}{(-s+a)} + \frac{1}{(-s-a)}\right]$$

$$= \frac{-1}{2}\left[\frac{1}{(-s+a)} + \frac{1}{(-s-a)}\right] = \frac{s}{s^2 - a^2},\ s > |a|$$

56. $f(t) = \sinh at$

$$F(s) = \int_0^\infty e^{-st} \sinh at \, dt = \int_0^\infty e^{-st}\left(\frac{e^{at} - e^{-at}}{2}\right) dt = \frac{1}{2}\int_0^\infty \left[e^{t(-s+a)} - e^{t(-s-a)}\right] dt$$

$$= \lim_{b\to\infty} \frac{1}{2}\left[\frac{1}{(-s+a)}e^{t(-s+a)} - \frac{1}{(-s-a)}e^{t(-s-a)}\right]_0^b = 0 - \frac{1}{2}\left[\frac{1}{(-s+a)} - \frac{1}{(-s-a)}\right]$$

$$= \frac{-1}{2}\left[\frac{1}{(-s+a)} - \frac{1}{(-s-a)}\right] = \frac{a}{s^2 - a^2}, \, s > |a|$$

57. (a) $A = \int_0^\infty e^{-x}\, dx$

$\qquad = \lim_{b\to\infty}\left[-e^{-x}\right]_0^b = 0 - (-1) = 1$

(b) **Disc:**

$$V = \pi \int_0^\infty (e^{-x})^2\, dx$$

$$= \lim_{b\to\infty} \pi\left[-\frac{1}{2}e^{-2x}\right]_0^b = \frac{\pi}{2}$$

(c) **Shell:**

$$V = 2\pi \int_0^\infty xe^{-x}\, dx$$

$$= \lim_{b\to\infty}\left\{2\pi\left[-e^{-x}(x+1)\right]_0^b\right\} = 2\pi$$

58. (a) $A = \int_1^\infty \frac{1}{x^2}\, dx = \left[-\frac{1}{x}\right]_1^\infty = 1$

(b) **Disc:**

$$V = \pi \int_1^\infty \frac{1}{x^4}\, dx = \lim_{b\to\infty}\left[-\frac{\pi}{3x^3}\right]_1^b = \frac{\pi}{3}$$

(c) **Shell:**

$$V = 2\pi \int_1^\infty x\left(\frac{1}{x^2}\right) dx = \lim_{b\to\infty}\left[2\pi(\ln x)\right]_1^b = \infty$$

Diverges

59. $x^{2/3} + y^{2/3} = 1$

$\dfrac{2}{3}x^{-1/3} + \dfrac{2}{3}y^{-1/3}y' = 0$

$y' = \dfrac{-y^{1/3}}{x^{1/3}}$

$\sqrt{1 + (y')^2} = \sqrt{1 + \dfrac{y^{2/3}}{x^{2/3}}} = \sqrt{\dfrac{1}{x^{2/3}}} = \dfrac{1}{x^{1/3}}$

$s = 4\int_0^1 \dfrac{1}{x^{1/3}}\, dx = \lim_{b\to 0^+}\left[4\left(\dfrac{3}{2}x^{2/3}\right)\right]_b^1 = 6$

60. $(x - 2)^2 + y^2 = 1$

$2(x - 2) + 2yy' = 0$

$y' = \dfrac{-(x - 2)}{y}$

$\sqrt{1 + (y')^2} = \sqrt{1 + [(x - 2)^2/y^2]} = \dfrac{1}{y}$ (Assume $y > 0$.)

$$S = 4\pi \int_1^3 \frac{x}{y}\, dx = 4\pi \int_1^3 \frac{x}{\sqrt{1 - (x-2)^2}}\, dx = 4\pi \int_1^3\left[\frac{x-2}{\sqrt{1-(x-2)^2}} + \frac{2}{\sqrt{1-(x-2)^2}}\right] dx$$

$$= \lim_{\substack{a\to 1^+ \\ b\to 3^-}}\left\{4\pi\left[-\sqrt{1-(x-2)^2} + 2\arcsin(x-2)\right]_a^b\right\} = 4\pi[0 + 2\arcsin(1) - 2\arcsin(-1)] = 8\pi^2$$

61. $\Gamma(n) = \displaystyle\int_0^\infty x^{n-1}e^{-x}\,dx$

(a) $\Gamma(1) = \displaystyle\int_0^\infty e^{-x}\,dx = \lim_{b\to\infty}\left[-e^{-x}\right]_0^b = 1$

$\Gamma(2) = \displaystyle\int_0^\infty xe^{-x}\,dx = \lim_{b\to\infty}\left[-e^{-x}(x+1)\right]_0^b = 1$

$\Gamma(3) = \displaystyle\int_0^\infty x^2e^{-x}\,dx = \lim_{b\to\infty}\left[-x^2e^{-x} - 2xe^{-x} - 2e^{-x}\right]_0^b = 2$

(b) $\Gamma(n+1) = \displaystyle\int_0^\infty x^ne^{-x}\,dx = \lim_{b\to\infty}\left[-x^ne^{-x}\right]_0^b + \lim_{b\to\infty}n\int_0^\infty x^{n-1}e^{-x}\,dx = 0 + n\Gamma(n) \quad (u = x^n,\, dv = e^{-x}\,dx)$

(c) $\Gamma(n) = (n-1)!$

62. (a) $F(x) = \dfrac{K}{x^2}, 5 = \dfrac{K}{(4000)^2}, K = 80,000,000$

$W = \displaystyle\int_{4000}^\infty \frac{80,000,000}{x^2}\,dx = \lim_{b\to\infty}\left[\frac{-80,000,000}{x}\right]_{4000}^b = 20,000 \text{ mi-ton}$

(b) $\dfrac{W}{2} = 10,000 = \left[\dfrac{-80,000,000}{x}\right]_{4000}^b = \dfrac{-80,000,000}{b} + 20,000$

$\dfrac{80,000,000}{b} = 10,000$

$b = 8000$

Therefore, 4000 miles *above* the earth's surface.

63. (a) $\displaystyle\int_{-\infty}^\infty \frac{1}{7}e^{-t/7}\,dt = \int_0^\infty \frac{1}{7}e^{-t/7}\,dt = \lim_{b\to\infty}\left[-e^{-t/7}\right]_0^b = 1$

(b) $\displaystyle\int_0^4 \frac{1}{7}e^{-t/7}\,dt = \left[-e^{-t/7}\right]_0^4 = -e^{-t/7} + 1$

$\approx 0.4353 = 43.53\%$

(c) $\displaystyle\int_0^\infty t\left[\frac{1}{7}e^{-t/7}\right]dt = \lim_{b\to\infty}\left[-te^{-t/7} - 7e^{-t/7}\right]_0^b$

$= 0 + 7 = 7$

64. (a) $\displaystyle\int_{-\infty}^\infty \frac{2}{5}e^{-2t/5}\,dt = \int_0^\infty \frac{2}{5}e^{-2t/5}\,dt = \lim_{b\to\infty}\left[-e^{-2t5}\right]_0^b = 1$

(b) $\displaystyle\int_0^4 \frac{2}{5}e^{-2t/5}\,dt = \left[-e^{-2t/5}\right]_0^4 = -e^{-8/5} + 1$

$\approx 0.7981 = 79.81\%$

(c) $\displaystyle\int_0^\infty t\left[\frac{2}{5}e^{-2t/5}\right]dt = \lim_{b\to\infty}\left[-te^{2t/5} - \frac{5}{2}e^{-2t/5}\right]_0^b = \frac{5}{2}$

65. (a) $C = 650,000 + \displaystyle\int_0^5 25,000\,e^{-0.06t}\,dt = 650,000 - \left[\frac{25,000}{0.06}e^{-0.06t}\right]_0^5 \approx \$757,992.41$

(b) $C = 650,000 + \displaystyle\int_0^{10} 25,000e^{-0.06t}\,dt \approx \$837,995.15$

(c) $C = 650,000 + \displaystyle\int_0^\infty 25,000e^{-0.06t}\,dt = 650,000 - \lim_{b\to\infty}\left[\frac{25,000}{0.06}e^{-0.06t}\right]_0^b \approx \$1,066,666.67$

66. (a) $C = 650,000 + \displaystyle\int_0^5 25,000(1 + 0.08t)e^{-0.06t}\,dt$

$= 650,000 + 25,000\left[-\frac{1}{0.06}e^{-0.06t} - 0.08\left(\frac{t}{0.06}e^{-0.06t} + \frac{1}{(0.06)^2}e^{-0.06t}\right)\right]_0^5 \approx \$778,512.58$

(b) $C = 650,000 + \displaystyle\int_0^{10} 25,000(1 + 0.08t)e^{-0.06t}\,dt$

$= 650,000 + 25,000\left[-\frac{1}{0.06}e^{-0.06t} - 0.08\left(\frac{t}{0.06}e^{-0.06t} + \frac{1}{(0.06)^2}e^{-0.06t}\right)\right]_0^{10} \approx \$905,718.14$

—CONTINUED—

66. —CONTINUED—

(c) $C = 650,000 + \int_0^\infty 25,000(1 + 0.08t)e^{-0.06t}\,dt$

$= 650,000 + 25,000 \lim_{b\to\infty}\left[-\dfrac{t}{0.06}e^{-0.06t} - 0.08\left(\dfrac{t}{0.06}e^{-0.06t} + \dfrac{1}{(0.06)^2}e^{-0.06t}\right)\right]_0^b \approx \$1,622,222.22$

67. Let $x = a\tan\theta$, $dx = a\sec^2\theta\,d\theta$, $\sqrt{a^2 + x^2} = a\sec\theta$.

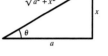

$\displaystyle\int \frac{1}{(a^2 + x^2)^{3/2}}\,dx = \int \frac{a\sec^2\theta\,d\theta}{a^3\sec^3\theta} = \frac{1}{a^2}\int \cos\theta\,d\theta$

$\displaystyle = \frac{1}{a^2}\sin\theta = \frac{1}{a^2}\frac{x}{\sqrt{a^2 + x^2}}$

Hence,

$\displaystyle P = k\int_1^\infty \frac{1}{(a^2 + x^2)^{3/2}}\,dx = \frac{k}{a^2}\lim_{b\to\infty}\left[\frac{x}{\sqrt{a^2 + x^2}}\right]_1^b$

$\displaystyle = \frac{k}{a^2}\left[1 - \frac{1}{\sqrt{a^2 + x^2}}\right] = \frac{k\sqrt{a^2 + x^2} - 1}{a^2\sqrt{a^2 + x^2}}$

68. (a) $\displaystyle\int_1^\infty \frac{1}{x}\,dx = \lim_{b\to\infty}\left[\ln|x|\right]_1^b = \infty$

$\displaystyle\int_1^\infty \frac{1}{x^2}\,dx = \lim_{b\to\infty}\left[-\frac{1}{x}\right]_1^b = 1$

$\displaystyle\int_1^\infty \frac{1}{x^n}\,dx$ will converge if $n > 1$ and will diverge if $n \le 1$.

(c) Let $dv = \sin x\,dx \implies v = -\cos x$

$\displaystyle u = \frac{1}{x} \implies du = -\frac{1}{x^2}\,dx$

$\displaystyle\int_1^\infty \frac{\sin x}{x}\,dx = \lim_{b\to 0}\left[-\frac{\cos x}{x}\right]_1^b - \int_1^\infty \frac{\cos x}{x^2}\,dx$

$\displaystyle = \cos 1 - \int_1^\infty \frac{\cos x}{x^2}\,dx$

Converges

(b) It would appear to converge.

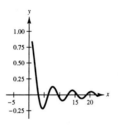

69. For $n = 1$,

$\displaystyle I_1 = \int_0^\infty \frac{x}{(x^2 + 1)^4}\,dx = \lim_{b\to\infty}\frac{1}{2}\int_0^b (x^2 + 1)^{-4}(2x\,dx) = \lim_{b\to\infty}\left[-\frac{1}{6}\frac{1}{(x^2 + 1)^3}\right]_0^b = \frac{1}{6}.$

For $n > 1$,

$\displaystyle I_n = \int_0^\infty \frac{x^{2n-1}}{(x^2 + 1)^{n+3}}\,dx = \lim_{b\to\infty}\left[\frac{-x^{2n-2}}{2(n + 2)(x^2 + 1)^n} + 2\right]_0^b + \frac{n - 1}{n + 2}\int_0^\infty \frac{x^{2n-3}}{(x^2 + 1)^{n+2}}\,dx = 0 + \frac{n - 1}{n + 2}(I_{n-1})$

$u = x^{2n-2}$, $du = (2n - 2)x^{2n-3}\,dx$, $dv = \dfrac{x}{(x^2 + 1)^{n+3}}\,dx$, $v = \dfrac{-1}{2(n + 2)(x^2 + 1)^{n+2}}$

(a) $\displaystyle\int_0^\infty \frac{x}{(x^2 + 1)^4}\,dx = \lim_{b\to\infty}\left[-\frac{1}{6(x^2 + 1)^3}\right]_0^b = \frac{1}{6}$

(b) $\displaystyle\int_0^\infty \frac{x^3}{(x^2 + 1)^5}\,dx = \frac{1}{4}\int_0^\infty \frac{x}{(x^2 + 1)^4}\,dx = \frac{1}{4}\left(\frac{1}{6}\right) = \frac{1}{24}$

(c) $\displaystyle\int_0^\infty \frac{x^5}{(x^2 + 1)^6} = \frac{2}{5}\int_0^\infty \frac{x^3}{(x^2 + 1)^5}\,dx = \frac{2}{5}\left(\frac{1}{24}\right) = \frac{1}{60}$

70. (a) $f(x) = \dfrac{1}{3\sqrt{2\pi}}e^{-(x-70)^2/18}$

$\displaystyle\int_{50}^{90} f(x)\,dx \approx 1.0$

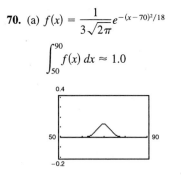

(b) $P(72 \le x < \infty) \approx 0.2525$

(c) $0.5 - P(70 \le x \le 72) \approx 0.5 - 0.2475 = 0.2525$

These are the same answers because by symmetry,

$$P(70 \le x < \infty) = 0.5$$

and

$$0.5 = P(70 \le x < \infty)$$
$$= P(70 \le x \le 72) + P(72 \le x < \infty).$$

71. False. $f(x) = 1/(x+1)$ is continuous on $[0, \infty)$, $\displaystyle\lim_{x \to \infty} 1/(x+1) = 0$, but

$$\int_0^\infty \frac{1}{x+1}\,dx = \lim_{b \to \infty}\left[\ln|x+1|\right]_0^b = \infty.$$

Diverges

72. False. This is equivalent to Exercise 71.

73. True

74. True

Review Exercises for Chapter 7

1. $\displaystyle\int e^{2x} \sin 3x\,dx = -\frac{1}{3}e^{2x}\cos 3x + \frac{2}{3}\int e^{2x}\cos 3x\,dx$

$$= -\frac{1}{3}e^{2x}\cos 3x + \frac{2}{3}\left(\frac{1}{3}e^{2x}\sin 3x - \frac{2}{3}\int e^{2x}\sin 3x\,dx\right)$$

$$\frac{13}{9}\int e^{2x}\sin 3x\,dx = -\frac{1}{3}e^{2x}\cos 3x + \frac{2}{9}e^{2x}\sin 3x$$

$$\int e^{2x}\sin 3x\,dx = \frac{e^{2x}}{13}(2\sin 3x - 3\cos 3x) + C$$

(1) $dv = \sin 3x\,dx \implies v = -\frac{1}{3}\cos 3x$

$\quad u = e^{2x} \qquad \implies du = 2e^{2x}\,dx$

(2) $dv = \cos 3x\,dx \implies v = \frac{1}{3}\sin 3x$

$\quad u = e^{2x} \qquad \implies du = 2e^{2x}\,dx$

2. $\displaystyle\int (x^2 - 1)e^x\,dx = (x^2 - 1)e^x - 2\int xe^x\,dx = (x^2 - 1)e^x - 2xe^x + 2\int e^x\,dx = e^x(x^2 - 2x + 1) + 1$

(1) $dv = e^x\,dx \implies v = e^x$

$\quad u = x^2 - 1 \implies du = 2x\,dx$

(2) $dv = e^x\,dx \implies v = e^x$

$\quad u = x \qquad \implies du = dx$

3. $\displaystyle\int \cos^3(\pi x - 1)\, dx = \int [1 - \sin^2(\pi x - 1)]\cos(\pi x - 1)\, dx$

$$= \frac{1}{\pi}\left[\sin(\pi x - 1) - \frac{1}{3}\sin^3(\pi x - 1)\right] + C$$

$$= \frac{1}{3\pi}\sin(\pi x - 1)[3 - \sin^2(\pi x - 1)] + C$$

$$= \frac{1}{3\pi}\sin(\pi x - 1)[3 - (1 - \cos^2(\pi x - 1))] + C$$

$$= \frac{1}{3\pi}\sin(\pi x - 1)[2 + \cos^2(\pi x - 1)] + C$$

4. $\displaystyle\int \sin^2\frac{\pi x}{2}\, dx = \int \frac{1}{2}(1 - \cos \pi x)\, dx = \frac{1}{2}\left[x - \frac{1}{\pi}\sin \pi x\right] + C = \frac{1}{2\pi}[\pi x - \sin \pi x] + C$

5. $\displaystyle\int \sec^4\!\left(\frac{x}{2}\right) dx = \int \left[\tan^2\!\left(\frac{x}{2}\right) + 1\right]\sec^2\!\left(\frac{x}{2}\right) dx$

$$= \int \tan^2\!\left(\frac{x}{2}\right)\sec^2\!\left(\frac{x}{2}\right) dx + \int \sec^2\!\left(\frac{x}{2}\right) dx$$

$$= \frac{2}{3}\tan^3\!\left(\frac{x}{2}\right) + 2\tan\!\left(\frac{x}{2}\right) + C = \frac{2}{3}\left[\tan^3\!\left(\frac{x}{2}\right) + 3\tan\!\left(\frac{x}{2}\right)\right] + C$$

6. $\displaystyle\int \tan \theta \sec^4 \theta\, d\theta = \int (\tan^3 \theta + \tan \theta)\sec^2 \theta\, d\theta = \frac{1}{4}\tan^4 \theta + \frac{1}{2}\tan^2 \theta + C_1$

or

$$\int \tan \theta \sec^4 \theta\, d\theta = \int \sec^3 \theta(\sec \theta \tan \theta)\, d\theta = \frac{1}{4}\sec^4 \theta + C_2$$

7. $\displaystyle\int \frac{-12}{x^2\sqrt{4 - x^2}}\, dx = \int \frac{-24 \cos \theta\, d\theta}{(4 \sin^2 \theta)(2 \cos \theta)}$

$$= -3\int \csc^2 \theta\, d\theta$$

$$= 3 \cot \theta + C$$

$$= \frac{3\sqrt{4 - x^2}}{x} + C$$

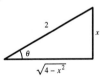

$x = 2 \sin \theta,\ dx = 2 \cos \theta\, d\theta,\ \sqrt{4 - x^2} = 2 \cos \theta$

8. $\displaystyle\int \frac{\sqrt{x^2 - 9}}{x}\, dx = \int \frac{3 \tan \theta}{3 \sec \theta}(3 \sec \theta \tan \theta\, d\theta)$

$$= 3\int \tan^2 \theta\, d\theta$$

$$= 3\int (\sec^2 \theta - 1)\, d\theta$$

$$= 3(\tan \theta - \theta) + C$$

$$= \sqrt{x^2 - 9} - 3 \arcsec\!\left(\frac{x}{3}\right) + C$$

$x = 3 \sec \theta,\ dx = 3 \sec \theta \tan \theta\, d\theta,\ \sqrt{x^2 - 9} = 3 \tan \theta$

9. $\dfrac{x^2 + 2x}{(x - 1)(x^2 + 1)} = \dfrac{A}{x - 1} + \dfrac{Bx + C}{x^2 + 1}$

$$x^2 + 2x = A(x^2 + 1) + (Bx + C)(x - 1)$$

Let $x = 1$: $3 = 2A \implies A = \dfrac{3}{2}$

Let $x = 0$: $0 = A - C \implies C = \dfrac{3}{2}$.

Let $x = 2$: $8 = 5A + 2B + C \implies B = -\dfrac{1}{2}$

$$\int \frac{x^2 + 2x}{x^3 - x^2 + x - 1} \, dx = \frac{3}{2} \int \frac{1}{x - 1} \, dx - \frac{1}{2} \int \frac{x - 3}{x^2 + 1} \, dx$$

$$= \frac{3}{2} \int \frac{1}{x - 1} \, dx - \frac{1}{4} \int \frac{2x}{x^2 + 1} \, dx + \frac{3}{2} \int \frac{1}{x^2 + 1} \, dx$$

$$= \frac{3}{2} \ln|x - 1| - \frac{1}{4} \ln|x^2 + 1| + \frac{3}{2} \arctan x + C$$

$$= \frac{1}{4} [6 \ln|x - 1| - \ln(x^2 + 1) + 6 \arctan x] + C$$

10. $\dfrac{4x - 2}{3(x - 1)^2} = \dfrac{A}{x - 1} + \dfrac{B}{(x - 1)^2}$

$$4x - 2 = 3A(x - 1) + 3B$$

Let $x = 1$: $2 = 3B \implies B = \dfrac{2}{3}$

Let $x = 2$: $6 = 3A + 3B \implies A = \dfrac{4}{3}$

$$\int \frac{4x - 2}{3(x - 1)^2} \, dx = \frac{4}{3} \int \frac{1}{x - 1} \, dx + \frac{2}{3} \int \frac{1}{(x - 1)^2} \, dx$$

$$= \frac{4}{3} \ln|x - 1| - \frac{2}{3(x - 1)} + C = \frac{2}{3} \left(2 \ln|x - 1| - \frac{1}{x - 1} \right) + C$$

11. $\dfrac{x^2}{x^2 + 2x - 15} = 1 + \dfrac{15 - 2x}{x^2 + 2x - 15}$

$\dfrac{15 - 2x}{(x - 3)(x + 5)} = \dfrac{A}{x - 3} + \dfrac{B}{x + 5}$

$$15 - 2x = A(x + 5) + B(x - 3)$$

Let $x = 3$: $\quad 9 = 8A \implies A = \dfrac{9}{8}$

Let $x = -5$: $25 = -8B \implies B = -\dfrac{25}{8}$

$$\int \frac{x^2}{x^2 + 2x - 15} \, dx = \int dx + \frac{9}{8} \int \frac{1}{x - 3} \, dx - \frac{25}{8} \int \frac{1}{x + 5} \, dx$$

$$= x + \frac{9}{8} \ln|x - 3| - \frac{25}{8} \ln|x + 5| + C$$

12. $\displaystyle\int \frac{3}{2x\sqrt{9x^2 - 1}} \, dx = \frac{3}{2} \int \frac{3}{3x\sqrt{9x^2 - 1}} \, dx = \frac{3}{2} \operatorname{arcsec}(3x) + C$

13. $\displaystyle\int \frac{1}{1 - \sin \theta}\, d\theta = \int \frac{1 + \sin \theta}{\cos^2 \theta}\, d\theta = \int (\sec^2 \theta + \sec \theta \tan \theta)\, d\theta = \tan \theta + \sec \theta + C$

14. $\displaystyle\int x^2 \sin 2x\, dx = -\frac{1}{2} x^2 \cos 2x + \int x \cos 2x\, dx$

$$= -\frac{1}{2} x^2 \cos 2x + \frac{1}{2} x \sin 2x - \frac{1}{2} \int \sin 2x\, dx$$

$$= -\frac{1}{2} x^2 \cos 2x + \frac{x}{2} \sin 2x + \frac{1}{4} \cos 2x + C$$

(1) $dv = \sin 2x\, dx \implies v = -\frac{1}{2} \cos 2x$

$\quad u = x^2 \qquad \implies du = 2x\, dx$

(2) $dv = \cos 2x\, dx \implies v = \frac{1}{2} \sin 2x$

$\quad u = x \qquad \implies du = dx$

15. $\displaystyle\int \frac{\ln 2x}{x^2}\, dx = \frac{-\ln 2x}{x} + \int \frac{1}{x^2}\, dx$

$$= \frac{-\ln 2x}{x} - \frac{1}{x} + C$$

$$= -\frac{1}{x}(1 + \ln 2x) + C$$

$dv = \dfrac{1}{x^2}\, dx \implies v = -\dfrac{1}{x}$

$u = \ln 2x \implies du = \dfrac{1}{x}\, dx$

16. $\displaystyle\int 2x\sqrt{2x - 3}\, dx = \int (u^4 + 3u^2)\, du = \frac{u^5}{5} + u^3 + C = \frac{2(2x - 3)^{3/2}}{5}(x + 1) + C$

$u = \sqrt{2x - 3},\, x = \dfrac{u^2 + 3}{2},\, dx = u\, du$

17. $\displaystyle\int \sqrt{4 - x^2}\, dx = \int (2 \cos \theta)(2 \cos \theta)\, d\theta$

$$= 2 \int (1 + \cos 2\theta)\, d\theta$$

$$= 2\left(\theta + \frac{1}{2} \sin 2\theta \right) + C$$

$$= 2(\theta + \sin \theta \cos \theta) + C$$

$$= 2\left[\arcsin\left(\frac{x}{2}\right) + \frac{x}{2}\left(\frac{\sqrt{4 - x^2}}{2} \right) \right] + C$$

$$= \frac{1}{2}\left[4 \arcsin\left(\frac{x}{2}\right) + x\sqrt{4 - x^2} \right] + C$$

$x = 2 \sin \theta,\, dx = 2 \cos \theta\, d\theta,\, \sqrt{4 - x^2} = 2 \cos \theta$

18. $\displaystyle\int \frac{\sec^2 \theta}{\tan \theta(\tan \theta - 1)}\, d\theta = \int \frac{1}{u(u - 1)}\, du = \int \frac{1}{u - 1}\, du - \int \frac{1}{u}\, du$

$$= \ln|u - 1| - \ln|u| + C = \ln\left| \frac{\tan \theta - 1}{\tan \theta} \right| + C = \ln|1 - \cot \theta| + C$$

$u = \tan \theta,\, du = \sec^2 \theta\, d\theta$

$\dfrac{1}{u(u - 1)} = \dfrac{A}{u} + \dfrac{B}{u - 1}$

$\quad 1 = A(u - 1) + Bu$

Let $u = 0$: $1 = -A \implies A = -1$

Let $u = 1$: $1 = B$

19. $\dfrac{3x^3 + 4x}{(x^2 + 1)^2} = \dfrac{Ax + B}{x^2 + 1} + \dfrac{Cx + D}{(x^2 + 1)^2}$

$3x^3 + 4x = (Ax + B)(x^2 + 1) + Cx + D$

$\qquad\qquad = Ax^3 + Bx^2 + (A + C)x + (B + D)$

$A = 3, B = 0, A + C = 4 \Rightarrow C = 1,$

$B + D = 0 \Rightarrow D = 0$

$\displaystyle\int \dfrac{3x^3 + 4x}{(x^2 + 1)^2}\, dx = 3\int \dfrac{x}{x^2 + 1}\, dx + \int \dfrac{x}{(x^2 + 1)^2}\, dx$

$\qquad\qquad = \dfrac{3}{2}\ln(x^2 + 1) - \dfrac{1}{2(x^2 + 1)} + C$

20. $\displaystyle\int \sqrt{\dfrac{x - 2}{x + 2}}\, dx = \int \dfrac{x - 2}{\sqrt{x^2 - 4}}\, dx$

$\qquad\qquad = \displaystyle\int \dfrac{x}{\sqrt{x^2 - 4}}\, dx - 2\int \dfrac{1}{\sqrt{x^2 - 4}}\, dx$

$\qquad\qquad = \sqrt{x^2 - 4} - 2\ln\left|x + \sqrt{x^2 - 4}\right| + C$

21. $\displaystyle\int \dfrac{16}{\sqrt{16 - x^2}}\, dx = 16\arcsin\left(\dfrac{x}{4}\right) + C$

22. $\displaystyle\int \dfrac{\sin\theta}{1 + 2\cos^2\theta}\, d\theta = \dfrac{-1}{\sqrt{2}}\int \dfrac{1}{1 + 2\cos^2\theta}\left(-\sqrt{2}\sin\theta\right) d\theta$

$\qquad\qquad = \dfrac{-1}{\sqrt{2}}\arctan\left(\sqrt{2}\cos\theta\right) + C$

$u = \sqrt{2}\cos\theta,\ du = -\sqrt{2}\sin\theta\, d\theta$

23. $\displaystyle\int \dfrac{x}{x^2 + 4x + 8}\, dx = \dfrac{1}{2}\int \dfrac{2x + 4 - 4}{x^2 + 4x + 8}\, dx$

$\qquad = \dfrac{1}{2}\displaystyle\int \dfrac{2x + 4}{x^2 + 4x + 8}\, dx - 2\int \dfrac{1}{(x + 2)^2 + 4}\, dx = \dfrac{1}{2}\ln|x^2 + 4x + 8| - \arctan\left(\dfrac{x + 2}{2}\right) + C$

24. $\displaystyle\int \dfrac{x}{x^2 - 4x + 8}\, dx = \dfrac{1}{2}\int \dfrac{2x - 4 + 4}{x^2 - 4x + 8}\, dx$

$\qquad = \dfrac{1}{2}\displaystyle\int \dfrac{2x - 4}{x^2 - 4x + 8}\, dx + 2\int \dfrac{1}{(x - 2)^2 + 4}\, dx = \dfrac{1}{2}\ln|x^2 - 4x + 8| + \arctan\left(\dfrac{x - 2}{2}\right) + C$

25. $\displaystyle\int \theta\sin\theta\cos\theta\, d\theta = \dfrac{1}{2}\int \theta\sin 2\theta\, d\theta$

$\qquad = -\dfrac{1}{4}\theta\cos 2\theta + \dfrac{1}{4}\displaystyle\int \cos 2\theta\, d\theta = -\dfrac{1}{4}\theta\cos 2\theta + \dfrac{1}{8}\sin 2\theta + C = \dfrac{1}{8}(\sin 2\theta - 2\theta\cos 2\theta) + C$

$dv = \sin 2\theta\, d\theta \Rightarrow v = -\dfrac{1}{2}\cos 2\theta$

$u = \theta \qquad \Rightarrow du = d\theta$

26. $\displaystyle\int \dfrac{\csc\sqrt{2x}}{\sqrt{x}}\, dx = \sqrt{2}\int \csc\sqrt{2x}\left(\dfrac{1}{\sqrt{2x}}\right) dx = -\sqrt{2}\ln\left|\csc\sqrt{2x} + \cot\sqrt{2x}\right| + C$

$u = \sqrt{2x},\ du = \dfrac{1}{\sqrt{2x}}\, dx$

27. $\displaystyle\int (\sin\theta + \cos\theta)^2\, d\theta = \int (\sin^2\theta + 2\sin\theta\cos\theta + \cos^2\theta)\, d\theta$

$\qquad\qquad = \displaystyle\int (1 + \sin 2\theta)\, d\theta = \theta - \dfrac{1}{2}\cos 2\theta + C = \dfrac{1}{2}(2\theta - \cos 2\theta) + C$

28. $\int \cos 2\theta(\sin \theta + \cos \theta)^2 \, d\theta = \int (\cos^2 \theta - \sin^2 \theta)(\sin \theta + \cos \theta)^2 \, d\theta$

$$= \int (\sin \theta + \cos \theta)^3(\cos \theta - \sin \theta) \, d\theta = \frac{1}{4}(\sin \theta + \cos\theta)^4 + C$$

29. $\int \sqrt{1 + \cos x} \, dx = \int \frac{\sin x}{\sqrt{1 - \cos x}} \, dx$

$$= \int (1 - \cos x)^{-1/2}(\sin x) \, dx$$

$$= 2\sqrt{1 - \cos x} + C$$

$u = 1 - \cos x, \, du = \sin x \, dx$

30. $\int \ln \sqrt{x^2 - 1} \, dx = \frac{1}{2} \int \ln(x^2 - 1) \, dx$

$$= \frac{1}{2}x \ln|x^2 - 1| - \int \frac{x^2}{x^2 - 1} \, dx$$

$$= \frac{1}{2}x \ln|x^2 - 1| - \int dx - \int \frac{1}{x^2 - 1} \, dx$$

$$= \frac{1}{2}x \ln|x^2 - 1| - x - \frac{1}{2} \ln\left|\frac{x - 1}{x + 1}\right| + C$$

$dv = dx \qquad \Longrightarrow \quad v = x$

$u = \ln(x^2 - 1) \Longrightarrow du = \frac{2x}{x^2 - 1} \, dx$

31. $\int \cos x \ln(\sin x) \, dx = \sin x \ln(\sin x) - \int \cos x \, dx$

$$= \sin x \ln(\sin x) - \sin x + C$$

$dv = \cos x \, dx \Longrightarrow \quad v = \sin x$

$u = \ln(\sin x) \Longrightarrow du = \frac{\cos x}{\sin x} \, dx$

32. $\dfrac{x^4 + 2x^2 + x + 1}{x^4 + 2x^2 + 1} = 1 + \dfrac{x}{(x^2 + 1)^2}$

$$\int \frac{x^4 + 2x^2 + x + 1}{(x^2 + 1)^2} \, dx = \int dx + \frac{1}{2} \int \frac{2x}{(x^2 + 1)^2} \, dx$$

$$= x - \frac{1}{2(x^2 + 1)} + C$$

33. $\int x \arcsin 2x \, dx = \dfrac{x^2}{2} \arcsin 2x - \int \dfrac{x^2}{\sqrt{1 - 4x^2}} \, dx$

$$= \frac{x^2}{2} \arcsin 2x - \frac{1}{8} \int \frac{2(2x)^2}{\sqrt{1 - (2x)^2}} \, dx$$

$$= \frac{x^2}{2} \arcsin 2x - \frac{1}{8}\left(\frac{1}{2}\right)\left[-(2x)\sqrt{1 - 4x^2} + \arcsin 2x\right] + C \quad \text{(by Formula 43 of Integration Tables)}$$

$$= \frac{1}{16}\left[(8x^2 - 1)\arcsin 2x + 2x\sqrt{1 - 4x^2}\right] + C$$

$dv = x \, dx \qquad \Longrightarrow \quad v = \dfrac{x^2}{2}$

$u = \arcsin 2x \Longrightarrow du = \dfrac{2}{\sqrt{1 - 4x^2}} \, dx$

34. $\int e^x \arctan(e^x) \, dx = e^x \arctan(e^x) - \int \dfrac{e^{2x}}{1 + e^{2x}} \, dx$

$$= e^x \arctan(e^x) - \frac{1}{2} \ln(1 + e^{2x}) + C$$

$dv = e^x \, dx \qquad \Longrightarrow \quad v = e^x$

$u = \arctan e^x \Longrightarrow du = \dfrac{e^x}{1 + e^{2x}} \, dx$

35. $\int \dfrac{x^{1/4}}{1 + x^{1/2}} \, dx = 4 \int \dfrac{u(u^3)}{1 + u^2} \, du$

$$= 4 \int \left(u^2 - 1 + \frac{1}{u^2 + 1}\right) du$$

$$= 4\left(\frac{1}{3}u^3 - u + \arctan u\right) + C$$

$$= \frac{4}{3}[x^{3/4} - 3x^{1/4} + 3 \arctan(x^{1/4})] + C$$

$y = \sqrt[4]{x}, x = u^4, dx = 4u^3 \, du$

36. $\int \sqrt{1 + \sqrt{x}}\, dx = \int u(4u^3 - 4u)\, du = \int (4u^4 - 4u^2)\, du = \dfrac{4u^5}{5} - \dfrac{4u^3}{3} + C = \dfrac{4}{15}\left(1 + \sqrt{x}\right)^{3/2}\left(3\sqrt{x} - 2\right) + C$

$u = \sqrt{1 + \sqrt{x}},\ x = u^4 - 2u^2 + 1,\ dx = (4u^3 - 4u)\, du$

37. $\int \dfrac{x^4}{(x - 1)^3}\, dx = \dfrac{1}{2}x^2 + 3x - \dfrac{1}{2(x - 1)^2} - \dfrac{4}{x - 1} + 6\ln|x - 1| + C$

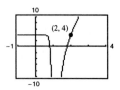

$(2, 4):\ 4 = 2 + 6 - \dfrac{1}{2} - 4 + C \Rightarrow C = \dfrac{1}{2}$

$y = 6\ln|x - 1| + \dfrac{1}{2}x^2 + 3x + \dfrac{1}{2} - \dfrac{1}{2(x - 1)^2} - \dfrac{4}{x - 1}$

38. $\int \dfrac{4x^2 - 1}{(2x)(x^2 + 2x + 1)}\, dx = \dfrac{-1}{2}\ln|x| + \dfrac{3}{2(x + 1)} + \dfrac{5}{2}\ln|x + 1| + C$

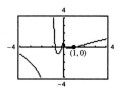

$(1, 0):\ \dfrac{3}{4} + \dfrac{5}{2}\ln 2 + C = 0 \Rightarrow C = \dfrac{-3}{4} - \dfrac{5}{3}\ln 2$

$y = \dfrac{-1}{2}\ln|x| + \dfrac{3}{2(x + 1)} + \dfrac{5}{2}\ln|x + 1| - \dfrac{3}{4} - \dfrac{5}{2}\ln 2$

39. $\int \dfrac{6x^2 - 3x + 14}{x^3 - 2x^2 + 4x - 8}\, dx = 4\ln|x - 2| + \ln(x^2 + 4) + \dfrac{1}{2}\arctan\left(\dfrac{1}{2}x\right) + C$

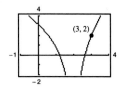

$(3, 2):\ \ln 13 + \dfrac{1}{2}\arctan\dfrac{3}{2} + C = 2 \Rightarrow C = 2 - \ln 13 - \dfrac{1}{2}\arctan\left(\dfrac{3}{2}\right)$

$y = \dfrac{1}{2}\arctan\dfrac{x}{2} + \ln(x^2 + 4) + 4\ln|x - 2| - \dfrac{1}{2}\arctan\dfrac{3}{2} - \ln 13 + 2$

40. $\int \dfrac{x^3 - 6x^2 + x - 4}{1 - x^4}\, dx = 2\ln|x - 1| - 3\ln|x + 1| + \arctan x + C$

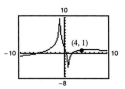

$(4, 1):\ 2\ln 3 - 3\ln 5 + \arctan 4 + C = 1 \Rightarrow C = 1 - 2\ln 3 + 3\ln 5 - \arctan 4$

$y = 2\ln|x - 1| - 3\ln|x + 1| + \arctan x + 1 - 2\ln 3 + 3\ln 5 - \arctan 4$

41. $\int \dfrac{d\theta}{2 - 3\sin\theta} = \dfrac{\sqrt{5}}{5}\ln\left|\dfrac{\cos\theta - \sqrt{5}(\sin\theta - 1)}{\cos\theta + \sqrt{5}(\sin\theta - 1)}\right| + C$

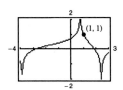

$(1, 1):\ \dfrac{\sqrt{5}}{5}\ln\left|\dfrac{\cos(1) - \sqrt{5}(\sin(1) - 1)}{\cos(1) + \sqrt{5}(\sin(1) - 1)}\right| + C = 1 \Rightarrow C \approx 0.297$

$\dfrac{\sqrt{5}}{5}\ln\left|\dfrac{\cos\theta - \sqrt{5}(\sin\theta - 1)}{\cos\theta + \sqrt{5}(\sin\theta - 1)}\right| + 0.297$

42. $\int \dfrac{d\theta}{\sec\theta - \tan\theta} = -\ln(1 - \sin\theta) + C$

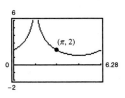

$(\pi, 2):\ 2 = -\ln(1 - \sin\pi) + C \Rightarrow C = 2$

$2 - \ln(1 - \sin\theta)$

43. $\displaystyle\int \frac{d\theta}{1 + \sin\theta + \cos\theta} = \ln\left|2 + 2\tan\left(\frac{\theta}{2}\right)\right| + C$

or $\displaystyle\ln\left|\frac{1 + \sin\theta + \cos\theta}{1 + \cos\theta}\right| + C$

$(0, 0)$: $\ln 1 + C = 0 \Rightarrow C = 0$

$$y = \ln\left|\frac{1 + \sin\theta + \cos\theta}{1 + \cos\theta}\right|$$

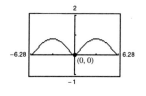

44. $\displaystyle\int \frac{1}{3 - 2\cos\theta}\, d\theta = \frac{2\sqrt{5}}{5}\arctan\left[\left(\tan\frac{x}{2}\right)\sqrt{5}\right] + C$

$(0, 0)$: $C = 0$

$$y = \frac{2\sqrt{5}}{5}\arctan\left[\sqrt{5}\tan\frac{x}{2}\right]$$

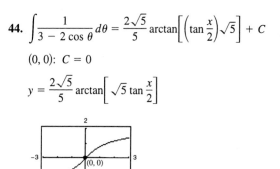

45. $\displaystyle\int \frac{\sin\theta}{3 - 2\cos\theta}\, d\theta = \frac{1}{2}\ln|3 - 2\cos\theta| + C$

$(0, 0)$: $\dfrac{1}{2}\ln 1 + C = 0 \Rightarrow C = 0$

$$y = \frac{1}{2}\ln(3 - 2\cos\theta)$$

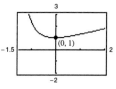

46. $\displaystyle\int \frac{\sin\theta}{1 + \sin\theta}\, d\theta = \frac{2}{\tan(\theta/2) + 1} + \theta + C$

$(0, 1)$: $\dfrac{2}{1} + 0 + C = 1 \Rightarrow C = -1$

$$y = \frac{2}{1 + \tan(\theta/2)} + \theta - 1$$

or, $\displaystyle\int \frac{\sin\theta}{1 + \sin\theta}\, d\theta = \frac{1 - \sin\theta}{\cos\theta} + \theta + C_1$

$(0, 1)$: $C_1 = 0$

$$y = \frac{1 - \sin\theta}{\cos\theta} + \theta$$

47. $\displaystyle y = \int \frac{9}{x^2 - 9}\, dx = \frac{3}{2}\ln\left|\frac{x - 3}{x + 3}\right| + C$ (by Formula 24 of Integration Tables)

48. $\displaystyle y = \int \frac{\sqrt{4 - x^2}}{2x}\, dx = \int \frac{2\cos\theta(2\cos\theta)\, d\theta}{4\sin\theta}$

$\displaystyle = \int (\csc\theta - \sin\theta)\, d\theta$

$= [-\ln|\csc\theta + \cos\theta| + \cos\theta] + C$

$\displaystyle = -\ln\left|\frac{2 + \sqrt{4 - x^2}}{x}\right| + \frac{\sqrt{4 - x^2}}{2} + C$

$x = 2\sin\theta,\ dx = 2\cos\theta\, d\theta,\ \sqrt{4 - x^2} = 2\cos\theta$

49. $\displaystyle y = \int \ln(x^2 + x)\, dx = x\ln|x^2 + x| - \int \frac{2x^2 + x}{x^2 + x}\, dx$

$\displaystyle = x\ln|x^2 + x| - \int \frac{2x + 1}{x + 1}\, dx$

$\displaystyle = x\ln|x^2 + x| - \int 2\, dx + \int \frac{1}{x + 1}\, dx$

$= x\ln|x^2 + x| - 2x + \ln|x + 1| + C$

$dv = dx \qquad \Rightarrow \quad v = x$

$u = \ln(x^2 + x) \Rightarrow du = \dfrac{2x + 1}{x^2 + x}\, dx$

50. $y = \displaystyle\int \sqrt{1 - \cos\theta}\, d\theta = \int \frac{\sin\theta}{\sqrt{1 + \cos\theta}}\, d\theta = -\int (1 + \cos\theta)^{-1/2}(-\sin\theta)d\theta = -2\sqrt{1 + \cos\theta} + C$

$u = 1 + \cos\theta,\, du = -\sin\theta\, d\theta$

51. (a) $\displaystyle\int \frac{1}{x^2\sqrt{4 + x^2}}\, dx = \frac{1}{4}\int \frac{\cos\theta}{\sin^2\theta}\, d\theta$

$\qquad = \dfrac{-1}{4\sin\theta} + C$

$\qquad = -\dfrac{1}{4}\csc\theta + C$

$\qquad = \dfrac{-\sqrt{4 + x^2}}{4x} + C$

$x = 2\tan\theta,\, dx = 2\sec^2\theta\, d\theta,\, \sqrt{4 + x^2} = 2\sec\theta$

(b) $\displaystyle\int \frac{1}{x^2\sqrt{4 + x^2}}\, dx = -\frac{1}{4}\int \frac{u}{\sqrt{1 + u^2}}\, du$

$\qquad = -\dfrac{1}{4}\sqrt{1 + u^2} + C$

$\qquad = -\dfrac{1}{4}\sqrt{1 + \dfrac{4}{x^2}}$

$\qquad = \dfrac{-\sqrt{4 + x^2}}{4x} + C$

$x = \dfrac{2}{u},\, dx = -\dfrac{2du}{u^2}$

52. (a) $\displaystyle\int \frac{1}{x\sqrt{4 + x^2}}\, dx = \frac{1}{2}\int \csc\theta\, d\theta$

$\qquad = -\dfrac{1}{2}\ln|\csc\theta + \cot\theta| + C$

$\qquad = -\dfrac{1}{2}\ln\left|\dfrac{\sqrt{x^2 + 4} + 2}{x}\right| + C$

$\qquad = \dfrac{1}{2}\ln\left|\dfrac{\sqrt{x^2 + 4} - 2}{x}\right| + C$

$x = 2\tan\theta,\, dx = 2\sec^2\theta\, d\theta$

(b) $\displaystyle\int \frac{1}{x\sqrt{4 + x^2}}\, dx = \int \frac{x\, dx}{x^2\sqrt{4 + x^2}}$

$\qquad = \displaystyle\int \dfrac{du}{u^2 - 4} = \dfrac{1}{4}\ln\left|\dfrac{u - 2}{u + 2}\right| + C$

$\qquad = \dfrac{1}{4}\ln\left|\dfrac{\sqrt{4 + x^2} - 2}{\sqrt{4 + x^2} + 2}\right| + C$

$\qquad = \dfrac{1}{4}\ln\left|\dfrac{\left(\sqrt{4 + x^2} - 2\right)^2}{x^2}\right| + C$

$\qquad = \dfrac{1}{2}\ln\left|\dfrac{\sqrt{4 + x^2} - 2}{x}\right| + C$

$u^2 = 4 + x^2,\, x^2 = u^2 - 4,\, x\, dx = u\, du$

53. (a) $\displaystyle\int \frac{x^3}{\sqrt{4 + x^2}}\, dx = 8\int \frac{\sin^3\theta}{\cos^4\theta}\, d\theta$

$\qquad = 8\displaystyle\int \left(\cos^{-4}\theta - \cos^{-2}\theta\right)\sin\theta\, d\theta$

$\qquad = \dfrac{8}{3}\sec\theta(\sec^2\theta - 3) + C$

$\qquad = \dfrac{\sqrt{4 + x^2}}{3}(x^2 - 8) + C$

$x = 2\tan\theta,\, dx = 2\sec^2\theta\, d\theta$

(b) $\displaystyle\int \frac{x^3}{\sqrt{4 + x^2}}\, dx = \int (u^2 - 4)\, du$

$\qquad = \dfrac{1}{3}u^3 - 4u + C$

$\qquad = \dfrac{u}{3}(u^2 - 12) + C$

$\qquad = \dfrac{\sqrt{4 + x^2}}{3}(x^2 - 8) + C$

$u^2 = 4 + x^2,\, 2u\, du = 2x\, dx$

(c) $\displaystyle\int \frac{x^3}{\sqrt{4 + x^2}}\, dx = x^2\sqrt{4 + x^2} - \int 2x\sqrt{4 + x^2}\, dx$

$\qquad = x^2\sqrt{4 + x^2} - \dfrac{2}{3}(4 + x^2)^{3/2} + C = \dfrac{\sqrt{4 + x^2}}{3}(x^2 - 8) + C$

$dv = \dfrac{x}{\sqrt{4 + x^2}}\, dx \implies v = \sqrt{4 + x^2}$

$u = x^2 \qquad\qquad \implies du = 2x\, dx$

54. (a) $\displaystyle\int x\sqrt{4+x}\,dx = 64\int \tan^3\theta \sec^3\theta\,d\theta$

$$= 64\int (\sec^4\theta - \sec^2\theta)\sec\theta\tan\theta\,d\theta$$

$$= \frac{64\sec^3\theta}{15}(3\sec^3\theta - 5) + C$$

$$= \frac{2(4+x)^{3/2}}{15}(3x-8) + C$$

$x = 4\tan^2\theta,\ dx = 8\tan\theta\sec^2\theta\,d\theta,$

$\sqrt{4+x} = 2\sec\theta$

(c) $\displaystyle\int x\sqrt{4+x}\,dx = \int (u^{3/2} - 4u^{1/2})\,du$

$$= \frac{2u^{3/2}}{15}(3u - 20) + C$$

$$= \frac{2(4+x)^{3/2}}{15}(3x-8) + C$$

$u = 4 + x,\ du = dx$

(b) $\displaystyle\int x\sqrt{4+x}\,dx = 2\int (u^4 - 4u^2)\,du$

$$= \frac{2u^3}{15}(3u^2 - 20) + C$$

$$= \frac{2(4+x)^{3/2}}{15}(3x-8) + C$$

$u^2 = 4 + x,\ dx = 2u\,du$

(d) $\displaystyle\int x\sqrt{4+x}\,dx = \frac{2x}{3}(4+x)^{3/2} - \frac{2}{3}\int (4+x)^{3/2}\,dx$

$$= \frac{2x}{3}(4+x)^{3/2} - \frac{4}{15}(4+x)^{5/2} + C$$

$$= \frac{2(4+x)^{3/2}}{15}(3x-8) + C$$

$dv = \sqrt{4+x}\,dx \implies v = \frac{2}{3}(4+x)^{3/2}$

$u = x \qquad\qquad \implies du = dx$

55. $\displaystyle\int_2^{\sqrt5} x(x^2-4)^{3/2}\,dx = \left[\frac{1}{5}(x^2-4)^{5/2}\right]_2^{\sqrt5} = \frac{1}{5}$

56. $\displaystyle\int_0^1 \frac{x}{(x-2)(x-4)}\,dx = \Big[2\ln|x-4| - \ln|x-2|\Big]_0^1$

$$= 2\ln 3 - 2\ln 4 + \ln 2$$

$$= \ln\frac{9}{8} \approx 0.118$$

57. $\displaystyle\int_1^4 \frac{\ln x}{x}\,dx = \left[\frac{1}{2}(\ln x)^2\right]_1^4 = \frac{1}{2}(\ln 4)^2 = 2(\ln 2)^2 \approx 0.961$

58. $\displaystyle\int_0^2 xe^{3x}\,dx = \left[\frac{e^{3x}}{9}(3x-1)\right]_0^2 = \frac{1}{9}(5e^6 + 1) \approx 224.238$

59. $\displaystyle\int_0^\pi x\sin x\,dx = \Big[-x\cos x + \sin x\Big]_0^\pi = \pi$

60. $\displaystyle\int_0^3 \frac{x}{\sqrt{1+x}}\,dx = \left[\frac{-2(2-x)}{3}\sqrt{1+x}\right]_0^3 = \frac{4}{3} + \frac{4}{3} = \frac{8}{3}$

61. $A = \displaystyle\int_0^4 x\sqrt{4-x}\,dx = \int_2^0 (4-u^2)u(-2u)\,du$

$$= \int_2^0 2(u^4 - 4u^2)\,du$$

$$= \left[2\left(\frac{u^5}{5} - \frac{4u^3}{3}\right)\right]_2^0 = \frac{128}{15}$$

$u = \sqrt{4-x},\ x = 4 - u^2,\ dx = -2u\,du$

62. $A = \displaystyle\int_0^4 \frac{1}{25 - x^2}\,dx$

$$= \left[-\frac{1}{10}\ln\left|\frac{x-5}{x+5}\right|\right]_0^4$$

$$= -\frac{1}{10}\ln\frac{1}{9} = \frac{1}{10}\ln 9 \approx 0.220$$

63. By symmetry, $\bar x = 0,\ A = \dfrac{1}{2}\pi.$

$$\bar y = \frac{2}{\pi}\left(\frac{1}{2}\right)\int_{-1}^1 (\sqrt{1-x^2})^2\,dx = \frac{1}{\pi}\left[x - \frac{1}{3}x^3\right]_{-1}^1 = \frac{4}{3\pi}$$

$$(\bar x, \bar y) = \left(0, \frac{4}{3\pi}\right)$$

64. By symmetry, $\bar y = 0.$

$$A = \pi + 4\pi = 5\pi$$

$$\bar x = \frac{1(\pi) + 4(4\pi)}{\pi + 4\pi}$$

$$= \frac{17\pi}{5\pi} = 3.4$$

$$(\bar x, \bar y) = (3.4, 0)$$

65. (a) $\displaystyle\int_0^1 e^x \, dx = \left[e^x \right]_0^1 = e - 1 \approx 1.72$ **(b)** $\displaystyle\int_0^1 xe^x \, dx = \left[e^x(x - 1) \right]_0^1 = 1.00$

(c) $\displaystyle\int_0^1 xe^{x^2} \, dx = \left[\frac{1}{2}e^{x^2} \right]_0^1 = \frac{1}{2}(e - 1) \approx 0.86$

(d) Simpson's Rule ($n = 8$)

$$\int_0^1 e^{x^2} \, dx = \frac{1}{24}\left[1 + 4e^{(1/8)^2} + 2e^{(1/4)^2} + 4e^{(3/8)^2} + 2e^{(1/2)^2} + 4e^{(5/8)^2} + 2e^{(3/4)^2} + 4e^{(7/8)^2} + e \right] \approx 1.46$$

66. (a) $\displaystyle\int_0^{\pi/2} \cos x \, dx = \left[\sin x \right]_0^{\pi/2} = 1$

(b) $\displaystyle\int_0^{\pi/2} \cos^2 x \, dx = \frac{1}{2}\int_0^{\pi/2} (1 + \cos 2x) \, dx = \frac{1}{2}\left[x + \sin x \cos x \right]_0^{\pi/2} = \frac{\pi}{4} \approx 0.78$

(c) Simpson's Rule ($n = 4$)

$$\int_0^{\pi/2} \cos(x^2) \, dx = \frac{\pi}{24}\left[1 + 4\cos\left(\frac{\pi}{8}\right)^2 + 2\cos\left(\frac{\pi}{4}\right)^2 + 4\cos\left(\frac{3\pi}{8}\right)^2 + \cos\left(\frac{\pi}{2}\right)^2 \right] \approx 0.85$$

(d) Simpson's Rule ($n = 4$)

$$\int_0^{\pi/2} \cos \sqrt{x} \, dx = \frac{\pi}{24}\left[1 + 4\cos\sqrt{\pi/8} + 2\cos\sqrt{\pi/4} + 4\cos\sqrt{3\pi/8} + \cos\sqrt{\pi/2} \right] \approx 1.01$$

67. $\displaystyle s = \int_0^\pi \sqrt{1 + \cos^2 x} \, dx \approx 3.82$ **68.** $\displaystyle s = \int_0^\pi \sqrt{1 + \sin^2 2x} \, dx \approx 3.82$

69. $\displaystyle\lim_{x \to 1} \left[\frac{(\ln x)^2}{x - 1} \right] = \lim_{x \to 1} \left[\frac{2(1/x)\ln x}{1} \right] = 0$ **70.** $\displaystyle\lim_{x \to 0} \frac{\sin \pi x}{\sin 2\pi x} = \lim_{x \to 0} \frac{\pi \cos \pi x}{2\pi \cos 2\pi x} = \frac{\pi}{2\pi} = \frac{1}{2}$

71. $\displaystyle\lim_{x \to \infty} \frac{e^{2x}}{x^2} = \lim_{x \to \infty} \frac{2e^{2x}}{2x} = \lim_{x \to \infty} \frac{4e^{2x}}{2} = \infty$ **72.** $\displaystyle\lim_{x \to \infty} xe^{-x^2} = \lim_{x \to \infty} \frac{x}{e^{x^2}} = \lim_{x \to \infty} \frac{1}{2xe^{x^2}} = 0$

73. $\displaystyle y = \lim_{x \to \infty} (\ln x)^{2/x}$

$\ln y = \displaystyle\lim_{x \to \infty} \frac{2 \ln(\ln x)}{x} = \lim_{x \to \infty} \left[\frac{2/(x \ln x)}{1} \right] = 0$

Since $\ln y = 0$, $y = 1$.

74. $\displaystyle y = \lim_{x \to 1^+} (x - 1)^{\ln x}$

$\ln y = \displaystyle\lim_{x \to 1^+} \left[(\ln x) \ln(x - 1) \right]$

$\displaystyle = \lim_{x \to 1^+} \left[\frac{\ln(x - 1)}{\dfrac{1}{\ln x}} \right] = \lim_{x \to 1^+} \left[\frac{\dfrac{1}{x - 1}}{\left(\dfrac{1}{x}\right)\dfrac{-1}{\ln^2 x}} \right] = \lim_{x \to 1^+} \left[\frac{-\ln^2 x}{x - 1} \right] = \lim_{x \to 1^+} \left[\frac{-2\left(\dfrac{1}{x}\right)(\ln x)}{\dfrac{1}{x^2}} \right]$

$\displaystyle = \lim_{x \to 1^+} 2x(\ln x) = 0$

Since $\ln y = 0$, $y = 1$.

75. $\displaystyle\lim_{n \to \infty} 1000\left(1 + \frac{0.09}{n}\right)^n = 1000 \lim_{n \to \infty} \left(1 + \frac{0.09}{n}\right)^n$

Let $y = \displaystyle\lim_{n \to \infty} \left(1 + \frac{0.09}{n}\right)^n$.

$\ln y = \displaystyle\lim_{n \to \infty} n \ln\left(1 + \frac{0.09}{n}\right) = \lim_{n \to \infty} \frac{\ln\left(1 + \dfrac{0.09}{n}\right)}{\dfrac{1}{n}} = \lim_{n \to \infty} \left(\frac{\dfrac{-0.09/n^2}{1 + (0.09/n)}}{-\dfrac{1}{n^2}}\right) = \lim_{n \to \infty} \frac{0.09}{1 + \left(\dfrac{0.09}{n}\right)} = 0.09$

Thus, $\ln y = 0.09 \implies y = e^{0.09}$ and $\displaystyle\lim_{n \to \infty} 1000\left(1 + \frac{0.09}{n}\right)^n = 1000e^{0.09} \approx 1094.17$.

76. $\displaystyle\lim_{x \to 1^+} \left(\frac{2}{\ln x} - \frac{2}{x - 1}\right) = \lim_{x \to 1^+} \left[\frac{2x - 2 - 2\ln x}{(\ln x)(x - 1)}\right]$

$= \displaystyle\lim_{x \to 1^+} \left[\frac{2 - (2/x)}{(x - 1)(1/x) + \ln x}\right]$

$= \displaystyle\lim_{x \to 1^+} \frac{2x - 2}{(x - 1) + x \ln x} = \lim_{x \to 1^+} \frac{2}{1 + 1 + \ln x} = 1$

77. $\displaystyle\int_0^{16} \frac{1}{\sqrt[4]{x}} \, dx = \lim_{b \to 0^+} \left[\frac{4}{3} x^{3/4}\right]_b^{16} = \frac{32}{3}$

Converges

78. $\displaystyle\int_0^1 \frac{6}{x - 1} \, dx = \lim_{b \to 1^-} \left[6 \ln|x - 1|\right]_0^b = -\infty$

Diverges

79. $\displaystyle\int_1^{\infty} x^2 \ln x \, dx = \lim_{b \to \infty} \left[\frac{x^3}{9}(-1 + 3\ln x)\right]_1^b = \infty$

Diverges

80. $\displaystyle\int_0^{\infty} \frac{e^{-1/x}}{x^2} \, dx = \lim_{\substack{a \to 0^+ \\ b \to \infty}} \left[e^{-1/x}\right]_a^b = 1 - 0 = 1$

81. $\displaystyle\int_0^{t_0} 500{,}000 e^{-0.05t} \, dt = \left[\frac{500{,}000}{-0.05} e^{-0.05t}\right]_0^{t_0}$

$= \dfrac{-500{,}000}{0.05}(e^{-0.05t_0} - 1)$

$= 10{,}000{,}000(1 - e^{-0.05t_0})$

(a) $t_0 = 20$: \$6,321,205.59

(b) $t_0 \to \infty$: \$10,000,000

82. $V = \pi \displaystyle\int_0^{\infty} (xe^{-x})^2 \, dx$

$= \pi \displaystyle\int_0^{\infty} x^2 e^{-2x} \, dx$

$= \displaystyle\lim_{b \to \infty} \left[-\frac{\pi e^{-2x}}{4}(2x^2 + 2x + 1)\right]_0^b = \frac{\pi}{4}$

83. (a) $P(13 \le x < \infty) = \dfrac{1}{0.95\sqrt{2\pi}} \displaystyle\int_{13}^{\infty} e^{-(x - 12.9)^2/2(0.95)^2} \, dx \approx 0.4581$

(b) $P(15 \le x < 20) = \dfrac{1}{0.95\sqrt{2\pi}} \displaystyle\int_{15}^{\infty} e^{-(x - 12.9)^2/2(0.95)^2} \, dx \approx 0.0135$

84. $\displaystyle\int_2^{\infty} \left[\frac{1}{x^5} + \frac{1}{x^{10}} + \frac{1}{x^{15}}\right] dx < \int_2^{\infty} \frac{1}{x^5 - 1} \, dx < \int_2^{\infty} \left[\frac{1}{x^5} + \frac{1}{x^{10}} + \frac{2}{x^{15}}\right] dx$

$\displaystyle\lim_{b \to \infty} \left[-\frac{1}{4x^4} - \frac{1}{9x^9} - \frac{1}{14x^{14}}\right]_2^b < \int_2^{\infty} \frac{1}{x^5 - 1} \, dx < \lim_{b \to \infty} \left[-\frac{1}{4x^4} - \frac{1}{9x^9} - \frac{1}{7x^{14}}\right]_2^b$

$0.015846 < \displaystyle\int_2^{\infty} \frac{1}{x^5 - 1} \, dx < 0.015851$

85. $dv = dx \implies v = x$

$u = (\ln x)^n \implies du = n(\ln x)^{n-1}\dfrac{1}{x}dx$

$\displaystyle\int (\ln x)^n \, dx = x(\ln x)^n - n\int (\ln x)^{n-1}\, dx$

86. $\displaystyle\int \tan^n x \, dx = \int \tan^{n-2}x(\sec^2 x - 1)\, dx$

$\qquad = \displaystyle\int \tan^{n-2} x \sec^2 x \, dx - \int \tan^{n-2} x \, dx$

$\qquad = \dfrac{1}{n-1}\tan^{n-1} x - \displaystyle\int \tan^{n-2} x \, dx$

87. False

$u = \ln x^2 = 2\ln x \implies du = \dfrac{2}{x}dx$

$\displaystyle\int \dfrac{\ln x^2}{x}\, dx = \dfrac{1}{2}\int (\ln x^2)\dfrac{2}{x}\, dx = \dfrac{1}{2}\int u \, du$

88. True

$u = \sqrt[3]{3x+1} \implies x = \dfrac{u^3 - 1}{3} \implies dx = u^2 \, du$

89. False

$\displaystyle\int_{-1}^{1} \sqrt{x^2 - x^3}\, dx = \int_{-1}^{1} \sqrt{x^2(1-x)}\, dx$

$\qquad = \displaystyle\int_{-1}^{1} |x|\sqrt{1-x}\, dx$

90. True

$\displaystyle\int_{1}^{\infty} x^{-p}\, dx = \lim_{b\to\infty}\left[\dfrac{x^{-p+1}}{-p+1}\right]_{1}^{b}$

$\qquad = 0 - \left(\dfrac{1}{-p+1}\right) = \dfrac{1}{p-1}$ if $p > 1$.

91. $\displaystyle\int_{x}^{1} \dfrac{1}{1+t^2}\, dt = \left[\arctan t\right]_{x}^{1} = \dfrac{\pi}{4} - \arctan x$

$\displaystyle\int_{1}^{1/x} \dfrac{1}{1+t^2}\, dt = \left[\arctan t\right]_{1}^{1/x} = \arctan\dfrac{1}{x} - \dfrac{\pi}{4}$

Since $\arctan x + \arctan\dfrac{1}{x} = \dfrac{\pi}{2}, x > 0$, we have:

$\arctan\dfrac{1}{x} = \dfrac{\pi}{4} + \dfrac{\pi}{4} - \arctan x$

$\arctan\dfrac{1}{x} - \dfrac{\pi}{4} = \dfrac{\pi}{4} - \arctan x$

Therefore, $\displaystyle\int_{1}^{1/x} \dfrac{1}{1+t^2}\, dt = \int_{x}^{1} \dfrac{1}{1+t^2}\, dt.$

C H A P T E R   8
Infinite Series

CHAPTER 8
Infinite Series

Section 8.1   Sequences

Solutions to Exercises

1. $a_n = 2^n$

$a_1 = 2^1 = 2$

$a_2 = 2^2 = 4$

$a_3 = 2^3 = 8$

$a_4 = 2^4 = 16$

$a_5 = 2^5 = 32$

2. $a_n = \dfrac{n}{n+1}$

$a_1 = \dfrac{1}{2}$

$a_2 = \dfrac{2}{3}$

$a_3 = \dfrac{3}{4}$

$a_4 = \dfrac{4}{5}$

$a_5 = \dfrac{5}{6}$

3. $a_n = \left(-\dfrac{1}{2}\right)^n$

$a_1 = \left(-\dfrac{1}{2}\right)^1 = -\dfrac{1}{2}$

$a_2 = \left(-\dfrac{1}{2}\right)^2 = \dfrac{1}{4}$

$a_3 = \left(-\dfrac{1}{2}\right)^3 = -\dfrac{1}{8}$

$a_4 = \left(-\dfrac{1}{2}\right)^4 = \dfrac{1}{16}$

$a_5 = \left(-\dfrac{1}{2}\right)^5 = -\dfrac{1}{32}$

4. $a_n = \sin\dfrac{n\pi}{2}$

$a_1 = \sin\dfrac{\pi}{2} = 1$

$a_2 = \sin\pi = 0$

$a_3 = \sin\dfrac{3\pi}{2} = -1$

$a_4 = \sin 2\pi = 0$

$a_5 = \sin\dfrac{5\pi}{2} = 1$

5. $a_n = \dfrac{(-1)^{n(n+1)/2}}{n^2}$

$a_1 = \dfrac{(-1)^1}{1^2} = -1$

$a_2 = \dfrac{(-1)^3}{2^2} = -\dfrac{1}{4}$

$a_3 = \dfrac{(-1)^6}{3^2} = \dfrac{1}{9}$

$a_4 = \dfrac{(-1)^{10}}{4^2} = \dfrac{1}{16}$

$a_5 = \dfrac{(-1)^{15}}{5^2} = -\dfrac{1}{25}$

6. $a_n = 5 - \dfrac{1}{n} + \dfrac{1}{n^2}$

$a_1 = 5 - 1 + 1 = 5$

$a_2 = 5 - \dfrac{1}{2} + \dfrac{1}{4} = \dfrac{19}{4}$

$a_3 = 5 - \dfrac{1}{3} + \dfrac{1}{9} = \dfrac{43}{9}$

$a_4 = 5 - \dfrac{1}{4} + \dfrac{1}{16} = \dfrac{77}{16}$

$a_5 = 5 - \dfrac{1}{5} + \dfrac{1}{25} = \dfrac{121}{25}$

7. $a_n = \dfrac{3^n}{n!}$

$a_1 = \dfrac{3}{1!} = 3$

$a_2 = \dfrac{3^2}{2!} = \dfrac{9}{2}$

$a_3 = \dfrac{3}{3!} = \dfrac{27}{6}$

$a_4 = \dfrac{3^4}{4!} = \dfrac{81}{24}$

$a_5 = \dfrac{3^5}{5!} = \dfrac{243}{120}$

8. $a_n = \dfrac{3n!}{(n-1)!} = 3n$

$a_1 = 3(1) = 3$

$a_2 = 3(2) = 6$

$a_3 = 3(3) = 9$

$a_4 = 3(4) = 12$

$a_5 = 3(5) = 15$

9. $a_1 = 3, a_{k+1} = 2(a_k - 1)$

$a_2 = 2(a_1 - 1)$

$\quad = 2(3 - 1) = 4$

$a_3 = 2(a_2 - 1)$

$\quad = 2(4 - 1) = 6$

$a_4 = 2(a_3 - 1)$

$\quad = 2(6 - 1) = 10$

$a_5 = 2(a_4 - 1)$

$\quad = 2(10 - 1) = 18$

10. $a_1 = 4, \ a_{k+1} = \left(\dfrac{k+1}{2}\right)a_k$

$a_2 = \left(\dfrac{2}{2}\right)a_1 = (1)(4) = 4$

$a_3 = \left(\dfrac{3}{2}\right)a_2 = \left(\dfrac{3}{2}\right)(4) = 6$

$a_4 = \left(\dfrac{4}{2}\right)a_3 = (2)(6) = 12$

$a_5 = \left(\dfrac{5}{2}\right)a_4 = \left(\dfrac{5}{2}\right)(12) = 30$

11. $a_1 = 32, \ a_{k+1} = \dfrac{1}{2}a_k$

$a_2 = \dfrac{1}{2}a_1 = \dfrac{1}{2}(32) = 16$

$a_3 = \dfrac{1}{2}a_2 = \dfrac{1}{2}(16) = 8$

$a_4 = \dfrac{1}{2}a_3 = \dfrac{1}{2}(8) = 4$

$a_5 = \dfrac{1}{2}a_4 = \dfrac{1}{2}(4) = 2$

12. $a_1 = 6, \ a_{k+1} = \dfrac{1}{3}a_k^2$

$a_2 = \dfrac{1}{3}a_1^2 = \dfrac{1}{3}(6^2) = 12$

$a_3 = \dfrac{1}{3}a_2^2 = \dfrac{1}{3}(12^2) = 48$

$a_4 = \dfrac{1}{3}a_3^2 = \dfrac{1}{3}(48^2) = 768$

$a_5 = \dfrac{1}{3}a_4^2 = \dfrac{1}{3}(768)^2 = 196{,}608$

13. Because $a_1 = 8/(1 + 1) = 4$ and $a_2 = 8/(2 + 1) = \frac{8}{3}$, the sequence matches graph (d).

14. Because the sequence tends to 8 as n tends to infinity, it matches (a).

15. This sequence decreases and $a_1 = 4, a_2 = 4(0.5) = 2$. Matches (c).

16. This sequence increases for a few terms, then decreases $a_2 = \frac{16}{2} = 8$. Matches (b).

17.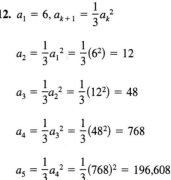

18.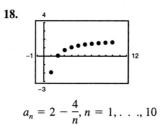

$a_n = 2 - \dfrac{4}{n}, \ n = 1, \ldots, 10$

19.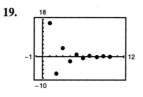

20.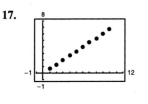

$a_n = 8(0.75)^{n-1}, \ n = 1, 2, \ldots, 10$

21.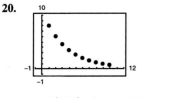

22.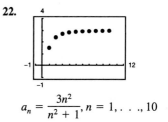

$a_n = \dfrac{3n^2}{n^2 + 1}, \ n = 1, \ldots, 10$

23. $a_n = 3n - 1$

$a_5 = 3(5) - 1 = 14$

$a_6 = 3(6) - 1 = 17$

24. $a_n = \dfrac{n + 6}{2}$

$a_5 = \dfrac{5 + 6}{2} = \dfrac{11}{2}$

$a_6 = \dfrac{6 + 6}{2} = 6$

25. $a_n = \dfrac{3}{(-2)^{n-1}}$

$a_n = \dfrac{3}{(-2)^4} = \dfrac{3}{16}$

$a_6 = \dfrac{3}{(-2)^5} = -\dfrac{3}{32}$

26. $a_{n+1} = 2a_n, \ a_1 = 5$

$a_5 = 2(40) = 80$

$a_6 = 2(80) = 160$

27. $\dfrac{10!}{8!} = \dfrac{8!(9)(10)}{8!}$

$= (9)(10) = 90$

28. $\dfrac{25!}{23!} = \dfrac{23!(24)(25)}{23!}$

$= (24)(25) = 600$

29. $\dfrac{(n + 1)!}{n!} = \dfrac{n!(n + 1)}{n!}$

$= n + 1$

30. $\dfrac{(n + 2)!}{n!} = \dfrac{n!(n + 1)(n + 2)}{n!}$

$= (n + 1)(n + 2)$

31. $\dfrac{(2n-1)!}{(2n+1)!} = \dfrac{(2n-1)!}{(2n-1)!(2n)(2n+1)}$

$\qquad = \dfrac{1}{2n(2n+1)}$

32. $\dfrac{(2n+2)!}{(2n)!} = \dfrac{(2n)!(2n+1)(2n+2)}{(2n)!}$

$\qquad = (2n+1)(2n+2)$

33. $a_n = 3n-2$

34. $a_n = 4n-1$

35. $a_n = n^2-2$

36. $a_n = \dfrac{(-1)^{n-1}}{n^2}$

37. $a_n = \dfrac{n+1}{n+2}$

38. $a_n = \dfrac{n+1}{2n-1}$

39. $a_n = \dfrac{(-1)^{n-1}}{2^{n-2}}$

40. $a_n = \dfrac{(-1)^n}{2}\left(\dfrac{2}{3}\right)^{n-1}$

41. $a_n = 1 + \dfrac{1}{n} = \dfrac{n+1}{n}$

42. $a_n = 1 + \dfrac{2^n-1}{2^n}$

$\qquad = \dfrac{2^{n+1}-1}{2^n}$

43. $a_n = \dfrac{n}{(n+1)(n+2)}$

44. $a_n = \dfrac{1}{n!}$

45. $a_n = \dfrac{(-1)^{n-1}}{1\cdot 3\cdot 5\cdots(2n-1)} = \dfrac{(-1)^{n-1}2^n n!}{(2n)!}$

46. $a_n = \dfrac{x^{n-1}}{(n-1)!}$

47.

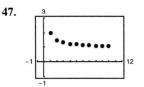

The graph seems to indicate that the sequence converges to 1. Analytically,

$$\lim_{n\to\infty} a_n = \lim_{n\to\infty}\frac{n+1}{n} = \lim_{x\to\infty}\frac{x+1}{x} = \lim_{x\to\infty} 1 = 1.$$

48.

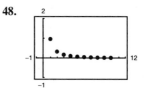

The graph seems to indicate that the sequence converges to 0. Analytically,

$$\lim_{n\to\infty} a_n = \lim_{n\to\infty}\frac{1}{n^{3/2}} = \lim_{x\to\infty}\frac{1}{x^{3/2}} = 0.$$

49.

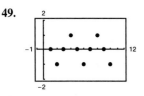

The graph seems to indicate that the sequence diverges. Analytically, the sequence is

$$\{a_n\} = \{0, -1, 0, 1, 0, -1, \ldots\}.$$

Hence, $\lim\limits_{n\to\infty} a_n$ does not exist.

50.

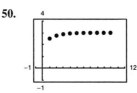

The graph seems to indicate that the sequence converges to 3. Analytically,

$$\lim_{n\to\infty} a_n = \lim_{n\to\infty}\left(3 - \frac{1}{2^n}\right) = 3 - 0 = 3.$$

51. $\lim\limits_{n\to\infty} (-1)^n\left(\dfrac{n}{n+1}\right)$

does not exist (oscillates between -1 and 1), diverges.

52. $\lim\limits_{n\to\infty} \left[1 + (-1)^n\right]$

does not exist, (alternates between 0 and 2), diverges.

53. $\lim\limits_{n\to\infty} \dfrac{3n^2-n+4}{2n^2+1} = \dfrac{3}{2}$, converges

54. $\lim\limits_{n\to\infty} \dfrac{\sqrt{n}}{\sqrt{n}+1} = 1$, converges

55. $\lim\limits_{n\to\infty} \dfrac{1+(-1)^n}{n} = 0$ converges

56. $\lim\limits_{n\to\infty} \dfrac{\ln(n^2)}{n} = \lim\limits_{n\to\infty} \dfrac{2/n}{1} = 0,$

converges (Use L'Hôpital's Rule.)

57. $\lim\limits_{n\to\infty} \left(\dfrac{3}{4}\right)^n = 0$, converges

58. $\lim\limits_{n\to\infty} (0.5)^n = 0$, converges

59. $\lim\limits_{n\to\infty} \dfrac{(n+1)!}{n!} = \lim\limits_{n\to\infty} (n+1) = \infty$, diverges

60. $\lim\limits_{n\to\infty} \dfrac{(n-2)!}{n!} = \lim\limits_{n\to\infty} \dfrac{1}{n(n-1)} = 0$, converges

61. $\lim\limits_{n\to\infty} \left(\dfrac{n-1}{n} - \dfrac{n}{n-1} \right) = \lim\limits_{n\to\infty} \dfrac{(n-1)^2 - n^2}{n(n-1)}$

$$= \lim\limits_{n\to\infty} \dfrac{1-2n}{n^2-n} = 0, \text{ converges}$$

63. $\lim\limits_{n\to\infty} \dfrac{n^p}{e^n} = 0$, converges

$(p > 0, n \geq 2)$

62. $\lim\limits_{n\to\infty} \left(\dfrac{n^2}{2n+1} - \dfrac{n^2}{2n-1} \right) = \lim\limits_{n\to\infty} \dfrac{-2n^2}{4n^2-1} = -\dfrac{1}{2}$,

converges

64. $a_n = n \sin \dfrac{1}{n}$

Let $f(x) = x \sin \dfrac{1}{x}$.

$$\lim\limits_{x\to\infty} x \sin \dfrac{1}{x} = \lim\limits_{x\to\infty} \dfrac{\sin(1/x)}{1/x} = \lim\limits_{x\to\infty} \dfrac{(-1/x^2)\cos(1/x)}{-1/x^2} = \lim\limits_{x\to\infty} \cos \dfrac{1}{x} = \cos 0 = 1 \text{ (L'Hôpital's Rule)}$$

or,

$$\lim\limits_{x\to\infty} \dfrac{\sin(1/x)}{1/x} = \lim\limits_{y\to 0^+} \dfrac{\sin(y)}{y} = 1. \text{ Therefore } \lim\limits_{n\to\infty} n \sin \dfrac{1}{n} = 1.$$

65. $a_n = \left(1 + \dfrac{k}{n} \right)^n$

$$\lim\limits_{n\to\infty} \left(1 + \dfrac{k}{n} \right)^n = \lim\limits_{u\to 0} \left[(1+u)^{1/u} \right]^k = e^k$$

where $u = \dfrac{k}{n}$, converges

66. $\lim\limits_{n\to\infty} 2^{1/n} = 2^0 = 1$, converges

67. $a_n = 4 - \dfrac{1}{n} < 4 - \dfrac{1}{n+1} = a_{n+1}$,

monotonic; $|a_n| < 4$ bounded.

68. Let $f(x) = \dfrac{4x}{x+1}$, then $f'(x) = \dfrac{4}{(x+1)^2}$.

Thus, f is increasing which implies $\{a_n\}$ is increasing; monotonic; $|a_n| < 4$ bounded.

69. $a_n = \dfrac{\cos n}{n}$

$a_1 = 0.5403$

$a_2 = -0.2081$

$a_3 = -0.3230$

$a_4 = -0.1634$

Not monotonic; $|a_n| \leq 1$, bounded

70. $a_n = ne^{-n/2}$

$a_1 = 0.6065$

$a_2 = 0.7358$

$a_3 = 0.6694$

Not monotonic; $|a_n| \leq 0.7358$, bounded

71. $a_n = (-1)^n \left(\dfrac{1}{n} \right)$

$a_1 = -1$

$a_2 = \dfrac{1}{2}$

$a_3 = -\dfrac{1}{3}$

Not monotonic; $|a_n| \leq 1$, bounded

72. $a_n = \left(-\dfrac{2}{3} \right)^n$

$a_1 = -\dfrac{2}{3}$

$a_2 = \dfrac{4}{9}$

$a_3 = -\dfrac{8}{27}$

Not monotonic; $|a_n| \leq \dfrac{2}{3}$, bounded

73. $a_n = \left(\dfrac{2}{3} \right)^n > \left(\dfrac{2}{3} \right)^{n+1} = a_{n+1}$

Monotonic; $|a_n| \leq \dfrac{2}{3}$, bounded

74. $a_n = \left(\frac{3}{2}\right)^n < \left(\frac{3}{2}\right)^{n+1} = a_{n+1}$

Monotonic; $\lim\limits_{n \to \infty} a_n = \infty$, not bounded

75. $a_n = \sin\left(\dfrac{n\pi}{6}\right)$

$a_1 = 0.500$

$a_2 = 0.8660$

$a_3 = 1.000$

$a_4 = 0.8660$

Not monotonic; $|a_n| \leq 1$, bounded

76. $\dfrac{n}{2^{n+2}} \overset{?}{\geq} \dfrac{n+1}{2^{(n+1)+2}}$

$2^{n+3}n \overset{?}{\geq} 2^{n+2}(n+1)$

$2n \overset{?}{\geq} n+1$

$n \geq 1$

Hence, $n \geq 1$

$2n \geq n+1$

$2^{n+3}n \geq 2^{n+2}(n+1)$

$\dfrac{n}{2^{n+2}} \geq \dfrac{n+1}{2^{(n+1)+2}}$

$a_n \geq a_{n+1}$

True; monotonic; $|a_n| \leq \frac{1}{8}$, bounded

77. (a) $a_n = 5 + \dfrac{1}{n}$

$\left|5 + \dfrac{1}{n}\right| \leq 6 \implies \{a_n\}$ bounded

$a_n = 5 + \dfrac{1}{n} > 5 + \dfrac{1}{n+1}$

$= a_{n+1} \implies \{a_n\}$ monotonic

Therefore, $\{a_n\}$ converges.

(b)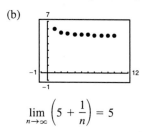

$\lim\limits_{n \to \infty} \left(5 + \dfrac{1}{n}\right) = 5$

78. (a) $a_n = 3 - \dfrac{4}{n}$

$\left|3 - \dfrac{4}{n}\right| < 3 \implies \{a_n\}$ bounded

$a_n = 3 - \dfrac{4}{n} < 3 - \dfrac{4}{n+1}$

$= a_{n+1} \implies \{a_n\}$ monotonic

Therefore, $\{a_n\}$ converges.

(b)

$\lim\limits_{n \to \infty} \left(3 - \dfrac{4}{n}\right) = 3$

79. (a) $a_n = \dfrac{1}{3}\left(1 - \dfrac{1}{3^n}\right)$

$\left|\dfrac{1}{3}\left(1 - \dfrac{1}{3^n}\right)\right| < \dfrac{1}{3} \implies \{a_n\}$ bounded

$a_n = \dfrac{1}{3}\left(1 - \dfrac{1}{3^n}\right) < \dfrac{1}{3}\left(1 - \dfrac{1}{3^{n+1}}\right)$

$= a_{n+1} \implies \{a_n\}$ monotonic

Therefore, $\{a_n\}$ converges.

(b)

$\lim\limits_{n \to \infty} \left[\dfrac{1}{3}\left(1 - \dfrac{1}{3^n}\right)\right] = \dfrac{1}{3}$

80. (a) $a_n = 4 + \dfrac{1}{2^n}$

$\left|4 + \dfrac{1}{2^n}\right| \leq 4.5 \implies \{a_n\}$ bounded

$a_n = 4 + \dfrac{1}{2^n} > 4 + \dfrac{1}{2^{n+1}}$

$= a_{n+1} \implies \{a_n\}$ monotonic

Therefore, $\{a_n\}$ converges.

(b)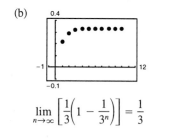

$\lim\limits_{n \to \infty} \left(4 + \dfrac{1}{2^n}\right) = 4$

81. (a) $a_n = 10 - \dfrac{1}{n}$

(b) Not possible; a bounded monotonic sequence must converge—see Theorem 8.5.

(c) $a_n = \dfrac{3n}{4n + 1}$

(d) Not possible; an unbounded sequence does not converge.

82. $a_n = \dfrac{3n}{8n + 1}$ (Many correct answers possible)

$\left| \dfrac{3n}{8n + 1} - \dfrac{3}{8} \right| < 0.001$ when $n = 47$.

83. $A_n = P \left[1 + \dfrac{r}{12} \right]^n$

(a) $\lim\limits_{n \to \infty} A_n = \infty$, divergent. The amount will grow arbitrarily large over time.

(b) $A_n = 9000 \left[1 + \dfrac{0.115}{12} \right]^n$

$A_1 = \$9086.25$	$A_6 = \$9530.06$
$A_2 = \$9173.33$	$A_7 = \$9621.39$
$A_3 = \$9261.24$	$A_8 = \$9713.59$
$A_4 = \$9349.99$	$A_9 = \$9806.68$
$A_5 = \$9439.60$	$A_{10} = \$9900.66$

84. $A_n = 100(101)[(1.01)^n - 1]$

(a) $A_1 = \$101.00$ (b) $A_{60} = \$8248.64$

$A_2 = \$203.01$ (c) $A_{240} = \$99,914.79$

$A_3 = \$306.04$

$A_4 = \$410.10$

$A_5 = \$515.20$

$A_6 = \$621.35$

85. (a) $A_n = (0.8)^n (2.5)$ billion

(b) $A_1 = \$2$ billion

$A_2 = \$1.6$ billion

$A_3 = \$1.28$ billion

$A_4 = \$1.024$ billion

(c) $\lim\limits_{n \to \infty} (0.8)^n (2.5) = 0$

86. $P_n = 16,000(1.045)^n$

$P_1 = \$16,720.00$

$P_2 = \$17,472.40$

$P_3 \approx \$18,258.66$

$P_4 \approx \$19,080.30$

$P_5 \approx \$19,938.91$

87. (a) $a_n = 59.69n + 697.32$

(b) For the year 2000, $n = 10$ and $a_{10} \approx \$1294$

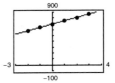

88. (a) $a_n = 6.742n^2 + 255.664n + 1109.909$

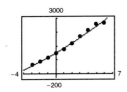

(b) For the year 2000, $n = 10$ and $a_{10} \approx \$4341$

89. $S_6 = 130 + 70 + 40 = 240$

$S_7 = 240 + 130 + 70 = 440$

$S_8 = 440 + 240 + 130 = 810$

$S_9 = 810 + 440 + 240 = 1490$

$S_{10} = 1490 + 810 + 440 = 2740$

90. $a_n = 0.905e^{0.134n}, \; n = 0, 1, 2, \ldots, 10$

$a_0 = 0.905$ trillion	$a_6 = 2.022$
$a_1 = 1.035$	$a_7 = 2.312$
$a_2 = 1.183$	$a_8 = 2.644$
$a_3 = 1.353$	$a_9 = 3.023$
$a_4 = 1.547$	$a_{10} = 3.456$
$a_5 = 1.769$	

91. $a_n = \dfrac{10^n}{n!}$

(a) $a_9 = a_{10} = \dfrac{10^9}{9!}$

$= \dfrac{1{,}000{,}000{,}000}{362{,}880}$

$= \dfrac{1{,}562{,}500}{567}$

(b) Decreasing

(c) Factorials increase more rapidly than exponentials.

92. $a_n = \left(1 + \dfrac{1}{n}\right)^n$

$a_1 = 2.0000$

$a_2 = 2.2500$

$a_3 \approx 2.3704$

$a_4 \approx 2.4414$

$a_5 \approx 2.4883$

$a_6 \approx 2.5216$

$\displaystyle\lim_{n\to\infty} \left(1 + \dfrac{1}{n}\right)^n = e$

93. $\{a_n\} = \left\{\sqrt[n]{n}\right\} = \{n^{1/n}\}$

$a_1 = 1^{1/1} = 1$

$a_2 = \sqrt{2} \approx 1.4142$

$a_3 = \sqrt[3]{3} \approx 1.4422$

$a_4 = \sqrt[4]{4} \approx 1.4142$

$a_5 = \sqrt[5]{5} \approx 1.3797$

$a_6 = \sqrt[6]{6} \approx 1.3480$

Let $y = \displaystyle\lim_{n\to\infty} n^{1/n}$.

$\ln y = \displaystyle\lim_{n\to\infty} \left(\dfrac{1}{n} \ln n\right)$

$= \displaystyle\lim_{n\to\infty} \dfrac{\ln n}{n} = \lim_{n\to\infty} \dfrac{1/n}{1} = 0$

Since $\ln y = 0$, we have $y = e^0 = 1$. Therefore,
$\displaystyle\lim_{n\to\infty} \sqrt[n]{n} = 1$.

94. Since

$\displaystyle\lim_{n\to\infty} s_n = L > 0,$

there exists for each $\epsilon > 0$, an integer N such that
$|s_n - L| < \epsilon$ for every $n > N$. Let $\epsilon = L > 0$ and we
have,

$|s_n - L| < L, \; -L < s_n - L < L, \text{ or } 0 < s_n < 2L$

for each $n > N$.

95. $a_{n+2} = a_n + a_{n+1}$

(a) $a_1 = 1$ $a_7 = 8 + 5 = 13$

$\quad\;\; a_2 = 1$ $a_8 = 13 + 8 = 21$

$\quad\;\; a_3 = 1 + 1 = 2$ $a_9 = 21 + 13 = 34$

$\quad\;\; a_4 = 2 + 1 = 3$ $a_{10} = 34 + 21 = 55$

$\quad\;\; a_5 = 3 + 2 = 5$ $a_{11} = 55 + 34 = 89$

$\quad\;\; a_6 = 5 + 3 = 8$ $a_{12} = 89 + 55 = 144$

(b) $b_n = \dfrac{a_{n+1}}{a_n}, \; n \geq 1$

$b_1 = \dfrac{1}{1} = 1 \qquad b_6 = \dfrac{13}{8}$

$b_2 = \dfrac{2}{1} = 2 \qquad b_7 = \dfrac{21}{13}$

$b_3 = \dfrac{3}{2} \qquad\qquad b_8 = \dfrac{34}{21}$

$b_4 = \dfrac{5}{3} \qquad\qquad b_9 = \dfrac{55}{34}$

$b_5 = \dfrac{8}{5} \qquad\qquad b_{10} = \dfrac{89}{55}$

—CONTINUED—

95. —CONTINUED—

(c) $1 + \dfrac{1}{b_{n-1}} = 1 + \dfrac{1}{a_n/a_{n-1}}$

$\qquad = 1 + \dfrac{a_{n-1}}{a_n}$

$\qquad = \dfrac{a_n + a_{n-1}}{a_n} = \dfrac{a_{n+1}}{a_n} = b_n$

(d) If $\lim\limits_{n\to\infty} b_n = \rho$, then $\lim\limits_{n\to\infty}\left(1 + \dfrac{1}{b_{n-1}}\right) = \rho$.

Since $\lim\limits_{n\to\infty} b_n = \lim\limits_{n\to\infty} b_{n-1}$ we have,

$\qquad 1 + (1/\rho) = \rho$.

$\qquad \rho + 1 = \rho^2$

$\qquad 0 = \rho^2 - \rho - 1$

$\qquad \rho = \dfrac{1 \pm \sqrt{1+4}}{2} = \dfrac{1 \pm \sqrt{5}}{2}$

Since a_n, and thus b_n, is positive,

$\qquad \rho = \left(1 + \sqrt{5}\right)/2 \approx 1.6180$.

96. If $\{a_n\}$ is bounded, monotonic and nonincreasing, then $a_1 \ge a_2 \ge a_3 \ge \cdots \ge a_n \ge \cdots$. Then

$\qquad -a_1 \le -a_2 \le -a_3 \le \cdots \le -a_n \le \cdots$

is a bounded, monotonic, nondecreasing sequence which converges by the first half of the theorem. Since $\{-a_n\}$ converges, then so does $\{a_n\}$.

97. True   **98.** True   **99.** True   **100.** True

101. $a_1 = \sqrt{2} \approx 1.4142$

$a_2 = \sqrt{2 + \sqrt{2}} \approx 1.8478$

$a_3 = \sqrt{2 + \sqrt{2 + \sqrt{2}}} \approx 1.9616$

$a_4 = \sqrt{2 + \sqrt{2 + \sqrt{2 + \sqrt{2}}}} \approx 1.9904$

$a_5 = \sqrt{2 + \sqrt{2 + \sqrt{2 + \sqrt{2 + \sqrt{2}}}}} \approx 1.9976$

$\{a_n\}$ is increasing and bounded by 2, and hence converges to L. Letting $\lim\limits_{n\to\infty} a_n = L$ implies that
$\sqrt{2 + L} = L \overset{n\to\infty}{\Longrightarrow} L = 2$. Hence, $\lim\limits_{n\to\infty} a_n = 2$.

102. $x_0 = 1, x_n = \dfrac{1}{2}x_{n-1} + \dfrac{1}{x_{n-1}}, n = 1, 2, \ldots$

$\begin{aligned} x_1 &= 1.5 & x_6 &= 1.414214 \\ x_2 &= 1.41667 & x_7 &= 1.414214 \\ x_3 &= 1.414216 & x_8 &= 1.414114 \\ x_4 &= 1.414214 & x_9 &= 1.414214 \\ x_5 &= 1.414214 & x_{10} &= 1.414214 \end{aligned}$

The limit of the sequence appears to be $\sqrt{2}$. In fact, this sequence is Newton's Method applied to $f(x) = x^2 - 2$.

Section 8.2   Series and Convergence

1. $S_1 = 1$

$S_2 = 1 + \dfrac{1}{4} = 1.2500$

$S_3 = 1 + \dfrac{1}{4} + \dfrac{1}{9} \approx 1.3611$

$S_4 = 1 + \dfrac{1}{4} + \dfrac{1}{9} + \dfrac{1}{16} \approx 1.4236$

$S_5 = 1 + \dfrac{1}{4} + \dfrac{1}{9} + \dfrac{1}{16} + \dfrac{1}{25} \approx 1.4636$

2. $S_1 = \dfrac{1}{6} \approx 0.1667$

$S_2 = \dfrac{1}{6} + \dfrac{1}{6} \approx 0.3333$

$S_3 = \dfrac{1}{6} + \dfrac{1}{6} + \dfrac{3}{20} \approx 0.4833$

$S_4 = \dfrac{1}{6} + \dfrac{1}{6} + \dfrac{3}{20} + \dfrac{2}{15} \approx 0.6167$

$S_5 = \dfrac{1}{6} + \dfrac{1}{6} + \dfrac{3}{20} + \dfrac{2}{15} + \dfrac{5}{42} \approx 0.7357$

3. $S_1 = 3$

$S_2 = 3 - \dfrac{9}{2} = -1.5$

$S_3 = 3 - \dfrac{9}{2} + \dfrac{27}{4} = 5.25$

$S_4 = 3 - \dfrac{9}{2} + \dfrac{27}{4} - \dfrac{81}{8} = -4.875$

$S_5 = 3 - \dfrac{9}{2} + \dfrac{27}{4} - \dfrac{81}{8} + \dfrac{243}{16} = 10.3125$

4. $S_1 = 1$

$S_2 = 1 + \dfrac{1}{3} \approx 1.3333$

$S_3 = 1 + \dfrac{1}{3} + \dfrac{1}{5} \approx 1.5333$

$S_4 = 1 + \dfrac{1}{3} + \dfrac{1}{5} + \dfrac{1}{9} \approx 1.6444$

$S_5 = 1 + \dfrac{1}{3} + \dfrac{1}{5} + \dfrac{1}{9} + \dfrac{1}{11} \approx 1.7354$

5. $S_1 = 3$

$S_2 = 3 + \frac{3}{2} = 4.5$

$S_3 = 3 + \frac{3}{2} + \frac{3}{4} = 5.250$

$S_4 = 3 + \frac{3}{2} + \frac{3}{4} + \frac{3}{8} = 5.625$

$S_5 = 3 + \frac{3}{2} + \frac{3}{4} + \frac{3}{8} + \frac{3}{16} = 5.8125$

6. $S_1 = 1$

$S_2 = 1 - \frac{1}{2} = 0.5$

$S_3 = 1 - \frac{1}{2} + \frac{1}{6} \approx 0.6667$

$S_4 = 1 - \frac{1}{2} + \frac{1}{6} - \frac{1}{24} \approx 0.6250$

$S_5 = 1 - \frac{1}{2} + \frac{1}{6} - \frac{1}{24} + \frac{1}{120} \approx 0.6333$

7. $\displaystyle\sum_{n=1}^{\infty} \frac{n}{n+1}$

$\displaystyle\lim_{n \to \infty} \frac{n}{n+1} = 1 \neq 0$

Diverges by Theorem 8.9

8. $\displaystyle\sum_{n=1}^{\infty} \frac{n}{2n+3}$

$\displaystyle\lim_{n \to \infty} \frac{n}{2n+3} = \frac{1}{2} \neq 0$

Diverges by Theorem 8.9

9. $\displaystyle\sum_{n=1}^{\infty} \frac{n^2}{n^2+1}$

$\displaystyle\lim_{n \to \infty} \frac{n^2}{n^2+1} = 1 \neq 0$

Diverges by Theorem 8.9

10. $\displaystyle\sum_{n=1}^{\infty} \frac{n}{\sqrt{n^2+1}}$

$\displaystyle\lim_{n \to \infty} \frac{n}{\sqrt{n^2+1}} = \lim_{n \to \infty} \frac{1}{\sqrt{1+(1/n^2)}} = 1 \neq 0$

Diverges by Theorem 8.9

11. $\displaystyle\sum_{n=0}^{\infty} 3\left(\frac{3}{2}\right)^n$ Geometric series

$r = \frac{3}{2} > 1$

Diverges by Theorem 8.6

12. $\displaystyle\sum_{n=0}^{\infty} \left(\frac{4}{3}\right)^n$ Geometric series

$r = \frac{4}{3} > 1$

Diverges by Theorem 8.6

13. $\displaystyle\sum_{n=0}^{\infty} 1000(1.055)^n$ Geometric series

$r = 1.055 > 1$

Diverges by Theorem 8.6

14. $\displaystyle\sum_{n=0}^{\infty} 2(-1.03)^n$ Geometric series

$|r| = 1.03 > 1$

Diverges by Theorem 8.6

15. $\displaystyle\sum_{n=0}^{\infty} \frac{2^n+1}{2^{n+1}}$

$\displaystyle\lim_{n \to \infty} \frac{2^n+1}{2^{n+1}} = \lim_{n \to \infty} \frac{1+2^{-n}}{2} = \frac{1}{2} \neq 0$

Diverges by Theorem 8.9

16. $\displaystyle\sum_{n=1}^{\infty} \frac{n!}{2^n}$

$\displaystyle\lim_{n \to \infty} \frac{n!}{2^n} = \infty$

Diverges by Theorem 8.9

17. $\displaystyle\sum_{n=0}^{\infty} 2\left(\frac{3}{4}\right)^n$

Geometric series with $r = \frac{3}{4} < 1$.

Converges by Theorem 8.6

18. $\displaystyle\sum_{n=0}^{\infty} 2\left(-\frac{1}{2}\right)^n$

Geometric series with $|r| = \left|-\frac{1}{2}\right| < 1$.

Converges by Theorem 8.6

19. $\displaystyle\sum_{n=0}^{\infty} (0.9)^n$

Geometric series with $r = 0.9 < 1$.

Converges by Theorem 8.6

20. $\displaystyle\sum_{n=0}^{\infty} (-0.6)^n$

Geometric series with $|r| = |-0.6| < 1$.

Converges by Theorem 8.6

21. $\displaystyle\sum_{n=1}^{\infty} \frac{1}{n(n+1)} = \sum_{n=1}^{\infty} \left(\frac{1}{n} - \frac{1}{n+1}\right) = \left(1 - \frac{1}{2}\right) + \left(\frac{1}{2} - \frac{1}{3}\right) + \left(\frac{1}{3} - \frac{1}{4}\right) + \left(\frac{1}{4} - \frac{1}{5}\right) + \cdots$

$\displaystyle\sum_{n=1}^{\infty} \frac{1}{n(n+1)} = \lim_{n \to \infty} S_n = \lim_{n \to \infty} \left(1 - \frac{1}{n+1}\right) = 1$

22. $\displaystyle\sum_{n=1}^{\infty} \frac{1}{n(n+2)} = \sum_{n=1}^{\infty}\left(\frac{1}{2n} - \frac{1}{2(n+2)}\right) = \left(\frac{1}{2} - \frac{1}{6}\right) + \left(\frac{1}{4} - \frac{1}{8}\right) + \left(\frac{1}{6} - \frac{1}{10}\right) + \left(\frac{1}{8} - \frac{1}{12}\right) + \left(\frac{1}{10} - \frac{1}{14}\right) + \cdots$

$\displaystyle\sum_{n=1}^{\infty} \frac{1}{n(n+2)} = \lim_{n\to\infty} S_n = \lim_{n\to\infty}\left[\frac{1}{2} + \frac{1}{4} - \frac{1}{2(n+1)} - \frac{1}{2(n+2)}\right] = \frac{1}{2} + \frac{1}{4} = \frac{3}{4}$

23. $\displaystyle\sum_{n=0}^{\infty} \frac{9}{4}\left(\frac{1}{4}\right)^n = \frac{9}{4}\left[1 + \frac{1}{4} + \frac{1}{16} + \cdots\right]$

$S_0 = \frac{9}{4}, \; S_1 = \frac{9}{4}\cdot\frac{5}{4} = \frac{45}{16}, \; S_2 = \frac{9}{4}\cdot\frac{21}{16} \approx 2.95, \ldots$

Matches graph (c).

Analytically, the series is geometric:

$$\sum_{n=0}^{\infty}\left(\frac{9}{4}\right)\left(\frac{1}{4}\right)^n = \frac{9/4}{1 - 1/4} = \frac{9/4}{3/4} = 3$$

24. $\displaystyle\sum_{n=0}^{\infty}\left(\frac{2}{3}\right)^n = 1 + \frac{2}{3} + \frac{4}{9} + \cdots$

$S_0 = 1, \; S_1 = \frac{5}{3}, \; S_2 \approx 2.11, \ldots$

Matches graph (b).

Analytically, the series is geometric:

$$\sum_{n=0}^{\infty}\left(\frac{2}{3}\right)^n = \frac{1}{1 - 2/3} = \frac{1}{1/3} = 3$$

25. $\displaystyle\sum_{n=0}^{\infty} \frac{15}{4}\left(-\frac{1}{4}\right)^n = \frac{15}{4}\left[1 - \frac{1}{4} + \frac{1}{16} - \cdots\right]$

$S_0 = \frac{15}{4}, \; S_1 = \frac{45}{16}, \; S_2 \approx 3.05, \ldots$

Matches graph (a).

Analytically, the series is geometric:

$$\sum_{n=0}^{\infty} \frac{15}{4}\left(-\frac{1}{4}\right)^n = \frac{15/4}{1 - (-1/4)} = \frac{15/4}{5/4} = 3$$

26. $\displaystyle\sum_{n=0}^{\infty} \frac{17}{3}\left(-\frac{8}{9}\right)^n = \frac{17}{3}\left[1 - \frac{8}{9} + \frac{64}{81} - \cdots\right]$

$S_0 = \frac{17}{3}, \; S_1 \approx 0.63, \; S_3 \approx 5.1, \ldots$

Matches (d).

Analytically, the series is geometric:

$$\sum_{n=0}^{\infty} \frac{17}{3}\left(-\frac{8}{9}\right)^n = \frac{17/3}{1 - (-8/9)} = \frac{17/3}{17/9} = 3$$

27. (a) $\displaystyle\sum_{n=1}^{\infty} 2(0.9)^{n-1} = \sum_{n=0}^{\infty} 2(0.9)^n = \frac{2}{1 - 0.9} = 20$

(b)

n	5	10	20	50	100
S_n	8.1902	13.0264	17.5685	19.8969	19.9995

(c)

(d) The terms of the series decrease in magnitude slowly. Thus, the sequence of partial sums approaches the sum slowly.

28. (a) $\displaystyle\sum_{n=1}^{\infty} \frac{4}{n(n+4)} = \sum_{n=1}^{\infty}\left(\frac{1}{n} - \frac{1}{n+4}\right)$

$= \left(1 - \frac{1}{5}\right) + \left(\frac{1}{2} - \frac{1}{6}\right) + \left(\frac{1}{3} - \frac{1}{7}\right) + \left(\frac{1}{4} - \frac{1}{8}\right) + \left(\frac{1}{5} - \frac{1}{9}\right) + \left(\frac{1}{6} - \frac{1}{10}\right) + \cdots$

$= 1 + \frac{1}{2} + \frac{1}{3} + \frac{1}{4} = \frac{25}{12} \approx 2.0833$

(b)

n	5	10	20	50	100
S_n	1.5377	1.7607	1.9051	2.0071	2.0443

(c)

(d) The terms of the series decrease in magnitude slowly. Thus, the sequence of partial sums approaches the sum slowly.

29. (a) $\displaystyle\sum_{n=1}^{\infty} 10(0.25)^{n-1} = \frac{10}{1-0.25} = \frac{40}{3} \approx 13.3333$

(c)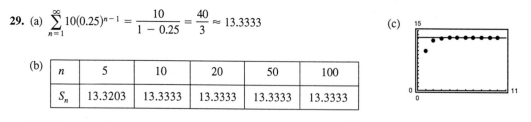

(b)

n	5	10	20	50	100
S_n	13.3203	13.3333	13.3333	13.3333	13.3333

(d) The terms of the series decrease in magnitude rapidly. Thus, the sequence of partial sums approaches the sum rapidly.

30. (a) $\displaystyle\sum_{n=1}^{\infty} 5\left(-\frac{1}{3}\right)^{n-1} = \frac{5}{1-(-1/3)} = \frac{15}{4} = 3.75$

(c)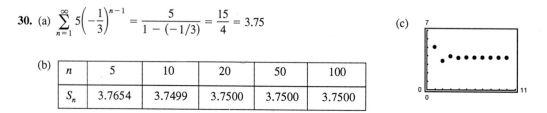

(b)

n	5	10	20	50	100
S_n	3.7654	3.7499	3.7500	3.7500	3.7500

(d) The terms of the series decrease in magnitude rapidly. Thus, the sequence of partial sums approaches the sum rapidly.

31. $\displaystyle\sum_{n=0}^{\infty} \left(\frac{1}{2}\right)^n = \frac{1}{1-(1/2)} = 2$

32. $\displaystyle\sum_{n=0}^{\infty} 2\left(\frac{2}{3}\right)^n = \frac{2}{1-(2/3)} = 6$

33. $\displaystyle\sum_{n=0}^{\infty} \left(-\frac{1}{2}\right)^n = \frac{1}{1-(-1/2)} = \frac{2}{3}$

34. $\displaystyle\sum_{n=0}^{\infty} 2\left(-\frac{2}{3}\right)^n = \frac{2}{1-(-2/3)} = \frac{6}{5}$

35. $\displaystyle\sum_{n=0}^{\infty} \left(\frac{1}{10}\right)^n = \frac{1}{1-(1/10)} = \frac{10}{9}$

36. $\displaystyle\sum_{n=0}^{\infty} 8\left(\frac{3}{4}\right)^n = \frac{8}{1-(3/4)} = 32$

37. $\displaystyle\sum_{n=0}^{\infty} 3\left(-\frac{1}{3}\right)^n = \frac{3}{1-(-1/3)} = \frac{9}{4}$

38. $\displaystyle\sum_{n=0}^{\infty} 4\left(-\frac{1}{2}\right)^n = \frac{4}{1-(-1/2)} = \frac{8}{3}$

39. $\displaystyle\sum_{n=2}^{\infty} \frac{1}{n^2-1} = \sum_{n=2}^{\infty}\left(\frac{1/2}{n-1} - \frac{1/2}{n+1}\right) = \frac{1}{2}\sum_{n=2}^{\infty}\left(\frac{1}{n-1} - \frac{1}{n+1}\right)$

$\displaystyle = \frac{1}{2}\left[\left(1-\frac{1}{3}\right) + \left(\frac{1}{2}-\frac{1}{4}\right) + \left(\frac{1}{3}-\frac{1}{5}\right) + \left(\frac{1}{4}-\frac{1}{6}\right) + \cdots\right]$

$\displaystyle = \frac{1}{2}\left(1+\frac{1}{2}\right) = \frac{3}{4}$

40. $\displaystyle\sum_{n=1}^{\infty} \frac{1}{n(n+1)} = \sum_{n=1}^{\infty}\left(\frac{1}{n} - \frac{1}{n+1}\right) = \left(1-\frac{1}{2}\right) + \left(\frac{1}{2}-\frac{1}{3}\right) + \left(\frac{1}{3}-\frac{1}{4}\right) + \cdots = 1$

41. $\displaystyle\sum_{n=1}^{\infty} \frac{4}{n(n+2)} = 2\sum_{n=1}^{\infty}\left(\frac{1}{n} - \frac{1}{n+2}\right) = 2\left[\left(1-\frac{1}{3}\right) + \left(\frac{1}{2}-\frac{1}{4}\right) + \left(\frac{1}{3}-\frac{1}{5}\right) + \cdots\right] = 2\left(1+\frac{1}{2}\right) = 3$

42. $\displaystyle\sum_{n=1}^{\infty} \frac{1}{(2n+1)(2n+3)} = \frac{1}{2}\sum_{n=1}^{\infty}\left(\frac{1}{2n+1} - \frac{1}{2n+3}\right) = \frac{1}{2}\left[\left(\frac{1}{3}-\frac{1}{5}\right) + \left(\frac{1}{5}-\frac{1}{7}\right) + \left(\frac{1}{7}-\frac{1}{9}\right) + \cdots\right] = \frac{1}{2}\left(\frac{1}{3}\right) = \frac{1}{6}$

43. $\displaystyle\sum_{n=0}^{\infty} \left(\frac{1}{2^n} - \frac{1}{3^n}\right) = \sum_{n=0}^{\infty}\left(\frac{1}{2}\right)^n - \sum_{n=0}^{\infty}\left(\frac{1}{3}\right)^n = \frac{1}{1-(1/2)} - \frac{1}{1-(1/3)} = 2 - \frac{3}{2} = \frac{1}{2}$

44. $\displaystyle\sum_{n=1}^{\infty} [(0.7)^n + (0.9)^n] = \sum_{n=0}^{\infty} \left(\frac{7}{10}\right)^n + \sum_{n=0}^{\infty} \left(\frac{9}{10}\right)^n - 2 = \frac{1}{1 - (7/10)} + \frac{1}{1 - (9/10)} - 2 = \frac{10}{3} + 10 - 2 = \frac{34}{3}$

45. $0.\overline{4} = \displaystyle\sum_{n=0}^{\infty} \frac{4}{10}\left(\frac{1}{10}\right)^n$

Geometric series with $a = \frac{4}{10}$ and $r = \frac{1}{10}$

$$S = \frac{a}{1 - r} = \frac{4/10}{1 - (1/10)} = \frac{4}{9}$$

46. $0.23\overline{23} = \displaystyle\sum_{n=0}^{\infty} \frac{23}{100}\left(\frac{1}{100}\right)^n$

Geometric series with $a = \frac{23}{100}$ and $r = \frac{1}{100}$

$$S = \frac{a}{1 - r} = \frac{23/100}{99/100} = \frac{23}{99}$$

47. $0.07\overline{5} = \displaystyle\sum_{n=0}^{\infty} \frac{3}{40}\left(\frac{1}{100}\right)^n$

Geometric series with $a = \frac{3}{40}$ and $r = \frac{1}{100}$

$$S = \frac{a}{1 - r} = \frac{3/40}{99/100} = \frac{5}{66}$$

48. $0.215\overline{15} = \dfrac{1}{5} + \displaystyle\sum_{n=0}^{\infty} \frac{3}{200}\left(\frac{1}{100}\right)^n$

Geometric series with $a = \frac{3}{200}$ and $r = \frac{1}{100}$

$$S = \frac{1}{5} + \frac{a}{1 - r} = \frac{1}{5} + \frac{3/200}{99/100} = \frac{71}{330}$$

49. $\displaystyle\sum_{n=1}^{\infty} \frac{n + 10}{10n + 1}$

$\displaystyle\lim_{n\to\infty} \frac{n + 10}{10n + 1} = \frac{1}{10} \neq 0$

Diverges by Theorem 8.9

50. $\displaystyle\sum_{n=1}^{\infty} \frac{n + 1}{2n - 1}$

$\displaystyle\lim_{n\to\infty} \frac{n + 1}{2n - 1} = \frac{1}{2} \neq 0$

Diverges by Theorem 8.9

51. $\displaystyle\sum_{n=1}^{\infty} \left(\frac{1}{n} - \frac{1}{n + 2}\right) = \left(1 - \frac{1}{3}\right) + \left(\frac{1}{2} - \frac{1}{4}\right) + \left(\frac{1}{3} - \frac{1}{5}\right) + \left(\frac{1}{4} - \frac{1}{6}\right) + \cdots = 1 + \frac{1}{2} = \frac{3}{2}$, converges

52. $\displaystyle\sum_{n=1}^{\infty} \frac{1}{n(n + 3)} = \frac{1}{3} \sum_{n=1}^{\infty} \left(\frac{1}{n} - \frac{1}{n + 3}\right)$

$$= \frac{1}{3}\left[\left(1 - \frac{1}{4}\right) + \left(\frac{1}{2} - \frac{1}{5}\right) + \left(\frac{1}{3} - \frac{1}{6}\right) + \left(\frac{1}{4} - \frac{1}{7}\right) + \left(\frac{1}{5} - \frac{1}{8}\right) + \left(\frac{1}{6} - \frac{1}{9}\right) + \cdots\right]$$

$$= \frac{1}{3}\left(1 + \frac{1}{2} + \frac{1}{3}\right) = \frac{11}{18}, \text{ converges}$$

53. $\displaystyle\sum_{n=1}^{\infty} \frac{3n - 1}{2n + 1}$

$\displaystyle\lim_{n\to\infty} \frac{3n - 1}{2n + 1} = \frac{3}{2} \neq 0$

Diverges by Theorem 8.9

54. $\displaystyle\sum_{n=1}^{\infty} \frac{2^n}{n^2}$

$\displaystyle\lim_{n\to\infty} \frac{2^n}{n^2} = \lim_{n\to\infty} \frac{(\ln 2)2^n}{2n}$

$$= \lim_{n\to\infty} \frac{(\ln 2)^2 2^n}{2} = \infty$$

(by L'Hôpital's Rule) Diverges by Theorem 8.9

55. $\displaystyle\sum_{n=0}^{\infty} \frac{4}{2^n} = 4 \sum_{n=0}^{\infty} \left(\frac{1}{2}\right)^n$

Geometric series with $r = \frac{1}{2}$

Converges by Theorem 8.6

56. $\displaystyle\sum_{n=0}^{\infty} \frac{1}{4^n}$

Geometric series with $r = \frac{1}{4}$

Converges by Theorem 8.6

57. $\displaystyle\sum_{n=0}^{\infty} (1.075)^n$

Geometric series with $r = 1.075$

Diverges by Theorem 8.6

58. $\displaystyle\sum_{n=1}^{\infty} \frac{2^n}{100}$

Geometric series with $r = 2$

Diverges by Theorem 8.6

59. $\displaystyle\sum_{n=2}^{\infty} \frac{n}{\ln n}$

$$\lim_{n \to \infty} \frac{n}{\ln n} = \lim_{n \to \infty} \frac{1}{1/n} = \infty$$

(by L'Hôpital's Rule) Diverges by Theorem 8.9

60. $\displaystyle\sum_{n=1}^{\infty} \left(1 + \frac{k}{n}\right)^n$

$$\lim_{n \to \infty} \left(1 + \frac{k}{n}\right)^n = e^k \neq 0$$

Diverges by Theorem 8.9

61. (a) x is the common ratio.

(c) $y_1 = \dfrac{1}{1-x}$

$\quad\ y_2 = 1 + x$

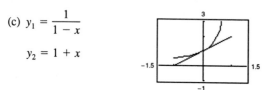

(b) $\displaystyle 1 + x + x^2 + \ldots = \sum_{n=0}^{\infty} x^n = \frac{1}{1-x}, |x| < 1$

Geometric series: $a = 1, r = x, |x| < 1$

62. (a) $(-x/2)$ is the common ratio.

(c) $y_1 = \dfrac{2}{2+x}$

$\quad\ y_2 = 1 - \dfrac{x}{2}$

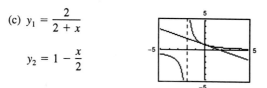

(b) $\displaystyle 1 - \frac{x}{2} + \frac{x^2}{4} - \frac{x^3}{8} = \sum_{n=0}^{\infty} \left(-\frac{x}{2}\right)^n = \frac{1}{1 - (-x/2)}$

$$= \frac{2}{2+x}, |x| < 2$$

Geometric series:

$$a = 1, r = -\frac{x}{2}, \left|-\frac{x}{2}\right| < 1 \implies |x| < 2$$

63. $f(x) = 3\left[\dfrac{1 - 0.5^x}{1 - 0.5}\right]$

Horizontal asymptote: $y = 6$

$$\sum_{n=0}^{\infty} 3\left(\frac{1}{2}\right)^n$$

$$S = \frac{3}{1 - (1/2)} = 6$$

The horizontal asymptote is the sum of the series. $f(n)$ is the nth partial sum.

64. $f(x) = 2\left[\dfrac{1 - 0.8^x}{1 - 0.8}\right]$

Horizontal asymptote: $y = 10$

$$\sum_{n=0}^{\infty} 2\left(\frac{4}{5}\right)^n$$

$$S = \frac{2}{1 - (4/5)} = 10$$

The horizontal asymptote is the sum of the series. $f(n)$ is the nth partial sum.

65. $\dfrac{1}{n(n+1)} < 0.001$

$\quad 10{,}000 < n^2 + n$

$\quad\quad 0 < n^2 + n - 10{,}000$

$\quad n = \dfrac{-1 \pm \sqrt{1^2 - 4(1)(-10{,}000)}}{2}$

Choosing the positive value for n we have $n \approx 99.5012$. The first *term* that is less than 0.001 is $n = 100$.

$\quad \left(\dfrac{1}{8}\right)^n < 0.001$

$\quad 10{,}000 < 8^n$

This inequality is true when $n = 5$. This series converges at a faster rate.

66. $\dfrac{1}{2^n} < 0.0001$

$10,000 < 2^n$

This inequality is true when $n = 14$.

$(0.01)^n < 0.0001$

$10,000 < 10^n$

This inequality is true when $n = 5$. This series converges at a faster rate.

67. $\displaystyle\sum_{i=0}^{n-1} 8000(0.9)^i = \dfrac{8000[1 - (0.9)^{(n-1)+1}]}{1 - 0.9}$

$\qquad\qquad\qquad = 80,000(1 - 0.9^n),\ n > 0$

68. $\displaystyle\sum_{i=0}^{n-1} 100(0.75)^i = \dfrac{100[1 - 0.75^{(n-1)+1}]}{1 - 0.75}$

$\qquad\qquad\qquad = 400(1 - 0.75^n)$ million dollars.

$\qquad\qquad$ Sum = 400 million dollars

69. $D_1 = 16$

$D_2 = \underbrace{0.81(16)}_{\text{up}} + \underbrace{0.81(16)}_{\text{down}} = 32(0.81)$

$D_3 = 16(0.81)^2 + 16(0.81)^2 = 32(0.81)^2$

$\vdots$

$D = 16 + 32(0.81) + 32(0.81)^2 + \ldots = -16 + \displaystyle\sum_{n=0}^{\infty} 32(0.81)^n = -16 + \dfrac{32}{1 - 0.81} = 152.42$ ft

70. The ball in Exercise 69 takes the following times for each fall.

$s_1 = -16t^2 + 16$ $\qquad\qquad$ $s_1 = 0$ if $t = 1$

$s_2 = -16t^2 + 16(0.81)$ $\qquad\qquad$ $s_2 = 0$ if $t = 0.9$

$s_3 = -16t^2 + 16(0.81)^2$ $\qquad\qquad$ $s_3 = 0$ if $t = (0.9)^2$

$\vdots$ $\qquad\qquad\qquad\qquad$ $\vdots$

$s_n = -16t^2 + 16(0.81)^{n-1}$ $\qquad\qquad$ $s_n = 0$ if $t = (0.9)^{n-1}$

Beginning with s_2, the ball takes the same amount of time to bounce up as it takes to fall. The total elapsed time before the ball comes to rest is

$$t = 1 + 2\sum_{n=1}^{\infty} (0.9)^n = -1 + 2\sum_{n=0}^{\infty} (0.9)^n = -1 + \dfrac{2}{1 - 0.9} = 19 \text{ seconds.}$$

71. $P(n) = \dfrac{1}{2}\left(\dfrac{1}{2}\right)^n$

$P(2) = \dfrac{1}{2}\left(\dfrac{1}{2}\right)^2 = \dfrac{1}{8}$

$\displaystyle\sum_{n=0}^{\infty} \dfrac{1}{2}\left(\dfrac{1}{2}\right)^n = \dfrac{1/2}{1 - (1/2)} = 1$

72. $P(n) = \dfrac{1}{3}\left(\dfrac{2}{3}\right)^n$

$P(2) = \dfrac{1}{3}\left(\dfrac{2}{3}\right)^2 = \dfrac{4}{27}$

$\displaystyle\sum_{n=0}^{\infty} \dfrac{1}{3}\left(\dfrac{2}{3}\right)^n = \dfrac{1/3}{1 - (2/3)} = 1$

73. (a) $64 + 32 + 16 + 8 + 4 + 2 = 126$ in.^{2}

(b) $\displaystyle\sum_{n=0}^{\infty} 64\left(\frac{1}{2}\right)^n = \frac{64}{1 - (1/2)} = 128$ in.^{2}

Note: This is one-half of the area of the original square!

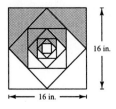

16 in.

16 in.

74. Surface area $= 4\pi(1)^2 + 9\left(4\pi\left(\frac{1}{3}\right)^2\right) + 9^2 \cdot 4\pi\left(\frac{1}{9}\right)^2 + \ldots = 4[\pi + \pi + \ldots] = \infty$

75. $w = \displaystyle\sum_{i=0}^{n-1} 0.01(2)^i = \frac{0.01(1 - 2^n)}{1 - 2} = 0.01(2^n - 1)$

(a) When $n = 29$: $w = \$5,368,709.11$

(b) When $n = 30$: $w = \$10,737,418.23$

(c) When $n = 31$: $w = \$21,474,836.47$

76. $\displaystyle\sum_{n=0}^{12t-1} P\left(1 + \frac{r}{12}\right)^n = \frac{P\left[1 - \left(1 + \dfrac{r}{12}\right)^{12t}\right]}{1 - \left(1 + \dfrac{r}{12}\right)}$

$= P\left(-\frac{12}{r}\right)\left[\left(1 - \left(1 + \frac{r}{12}\right)^{12t}\right)\right]$

$= P\left(\frac{12}{r}\right)\left[\left(1 + \frac{r}{12}\right)^{12t} - 1\right]$

$\displaystyle\sum_{n=0}^{12t-1} P(e^{r/12})^n = \frac{P(1 - (e^{r/12})^{12t})}{1 - e^{r/12}} = \frac{P(e^{rt} - 1)}{e^{r/12} - 1}$

77. $P = 50, r = 0.03, t = 20$

(a) $A = 50\left(\frac{12}{0.03}\right)\left[\left(1 + \frac{0.03}{12}\right)^{12(20)} - 1\right] \approx \$16,415.10$

(b) $A = \dfrac{50 - (e^{0.03(20)} - 1)}{e^{0.03/12} - 1} \approx \$16,421.83$

78. $P = 75, r = 0.05, t = 25$

(a) $A = 75\left(\frac{12}{0.05}\right)\left[\left(1 + \frac{0.05}{12}\right)^{12(25)} - 1\right] \approx \$44,663.23$

(b) $A = \dfrac{75(e^{0.05(25)} - 1)}{e^{0.05/12} - 1} \approx \$44,732.85$

79. $P = 100, r = 0.04, t = 40$

(a) $A = 100\left(\frac{12}{0.04}\right)\left[\left(1 + \frac{0.04}{12}\right)^{12(40)} - 1\right] \approx \$118,196.13$

(b) $A = \dfrac{100(e^{0.04(40)} - 1)}{e^{0.04/12} - 1} \approx \$118,393.43$

80. $P = 20, r = 0.06, t = 50$

(a) $A = 20\left(\frac{12}{0.06}\right)\left[\left(1 + \frac{0.06}{12}\right)^{12(50)} - 1\right] \approx \$75,743.82$

(b) $A = \dfrac{20(e^{0.06(50)} - 1)}{e^{0.06/12} - 1} \approx \$76,151.45$

81. $x = 0.749999\ldots = 0.74 + \displaystyle\sum_{n=0}^{\infty} 0.009(0.1)^n$

$= 0.74 + \dfrac{0.009}{1 - 0.1}$

$= 0.74 + 0.01 = 0.75$

82. $x = 0.a_1a_2a_3 \ldots a_k\overline{a_1a_2a_3 \ldots a_k}$

$= 0.a_1a_2a_3 \ldots a_k\left[1 + \frac{1}{10^k} + \left(\frac{1}{10^k}\right)^2 + \left(\frac{1}{10^k}\right)^3 + \ldots\right]$

$= 0.a_1a_2a_3 \ldots a_k\displaystyle\sum_{n=0}^{\infty}\left(\frac{1}{10^k}\right)^n$

$= 0.a_1a_2a_3 \ldots a_k\left[\dfrac{1}{1 - (1/10^k)}\right] = $ a rational number

83. By letting $S_0 = 0$, we have $a_n = \displaystyle\sum_{k=1}^{n} a_k - \sum_{k=1}^{n-1} a_k = S_n - S_{n-1}$. Thus,

$\displaystyle\sum_{n=1}^{\infty} a_n = \sum_{n=1}^{\infty} (S_n - S_{n-1}) = \sum_{n=1}^{\infty} (S_n - S_{n-1} + c - c) = \sum_{n=1}^{\infty} [(c - S_{n-1}) - (c - S_n)].$

84. Let $\{S_n\}$ be the sequence of partial sums for the convergent series $\sum\limits_{n=1}^{\infty} a_n = L$. Then

$$\lim_{n\to\infty} S_n = L \text{ and since } R_n = \sum_{k=n+1}^{\infty} a_k = L - S_n,$$

we have

$$\lim_{n\to\infty} R_n = \lim_{n\to\infty}(L - S_n) = \lim_{n\to\infty} L - \lim_{n\to\infty} S_n = L - L = 0.$$

85. $\sum\limits_{n=0}^{39} 30{,}000(1.05)^n = \dfrac{30{,}000(1 - 1.05^{40})}{1 - 1.05} \approx \$3{,}623{,}993.23$

86. $\sum\limits_{n=0}^{9} 167.5(e^{0.12})^n = \dfrac{167.5(1 - e^{0.12(10)})}{1 - e^{0.12}}$

$\approx \$3048.1$ million

87. Let $\sum a_n = \sum\limits_{n=0}^{\infty} 1$ and $\sum b_n = \sum\limits_{n=0}^{\infty}(-1)$.

Both are divergent series.

$$\sum(a_n + b_n) = \sum_{n=0}^{\infty}[1 + (-1)] = \sum_{n=0}^{\infty}[1 - 1] = 0$$

88. If $\sum(a_n + b_n)$ converged, then $\sum(a_n + b_n) - \sum a_n = \sum b_n$ would converge, which is a contradiction. Thus, $\sum(a_n + b_n)$ diverges.

89. False. $\lim\limits_{n\to\infty} \dfrac{1}{n} = 0$, but $\sum\limits_{n=1}^{\infty} \dfrac{1}{n}$ diverges.

90. True

91. False

$$\sum_{n=1}^{\infty} ar^n = \left(\frac{a}{1 - r}\right) - a$$

The formula requires that the geometric series begins with $n = 0$.

92. True

$$\lim_{n\to\infty} \frac{n}{1000(n + 1)} = \frac{1}{1000} \neq 0$$

93. Let H represent the half-life of the drug. If a patient receives n equal doses of P units each of this drug, administered at equal time interval of length t, the total amount of the drug in the patient's system at the time the last dose is administered is given by

$$T_n = P + Pe^{kt} + Pe^{2kt} + \cdots + Pe^{(n-1)kt}$$

where $k = -(\ln 2)/H$. One time interval *after* the last dose is administered is given by

$$T_{n+1} = Pe^{kt} + Pe^{2kt} + Pe^{3kt} + \cdots + Pe^{nkt}.$$

Two time intervals *after* the last dose is administered is given by

$$T_{n+1} = Pe^{2kt} + Pe^{3kt} + Pe^{4kt} + \cdots + Pe^{(n+1)kt}$$

and so on. Since $k < 0$, $T_n \to 0$ as $n \to \infty$.

94. $\dfrac{1}{r} + \dfrac{1}{r^2} + \dfrac{1}{r^3} + \cdots = \sum\limits_{n=0}^{\infty} \dfrac{1}{r}\left(\dfrac{1}{r}\right)^n = \dfrac{1/r}{1 - (1/r)} = \dfrac{1}{r - 1}$

This is a geometric series which converges if $\left|\dfrac{1}{r}\right| < 1 \implies |r| > 1$.

Section 8.3 The Integral Test and *p*-series

1. $\displaystyle\sum_{n=1}^{\infty} \frac{1}{n+1}$

Let $f(x) = \dfrac{1}{x+1}$.

f is positive, continuous and decreasing for $x \geq 1$.

$$\int_1^{\infty} \frac{1}{x+1}\, dx = \left[\ln(x+1) \right]_1^{\infty} = \infty$$

Diverges by Theorem 8.10

2. $\displaystyle\sum_{n=1}^{\infty} ne^{-n}$

Let $f(x) = \dfrac{x}{e^x}$.

f is positive, continuous, and decreasing for $x > 1$ since

$$f'(x) = \frac{1-x}{e^x} < 0 \text{ for } x > 1.$$

$$\int_1^{\infty} xe^{-x}\, dx = \left[-e^{-x}(x+1) \right]_1^{\infty} = \frac{2}{e}$$

Converges by Theorem 8.10

3. $\displaystyle\sum_{n=1}^{\infty} e^{-n}$

Let $f(x) = e^{-x}$.

f is positive, continuous, and decreasing for $x \geq 1$.

$$\int_1^{\infty} e^{-x}\, dx = \left[-e^{-x} \right]_1^{\infty} = \frac{1}{e}$$

Converges by Theorem 8.10

4. $\displaystyle\sum_{n=1}^{\infty} \frac{1}{4n+1}$

Let $f(x) = \dfrac{1}{4x+1}$.

f is positive, continuous, and decreasing for $x \geq 1$.

$$\int_1^{\infty} \frac{1}{4x+1}\, dx = \left[\frac{1}{4} \ln|4x+1| \right]_1^{\infty} = \infty$$

Diverges by Theorem 8.10

5. $\displaystyle\sum_{n=1}^{\infty} \frac{1}{n^2+1}$

Let $f(x) = \dfrac{1}{x^2+1}$.

f is positive, continuous, and decreasing for $x \geq 1$.

$$\int_1^{\infty} \frac{1}{x^2+1}\, dx = \left[\arctan x \right]_1^{\infty} = \frac{\pi}{4}$$

Converges by Theorem 8.10

6. $\displaystyle\sum_{n=1}^{\infty} \frac{1}{2n+1}$

Let $f(x) = \dfrac{1}{2x+1}$.

f is positive, continuous, and decreasing for $x \geq 1$.

$$\int_1^{\infty} \frac{1}{2x+1}\, dx = \left[\ln\sqrt{2x+1} \right]_1^{\infty} = \infty$$

Diverges by Theorem 8.10

7. $\displaystyle\sum_{n=1}^{\infty} \frac{\ln(n+1)}{n+1}$

Let $f(x) = \dfrac{\ln(x+1)}{x+1}$

f is positive, continuous, and decreasing for $x \geq 2$ since

$$f'(x) = \frac{1 - \ln(x+1)}{(x+1)^2} < 0 \text{ for } x \geq 2.$$

$$\int_1^{\infty} \frac{\ln(x+1)}{x+1}\, dx = \left[\frac{\ln^2(x+1)}{2} \right]_1^{\infty} = \infty$$

Diverges by Theorem 8.10

8. $\displaystyle\sum_{n=1}^{\infty} \frac{n}{n^2+3}$

Let $f(x) = \dfrac{x}{x^2+3}$.

$f(x)$ is positive, continuous, and decreasing for $x \geq 2$ since

$$f'(x) = \frac{3-x^2}{(x^2+3)} < 0 \text{ for } x \geq 2.$$

$$\int_1^{\infty} \frac{x}{x^2+3}\, dx = \left[\ln\sqrt{x^2+3} \right]_1^{\infty} = \infty$$

Diverges by Theorem 8.10

9. $\displaystyle\sum_{n=1}^{\infty} \frac{n^{k-1}}{n^k + c}$

Let $f(x) = \dfrac{x^{k-1}}{x^k + c}$.

f is positive, continuous, and decreasing for $x > \sqrt[k]{c(k-1)}$ since

$$f'(x) = \frac{x^{k-2}[c(k-1) - x^k]}{(x^k + c)^2} < 0$$

for $x > \sqrt[k]{c(k-1)}$.

$$\int_1^\infty \frac{x^{k-1}}{x^k + c}\, dx = \left[\frac{1}{k}\ln(x^k + c)\right]_1^\infty = \infty$$

Diverges by Theorem 8.10

10. $\displaystyle\sum_{n=1}^{\infty} n^k e^{-n}$

Let $f(x) = \dfrac{x^k}{e^x}$.

f is positive, continuous and decreasing for $x > k$ since

$$f'(x) = \frac{x^{k-1}(k - x)}{e^x} < 0$$

for $x > k$. We use integration by parts.

$$\int_1^\infty x^k e^{-x}\, dx = \left[-x^k e^{-x}\right]_1^\infty + k\int_1^\infty x^{k-1} e^{-x}\, dx$$
$$= \frac{1}{e} + \frac{k}{e} + \frac{k(k-1)}{e} + \cdots + \frac{k!}{e}$$

Converges by Theorem 8.10

11. $\displaystyle\sum_{n=1}^{\infty} \frac{1}{n^3}$

Let $f(x) = \dfrac{1}{x^3}$.

f is positive, continuous, and decreasing for $x \geq 1$.

$$\int_1^\infty \frac{1}{x^3}\, dx = \left[-\frac{1}{2x^2}\right]_1^\infty = \frac{1}{2}$$

Converges by Theorem 8.10

12. $\displaystyle\sum_{n=1}^{\infty} \frac{1}{n^{1/3}}$

Let $f(x) = \dfrac{1}{x^{1/3}}$.

f is positive, continuous, and decreasing for $x \geq 1$.

$$\int_1^\infty \frac{1}{x^{1/3}}\, dx = \left[\frac{3}{2}x^{2/3}\right]_1^\infty = \infty$$

Diverges by Theorem 8.10

13. $\displaystyle\sum_{n=2}^{\infty} \frac{1}{n(\ln n)^p}$

If $p = 1$, then the series diverges by the Integral Test. If $p \neq 1$,

$$\int_2^\infty \frac{1}{x(\ln x)^p}\, dx = \int_2^\infty (\ln x)^{-p}\frac{1}{x}\, dx = \left[\frac{(\ln x)^{-p+1}}{-p+1}\right]_2^\infty .$$

Converges for $-p + 1 < 0$ or $p > 1$.

14. $\displaystyle\sum_{n=2}^{\infty} \frac{\ln n}{n^p}$

If $p = 1$, then the series diverges by the Integral Test. If $p \neq 1$,

$$\int_2^\infty \frac{\ln x}{x^p}\, dx = \int_2^\infty x^{-p} \ln x\, dx = \left[\frac{x^{-p+1}}{(-p+1)^2}\left[-1 + (-p+1)\ln x\right]\right]_2^\infty . \text{ (Use Integration by Parts.)}$$

Converges for $-p + 1 < 0$ or $p > 1$.

15. $\displaystyle\sum_{n=1}^{\infty} \frac{1}{\sqrt[5]{n}} = \sum_{n=1}^{\infty} \frac{1}{n^{1/5}}$

Divergent p-series with $p = \frac{1}{5} < 1$

16. $\displaystyle\sum_{n=1}^{\infty} \frac{1}{n^{4/3}}$

Convergent p-series with $p = \frac{4}{3} > 1$

17. $\displaystyle\sum_{n=1}^{\infty} \frac{1}{n^{1/2}}$

Divergent p-series with $p = \frac{1}{2} < 1$

18. $\displaystyle\sum_{n=1}^{\infty} \frac{1}{n^2}$

Convergent p-series with $p = 2 > 1$

19. $\sum_{n=1}^{\infty} \dfrac{1}{n^{3/2}}$

Convergent *p*-series with $p = \frac{3}{2} > 1$

20. $\sum_{n=1}^{\infty} \dfrac{1}{n^{2/3}}$

Divergent *p*-series with $p = \frac{2}{3} < 1$

21. $\sum_{n=1}^{\infty} \dfrac{1}{n^{1.04}}$

Convergent *p*-series with $p = 1.04 > 1$

22. $\sum_{n=1}^{\infty} \dfrac{1}{n^{\pi}}$

Convergent *p*-series with $p = \pi > 1$

23. $\sum_{n=1}^{\infty} \dfrac{2}{\sqrt[4]{n^3}} = \dfrac{2}{1} + \dfrac{2}{2^{3/4}} + \dfrac{2}{3^{3/4}} + \cdots$

$S_1 = 2$

$S_2 \approx 3.189$

$S_3 \approx 4.067$

Matches (a)

Diverges—*p*-series with $p = \frac{3}{4} < 1$

24. $\sum_{n=1}^{\infty} \dfrac{2}{n} = \dfrac{2}{1} + \dfrac{2}{2} + \dfrac{2}{3} + \cdots$

$S_1 = 2$

$S_2 = 3$

$S_3 \approx 3.67$

Matches (d)

Diverges—harmonic series

25. $\sum_{n=1}^{\infty} \dfrac{2}{n\sqrt{n}} = 2 + 2/2^{3/2} + 2/3^{3/2} + \cdots$

$S_1 = 2$

$S_2 \approx 2.707$

$S_3 \approx 3.092$

Matches (b)

Converges—*p*-series with $p = 3/2 > 1$

26. $\sum_{n=1}^{\infty} \dfrac{2}{n^2} = 2 + \dfrac{2}{2^2} + \dfrac{2}{3^2} + \cdots$

$S_1 = 2$

$S_2 = 2.5$

$S_3 \approx 2.722$

Matches (c)

Converges—*p*-series with $p = 2 > 1$.

27. No. Theorem 8.9 says that if the series converges, then the terms a_n tend to zero. Some of the series in Exercises 23-26 converge because the terms tend to 0 very rapidly.

28. (a)

n	5	10	20	50	100
S_n	3.7488	3.75	3.75	3.75	3.75

The partial sums approach the sum 3.75 very rapidly.

(b)

n	5	10	20	50	100
S_n	1.4636	1.5498	1.5962	1.6251	1.635

The partial sums approach the sum $\dfrac{\pi^2}{6} \approx 1.6449$ slower than the series in part (a).

29. $\sum_{n=1}^{N} \frac{1}{n} = 1 + \frac{1}{2} + \frac{1}{3} + \frac{1}{4} + \cdots + \frac{1}{N} > M$

(a)

M	2	4	6	8
N	4	31	227	1674

(b) No. Since the terms are decreasing (approaching zero), more and more terms are required to increase the partial sum by 2.

30. $\xi(x) = \sum_{n=1}^{\infty} n^{-x} = \sum_{n=1}^{\infty} \frac{1}{n^x}$

Converges for $x > 1$ by Theorem 8.11.

31. Since f is positive, continuous, and decreasing for $x \geq 1$ and $a_n = f(n)$, we have,

$$R_N = S - S_N = \sum_{n=1}^{\infty} a_n - \sum_{n=1}^{N} a_n = \sum_{n=N+1}^{\infty} a_n > 0.$$

Also, $R_N = S - S_N = \sum_{n=N+1}^{\infty} a_n \leq a_{N+1} + \int_{N+1}^{\infty} f(x)\,dx \leq \int_{N}^{\infty} f(x)\,dx.$ Thus,

$$0 \leq R_N \leq \int_{N}^{\infty} f(x)\,dx.$$

32. From Exercise 31, we have:

$$0 \leq S - S_N \leq \int_{N}^{\infty} f(x)\,dx$$

$$S_N \leq S \leq S_N + \int_{N}^{\infty} f(x)\,dx$$

$$\sum_{n=1}^{N} a_n \leq S \leq \sum_{n=1}^{N} a_n + \int_{N}^{\infty} f(x)\,dx$$

33. $S_6 = 1 + \frac{1}{2^4} + \frac{1}{3^4} + \frac{1}{4^4} + \frac{1}{5^4} + \frac{1}{6^4} \approx 1.0811$

$$R_6 \leq \int_{6}^{\infty} \frac{1}{x^4}\,dx = \left[-\frac{1}{3x^3} \right]_{6}^{\infty} \approx 0.0015$$

$$1.0811 \leq \sum_{n=1}^{\infty} \frac{1}{n^4} \leq 1.0811 + 0.0015 = 1.0826$$

34. $S_4 = 1 + \frac{1}{2^5} + \frac{1}{3^5} + \frac{1}{4^5} \approx 1.0363$

$$R_4 \leq \int_{4}^{\infty} \frac{1}{x^5}\,dx = \left[-\frac{1}{4x^4} \right]_{4}^{\infty} \approx 0.0010$$

$$1.0363 \leq \sum_{n=1}^{\infty} \frac{1}{n^5} \leq 1.0363 + 0.0010 = 1.0373$$

35. $S_{10} = \frac{1}{2} + \frac{1}{5} + \frac{1}{10} + \frac{1}{17} + \frac{1}{26} + \frac{1}{37} + \frac{1}{50} + \frac{1}{65} + \frac{1}{82} + \frac{1}{101} \approx 0.9818$

$$R_{10} = \leq \int_{10}^{\infty} \frac{1}{x^2 + 1}\,dx = \left[\arctan x \right]_{10}^{\infty} = \frac{\pi}{2} - \arctan 10 \approx 0.0997$$

$$0.9818 \leq \sum_{n=1}^{\infty} \frac{1}{n^5} \leq 0.9818 + 0.0997 = 1.0815$$

36. $S_{10} = \frac{1}{2(\ln 2)^3} + \frac{1}{3(\ln 3)^3} + \frac{1}{4(\ln 4)^3} + \cdots + \frac{1}{11(\ln 11)^3} \approx 1.9821$

$$R_{10} \leq \int_{10}^{\infty} \frac{1}{(x+1)[\ln(x+1)]^3}\,dx = \left[-\frac{1}{2[\ln(x+1)]^2} \right]_{10}^{\infty} = \frac{1}{2(\ln 11)^3} \approx 0.0870$$

$$1.9821 \leq \sum_{n=1}^{\infty} \frac{1}{(n+1)[\ln(n+1)]^3} \leq 1.9821 + 0.0870 = 2.0691$$

37. $S_4 = \dfrac{1}{e} + \dfrac{2}{e^4} + \dfrac{3}{e^9} + \dfrac{4}{e^{16}} \approx 0.4049$

$R_4 \leq \displaystyle\int_4^\infty x e^{-x^2}\, dx = \left[-\dfrac{1}{2} e^{-x^2} \right]_4^\infty = 5.6 \times 10^{-8}$

$0.4049 \leq \displaystyle\sum_{n=1}^\infty n e^{-n^2} \leq 0.4049 + 5.6 \times 10^{-8}$

38. $S_4 = \dfrac{1}{e} + \dfrac{1}{e^2} + \dfrac{1}{e^3} + \dfrac{1}{e^4} \approx 0.5713$

$R_4 \leq \displaystyle\int_4^\infty e^{-x}\, dx = \left[-e^{-x} \right]_4^\infty \approx 0.0183$

$0.5713 \leq \displaystyle\sum_{n=0}^\infty e^{-n} \leq 0.5713 + 0.0183 = 0.5896$

39. $0 < R_N < \displaystyle\int_N^\infty \dfrac{1}{x^4}\, dx = \left[-\dfrac{1}{3x^3} \right]_N^\infty = \dfrac{1}{3N^3} < 0.001$

$\dfrac{1}{N^3} < 0.003$

$N^3 > 333.33$

$N > 6.93$

$N \geq 7$

40. $0 < R_N < \displaystyle\int_N^\infty \dfrac{1}{x^{3/2}}\, dx = \left[-\dfrac{2}{x^{1/2}} \right]_N^\infty = \dfrac{2}{\sqrt{N}} < 0.001$

$N^{-1/2} < 0.0005$

$\sqrt{N} > 2000$

$N \geq 4{,}000{,}000$

41. $R_N < \displaystyle\int_N^\infty e^{-5x}\, dx = \left[-\dfrac{1}{5} e^{-5x} \right]_N^\infty = \dfrac{e^{-5N}}{5} < 0.001$

$\dfrac{1}{e^{5N}} < 0.005$

$e^{5N} > 200$

$5N > \ln 200$

$N > \dfrac{\ln 200}{5}$

$N > 1.0597$

$N \geq 2$

42. $R_N < \displaystyle\int_N^\infty \dfrac{1}{x^2 + 1}\, dx = \left[\arctan x \right]_N^\infty$

$= \dfrac{\pi}{2} - \arctan N < 0.001$

$-\arctan N < -1.5698$

$\arctan N > 1.5698$

$N > \tan 1.5698$

$N \geq 1004$

43. Your friend is not correct. The series

$$\sum_{n=10{,}000}^\infty \dfrac{1}{n} = \dfrac{1}{10{,}000} + \dfrac{1}{10{,}001} + \cdots$$

is the harmonic series, starting with the 10,000th term, and hence diverges.

44. (a) $\displaystyle\int_{10}^\infty \dfrac{1}{x^p}\, dx = \left[\dfrac{x^{-p+1}}{-p + 1} \right]_{10}^\infty = \dfrac{1}{(p-1)10^{p-1}}, p > 1$

(b) $f(x) = \dfrac{1}{x^p}$

$R_{10}(p) = \displaystyle\sum_{n=11}^\infty \dfrac{1}{n^p} \leq$ Area under the graph of f over the interval $[10, \infty)$

(c) The horizontal asymptote is $y = 0$. As n increases, the error decreases.

45. (a) $\displaystyle\sum_{n=2}^\infty \dfrac{1}{n^{1.1}}$. This is a convergent p-series with $p = 1.1 > 1$.

$\displaystyle\sum_{n=2}^\infty \dfrac{1}{n \ln n}$ is a divergent series. Use the Integral Test.

$\displaystyle\int_2^\infty \dfrac{1}{x \ln x}\, dx = \left[\ln|\ln x| \right]_2^\infty = \infty$

—CONTINUED—

45. —CONTINUED—

(b) $\displaystyle\sum_{n=2}^{6} \frac{1}{n^{1.1}} = \frac{1}{2^{1.1}} + \frac{1}{3^{1.1}} + \frac{1}{4^{1.1}} + \frac{1}{5^{1.1}} + \frac{1}{6^{1.1}} \approx 0.4665 + 0.2987 + 0.2176 + 0.1703 + 0.1393$

$\displaystyle\sum_{n=2}^{6} \frac{1}{n \ln n} = \frac{1}{2 \ln 2} + \frac{1}{3 \ln 3} + \frac{1}{4 \ln 4} + \frac{1}{5 \ln 5} + \frac{1}{6 \ln 6} \approx 0.7213 + 0.3034 + 0.1803 + 0.1243 + 0.0930$

The terms of the convergent series **seem** to be larger than those of the divergent series!

(c) $\dfrac{1}{n^{1.1}} < \dfrac{1}{n \ln n}$

$n \ln n < n^{1.1}$

$\ln n < n^{0.1}$

This inequality holds when $n \geq 3.5 \times 10^{15}$. Or, $n > e^{40}$. Then $\ln e^{40} = 40 < (e^{40})^{0.1} = e^4 \approx 55$.

46. (a) Let $f(x) = 1/x$. f is positive, continuous, and decreasing on $[1, \infty)$.

$S_n - 1 \leq \displaystyle\int_{1}^{n} \frac{1}{x}\, dx$

$S_n - 1 \leq \ln n$

Hence, $S_n \leq 1 + \ln n$. Similarly,

$S_n \geq \displaystyle\int_{1}^{n+1} \frac{1}{x}\, dx = \ln(n+1).$

Thus, $\ln(n+1) \leq S_n \leq 1 + \ln n$.

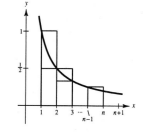

(b) Since $\ln(n+1) \leq S_n \leq 1 + \ln n$, we have $\ln(n+1) - \ln n \leq S_n - \ln n \leq 1$. Also, since $\ln x$ is an increasing function, $\ln(n+1) - \ln n > 0$ for $n \geq 1$. Thus, $0 \leq S_n - \ln n \leq 1$ and the sequence $\{a_n\}$ is bounded.

(c) $a_n - a_{n+1} = [S_n - \ln n] - [S_{n+1} - \ln(n+1)] = \displaystyle\int_{n}^{n+1} \frac{1}{x}\, dx - \frac{1}{n+1} \geq 0$

Thus, $a_n \geq a_{n+1}$ and the sequence is decreasing.

(d) Since the sequence is bounded and monotonic, it converges to a limit, γ.

(e) $a_{100} = S_{100} - \ln 100 \approx 0.5822$ (Actually $\gamma \approx 0.577216$.)

47. The harmonic series $\displaystyle\sum_{n=1}^{\infty} \frac{1}{n}$.

48. $\displaystyle\sum_{n=2}^{\infty} \ln\left(1 - \frac{1}{n^2}\right) = \sum_{n=2}^{\infty} \ln\left(\frac{n^2 - 1}{n^2}\right) = \sum_{n=2}^{\infty} \ln \frac{(n+1)(n-1)}{n^2} = \sum_{n=2}^{\infty} \left[\ln(n+1) + \ln(n-1) - 2 \ln n\right]$

$= (\ln 3 + \ln 1 - 2 \ln 2) + (\ln 4 + \ln 2 - 2 \ln 3) + (\ln 5 + \ln 3 - 2 \ln 4) + (\ln 6 + \ln 4 - 2 \ln 5)$

$+ (\ln 7 + \ln 5 - 2 \ln 6) + (\ln 8 + \ln 6 - 2 \ln 7) + (\ln 9 + \ln 7 - 2 \ln 8) + \ldots = -\ln 2$

49. $\displaystyle\sum_{n=1}^{\infty} \frac{1}{2n - 1}$

Let $f(x) = \dfrac{1}{2x - 1}$.

f is positive, continuous, and decreasing for $x \geq 1$.

$\displaystyle\int_{1}^{\infty} \frac{1}{2x - 1}\, dx = \left[\ln \sqrt{2x - 1}\,\right]_{1}^{\infty} = \infty$

Diverges by Theorem 8.10

50. $\displaystyle\sum_{n=2}^{\infty} \frac{1}{n\sqrt{n^2 - 1}}$

Let $f(x) = \dfrac{1}{x\sqrt{x^2 - 1}}$.

f is positive, continuous, and decreasing for $x \geq 2$.

$\displaystyle\int_{2}^{\infty} \frac{1}{x\sqrt{x^2 - 1}}\, dx = \left[\operatorname{arcsec} x\right]_{2}^{\infty} = \frac{\pi}{2} - \frac{\pi}{3}$

Converges by Theorem 8.10

51. $\sum_{n=1}^{\infty} \frac{1}{n\sqrt[4]{n}} = \sum_{n=1}^{\infty} \frac{1}{n^{5/4}}$

p-series with $p = \frac{5}{4}$

Converges by Theorem 8.11

52. $3\sum_{n=1}^{\infty} \frac{1}{n^{0.95}}$

p-series with $p = 0.95$

Diverges by Theorem 8.11

53. $\sum_{n=0}^{\infty} \left(\frac{2}{3}\right)^n$

Geometric series with $r = \frac{2}{3}$

Converges by Theorem 8.6

54. $\sum_{n=0}^{\infty} (1.075)^n$

Geometric series with $r = 1.075$

Diverges by Theorem 8.6

55. $\sum_{n=1}^{\infty} \frac{n}{\sqrt{n^2 + 1}}$

$\lim_{n\to\infty} \frac{n}{\sqrt{n^2 + 1}} = \lim_{n\to\infty} \frac{1}{\sqrt{1 + (1/n^2)}} = 1 \neq 0$

Diverges by Theorem 8.9

56. $\sum_{n=1}^{\infty} \left(\frac{1}{n^2} - \frac{1}{n^3}\right) = \sum_{n=1}^{\infty} \frac{1}{n^2} - \sum_{n=1}^{\infty} \frac{1}{n^3}$

Since these are both convergent *p*-series, the difference is convergent.

57. $\sum_{n=1}^{\infty} \left(1 + \frac{1}{n}\right)^n$

$\lim_{n\to\infty} \left(1 + \frac{1}{n}\right)^n = e \neq 0$

Fails *n*th Term Test

Diverges by Theorem 8.9

58. $\sum_{n=2}^{\infty} \ln(n)$

$\lim_{n\to\infty} \ln(n) = \infty$

Diverges by Theorem 8.9

59. $\sum_{n=2}^{\infty} \frac{1}{n(\ln n)^3}$

Let $f(x) = \frac{1}{x(\ln x)^3}$.

f is positive, continuous and decreasing for $x \geq 2$.

$\int_2^{\infty} \frac{1}{x(\ln x)^3} dx = \int_2^{\infty} (\ln x)^{-3} \frac{1}{x} dx = \left[\frac{(\ln x)^{-2}}{-2}\right]_2^{\infty} = \left[-\frac{1}{2(\ln x)^2}\right]_2^{\infty} = \frac{1}{2(\ln 2)^2}$

Converges by Theorem 8.10. See Exercise 13.

60. $\sum_{n=2}^{\infty} \frac{\ln n}{n^3}$

Let $f(x) = \frac{\ln x}{x^3}$.

f is positive, continuous, and decreasing for $x \geq 2$ since $f'(x) = \frac{1 - 3\ln x}{x^4} < 0$ for $x \geq 2$.

$\int_2^{\infty} \frac{\ln x}{x^3} dx = \left[-\frac{\ln x}{2x^2}\right]_2^{\infty} + \frac{1}{2} \int_2^{\infty} \frac{1}{x^3} dx = \frac{\ln 2}{8} + \left[-\frac{1}{4x^2}\right]_2^{\infty} = \frac{\ln 2}{8} + \frac{1}{16}$ (Use Integration by Parts.)

Converges by Theorem 8.10. See Exercise 14.

Section 8.4   Comparisons of Series

1. (a) $\displaystyle\sum_{n=1}^{\infty} \frac{6}{n^{3/2}} = \frac{6}{1} + \frac{6}{2^{3/2}} + \cdots \quad S_1 = 6$

$\displaystyle\sum_{n=1}^{\infty} \frac{6}{n^{3/2} + 3} = \frac{6}{4} + \frac{6}{2^{3/2} + 3} + \cdots \quad S_1 = \frac{3}{2}$

$\displaystyle\sum_{n=1}^{\infty} \frac{6}{n\sqrt{n^2 + 0.5}} = \frac{6}{1\sqrt{1.5}} + \frac{6}{2\sqrt{4.5}} + \cdots \quad S_1 = \frac{6}{\sqrt{1.5}} \approx 4.9$

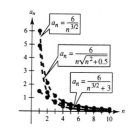

(b) The first series is a *p*-series. It converges ($p = 3/2 > 1$).

(c) The magnitude of the terms of the other two series are less than the corresponding terms at the convergent *p*-series. Hence, the other two series converge.

(d) The smaller the magnitude of the terms, the smaller the magnitude of the terms of the sequence of partial sums.

2. (a) $\displaystyle\sum_{n=1}^{\infty} \frac{2}{\sqrt{n}} = 2 + \frac{2}{\sqrt{2}} + \cdots \quad S_1 = 2$

$\displaystyle\sum_{n=1}^{\infty} \frac{2}{\sqrt{n} - 0.5} = \frac{2}{0.5} + \frac{2}{\sqrt{2} - 0.5} + \cdots \quad S_1 = 4$

$\displaystyle\sum_{n=1}^{\infty} \frac{4}{\sqrt{n} + 0.5} = \frac{4}{\sqrt{1.5}} + \frac{4}{\sqrt{2.5}} + \cdots \quad S_1 \approx 3.3$

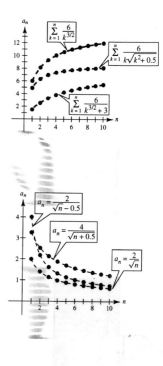

(b) The first series is a *p*-series. It diverges $\left(p = \frac{1}{2} < 1\right)$.

(c) The magnitude of the terms of the other two series are greater than the corresponding terms of the divergent *p*-series. Hence, the other two series diverge.

(d) The larger the magnitude of the terms, the larger the magnitude of the terms of the sequence of partial sums.

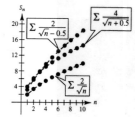

3. $\dfrac{1}{n^2 + 1} < \dfrac{1}{n^2}$

Therefore,

$$\sum_{n=1}^{\infty} \frac{1}{n^2 + 1}$$

converges by comparison with the convergent *p*-series

$$\sum_{n=1}^{\infty} \frac{1}{n^2}.$$

4. $\dfrac{1}{3n^2 + 2} < \dfrac{1}{3n^2}$

Therefore,

$$\sum_{n=1}^{\infty} \frac{1}{3n^2 + 2}$$

converges by comparison with the **convergent *p*-series**

$$\frac{1}{3}\sum_{n=1}^{\infty} \frac{1}{n^2}.$$

5. $\dfrac{1}{n-1} > \dfrac{1}{n}$ for $n \geq 2$

Therefore,

$$\sum_{n=2}^{\infty} \frac{1}{n-1}$$

diverges by comparison with the divergent p-series

$$\sum_{n=2}^{\infty} \frac{1}{n}.$$

6. $\dfrac{1}{\sqrt{n-1}} > \dfrac{1}{\sqrt{n}}$ for $n \geq 2$.

Therefore,

$$\sum_{n=2}^{\infty} \frac{1}{\sqrt{n-1}}$$

diverges by comparison with the divergent p-series

$$\sum_{n=2}^{\infty} \frac{1}{\sqrt{n}}.$$

7. $\dfrac{1}{3^n+1} < \dfrac{1}{3^n}$

Therefore,

$$\sum_{n=0}^{\infty} \frac{1}{3^n+1}$$

converges by comparison with the convergent geometric series

$$\sum_{n=0}^{\infty} \left(\frac{1}{3}\right)^n.$$

8. $\dfrac{2^n}{3^n+5} < \left(\dfrac{2}{3}\right)^n$

Therefore,

$$\sum_{n=1}^{\infty} \frac{2^n}{3^n+5}$$

converges by comparison with the convergent geometric series

$$\sum_{n=0}^{\infty} \left(\frac{2}{3}\right)^n.$$

9. For $n \geq 3$, $\dfrac{\ln n}{n+1} > \dfrac{1}{n+1}$.

Therefore,

$$\sum_{n=1}^{\infty} \frac{\ln n}{n+1}$$

diverges by comparison with the divergent series

$$\sum_{n=1}^{\infty} \frac{1}{n+1}.$$

Note: $\displaystyle\sum_{n=1}^{\infty} \frac{1}{n+1}$ diverges by the integral test.

10. $\dfrac{1}{\sqrt{n^3+1}} < \dfrac{1}{n^{3/2}}$

Therefore,

$$\sum_{n=1}^{\infty} \frac{1}{\sqrt{n^3+1}}$$

converges by comparison with the convergent p-series

$$\sum_{n=1}^{\infty} \frac{1}{n^{3/2}}.$$

11. For $n > 3$, $\dfrac{1}{n^2} > \dfrac{1}{n!}$.

Therefore,

$$\sum_{n=0}^{\infty} \frac{1}{n!}$$

converges by comparison with the convergent p-series

$$\sum_{n=1}^{\infty} \frac{1}{n^2}.$$

12. $\dfrac{1}{3\sqrt[4]{n}-1} > \dfrac{1}{3\sqrt[4]{n}}$

Therefore,

$$\sum_{n=1}^{\infty} \frac{1}{3\sqrt[4]{n}-1}$$

diverges by comparison with the divergent p-series

$$\frac{1}{3}\sum_{n=1}^{\infty} \frac{1}{\sqrt[4]{n}}.$$

13. $\dfrac{1}{e^{n^2}} \le \dfrac{1}{e^n}$

Therefore,

$$\sum_{n=0}^{\infty} \frac{1}{e^{n^2}}$$

converges by comparison with the convergent geometric series

$$\sum_{n=0}^{\infty} \left(\frac{1}{e}\right)^n.$$

14. $\dfrac{4^n}{3^n - 1} > \dfrac{4^n}{3^n}$

Therefore,

$$\sum_{n=1}^{\infty} \frac{4^n}{3^n - 1}$$

diverges by comparison with the divergent geometric series

$$\sum_{n=1}^{\infty} \left(\frac{4}{3}\right)^n.$$

15. $\displaystyle\lim_{n \to \infty} \frac{n/(n^2 + 1)}{1/n} = \lim_{n \to \infty} \frac{n^2}{n^2 + 1} = 1$

Therefore,

$$\sum_{n=1}^{\infty} \frac{n}{n^2 + 1}$$

diverges by a limit comparison with the divergent p-series

$$\sum_{n=1}^{\infty} \frac{1}{n}.$$

16. $\displaystyle\lim_{n \to \infty} \frac{1/\sqrt{n^2 - 1}}{1/n} = \lim_{n \to \infty} \frac{n}{\sqrt{n^2 - 1}} = 1$

Therefore,

$$\sum_{n=2}^{\infty} \frac{1}{\sqrt{n^2 - 1}}$$

diverges by a limit comparison with the divergent p-series

$$\sum_{n=2}^{\infty} \frac{1}{n}.$$

17. $\displaystyle\lim_{n \to \infty} \frac{1/\sqrt{n^2 + 1}}{1/n} = \lim_{n \to \infty} \frac{n}{\sqrt{n^2 + 1}} = 1$

Therefore,

$$\sum_{n=0}^{\infty} \frac{1}{\sqrt{n^2 + 1}}$$

diverges by a limit comparison with the divergent p-series

$$\sum_{n=1}^{\infty} \frac{1}{n}.$$

18. $\displaystyle\lim_{n \to \infty} \frac{1/(2^n - 5)}{1/2^n} = \lim_{n \to \infty} \frac{2^n}{2^n - 5} = 1$

Therefore,

$$\sum_{n=1}^{\infty} \frac{1}{2^n - 5}$$

converges by a limit comparison with the convergent geometric series

$$\sum_{n=0}^{\infty} \left(\frac{1}{2}\right)^n.$$

19. $\displaystyle\lim_{n \to \infty} \frac{\dfrac{2n^2 - 1}{3n^5 + 2n + 1}}{1/n^3} = \lim_{n \to \infty} \frac{2n^5 - n^3}{3n^5 + 2n + 1} = \frac{2}{3}$

Therefore,

$$\sum_{n=1}^{\infty} \frac{2n^2 - 1}{3n^5 + 2n + 1}$$

converges by a limit comparison with the convergent p-series

$$\sum_{n=1}^{\infty} \frac{1}{n^3}.$$

20. $\displaystyle\lim_{n \to \infty} \frac{\dfrac{5n - 3}{n^2 - 2n + 5}}{1/n} = \lim_{n \to \infty} \frac{5n^2 - 3n}{n^2 - 2n + 5} = 5$

Therefore,

$$\sum_{n=1}^{\infty} \frac{5n - 3}{n^2 - 2n + 5}$$

diverges by a limit comparison with the divergent p-series

$$\sum_{n=1}^{\infty} \frac{1}{n}.$$

21. $\lim\limits_{n\to\infty} \dfrac{\dfrac{n+3}{n(n+2)}}{1/n} = \lim\limits_{n\to\infty} \dfrac{n^2+3n}{n^2+2n} = 1$

Therefore,

$$\sum_{n=1}^{\infty} \frac{n+3}{n(n+2)}$$

diverges by a limit comparison with the divergent p-series

$$\sum_{n=1}^{\infty} \frac{1}{n}.$$

22. $\lim\limits_{n\to\infty} \dfrac{\dfrac{1}{n(n^2+1)}}{1/n^3} = \lim\limits_{n\to\infty} \dfrac{n^3}{n^3+n} = 1$

Therefore,

$$\sum_{n=1}^{\infty} \frac{1}{n(n^2+1)}$$

converges by a limit comparison with the convergent p-series

$$\sum_{n=1}^{\infty} \frac{1}{n^3}.$$

23. $\lim\limits_{n\to\infty} \dfrac{1/\left(n\sqrt{n^2+1}\right)}{1/n^2} = \lim\limits_{n\to\infty} \dfrac{n^2}{n\sqrt{n^2+1}} = 1$

Therefore,

$$\sum_{n=1}^{\infty} \frac{1}{n\sqrt{n^2+1}}$$

converges by a limit comparison with the convergent p-series

$$\sum_{n=1}^{\infty} \frac{1}{n^2}.$$

24. $\lim\limits_{n\to\infty} \dfrac{n/[(n+1)2^{n-1}]}{1/(2^{n-1})} = \lim\limits_{n\to\infty} \dfrac{n}{n+1} = 1$

Therefore,

$$\sum_{n=1}^{\infty} \frac{n}{(n+1)2^{n-1}}$$

converges by a limit comparison with the convergent geometric series

$$\sum_{n=1}^{\infty} \left(\frac{1}{2}\right)^{n-1}.$$

25. $\lim\limits_{n\to\infty} \dfrac{(n^{k-1})/(n^k+1)}{1/n} = \lim\limits_{n\to\infty} \dfrac{n^k}{n^k+1} = 1$

Therefore,

$$\sum_{n=1}^{\infty} \frac{n^{k-1}}{n^k+1}$$

diverges by a limit comparison with the divergent p-series

$$\sum_{n=1}^{\infty} \frac{1}{n}.$$

26. $\lim\limits_{n\to\infty} \dfrac{1/\left(n+\sqrt{n^2+1}\right)}{1/n} = \lim\limits_{n\to\infty} \dfrac{n}{n+\sqrt{n^2+1}} = \dfrac{1}{2}$

Therefore,

$$\sum_{n=1}^{\infty} \frac{1}{n+\sqrt{n^2+1}}$$

diverges by a limit comparison with the divergent p-series

$$\sum_{n=1}^{\infty} \frac{1}{n}.$$

27. $\lim\limits_{n\to\infty} \dfrac{\sin(1/n)}{1/n} = \lim\limits_{n\to\infty} \dfrac{(-1/n^2)\cos(1/n)}{-1/n^2}$

$$= \lim\limits_{n\to\infty} \cos\left(\frac{1}{n}\right) = 1$$

Therefore,

$$\sum_{n=1}^{\infty} \sin\left(\frac{1}{n}\right)$$

diverges by a limit comparison with the divergent p-series

$$\sum_{n=1}^{\infty} \frac{1}{n}.$$

28. $\lim\limits_{n\to\infty} \dfrac{\tan(1/n)}{1/n} = \lim\limits_{n\to\infty} \dfrac{(-1/n^2)\sec^2(1/n)}{-1/n^2}$

$$= \lim\limits_{n\to\infty} \sec^2\left(\frac{1}{n}\right) = 1$$

Therefore,

$$\sum_{n=1}^{\infty} \tan\left(\frac{1}{n}\right)$$

diverges by a limit comparison with the divergent p-series

$$\sum_{n=1}^{\infty} \frac{1}{n}.$$

29. $\displaystyle\sum_{n=1}^{\infty} \frac{\sqrt{n}}{n} = \sum_{n=1}^{\infty} \frac{1}{\sqrt{n}}$

Diverges

p-series with $p = \frac{1}{2}$

30. $\displaystyle\sum_{n=0}^{\infty} 5\left(-\frac{1}{5}\right)^n$

Converges

Geometric series with $r = -\frac{1}{5}$

31. $\displaystyle\sum_{n=1}^{\infty} \frac{1}{3^n+2}$

Converges

Direct comparison with $\displaystyle\sum_{n=1}^{\infty} \left(\frac{1}{3}\right)^n$

32. $\displaystyle\sum_{n=4}^{\infty} \frac{1}{3n^2 - 2n - 15}$

Converges

Limit comparison with $\displaystyle\sum_{n=1}^{\infty} \frac{1}{n^2}$

33. $\displaystyle\sum_{n=1}^{\infty} \frac{n}{2n + 3}$

Diverges; nth Term Test

$$\lim_{n \to \infty} \frac{n}{2n + 3} = \frac{1}{2} \neq 0$$

34. $\displaystyle\sum_{n=1}^{\infty} \left(\frac{1}{n+1} - \frac{1}{n+2}\right) = \left(\frac{1}{2} - \frac{1}{3}\right) + \left(\frac{1}{3} - \frac{1}{4}\right) + \left(\frac{1}{4} - \frac{1}{5}\right) + \cdots = \frac{1}{2}$

Converges; telescoping series

35. $\displaystyle\sum_{n=1}^{\infty} \frac{n}{(n^2 + 1)^2}$

Converges; integral test

36. $\displaystyle\sum_{n=1}^{\infty} \frac{3}{n(n+3)}$

Converges; telescoping series

$$\sum_{n=1}^{\infty} \left(\frac{1}{n} - \frac{1}{n+3}\right)$$

37. $\displaystyle\lim_{n \to \infty} \frac{a_n}{1/n} = \lim_{n \to \infty} na_n \neq 0$

Therefore,

$$\sum_{n=1}^{\infty} a_n$$

diverges by a limit comparison with the p-series

$$\sum_{n=1}^{\infty} \frac{1}{n}.$$

38. If $j < k - 1$, then $k - j > 1$. The p-series with $p = k - j$ converges and since

$$\lim_{n \to \infty} \frac{P(n)/Q(n)}{1/n^{k-j}} = L > 0,$$

the series $\displaystyle\sum_{n=1}^{\infty} \frac{P(n)}{Q(n)}$ converges by the limit comparison test. Similarly, if $j \geq k - 1$, then $k - j \leq 1$ which implies that

$$\sum_{n=1}^{\infty} \frac{P(n)}{Q(n)}$$

diverges by the limit comparison test.

39. $\dfrac{1}{2} + \dfrac{2}{5} + \dfrac{3}{10} + \dfrac{4}{17} + \dfrac{5}{26} + \cdots = \displaystyle\sum_{n=1}^{\infty} \frac{n}{n^2 + 1}$,

which diverges since the degree of the numerator is only one less than the degree of the denominator.

40. $\dfrac{1}{3} + \dfrac{1}{8} + \dfrac{1}{15} + \dfrac{1}{24} + \dfrac{1}{35} + \cdots = \displaystyle\sum_{n=2}^{\infty} \frac{1}{n^2 - 1}$,

which converges since the degree of the numerator is two less than the degree of the denominator.

41. $\displaystyle\sum_{n=1}^{\infty} \frac{1}{n^3 + 1}$

converges since the degree of the numerator is three less than the degree of the denominator.

42. $\displaystyle\sum_{n=1}^{\infty} \frac{n^2}{n^3 + 1}$

diverges since the degree of the numerator is only one less than the degree of the denominator.

43. $\lim\limits_{n\to\infty} n\left(\dfrac{n^3}{5n^4+3}\right) = \lim\limits_{n\to\infty} \dfrac{n^4}{5n^4+3} = \dfrac{1}{5} \neq 0$

Therefore,

$$\sum_{n=1}^{\infty} \frac{n^3}{5n^4+3} \text{ diverges.}$$

44. $\lim\limits_{n\to\infty} \dfrac{n}{\ln n} = \lim\limits_{n\to\infty} \dfrac{1}{1/n} = \lim\limits_{n\to\infty} n = \infty \neq 0$

Therefore,

$$\sum_{n=2}^{\infty} \frac{1}{\ln n} \text{ diverges.}$$

45. (a) $\sum\limits_{n=1}^{\infty} \dfrac{1}{(2n-1)^2} = \sum\limits_{n=1}^{\infty} \dfrac{1}{4n^2-4n+1}$

converges since the degree of the numerator is two less than the degree of the denominator. (See Exercise 38.)

(b)

n	5	10	20	50	100
S_n	1.1839	1.02087	1.2212	1.2287	1.2312

(c) $\sum\limits_{n=3}^{\infty} \dfrac{1}{(2n-1)^2} = \dfrac{\pi^2}{8} - S_2 \approx 0.1226$

(d) $\sum\limits_{n=10}^{\infty} \dfrac{1}{(2n-1)^2} = \dfrac{\pi^2}{8} - S_9 \approx 0.0277$

46. $\dfrac{1}{1000} + \dfrac{1}{1001} + \dfrac{1}{1002} + \dfrac{1}{1003} + \cdots = \sum\limits_{n=1}^{\infty} \dfrac{1}{n+999}$

which diverges by limit comparison to

$$\sum_{n=1}^{\infty} \frac{1}{n}.$$

This series diverges very slowly. You cannot conclude convergence or divergence of a series by comparing the first finite number of terms. In fact,

$$\frac{1}{n+999} > \frac{1}{n^2} \text{ for } n \geq 33.$$

47. False. Let $a_n = 1/n^3$ and $b_n = 1/n^2$. $0 < a_n \leq b_n$ and both

$$\sum_{n=1}^{\infty} \frac{1}{n^3} \text{ and } \sum_{n=1}^{\infty} \frac{1}{n^2}$$

converge.

48. True

49. True

50. False. Let $a_n = 1/n$, $b_n = 1/n$, $c_n = 1/n^2$. Then, $a_n \leq b_n + c_n$, but $\sum\limits_{n=1}^{\infty} c_n$ converges.

51. Since $\sum\limits_{n=1}^{\infty} b_n$ converges, $\lim\limits_{n\to\infty} b_n = 0$. There exists N such that $b_n < 1$ for $n > N$. Thus,

$$a_n b_n < a_n \text{ for } n > N \text{ and } \sum_{n=1}^{\infty} a_n b_n$$

converges by comparison to the convergent series $\sum\limits_{i=1}^{\infty} a_n$.

52. Since $\sum\limits_{n=1}^{\infty} a_n$ converges, then $\sum\limits_{n=1}^{\infty} a_n a_n = \sum\limits_{n=1}^{\infty} a_n^2$ converges by Exercise 51.

53. $\sum \dfrac{1}{n^2}$ and $\sum \dfrac{1}{n^3}$ both converge, and hence so does

$$\sum\left(\frac{1}{n^2}\right)\left(\frac{1}{n^3}\right) = \sum \frac{1}{n^5}$$

54. $\sum \dfrac{1}{n^2}$ converge, and hence so does $\sum\left(\dfrac{1}{n^2}\right)^2 = \sum \dfrac{1}{n^4}$.

55. (a) Suppose Σb_n converges and Σa_n diverges. Then there exists N such that $0 < b_n < a_n$ for $n \geq N$. This means that $1 < a_n/b_n$ for $n \geq N$. Therefore, $\lim\limits_{n\to\infty} a_n/b_n \neq 0$. Thus, Σa_n must also converge.

(b) Suppose Σb_n diverges and Σa_n converges. Then there exists N such that $0 < a_n < b_n$ for $n \geq N$. This means that $0 < a_n/b_n < 1$ for $n \geq N$. Therefore, $\lim\limits_{n\to\infty} a_n/b_n \neq \infty$. Thus, Σa_n must also diverge.

56. (a) $\sum a_n = \sum \dfrac{1}{n^3}$ and $\sum b_n = \sum \dfrac{1}{n^2}$. Since

$$\lim_{n\to\infty} \frac{a_n}{b_n} = \lim_{n\to\infty} \frac{1/n^3}{1/n^2} = \lim_{n\to\infty} \frac{1}{n} = 0 \quad \text{and} \quad \sum \frac{1}{n^2}$$

converges, so does $\sum \dfrac{1}{n^3}$.

(b) $\sum a_n = \sum \dfrac{1}{\sqrt{n}}$ and $\sum b_n = \sum \dfrac{1}{n}$. Since

$$\lim_{n\to\infty} \frac{a_n}{b_n} = \lim_{n\to\infty} \frac{1/\sqrt{n}}{1/n} = \lim_{n\to\infty} \sqrt{n} = \infty \quad \text{and} \quad \sum \frac{1}{n}$$

diverges, so does $\sum \dfrac{1}{\sqrt{n}}$.

57. Since $0 < a_n < 1$, $0 < a_n^2 < a_n < 1$. The squared terms will be below the others.

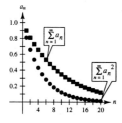

58. Start with one triangle whose sides have length 9. At the nth step, each side is replaced by four smaller line segments each having $\frac{1}{3}$ the length of the original side.

#Sides	Length of sides
3	9
$3 \cdot 4$	$9\left(\frac{1}{3}\right)$
$3 \cdot 4^2$	$9\left(\frac{1}{3}\right)^2$
$\vdots$	
$3 \cdot 4^n$	$9\left(\frac{1}{3}\right)^n$

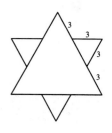

At the nth step there are $3 \cdot 4^n$ sides, each of length $9\left(\frac{1}{3}\right)^n$. At the next step, there are $3 \cdot 4^n$ new triangles of side $9\left(\frac{1}{3}\right)^{n+1}$. The area of an equilateral triangle of side x is $\frac{1}{4}\sqrt{3}\, x^2$. Thus, the new triangles each have area

$$9\,\frac{\sqrt{3}}{4}\left(\frac{1}{3^{n+1}}\right)^2 = \frac{\sqrt{3}}{4}\frac{1}{3^{2n}}.$$

The area of the $3 \cdot 4^n$ new triangles is

$$(3 \cdot 4^n)\left(\frac{\sqrt{3}}{4}\frac{1}{3^{2n}}\right) = \frac{3\sqrt{3}}{4}\left(\frac{4}{9}\right)^n.$$

The total area is the infinite sum

$$\frac{9\sqrt{3}}{4} + \sum_{n=0}^{\infty} \frac{3\sqrt{3}}{4}\left(\frac{4}{9}\right)^n = \frac{9\sqrt{3}}{4} + \frac{3\sqrt{3}}{4}\left(\frac{1}{1-4/9}\right) = \frac{9\sqrt{3}}{4} + \frac{3\sqrt{3}}{4}\left(\frac{9}{5}\right) = \frac{18\sqrt{3}}{5}.$$

The perimeter is infinite, since at step n there are $3 \cdot 4^n$ sides of length $9\left(\frac{1}{3}\right)^n$. Thus, the perimeter at step n is $27\left(\frac{4}{3}\right)^n \to \infty$.

Section 8.5 Alternating Series

1. $\displaystyle\sum_{n=1}^{\infty} \frac{6}{n^2} = \frac{6}{1} + \frac{6}{4} + \frac{6}{9} + \cdots$

$S_1 = 6, S_2 = 7.5$

Matches (b)

2. $\displaystyle\sum_{n=1}^{\infty} \frac{(-1)^{n-1} 6}{n^2} = \frac{6}{1} - \frac{6}{4} + \frac{6}{9} - \cdots$

$S_1 = 6, S_2 = 4.5$

Matches (d)

3. $\displaystyle\sum_{n=1}^{\infty} \frac{10}{n2^n} = \frac{10}{2} + \frac{10}{8} + \cdots$

$S_1 = 5, S_2 = 6.25$

Matches (c)

4. $\displaystyle\sum_{n=1}^{\infty} \frac{(-1)^{n-1} 10}{n2^n} = \frac{10}{2} - \frac{10}{8} + \cdots$

$S_1 = 5, S_2 = 3.75$

Matches (a)

5. $\displaystyle\sum_{n=1}^{\infty} \frac{(-1)^{n-1}}{2n - 1} = \frac{\pi}{4} \approx 0.7854$

(a)

n	1	2	3	4	5	6	7	8	9	10
S_n	1	0.6667	0.8667	0.7238	0.8349	0.7440	0.8209	0.7543	0.8131	0.7605

(b)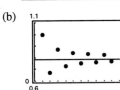

(c) The points alternate sides of the horizontal line that represents the sum of the series. The distance between successive points and the line decreases.

(d) The distance in part (c) is always less than the magnitude of the next term of the series.

6. $\displaystyle\sum_{n=1}^{\infty} \frac{(-1)^{n-1}}{(n - 1)!} = \frac{1}{e} \approx 0.3679$

(a)

n	1	2	3	4	5	6	7	8	9	10
S_n	1	0	0.5	0.3333	0.375	0.3667	0.3681	0.3679	0.3679	0.3679

(b)

(c) The points alternate sides of the horizontal line that represents the sum of the series. The distance between successive points and the line decreases.

(d) The distance in part (c) is always less than the magnitude of the next series.

7. $\displaystyle\sum_{n=1}^{\infty} \frac{(-1)^{n-1}}{n^2} = \frac{\pi^2}{12} \approx 0.8225$

(a)

n	1	2	3	4	5	6	7	8	9	10
S_n	1	0.75	0.8611	0.7986	0.8386	0.8108	0.8312	0.8156	0.8280	0.8180

—CONTINUED—

7. —CONTINUED—

(b)

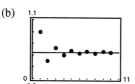

(c) The points alternate sides of the horizontal line that represents the sum of the series. The distance between successive points and the line decreases.

(d) The distance in part (c) is always less than the magnitude of the next series.

8. $\displaystyle\sum_{n=1}^{\infty} \frac{(-1)^{n-1}}{(2n-1)!} = \sin(1) \approx 0.8415$

(a)

n	1	2	3	4	5	6	7	8	9	10
S_n	1	0.8333	0.8417	0.8415	0.8415	0.8415	0.8415	0.8415	0.8415	0.8415

(b)

(c) The points alternate sides of the horizontal line that represents the sum of the series. The distance between successive points and the line decreases.

(d) The distance in part (c) is always less than the magnitude of the next series.

9. $\displaystyle\sum_{n=1}^{\infty} \frac{(-1)^{n+1}}{n}$

$a_{n+1} = \dfrac{1}{n+1} < \dfrac{1}{n} = a_n$

$\displaystyle\lim_{n\to\infty} \frac{1}{n} = 0$

Converges by Theorem 8.14.

10. $\displaystyle\sum_{n=1}^{\infty} \frac{(-1)^{n+1}\,n}{2n-1}$

$\displaystyle\lim_{n\to\infty} \frac{n}{2n-1} = \frac{1}{2}$

Diverges by the nth Term Test.

11. $\displaystyle\sum_{n=1}^{\infty} \frac{(-1)^{n+1}}{2n-1}$

$a_{n+1} = \dfrac{1}{2(n+1)-1} < \dfrac{1}{2n-1} = a_n$

$\displaystyle\lim_{n\to\infty} \frac{1}{2n-1} = 0$

Converges by Theorem 8.14

12. $\displaystyle\sum_{n=2}^{\infty} \frac{(-1)^{n}}{\ln n}$

$a_{n+1} = \dfrac{1}{\ln(n+1)} < \dfrac{1}{\ln n} = a_n$

$\displaystyle\lim_{n\to\infty} \frac{1}{\ln n} = 0$

Converges by Theorem 8.14

13. $\displaystyle\sum_{n=1}^{\infty} \frac{(-1)^{n}\,n^2}{n^2+1}$

$\displaystyle\lim_{n\to\infty} \frac{n^2}{n^2+1} = 1$

Diverges by the nth Term Test

14. $\displaystyle\sum_{n=1}^{\infty} \frac{(-1)^{n+1}\,n}{n^2+1}$

$a_{n+1} = \dfrac{n+1}{(n+1)^2+1} < \dfrac{n}{n^2+1} = a_n$

$\displaystyle\lim_{n\to\infty} \frac{n}{n^2+1} = 0$

Converges by Theorem 8.14

15. $\displaystyle\sum_{n=1}^{\infty} \frac{(-1)^{n}}{\sqrt{n}}$

$a_{n+1} = \dfrac{1}{\sqrt{n+1}} < \dfrac{1}{\sqrt{n}} = a_n$

$\displaystyle\lim_{n\to\infty} \frac{1}{\sqrt{n}} = 0$

Converges by Theorem 8.14

16. $\displaystyle\sum_{n=1}^{\infty} \frac{(-1)^{n+1}\,n^3}{n^3+6}$

$\displaystyle\lim_{n\to\infty} \frac{n^3}{n^3+6} = 1$

Diverges by the nth Term Test

17. $\displaystyle\sum_{n=1}^{\infty} \frac{(-1)^{n+1}(n+1)}{\ln(n+1)}$

$$\lim_{n\to\infty} \frac{n+1}{\ln(n+1)} = \lim_{n\to\infty} \frac{1}{1/(n+1)} = \lim_{n\to\infty} (n+1) = \infty$$

Diverges by the *n*th Term Test

18. $\displaystyle\sum_{n=1}^{\infty} \frac{(-1)^{n+1} \ln(n+1)}{n+1}$

$$a_{n+1} = \frac{\ln[(n+1)+1]}{(n+1)+1} < \frac{\ln(n+1)}{n+1} \text{ for } n \geq 2$$

$$\lim_{n\to\infty} \frac{\ln(n+1)}{n+1} = \lim_{n\to\infty} \frac{1/(n+1)}{1} = 0$$

Converges by Theorem 8.14

19. $\displaystyle\sum_{n=1}^{\infty} \sin\left[\frac{(2n-1)\pi}{2}\right] = \sum_{n=1}^{\infty} (-1)^{n+1}$

Diverges by the *n*th Term Test

20. $\displaystyle\sum_{n=1}^{\infty} \cos n\pi = \sum_{n=1}^{\infty} (-1)^n$

Diverges by the *n*th Term Test

21. $\displaystyle\sum_{n=1}^{\infty} \frac{1}{n} \sin\left[\frac{(2n-1)\pi}{2}\right] = \sum_{n=1}^{\infty} \frac{(-1)^{n+1}}{n}$

Converges; (see Exercise 9)

22. $\displaystyle\sum_{n=1}^{\infty} \frac{1}{n} \cos n\pi = \sum_{n=1}^{\infty} \frac{(-1)^n}{n}$

Converges; (see Exercise 9)

23. $\displaystyle\sum_{n=0}^{\infty} \frac{(-1)^n}{n!}$

$$a_{n+1} = \frac{1}{(n+1)!} < \frac{1}{n!} = a_n$$

$$\lim_{n\to\infty} \frac{1}{n!} = 0$$

Converges by Theorem 8.14

24. $\displaystyle\sum_{n=0}^{\infty} \frac{(-1)^n}{(2n)!}$

$$a_{n+1} = \frac{1}{(2n+2)!} < \frac{1}{(2n)!} = a_n$$

$$\lim_{n\to\infty} \frac{1}{(2n)!} = 0$$

Converges by Theorem 8.14

25. $\displaystyle\sum_{n=1}^{\infty} \frac{(-1)^{n+1}\sqrt{n}}{n+2}$

$$a_{n+1} = \frac{\sqrt{n+1}}{(n+1)+2} < \frac{\sqrt{n}}{n+2} \text{ for } n \geq 2$$

$$\lim_{n\to\infty} \frac{\sqrt{n}}{n+2} = 0$$

Converges by Theorem 8.14

26. $\displaystyle\sum_{n=1}^{\infty} \frac{(-1)^{n+1}\sqrt{n}}{\sqrt[3]{n}}$

$$\lim_{n\to\infty} \frac{n^{1/2}}{n^{1/3}} = \lim_{n\to\infty} n^{1/6} = \infty$$

Diverges by the *n*th Term Test

27. $\displaystyle\sum_{n=1}^{\infty} \frac{(-1)^{n+1}(2)}{e^n - e^{-n}} = \sum_{n=1}^{\infty} \frac{(-1)^{n+1}(2e^n)}{e^{2n} - 1}$

Let $f(x) = \dfrac{2e^x}{e^{2x} - 1}$. Then

$$f'(x) = \frac{-2e^x(e^{2x}+1)}{(e^{2x}-1)^2} < 0.$$

Thus, $f(x)$ is decreasing. Therefore, $a_{n+1} < a_n$, and

$$\lim_{n\to\infty} \frac{2e^n}{e^{2n}-1} = \lim_{n\to\infty} \frac{2e^n}{2e^{2n}} = \lim_{n\to\infty} \frac{1}{e^n} = 0.$$

The series converges by Theorem 8.14.

28. $\displaystyle\sum_{n=1}^{\infty} \frac{2(-1)^{n+1}}{e^n + e^{-n}} = \sum_{n=1}^{\infty} \frac{(-1)^{n+1}(2e^n)}{e^{2n} + 1}$

Let $f(x) = \dfrac{2e^x}{e^{2x} + 1}$. Then

$$f'(x) = \frac{2e^{2x}(1 - e^{2x})}{(e^{2x}+1)^2} < 0 \text{ for } x > 0.$$

Thus, $f(x)$ is decreasing for $x > 0$ which implies $a_{n+1} < a_n$.

$$\lim_{n\to\infty} \frac{2e^n}{e^{2n}+1} = \lim_{n\to\infty} \frac{2e^n}{2e^{2n}} = \lim_{n\to\infty} \frac{1}{e^n} = 0$$

The series converges by Theorem 8.14.

29. $\displaystyle\sum_{n=0}^{\infty} \frac{(-1)^n}{n!}$

(a) By Theorem 8.15,

$$|R_N| \le a_{N+1} = \frac{1}{(N+1)!} < 0.001.$$

This inequality is valid when $N = 6$.

(b) We may approximate the series by

$$\sum_{n=0}^{6} \frac{(-1)^n}{n!} = 1 - 1 + \frac{1}{2} - \frac{1}{6} + \frac{1}{24} - \frac{1}{120} + \frac{1}{720}$$
$$\approx 0.368.$$

(7 terms. Note that the sum begins with $n = 0$.)

30. $\displaystyle\sum_{n=0}^{\infty} \frac{(-1)^n}{2^n n!}$

(a) By Theorem 8.15,

$$|R_n| \le a_{N+1} = \frac{1}{2^{N+1}(N+1)!} < 0.001.$$

This inequality is valid when $N = 4$.

(b) We may approximate the series by

$$\sum_{n=0}^{4} \frac{(-1)^n}{2^n n!} = 1 - \frac{1}{2} + \frac{1}{8} - \frac{1}{48} + \frac{1}{348} \approx 0.607.$$

(5 terms. Note that the sum begins with $n = 0$.)

31. $\displaystyle\sum_{n=0}^{\infty} \frac{(-1)^n}{(2n+1)!}$

(a) By Theorem 8.15,

$$|R_N| \le a_{N+1} = \frac{1}{[2(N+1)+1]!} < 0.001.$$

This inequality is valid when $N = 2$.

(b) We may approximate the series by

$$\sum_{n=0}^{2} \frac{(-1)^n}{(2n+1)!} = 1 - \frac{1}{6} + \frac{1}{120} \approx 0.842.$$

(3 terms. Note that the sum begins with $n = 0$.)

32. $\displaystyle\sum_{n=0}^{\infty} \frac{(-1)^n}{(2n)!}$

(a) By Theorem 8.15,

$$|R_N| \le a_{N+1} = \frac{1}{(2N+2)!} < 0.001.$$

This inequality is valid when $N = 3$.

(b) We may approximate the series by

$$\sum_{n=0}^{3} \frac{(-1)^n}{(2n)!} = 1 - \frac{1}{2} + \frac{1}{24} - \frac{1}{720} \approx 0.540.$$

(4 terms. Note that the sum begins with $n = 0$.)

33. $\displaystyle\sum_{n=1}^{\infty} \frac{(-1)^{n+1}}{n}$

(a) By Theorem 8.15,

$$|R_N| \le a_{N+1} = \frac{1}{N+1} < 0.001.$$

This inequality is valid when $N = 1000$.

(b) We may approximate the series by

$$\sum_{n=1}^{1000} \frac{(-1)^{n+1}}{n} = 1 - \frac{1}{2} + \frac{1}{3} - \frac{1}{4} + \cdots - \frac{1}{1000}$$
$$\approx 0.693.$$

(1000 terms)

34. $\displaystyle\sum_{n=1}^{\infty} \frac{(-1)^{n+1}}{4^n n}$

(a) By Theorem 8.15,

$$|R_N| \le a_{N+1} = \frac{1}{4^{N+1}(N+1)} < 0.001.$$

This inequality is valid when $N = 3$.

(b) We may approximate the series by

$$\sum_{n=1}^{3} \frac{(-1)^{n+1}}{4^n n} = \frac{1}{4} - \frac{1}{32} + \frac{1}{192} \approx 0.224.$$

(3 terms)

35. $\displaystyle\sum_{n=1}^{\infty} \frac{(-1)^{n+1}}{2n^3 - 1}$

By Theorem 8.15,

$$|R_N| \le a_{N+1} = \frac{1}{2(N+1)^3 - 1} < 0.001.$$

This inequality is valid when $N = 7$.

36. $\displaystyle\sum_{n=1}^{\infty} \frac{(-1)^{n+1}}{n^4}$

By Theorem 8.15, $|R_N| \le a_{N+1} = \dfrac{1}{(N+1)^4} < 0.001.$

This inequality is valid when $N = 5$.

37. $\displaystyle\sum_{n=1}^{\infty} \frac{(-1)^{n+1}}{(n+1)^2}$

$\displaystyle\sum_{n=1}^{\infty} \frac{1}{(n+1)^2}$ converges by comparison to the p-series

$$\sum_{n=1}^{\infty} \frac{1}{n^2}.$$

Therefore, the given series converge absolutely.

38. $\displaystyle\sum_{n=1}^{\infty} \frac{(-1)^{n+1}}{n+1}$

The given series converges by the Alternating Series Test, but does not converge absolutely since the series

$$\sum_{n=1}^{\infty} \frac{1}{n+1}$$

diverges by the Integral Test. Therefore, the series converge conditionally.

39. $\displaystyle\sum_{n=1}^{\infty} \frac{(-1)^{n+1}}{\sqrt{n}}$

The given series converges by the Alternating Series Test, but does not converge absolutely since

$$\sum_{n=1}^{\infty} \frac{1}{\sqrt{n}}$$

is a divergent p-series. Therefore, the series converges conditionally.

40. $\displaystyle\sum_{n=1}^{\infty} \frac{(-1)^{n+1}}{n\sqrt{n}}$

$\displaystyle\sum_{n=1}^{\infty} \frac{1}{n\sqrt{n}} = \sum_{n=1}^{\infty} \frac{1}{n^{3/2}}$ which is a convergent p-series.

Therefore, the given series converges absolutely.

41. $\displaystyle\sum_{n=1}^{\infty} \frac{(-1)^{n+1} n^2}{(n+1)^2}$

$\displaystyle\lim_{n\to\infty} \frac{n^2}{(n+1)^2} = 1$ Therefore, the series diverges by the nth Term Test.

42. $\displaystyle\sum_{n=1}^{\infty} \frac{(-1)^{n+1}(2n+3)}{n+10}$

$\displaystyle\lim_{n\to\infty} \frac{2n+3}{n+10} = 2$ Therefore, the series diverges by the nth Term Test.

43. $\displaystyle\sum_{n=2}^{\infty} \frac{(-1)}{\ln(n)}$

The given series converges by the Alternating Series Test, but does not converge absolutely since the series

$$\sum_{n=2}^{\infty} \frac{1}{\ln n}$$

diverges by comparison to the harmonic series

$$\sum_{n=1}^{\infty} \frac{1}{n}.$$

Therefore, the series converges conditionally.

44. $\displaystyle\sum_{n=0}^{\infty} \frac{(-1)^n}{e^{n^2}}$

$$\sum_{n=0}^{\infty} \frac{1}{e^{n^2}}$$

converges by a comparison to the convergent geometric series

$$\sum_{n=0}^{\infty} \left(\frac{1}{e}\right)^n.$$

Therefore, the given series converges absolutely.

45. $\displaystyle\sum_{n=2}^{\infty} \frac{(-1)^n n}{n^3 - 1}$

$$\sum_{n=2}^{\infty} \frac{n}{n^3 - 1}$$

converges by a limit comparison to the convergent p-series

$$\sum_{n=2}^{\infty} \frac{1}{n^2}.$$

Therefore, the given series converges absolutely.

46. $\displaystyle\sum_{n=1}^{\infty} \frac{(-1)^{n+1}}{n^{1.5}}$

$\displaystyle\sum_{n=1}^{\infty} \frac{1}{n^{1.5}}$ is a convergent p-series. Therefore, the given series converge absolutely.

47. $\displaystyle\sum_{n=0}^{\infty} \frac{(-1)^n}{(2n+1)!}$

$$\sum_{n=0}^{\infty} \frac{1}{(2n+1)!}$$

is convergent by comparison to the convergent geometric series

$$\sum_{n=0}^{\infty} \left(\frac{1}{2}\right)^n$$

since

$$\frac{1}{(2n+1)!} < \frac{1}{2^n} \text{ for } n > 0.$$

Therefore, the given series converges absolutely.

49. $\displaystyle\sum_{n=0}^{\infty} \frac{\cos n\pi}{n+1} = \sum_{n=0}^{\infty} \frac{(-1)^n}{n+1}$

The given series converges by the Alternating Series Test, but

$$\sum_{n=0}^{\infty} \frac{|\cos n\pi|}{n+1} = \sum_{n=0}^{\infty} \frac{1}{n+1}$$

diverges by a limit comparison to the divergent harmonic series,

$$\sum_{n=1}^{\infty} \frac{1}{n}.$$

$$\lim_{n\to\infty} \frac{|\cos n\pi|/(n+1)}{1/n} = 1, \text{ therefore the series}$$

converges conditionally.

51. $\displaystyle\sum_{n=1}^{\infty} \frac{\cos n\pi}{n^2} = \sum_{n=1}^{\infty} \frac{(-1)^n}{n^2}$

$\displaystyle\sum_{n=1}^{\infty} \frac{1}{n^2}$ is a convergent p-series. Therefore, the given

series converges absolutely.

53. $\displaystyle\sum_{n=1}^{\infty} \frac{(-1)^n}{n^p}$

If $p = 0$, then

$$\lim_{n\to\infty} \frac{1}{n^p} = 1$$

and the series diverges. If $p > 0$, then

$$\lim_{n\to\infty} \frac{1}{n^p} = 0 \text{ and } \frac{1}{(n+1)^p} < \frac{1}{n^p}.$$

Therefore, the series converge by the Alternating Series Test.

48. $\displaystyle\sum_{n=0}^{\infty} \frac{(-1)^n}{\sqrt[3]{n+1}}$

The given series converges by the Alternating Series Test, but

$$\sum_{n=0}^{\infty} \frac{1}{\sqrt[3]{n+1}}$$

diverges by a limit comparison to the divergent p-series

$$\sum_{n=1}^{\infty} \frac{1}{\sqrt[3]{n}}.$$

Therefore, the series converges conditionally.

50. $\displaystyle\sum_{n=1}^{\infty} (-1)^{n+1} \arctan n$

$$\lim_{n\to\infty} \arctan n = \frac{\pi}{2} \neq 0 \text{ Therefore, the series diverges by}$$

the nth Term Test.

52. $\displaystyle\sum_{n=1}^{\infty} \frac{\sin[(2n-1)\pi/2]}{n} = \sum_{n=1}^{\infty} \frac{(-1)^{n+1}}{n}$

The given series converges by the Alternating Series Test, but

$$\sum_{n=1}^{\infty} \left|\frac{\sin[(2n-1)\pi/2]}{n}\right| = \sum_{n=1}^{\infty} \frac{1}{n}$$

is a divergent p-series. Therefore, the series converges conditionally.

54. Since $\displaystyle\sum_{n=1}^{\infty} |a_n|$ converges we have

$$\lim_{n\to\infty} |a_n| = 0.$$

Thus, there must exist an $N > 0$ such that $|a_N| < 1$ for all $n > N$ and it follows that $a_n{}^2 \le |a_n|$ for all $n > N$. Hence, by the Comparison Test,

$$\sum_{n=1}^{\infty} a_n{}^2$$

converges. Let $a_n = 1/n$ to see that the converse is false.

55. $\displaystyle\sum_{n=1}^{\infty} \frac{(-1)^{n-1}}{n}$ converges, but $\displaystyle\sum_{n=1}^{\infty} \frac{1}{n}$ diverges

56. $\displaystyle\sum_{n=1}^{\infty} \frac{1}{n^2}$ converges, hence so does $\displaystyle\sum_{n=1}^{\infty} \frac{1}{n^4}$.

57. (a) $\displaystyle\sum_{n=1}^{\infty} \frac{x^n}{n}$

converges absolutely (by comparison) for

$$-1 < x < 1,$$

since

$$\left|\frac{x^n}{n}\right| < |x^n| \text{ and } \sum x^n$$

is a convergent geometric series for $-1 < x < 1$.

(b) When $x = -1$, we have the convergent alternating series

$$\sum_{n=1}^{\infty} \frac{(-1)^n}{n}.$$

When $x = 1$, we have the divergent harmonic series

$$\frac{1}{n}.$$

Therefore,

$$\sum_{n=1}^{\infty} \frac{x^n}{n}$$

converges conditionally for $x = -1$.

58. The first term of the series is zero, not one. You cannot regroup series terms arbitrarily.

59. False

Let $a_n = \dfrac{(-1)^n}{n}$.

60. True, equivalent to Theorem 8.16

Section 8.6   The Ratio and Root Tests

1. $\dfrac{(n+1)!}{(n-2)!} = \dfrac{(n+1)(n)(n-1)(n-2)!}{(n-2)!}$

$= (n+1)(n)(n-1)$

2. $\dfrac{(2k-2)!}{(2k)!} = \dfrac{(2k-2)!}{(2k)(2k-1)(2k-2)!} = \dfrac{1}{(2k)(2k-1)}$

3. Use the Principle of Mathematical Induction. When $k = 1$, the formula is valid since $1 = \dfrac{(2(1))!}{2^1 \cdot 1!}$. Assume that

$$1 \cdot 3 \cdot 5 \cdots (2n-1) = \frac{(2n)!}{2^n n!}$$

and show that

$$1 \cdot 3 \cdot 5 \cdots (2n-1)(2n+1) = \frac{(2n+2)!}{2^{n+1}(n+1)!}.$$

To do this, note that:

$$1 \cdot 3 \cdot 5 \cdots (2n-1)(2n+1) = [1 \cdot 3 \cdot 5 \cdots (2n-1)](2n+1)$$

$$= \frac{(2n)!}{2^n n!} \cdot (2n+1)$$

$$= \frac{(2n)!(2n+1)}{2^n n!} \cdot \frac{(2n+2)}{2(n+1)}$$

$$= \frac{(2n)!(2n+1)(2n+2)}{2^{n+1}n!(n+1)}$$

$$= \frac{(2n+2)!}{2^{n+1}(n+1)}$$

The formula is valid for all $n \geq 1$.

4. Use the Principle of Mathematical Induction. When $k = 3$, the formula is valid since $\dfrac{1}{1} = \dfrac{2^3 3!(3)(5)}{6!} = 1$. Assume that

$$\frac{1}{1 \cdot 3 \cdot 5 \cdots (2n - 5)} = \frac{2^n n!(2n - 3)(2n - 1)}{(2n)!}$$

and show that

$$\frac{1}{1 \cdot 3 \cdot 5 \cdots (2n - 5)(2n - 3)} = \frac{2^{n+1}(n + 1)!(2n - 1)(2n + 1)}{(2n + 2)!}.$$

To do this, note that:

$$\frac{1}{1 \cdot 3 \cdot 5 \cdots (2n - 5)(2n - 3)} = \frac{1}{1 \cdot 3 \cdot 5 \cdots (2n - 5)} \cdot \frac{1}{(2n - 3)}$$

$$= \frac{2^n n!(2n - 3)(2n - 1)}{(2n)!} \cdot \frac{1}{(2n - 3)}$$

$$= \frac{2^n n!(2n - 1)}{(2n)!} \cdot \frac{(2n + 1)(2n + 2)}{(2n + 1)(2n + 2)}$$

$$= \frac{2^n (2)(n + 1)n!(2n - 1)(2n + 1)}{(2n)!(2n + 1)(2n + 2)}$$

$$= \frac{2^{n+1}(n + 1)!(2n - 1)(2n + 1)}{(2n + 2)!}$$

The formula is valid for all $n \geq 3$.

5. $\displaystyle\sum_{n=1}^{\infty} n\left(\frac{3}{4}\right)^n = 1\left(\frac{3}{4}\right) + 2\left(\frac{9}{16}\right) + \cdots$

$S_1 = \dfrac{3}{4}, S_2 \approx 1.875$

Matches (d)

6. $\displaystyle\sum_{n=1}^{\infty} \left(\frac{3}{4}\right)^n \left(\frac{1}{n!}\right) = \frac{3}{4} + \frac{9}{16}\left(\frac{1}{2}\right) + \cdots$

$S_1 = \dfrac{3}{4}, S_2 \approx 1.03$

Matches (c)

7. $\displaystyle\sum_{n=1}^{\infty} \frac{(-3)^{n+1}}{n!} = 9 - \frac{3^3}{2} + \cdots$

$S_1 = 9$

Matches (a)

8. $\displaystyle\sum_{n=1}^{\infty} \frac{(-1)^{n-1} 4}{(2n)!} = \frac{4}{2} - \frac{4}{24} + \cdots$

$S_1 = 2$

Matches (b)

9. (a) Ratio Test: $\displaystyle\lim_{n \to \infty} \left|\frac{a_{n+1}}{a_n}\right| = \lim_{n \to \infty} \frac{(n + 1)^2 (5/8)^{n+1}}{n^2 (5/8)^n} = \lim_{n \to \infty} \left(\frac{n + 1}{n}\right)^2 \frac{5}{8} = \frac{5}{8} < 1$. Converges

(b)

n	5	10	15	20	25
S_n	9.2104	16.7598	18.8016	19.1878	19.2491

(c)

(d) The sum is approximately 19.26.

(e) The more rapidly the terms of the series approach 0, the more rapidly the sequence of the partial sums approaches the sum of the series.

10. (a) Ratio Test: $\lim\limits_{n\to\infty}\left|\dfrac{a_{n+1}}{a_n}\right| = \lim\limits_{n\to\infty}\dfrac{\dfrac{(n+1)^2+1}{(n+1)!}}{\dfrac{n^2+1}{n!}} = \lim\limits_{n\to\infty}\left(\dfrac{n^2+2n+2}{n^2+1}\right)\left(\dfrac{1}{n+1}\right) = 0 < 1.$ Converges

(b)

n	5	10	15	20	25
S_n	7.0917	7.1548	7.1548	7.1548	7.1548

(c)

(d) The sum is approximately 7.15485

(e) The more rapidly the terms of the series approach 0, the more rapidly the sequence of the partial sums approaches the sum of the series.

11. $\sum\limits_{n=0}^{\infty}\dfrac{n!}{3^n}$

$\lim\limits_{n\to\infty}\left|\dfrac{a_{n+1}}{a_n}\right| = \lim\limits_{n\to\infty}\left|\dfrac{(n+1)!}{3^{n+1}}\cdot\dfrac{3^n}{n!}\right|$

$= \lim\limits_{n\to\infty}\dfrac{n+1}{3} = \infty$

Therefore, by the Ratio Test, the series diverges.

12. $\sum\limits_{n=1}^{\infty} n\left(\dfrac{2}{3}\right)^n$

$\lim\limits_{n\to\infty}\left|\dfrac{a_{n+1}}{a_n}\right| = \lim\limits_{n\to\infty}\left|\dfrac{(n+1)(2/3)^{n+1}}{n(2/3)^n}\right|$

$= \lim\limits_{n\to\infty}\dfrac{2(n+1)}{3n} = \dfrac{2}{3}$

Therefore, by the Ratio Test, the series converges.

13. $\sum\limits_{n=0}^{\infty}\dfrac{3^n}{n!}$

$\lim\limits_{n\to\infty}\left|\dfrac{a_{n+1}}{a_n}\right| = \lim\limits_{n\to\infty}\left|\dfrac{3^{n+1}}{(n+1)!}\cdot\dfrac{n!}{3^n}\right|$

$= \lim\limits_{n\to\infty}\dfrac{3}{n+1} = 0$

Therefore, by the Ratio Test, the series converges.

14. $\sum\limits_{n=1}^{\infty} n\left(\dfrac{3}{2}\right)^n$

$\lim\limits_{n\to\infty}\left|\dfrac{a_{n+1}}{a_n}\right| = \lim\limits_{n\to\infty}\left|\dfrac{(n+1)3^{n+1}}{2^{n+1}}\cdot\dfrac{2^n}{n3^n}\right|$

$= \lim\limits_{n\to\infty}\dfrac{3(n+1)}{2n} = \dfrac{3}{2}$

Therefore, by the Ratio Test, the series diverges.

15. $\sum\limits_{n=1}^{\infty}\dfrac{n}{2^n}$

$\lim\limits_{n\to\infty}\left|\dfrac{a_{n+1}}{a_n}\right| = \lim\limits_{n\to\infty}\left|\dfrac{n+1}{2^{n+1}}\cdot\dfrac{2^n}{n}\right|$

$= \lim\limits_{n\to\infty}\dfrac{n+1}{2n} = \dfrac{1}{2}$

Therefore, by the Ratio Test, the series converges.

16. $\sum\limits_{n=1}^{\infty}\dfrac{n^2}{2^n}$

$\lim\limits_{n\to\infty}\left|\dfrac{a_{n+1}}{a_n}\right| = \lim\limits_{n\to\infty}\left|\dfrac{n^2+2n+1}{2^{n+1}}\cdot\dfrac{2^n}{n}\right|$

$= \lim\limits_{n\to\infty}\dfrac{n^2+2n+1}{2n^2} = \dfrac{1}{2}$

Therefore, by the Ratio Test, the series converges.

17. $\sum\limits_{n=1}^{\infty}\dfrac{2^n}{n^2}$

$\lim\limits_{n\to\infty}\left|\dfrac{a_{n+1}}{a_n}\right| = \lim\limits_{n\to\infty}\left|\dfrac{2^{n+1}}{(n+1)^2}\cdot\dfrac{n^2}{2^n}\right|$

$= \lim\limits_{n\to\infty}\dfrac{2n^2}{(n+1)^2} = 2$

Therefore, by the Ratio Test, the series diverges.

18. $\sum\limits_{n=1}^{\infty}\dfrac{(-1)^{n+1}(n+2)}{n(n+1)}$

$a_{n+1} = \dfrac{n+3}{(n+1)(n+2)} \leq \dfrac{n+2}{n(n+1)} = a_n$

$\lim\limits_{n\to\infty}\dfrac{n+2}{n(n+1)} = 0$

Therefore, by Theorem 8.14, the series converges.

Note: The Ratio Test is inconclusive since $\lim\limits_{n\to\infty}\left|\dfrac{a_{n+1}}{a_n}\right| = 1$. The series converges conditionally.

19. $\displaystyle\sum_{n=0}^{\infty} \frac{(-1)^n 2^n}{n!}$

$$\lim_{n\to\infty}\left|\frac{a_{n+1}}{a_n}\right| = \lim_{n\to\infty}\left|\frac{2^{n+1}}{(n+1)!}\cdot\frac{n!}{2^n}\right|$$

$$= \lim_{n\to\infty}\frac{2}{n+1} = 0$$

Therefore, by the Ratio Test, the series converges.

20. $\displaystyle\sum_{n=1}^{\infty} \frac{(-1)^{n-1}(3/2)^n}{n^2}$

$$\lim_{n\to\infty}\left|\frac{a_{n+1}}{a_n}\right| = \lim_{n\to\infty}\left|\frac{(3/2)^{n+1}}{n^2+2n+1}\cdot\frac{n^2}{(3/2)^n}\right|$$

$$= \lim_{n\to\infty}\frac{3n^2}{2(n^2+2n+1)} = \frac{3}{2} > 1$$

Therefore, by the Ratio Test, the series diverges.

21. $\displaystyle\sum_{n=1}^{\infty} \frac{n!}{n3^n}$

$$\lim_{n\to\infty}\left|\frac{a_{n+1}}{a_n}\right| = \lim_{n\to\infty}\left|\frac{(n+1)!}{(n+1)3^{n+1}}\cdot\frac{n3^n}{n!}\right|$$

$$= \lim_{n\to\infty}\frac{n}{3} = \infty$$

Therefore, by the Ratio Test, the series diverges.

22. $\displaystyle\sum_{n=1}^{\infty} \frac{(2n)!}{n^5}$

$$\lim_{n\to\infty}\left|\frac{a_{n+1}}{a_n}\right| = \lim_{n\to\infty}\left|\frac{(2n+2)!}{(n+1)^5}\cdot\frac{n^5}{(2n)!}\right|$$

$$= \lim_{n\to\infty}\frac{(2n+2)(2n+1)n^5}{(n+1)^5} = \infty$$

Therefore, by the Ratio Test, the series diverges.

23. $\displaystyle\sum_{n=0}^{\infty} \frac{4^n}{n!}$

$$\lim_{n\to\infty}\left|\frac{a_{n+1}}{a_n}\right| = \lim_{n\to\infty}\left|\frac{4^{n+1}}{(n+1)!}\cdot\frac{n!}{4^n}\right|$$

$$= \lim_{n\to\infty}\frac{4}{n+1} = 0$$

Therefore, by the Ratio Test, the series converges.

24. $\displaystyle\sum_{n=1}^{\infty} \frac{n^n}{n!}$

$$\lim_{n\to\infty}\left|\frac{a_{n+1}}{a_n}\right| = \lim_{n\to\infty}\left|\frac{(n+1)^{n+1}}{(n+1)!}\cdot\frac{n!}{n^n}\right|$$

$$= \lim_{n\to\infty}\frac{(n+1)(n+1)^n n!}{(n+1)n!n^n}$$

$$= \lim_{n\to\infty}\left(\frac{n+1}{n}\right)^n = e > 1$$

Therefore, by the Ratio Test, the series diverges.

25. $\displaystyle\sum_{n=0}^{\infty} \frac{3^n}{(n+1)^n}$

$$\lim_{n\to\infty}\left|\frac{a_{n+1}}{a_n}\right| = \lim_{n\to\infty}\left|\frac{3^{n+1}}{(n+2)^{n+1}}\cdot\frac{(n+1)^n}{3^n}\right| = \lim_{n\to\infty}\frac{3(n+1)^n}{(n+2)^{n+1}} = \lim_{n\to\infty}\frac{3}{n+2}\left(\frac{n+1}{n+2}\right)^n = (0)\left(\frac{1}{e}\right) = 0$$

To find $\displaystyle\lim_{n\to\infty}\left(\frac{n+1}{n+2}\right)^n$, let $\displaystyle y = \lim_{n\to\infty}\left(\frac{n+1}{n+2}\right)^n$. Then,

$$\ln y = \lim_{n\to\infty} n \ln\left(\frac{n+1}{n+2}\right) = \lim_{n\to\infty}\frac{\ln[(n+1)/(n+2)]}{1/n} = \frac{0}{0}$$

$$\ln y = \lim_{n\to\infty}\frac{[(1)/(n+1)] - [(1)/(n+2)]}{-(1/n^2)} = -1 \text{ by L'Hôpital's Rule}$$

$$y = e^{-1} = \frac{1}{e}.$$

Therefore, by the Ratio Test, the series converges.

26. $\displaystyle\sum_{n=0}^{\infty} \frac{(n!)^2}{(3n)!}$

$$\lim_{n\to\infty}\left|\frac{a_{n+1}}{a_n}\right| = \lim_{n\to\infty}\left|\frac{[(n+1)!]^2}{(3n+3)!}\cdot\frac{(3n)!}{(n!)^2}\right| = \lim_{n\to\infty}\frac{(n+1)^2}{(3n+3)(3n+2)(3n+1)} = 0$$

Therefore, by the Ratio Test, the series converges.

27. $\displaystyle\sum_{n=0}^{\infty} \frac{4^n}{3^n + 1}$

$$\lim_{n\to\infty} \left|\frac{a_{n+1}}{a_n}\right| = \lim_{n\to\infty} \left|\frac{4^{n+1}}{3^{n+1} + 1} \cdot \frac{3^n + 1}{4^n}\right| = \lim_{n\to\infty} \frac{4(3^n + 1)}{3^{n+1} + 1} = \lim_{n\to\infty} \frac{4(1 + 1/3^n)}{3 + 1/3^n} = \frac{4}{3}$$

Therefore, by the Ratio Test, the series diverges.

28. $\displaystyle\sum_{n=0}^{\infty} \frac{(-1)^n 2^{4n}}{(2n + 1)!}$

$$\lim_{n\to\infty} \left|\frac{a_{n+1}}{a_n}\right| = \lim_{n\to\infty} \left|\frac{2^{4n+4}}{(2n + 3)!} \cdot \frac{(2n + 1)!}{2^{4n}}\right| = \lim_{n\to\infty} \frac{2^4}{(2n + 3)(2n + 2)} = 0$$

Therefore, by the Ratio Test, the series converges.

29. $\displaystyle\sum_{n=0}^{\infty} \frac{(-1)^{n+1}n!}{1 \cdot 3 \cdot 5 \cdots (2n + 1)}$

$$\lim_{n\to\infty} \left|\frac{a_{n+1}}{a_n}\right| = \lim_{n\to\infty} \left|\frac{(n + 1)!}{1 \cdot 3 \cdot 5 \cdots (2n + 1)(2n + 3)} \cdot \frac{1 \cdot 3 \cdot 5 \cdots (2n + 1)}{n!}\right| = \lim_{n\to\infty} \frac{n + 1}{2n + 3} = \frac{1}{2}$$

Therefore, by the Ratio Test, the series converges.

Note: The first few terms of this series are $-1 + \dfrac{1}{1 \cdot 3} - \dfrac{2!}{1 \cdot 3 \cdot 5} + \dfrac{3!}{1 \cdot 3 \cdot 5 \cdot 7} - \cdots$

30. $\displaystyle\sum_{n=1}^{\infty} \frac{(-1)^n 2 \cdot 4 \cdot 6 \cdots 2n}{2 \cdot 5 \cdot 8 \cdots (3n - 1)}$

$$\lim_{n\to\infty} \left|\frac{a_{n+1}}{a_n}\right| = \lim_{n\to\infty} \left|\frac{2 \cdot 4 \cdots 2n(2n + 2)}{2 \cdot 5 \cdots (3n - 1)(3n + 2)} \cdot \frac{2 \cdot 5 \cdots (3n - 1)}{2 \cdot 4 \cdots 2n}\right| = \lim_{n\to\infty} \frac{2n + 2}{3n + 2} = \frac{2}{3}$$

Therefore, by the Ratio Test, the series converges.

Note: The first few terms of this series are $-\dfrac{2}{2} + \dfrac{2 \cdot 4}{2 \cdot 5} - \dfrac{2 \cdot 4 \cdot 6}{2 \cdot 5 \cdot 8} + \cdots$

31. (a) $\displaystyle\sum_{n=1}^{\infty} \frac{1}{n^{3/2}}$

$$\lim_{n\to\infty} \left|\frac{a_{n+1}}{a_n}\right| = \lim_{n\to\infty} \left|\frac{1}{(n + 1)^{3/2}} \cdot \frac{n^{3/2}}{1}\right| = \lim_{n\to\infty} \left(\frac{n}{n + 1}\right)^{3/2} = 1$$

(b) $\displaystyle\sum_{n=1}^{\infty} \frac{1}{n^{1/2}}$

$$\lim_{n\to\infty} \left|\frac{a_{n+1}}{a_n}\right| = \lim_{n\to\infty} \left|\frac{1}{(n + 1)^{1/2}} \cdot \frac{n^{1/2}}{1}\right| = \lim_{n\to\infty} \left(\frac{n}{n + 1}\right)^{1/2} = 1$$

32. (a) $\displaystyle\sum_{n=1}^{\infty} \frac{1}{n^3}$

$$\lim_{n\to\infty} \left|\frac{a_{n+1}}{a_n}\right| = \lim_{n\to\infty} \left|\frac{1}{(n + 1)^3} \cdot \frac{n^3}{1}\right| = \lim_{n\to\infty} \left(\frac{n}{n + 1}\right)^3 = 1$$

(b) $\displaystyle\sum_{n=1}^{\infty} \frac{1}{n^p}$

$$\lim_{n\to\infty} \left|\frac{a_{n+1}}{a_n}\right| = \lim_{n\to\infty} \left|\frac{1}{(n + 1)^p} \cdot \frac{n^p}{1}\right| = \lim_{n\to\infty} \left(\frac{n}{n + 1}\right)^p = 1$$

33. $\displaystyle\sum_{n=1}^{\infty} \left(\frac{n}{2n+1}\right)^n$

$$\lim_{n\to\infty} \sqrt[n]{|a_n|} = \lim_{n\to\infty} \sqrt[n]{\left(\frac{n}{2n+1}\right)^n}$$

$$= \lim_{n\to\infty} \frac{n}{2n+1} = \frac{1}{2}$$

Therefore, by the Root Test, the series converges.

34. $\displaystyle\sum_{n=1}^{\infty} \left(\frac{2n}{n+1}\right)^n$

$$\lim_{n\to\infty} \sqrt[n]{|a_n|} = \lim_{n\to\infty} \sqrt[n]{\left(\frac{2n}{n+1}\right)^n}$$

$$= \lim_{n\to\infty} \frac{2n}{n+1} = 2$$

Therefore, by the Root Test, the series diverges.

35. $\displaystyle\sum_{n=2}^{\infty} \frac{(-1)^n}{(\ln n)^n}$

$$\lim_{n\to\infty} \sqrt[n]{|a_n|} = \lim_{n\to\infty} \sqrt[n]{\left|\frac{(-1)^n}{(\ln n)^n}\right|}$$

$$= \lim_{n\to\infty} \frac{1}{|\ln n|} = 0$$

Therefore, by the Root Test, the series converges.

36. $\displaystyle\sum_{n=1}^{\infty} \left(\frac{-2n}{3n+1}\right)^{3n}$

$$\lim_{n\to\infty} \sqrt[n]{|a_n|} = \lim_{n\to\infty} \sqrt[n]{\left|\left(\frac{-2n}{3n+1}\right)^{3n}\right|}$$

$$= \lim_{n\to\infty} \left(\frac{2n}{3n+1}\right)^3 = \left(\frac{2}{3}\right)^3 = \frac{8}{27}$$

Therefore, by the Root Test, the series converges.

37. $\displaystyle\sum_{n=1}^{\infty} \left(2\sqrt[n]{n}+1\right)^n$

$$\lim_{n\to\infty} \sqrt[n]{|a_n|} = \lim_{n\to\infty} \sqrt[n]{\left(2\sqrt[n]{n}+1\right)^n} = \lim_{n\to\infty} \left(2\sqrt[n]{n}+1\right)$$

To find $\displaystyle\lim_{n\to\infty} \sqrt[n]{n}$, let $y = \displaystyle\lim_{n\to\infty} \sqrt[x]{x}$. Then

$$\ln y = \lim_{n\to\infty} \left(\ln \sqrt[x]{x}\right) = \lim_{n\to\infty} \frac{1}{x} \ln x = \lim_{n\to\infty} \frac{\ln x}{x} = \lim_{n\to\infty} \frac{1/x}{1} = 0.$$

Thus, $\ln y = 0$, so $y = e^0 = 1$ and $\displaystyle\lim_{n\to\infty} \left(2\sqrt[n]{n}+1\right) = 2(1)+1 = 3$. Therefore, by the Root Test, the series diverges.

38. $\displaystyle\sum_{n=0}^{\infty} e^{-n}$

$$\lim_{n\to\infty} \sqrt[n]{|a_n|} = \lim_{n\to\infty} \sqrt[n]{\frac{1}{e^n}} = \frac{1}{e}$$

Therefore, by the Root Test, the series converges.

39. $\displaystyle\sum_{n=3}^{\infty} \frac{1}{(\ln n)^n}$

$$\lim_{n\to\infty} \sqrt[n]{|a_n|} = \lim_{n\to\infty} \sqrt[n]{\frac{1}{(\ln n)^n}} = \lim_{n\to\infty} \frac{1}{\ln n} = 0$$

Therefore, by the Root Test, the series converges.

40. $\displaystyle\sum_{n=0}^{\infty} \frac{n+1}{3^n}$

$$\lim_{n\to\infty} \sqrt[n]{|a_n|} = \lim_{n\to\infty} \sqrt[n]{\frac{n+1}{3^n}} = \lim_{n\to\infty} \frac{\sqrt[n]{n+1}}{3}$$

$$\text{Let} \quad y = \lim_{n\to\infty} \sqrt[x]{x+1}$$

$$\ln y = \lim_{n\to\infty} \left(\ln \sqrt[x]{x+1}\right)$$

$$= \lim_{n\to\infty} \frac{1}{x} \ln(x+1)$$

$$= \lim_{n\to\infty} \frac{\ln(x+1)}{x} = \frac{1}{x+1} = 0.$$

Since $\ln y = 0$, $y = e^0 = 1$, so

$$\lim_{n\to\infty} \frac{\sqrt[n]{n+1}}{3} = \frac{1}{3}.$$

Therefore, by the Root Test, the series converges.

41. $\displaystyle\sum_{n=1}^{\infty} \frac{(-1)^{n+1}5}{n}$

$$a_{n+1} = \frac{5}{n+1} < \frac{5}{n} = a_n$$

$$\lim_{n\to\infty} \frac{5}{n} = 0$$

Therefore, by the Alternating Series Test, the series converges (conditional convergence).

42. $\displaystyle\sum_{n=1}^{\infty} \frac{5}{n} = 5\sum_{n=1}^{\infty} \frac{1}{n}$

This is the divergent harmonic series.

43. $\displaystyle\sum_{n=1}^{\infty} \frac{3}{n\sqrt{n}} = 3\sum_{n=1}^{\infty} \frac{1}{n^{3/2}}$

This is convergent *p*-series.

44. $\displaystyle\sum_{n=1}^{\infty} \left(\frac{\pi}{4}\right)^n$

Since $\pi/4 < 1$, this is convergent geometric series.

45. $\displaystyle\sum_{n=1}^{\infty} \frac{2n}{n+1}$

$$\lim_{n\to\infty} \frac{2n}{n+1} = 2 \neq 0$$

This diverges by the *n*th Term Test for Divergence.

46. $\displaystyle\sum_{n=1}^{\infty} \frac{n}{2n^2+1}$

$$\lim_{n\to\infty} \frac{n/(2n^2+1)}{1/n} = \lim_{n\to\infty} \frac{n^2}{2n^2+1} = \frac{1}{2} > 0$$

This series diverges by limit comparison to the divergent harmonic series

$$\sum_{n=1}^{\infty} \frac{1}{n}.$$

47. $\displaystyle\sum_{n=1}^{\infty} \frac{(-1)^n 3^{n-2}}{2^n} = \sum_{n=1}^{\infty} \frac{(-1)^n 3^n 3^{-2}}{2^n} = \sum_{n=1}^{\infty} \frac{1}{9}\left(-\frac{3}{2}\right)^n$

Since $|r| = \frac{3}{2} > 1$, this is a divergent geometric series.

48. $\displaystyle\sum_{n=1}^{\infty} \frac{10}{3\sqrt{n^3}}$

$$\lim_{n\to\infty} \frac{10/3n^{3/2}}{1/n^{3/2}} = \frac{10}{3}$$

Therefore, the series converges by a limit comparison test with the *p*-series

$$\sum_{n=1}^{\infty} \frac{1}{n^{3/2}}.$$

49. $\displaystyle\sum_{n=1}^{\infty} \frac{10n+3}{n2^n}$

$$\lim_{n\to\infty} \frac{(10n+3)/n2^n}{1/2^n} = \lim_{n\to\infty} \frac{10n+3}{n} = 10$$

Therefore, the series converges by a limit comparison test with the geometric series

$$\sum_{n=0}^{\infty} \left(\frac{1}{2}\right)^n.$$

50. $\displaystyle\sum_{n=1}^{\infty} \frac{2^n}{4n^2-1}$

$$\lim_{n\to\infty} \frac{2^n}{4n^2-1} = \lim_{n\to\infty} \frac{(\ln 2)2^n}{8n} = \lim_{n\to\infty} \frac{(\ln 2)^2 2^n}{8} = \infty$$

Therefore, the series diverges by the *n*th Term Test for Divergence.

51. $\displaystyle\sum_{n=1}^{\infty} \frac{\cos(n)}{2^n}$

$$\left|\frac{\cos(n)}{2^n}\right| \leq \frac{1}{2^n}$$

Therefore, the series

$$\sum_{n=1}^{\infty} \left|\frac{\cos(n)}{2^n}\right|$$

converges by comparison with the geometric series

$$\sum_{n=0}^{\infty} \left(\frac{1}{2}\right)^n.$$

52. $\displaystyle\sum_{n=2}^{\infty} \frac{(-1)^n}{n\ln(n)}$

$$a_{n+1} = \frac{1}{(n+1)\ln(n+1)} \leq \frac{1}{n\ln(n)} = a_n$$

$$\lim_{n\to\infty} \frac{1}{n\ln(n)} = 0$$

Therefore, by the Alternating Series Test, the series converges.

53. $\displaystyle\sum_{n=1}^{\infty} \frac{n7^n}{n!}$

$$\lim_{n\to\infty} \left|\frac{a_{n+1}}{a_n}\right| = \lim_{n\to\infty} \left|\frac{(n+1)7^{n+1}}{(n+1)!} \cdot \frac{n!}{n7^n}\right| = \lim_{n\to\infty} \frac{7}{n} = 0$$

Therefore, by the Ratio Test, the series converges.

54. $\displaystyle\sum_{n=1}^{\infty} \frac{\ln(n)}{n^2}$

$$\frac{\ln(n)}{n^2} \le \frac{1}{n^{3/2}}$$

Therefore, the series converges by comparison with the *p*-series

$$\sum_{n=1}^{\infty} \frac{1}{n^{3/2}}.$$

56. $\displaystyle\sum_{n=1}^{\infty} \frac{(-1)^n \, 3^n}{n2^n}$

$$\lim_{n\to\infty} \left| \frac{a_{n+1}}{a_n} \right| = \lim_{n\to\infty} \left| \frac{3^{n+1}}{(n+1)2^{n+1}} \cdot \frac{n2^n}{3^n} \right| = \lim_{n\to\infty} \frac{3n}{2(n+1)} = \frac{3}{2}$$

Therefore, by the Ratio Test, the series diverges.

57. $\displaystyle\sum_{n=1}^{\infty} \frac{(-3)^n}{3 \cdot 5 \cdot 7 \cdots (2n+1)}$

$$\lim_{n\to\infty} \left| \frac{a_{n+1}}{a_n} \right| = \lim_{n\to\infty} \left| \frac{(-3)^{n+1}}{3 \cdot 5 \cdot 7 \cdots (2n+1)(2n+3)} \cdot \frac{3 \cdot 5 \cdot 7 \cdots (2n+1)}{(-3)^n} \right| = \lim_{n\to\infty} \frac{3}{2n+3} = 0$$

Therefore, by the Ratio Test, the series converges.

58. $\displaystyle\sum_{n=1}^{\infty} \frac{3 \cdot 5 \cdot 7 \cdots (2n+1)}{18^n(2n-1)n!}$

$$\lim_{n\to\infty} \left| \frac{a_{n+1}}{a_n} \right| = \lim_{n\to\infty} \left| \frac{3 \cdot 5 \cdot 7 \cdots (2n+1)(2n+3)}{18^{n+1}(2n+1)(2n-1)n!} \cdot \frac{18^n(2n-1)n!}{3 \cdot 5 \cdot 7 \cdots (2n+1)} \right| = \lim_{n\to\infty} \frac{(2n+3)(2n-1)}{18(2n+1)(2n-1)} = \frac{2}{18} = \frac{1}{9}$$

Therefore, by the Ratio Test, the series converge.

55. $\displaystyle\sum_{n=1}^{\infty} \frac{(-1)^n \, 3^{n-1}}{n!}$

$$\lim_{n\to\infty} \left| \frac{a_{n+1}}{a_n} \right| = \lim_{n\to\infty} \left| \frac{3^n}{(n+1)!} \cdot \frac{n!}{3^{n-1}} \right| = \lim_{n\to\infty} \frac{3}{n+1} = 0$$

Therefore, by the Ratio Test, the series converges.

59. (a) and (c)

$$\sum_{n=1}^{\infty} \frac{n5^n}{n!} = \sum_{n=0}^{\infty} \frac{(n+1)5^{n+1}}{(n+1)!}$$

$$= 5 + \frac{(2)(5)^2}{2!} + \frac{(3)(5)^3}{3!} + \frac{(4)(5)^4}{4!} + \cdots$$

60. (b) and (c)

$$\sum_{n=0}^{\infty} (n+1)\left(\frac{3}{4}\right)^n = \sum_{n=1}^{\infty} n\left(\frac{3}{4}\right)^{n-1}$$

$$= 1 + 2\left(\frac{3}{4}\right) + 3\left(\frac{3}{4}\right)^2 + 4\left(\frac{3}{4}\right)^3 + \cdots$$

61. (a) and (b) are the same.

62. (a) and (b) are the same.

$$\sum_{n=2}^{\infty} \frac{(-1)^n}{(n-1)2^{n-1}} = \frac{1}{2} - \frac{1}{2 \cdot 2^2} + \frac{1}{3 \cdot 2^3} - \cdots$$

$$\sum_{n=1}^{\infty} \frac{(-1)^{n+1}}{n2^n} = \frac{1}{2} - \frac{1}{2 \cdot 2^2} + \frac{1}{3 \cdot 2^3} - \cdots$$

63. Replace *n* with *n* + 1.

$$\sum_{n=1}^{\infty} \frac{n}{4^n} = \sum_{n=0}^{\infty} \frac{n+1}{4^{n+1}}$$

64. Replace *n* with *n* + 2.

$$\sum_{n=2}^{\infty} \frac{2^n}{(n-2)!} = \sum_{n=0}^{\infty} \frac{2^{n+2}}{n!}$$

65. Since

$$\frac{3^{10}}{2^{10} \, 10!} = 1.59 \times 10^{-5},$$

use 9 terms.

$$\sum_{k=1}^{9} \frac{(-3)^k}{2^k \, k!} \approx -0.7769$$

66. $\displaystyle\sum_{k=0}^{\infty} \frac{(-3)^k}{1 \cdot 3 \cdot 5 \ldots (2k+1)} = \sum_{k=0}^{\infty} \frac{(-3)^k 2^k k!}{(2k)!(2k+1)}$

$\displaystyle\qquad\qquad\qquad\qquad = \sum_{k=0}^{\infty} \frac{(-6)^k k!}{(2k+1)!}$

$\displaystyle\qquad\qquad\qquad\qquad \approx 0.40967$

(See Exercise 3 and use 10 terms, $k = 9$.)

67. No. Let $a_n = \dfrac{1}{n + 10,000}$.

The series $\displaystyle\sum_{n=1}^{\infty} \frac{1}{n + 10,000}$ diverges.

68. Assume that

$$\lim_{n \to \infty} |a_{n+1}/a_n| = L > 1 \text{ or that } \lim_{n \to \infty} |a_{n+1}/a_n| = \infty.$$

Then there exists $N > 0$ such that $|a_{n+1}/a_n| > 1$ for all $n > N$. Therefore,

$$|a_{n+1}| > |a_n|, \quad n > N$$

$$\Rightarrow \lim_{n \to \infty} a_n \neq 0 \Rightarrow \sum a_n \text{ diverges}$$

69. First, let

$$\lim_{n \to \infty} \sqrt[n]{|a_n|} = r < 1$$

and choose R such that $0 \leq r < R < 1$. There must exist some $N > 0$ such that $\sqrt[n]{|a_n|} < R$ for all $n > N$. Thus, for $n > N$, we $|a_n| < R^n$ and since the geometric series

$$\sum_{n=0}^{\infty} R^n$$

converges, we can apply the Comparison Test to conclude that

$$\sum_{n=1}^{\infty} |a_n|$$

converges which in turn implies that $\displaystyle\sum_{n=1}^{\infty} a_n$ converges.

Second, let

$$\lim_{n \to \infty} \sqrt[n]{|a_n|} = r > R > 1.$$

Then there must exist some $M > 0$ such that $\sqrt[n]{|a_n|} > R$ for all $n > M$. Thus, for $n > M$, we have $|a_n| > R_n > 1$ which implies that $\lim_{n \to \infty} a_n \neq 0$ which in turn implies that

$$\sum_{n=1}^{\infty} a_n \text{ diverges}.$$

70. The differentiation test states that if

$$\sum_{n=1}^{\infty} U_n$$

is an infinite series with real terms and $f(x)$ is a real function such that $f(1/n) = U_n$ for all positive integers n and $d^2 f/dx^2$ exists at $x = 0$, then

$$\sum_{n=1}^{\infty} U_n$$

converges absolutely if $f(0) = f'(0) = 0$ and diverges otherwise. Below are some examples.

Convergent Series	Divergent Series
$\displaystyle\sum \frac{1}{n^3}, f(x) = x^3$	$\displaystyle\sum \frac{1}{n}, f(x) = x$
$\displaystyle\sum \left(1 - \cos \frac{1}{n}\right), f(x) = 1 - \cos x$	$\displaystyle\sum \sin \frac{1}{n}, f(x) = \sin x$

Section 8.7   Taylor Polynomials and Approximations

1. $y = -\frac{1}{2}x^2 + 1$

Parabola

Matches (d)

2. $y = \frac{1}{8}x^4 - \frac{1}{2}x^2 + 1$

y-axis symmetry

Three relative extrema

Matches (c)

3. $y = e^{-1/2}[(x + 1) + 1]$

Linear

Matches (a)

4. $y = e^{-1/2}\left[\frac{1}{3}(x - 1)^3 - (x - 1) + 1\right]$

Cubic

Matches (b)

5. $f(x) = \cos x$

$P_2(x) = 1 - \frac{1}{2}x^2$

$P_4(x) = 1 - \frac{1}{2}x^2 + \frac{1}{24}x^4$

$P_6(x) = 1 - \frac{1}{2}x^2 + \frac{1}{24}x^4 - \frac{1}{720}x^6$

(a)

(c) In general, $f^{(n)}(0) = P_n^{(n)}(0)$ for all n.

(b) $\quad f'(x) = -\sin x \qquad P_2'(x) = -x$

$\qquad f''(x) = -\cos x \qquad P_2''(x) = -1$

$\qquad f''(0) = P_2''(0) = -1$

$\qquad f'''(x) = \sin x \qquad P_4'''(x) = x$

$\qquad f^{(4)}(x) = \cos x \qquad P_4^{(4)}(x) = 1$

$\qquad f^{(4)}(0) = 1 = P_4^{(4)}(0)$

$\qquad f^{(5)}(x) = -\sin x \qquad P_6^{(5)}(x) = -x$

$\qquad f^{(6)}(x) = -\cos x \qquad P^{(6)}(x) = -1$

$\qquad f^{(6)}(0) = -1 = P_6^{(6)}(0)$

6. $f(x) = x^2 e^x, f(0) = 0$

(a) $f'(x) = (x^2 + 2x)e^x \qquad f'(0) = 0$

$\quad f''(x) = (x^2 + 4x + 2)e^x \qquad f''(0) = 2$

$\quad f'''(x) = (x^2 + 6x + 6)e^x \qquad f'''(0) = 6$

$\quad f^{(4)}(x) = (x^2 + 8x + 12)e^x \qquad f^{(4)}(0) = 12$

$\quad P_2(x) = \dfrac{2x^2}{2!} = x^2$

$\quad P_3(x) = x^2 + \dfrac{6x^3}{3!} = x^2 + x^3$

$\quad P_4(x) = x^2 + x^3 + \dfrac{12x^4}{4!} = x^2 + x^3 + \dfrac{x^4}{2}$

(b)

(c) $f''(0) = 2 = P_2''(0)$

$\qquad f'''(0) = 6 = P_3'''(0)$

$\qquad f^{(4)}(0) = 12 = P_4^{(4)}(0)$

(d) $f^{(n)}(0) = P_n^{(n)}(0)$

7. $f(x) = e^{-x} \qquad f(0) = 1$

$\quad f'(x) = -e^{-x} \qquad f'(0) = -1$

$\quad f''(x) = e^{-x} \qquad f''(0) = 1$

$\quad f'''(x) = -e^{-x} \qquad f'''(0) = -1$

$\quad P_3(x) = f(0) + f'(0)x + \dfrac{f''(0)}{2!}x^2 + \dfrac{f'''(0)}{3!}x^3 = 1 - x + \dfrac{x^2}{2} - \dfrac{x^3}{6}$

8. $f(x) = e^{-x}$ $f(0) = 1$

$f'(x) = -e^{-x}$ $f'(0) = -1$

$f''(x) = e^{-x}$ $f''(0) = 1$

$f'''(x) = -e^{-x}$ $f'''(0) = -1$

$f^{(4)}(x) = e^{-x}$ $f^{(4)}(0) = 1$

$f^{(5)}(x) = -e^{-x}$ $f^{(5)}(0) = -1$

$$P_5(x) = f(0) + f'(0)x + \frac{f'(0)}{2!}x^2 + \frac{f'''(0)}{3!}x^3 + \frac{f^{(4)}(0)}{4!}x^4 + \frac{f^{(5)}(0)}{5!}x^5 = 1 - x + \frac{x^2}{2} - \frac{x^3}{6} + \frac{x^4}{24} - \frac{x^5}{120}$$

9. $f(x) = e^{2x}$ $f(0) = 1$

$f'(x) = 2e^{2x}$ $f'(0) = 2$

$f''(x) = 4e^{2x}$ $f''(0) = 4$

$f'''(x) = 8e^{2x}$ $f'''(0) = 8$

$f^{(4)}(x) = 16^{2x}$ $f^{(4)}(0) = 16$

$$P_4(x) = 1 + 2x + \frac{4}{2!}x^2 + \frac{8}{3!}x^3 + \frac{16}{4!}x^4$$

$$= 1 + 2x + 2x^2 + \frac{4}{3}x^3 + \frac{2}{3}x^4$$

10. $f(x) = e^{3x}$ $f(0) = 1$

$f'(x) = 3e^{3x}$ $f'(0) = 3$

$f''(x) = 9e^{3x}$ $f''(0) = 9$

$f'''(x) = 27e^{3x}$ $f'''(0) = 27$

$f^{(4)}(x) = 81e^{3x}$ $f^{(4)}(0) = 81$

$$P_4(x) = 1 + 3x + \frac{9}{2!}x^2 + \frac{27}{3!}x^3 + \frac{81}{4!}x^4$$

$$= 1 + 3x + \frac{9}{2}x^2 + \frac{9}{2}x^3 + \frac{27}{8}x^4$$

11. $f(x) = \sin x$ $f(0) = 0$

$f'(x) = \cos x$ $f'(0) = 1$

$f''(x) = -\sin x$ $f''(0) = 0$

$f'''(x) = -\cos x$ $f'''(0) = -1$

$f^{(4)}(x) = \sin x$ $f^{(4)}(0) = 0$

$f^{(5)}(x) = \cos x$ $f^{(5)}(0) = 1$

$$P_5(x) = 0 + (1)x + \frac{0}{2!}x^2 + \frac{-1}{3!}x^3 + \frac{0}{4!}x^4 + \frac{1}{5!}x^5$$

$$= x - \frac{1}{6}x^3 + \frac{1}{120}x^5$$

12. $f(x) = \sin \pi x$ $f(0) = 0$

$f'(x) = \pi \cos \pi x$ $f'(0) = \pi$

$f''(x) = -\pi^2 \sin \pi x$ $f''(0) = 0$

$f'''(x) = -\pi^3 \cos \pi x$ $f'''(0) = -\pi^3$

$$P_3(x) = 0 + \pi x + \frac{0}{2!}x^2 + \frac{-\pi^3}{3!}x^3 = \pi x - \frac{\pi^3}{6}x^3$$

13. $f(x) = xe^x$ $f(0) = 0$

$f'(x) = xe^x + e^x$ $f'(0) = 1$

$f''(x) = xe^x + 2e^x$ $f''(0) = 2$

$f'''(x) = xe^x + 3e^x$ $f'''(0) = 3$

$f^{(4)}(x) = xe^x + 4e^x$ $f^{(4)}(0) = 4$

$$P_4(x) = 0 + x + \frac{2}{2!}x^2 + \frac{3}{3!}x^3 + \frac{4}{4!}x^4$$

$$= x + x^2 + \frac{1}{2}x^3 + \frac{1}{6}x^4$$

14. $f(x) = x^2 e^{-x}$ $f(0) = 0$

$f'(x) = 2xe^{-x} - x^2 e^{-x}$ $f'(0) = 0$

$f''(x) = 2e^{-x} - 4xe^{-x} + x^2 e^{-x}$ $f''(0) = 2$

$f'''(x) = -6e^{-x} + 6xe^{-x} - x^2 e^{-x}$ $f'''(0) = -6$

$f^{(4)}(x) = 12e^{-x} - 8xe^{-x} + x^2 e^{-x}$ $f^{(4)}(0) = 12$

$$P_4(x) = 0 + 0x + \frac{2}{2!}x^2 + \frac{-6}{3!}x^3 + \frac{12}{4!}x^4$$

$$= x^2 - x^3 + \frac{1}{2}x^4$$

15. $f(x) = \dfrac{1}{x+1}$ $f(0) = 1$

     $f'(x) = -\dfrac{1}{(x+1)^2}$ $f'(0) = -1$

     $f''(x) = \dfrac{2}{(x+1)^2}$ $f''(0) = 2$

     $f'''(x) = \dfrac{-6}{(x+1)^4}$ $f'''(0) = -6$

     $f^{(4)}(x) = \dfrac{24}{(x+1)^5}$ $f^{(4)}(0) = 24$

     $P_4(x) = 1 - x + \dfrac{2}{2!}x^2 + \dfrac{-6}{3!}x^3 + \dfrac{24}{4!}x^4$

           $= 1 - x + x^2 - x^3 + x^4$

16. $f(x) = \sec x$ $f(0) = 1$

     $f'(x) = \sec x \tan x$ $f'(0) = 0$

     $f''(x) = \sec^3 x + \sec x \tan^2 x$ $f''(0) = 1$

     $P_2(x) = 1 + 0x + \dfrac{1}{2!}x^2 = 1 + \dfrac{1}{2}x^2$

17. $f(x) = \dfrac{1}{x}$ $f(1) = 1$

     $f'(x) = -\dfrac{1}{x^2}$ $f'(1) = -1$

     $f''(x) = \dfrac{2}{x^3}$ $f''(1) = 2$

     $f'''(x) = -\dfrac{6}{x^4}$ $f'''(1) = -6$

     $f^{(4)}(x) = \dfrac{24}{x^5}$ $f^{(4)}(1) = 24$

     $P_4(x) = 1 - (x-1) + \dfrac{2}{2!}(x-1)^2 + \dfrac{-6}{3!}(x-1)^3 + \dfrac{24}{4!}(x-1)^4$

           $= 1 - (x-1) + (x-1)^2 - (x-1)^3 + (x-1)^4$

18. $f(x) = \sqrt{x}$ $f(4) = 2$

     $f'(x) = \dfrac{1}{2\sqrt{x}}$ $f'(4) = \dfrac{1}{4}$

     $f''(x) = -\dfrac{1}{4x\sqrt{x}}$ $f''(4) = -\dfrac{1}{32}$

     $f'''(x) = \dfrac{3}{8x^2\sqrt{x}}$ $f'''(4) = \dfrac{3}{256}$

     $f^{(4)}(x) = -\dfrac{15}{16x^3\sqrt{x}}$ $f^{(4)}(4) = -\dfrac{15}{2048}$

     $P_4(x) = 2 + \dfrac{1}{4}(x-4) + \dfrac{-1/32}{2!}(x-4)^2 + \dfrac{3/256}{3!}(x-4)^3 + \dfrac{-15/2048}{4!}(x-4)^4$

           $= 2 + \dfrac{1}{4}(x-4) - \dfrac{1}{64}(x-4)^2 + \dfrac{1}{512}(x-4)^3 - \dfrac{5}{16,384}(x-4)^4$

19. $f(x) = \ln x$ $f(1) = 0$

$$f'(x) = \frac{1}{x} \qquad f'(1) = 1$$

$$f''(x) = -\frac{1}{x^2} \qquad f''(1) = -1$$

$$f'''(x) = \frac{2}{x^3} \qquad f'''(1) = 2$$

$$f^{(4)}(x) = -\frac{6}{x^4} \qquad f^{(4)}(1) = -6$$

$$P_4(x) = 0 + (x - 1) - \frac{1}{2}(x - 1)^2 + \frac{1}{3}(x - 1)^3 - \frac{1}{4}(x - 1)^4$$

20. $f(x) = x^2 \cos x$ $f(\pi) = -\pi^2$

$$f'(x) = 2x \cos x - x^2 \sin x \qquad f'(\pi) = -2\pi$$

$$f''(x) = 2 \cos x - 4x \sin x - x^2 \cos x \qquad f''(\pi) = -2 + \pi^2$$

$$P_2(x) = -\pi^2 - 2\pi(x - \pi) + \frac{(\pi^2 - 2)}{2}(x - \pi)^2$$

21. $f(x) = \tan x$

$$f'(x) = \sec^2 x$$

$$f''(x) = 2 \sec^2 x \tan x$$

$$f'''(x) = 4 \sec^2 x \tan^2 x + 2 \sec^4 x$$

$$f^{(4)}(x) = 8 \sec^2 x \tan^3 x + 16 \sec^4 x \tan x$$

$$f^{(5)}(x) = 16 \sec^2 x \tan^4 x + 88 \sec^4 x \tan^2 x + 16 \sec^6 x$$

(a) $n = 3, c = 0$

$$P_3(x) = 0 + x + \frac{0}{2!}x^2 + \frac{2}{3!}x^3 = x + \frac{1}{3}x^3$$

(b) $n = 5, c = 0$

$$P_5(x) = 0 + x + \frac{0}{2!}x^2 + \frac{2}{3!}x^3 + \frac{0}{4!}x^4 + \frac{16}{5!}x^5$$

$$= x + \frac{1}{3}x^3 + \frac{2}{15}x^5$$

(c) $n = 3, c = \frac{\pi}{4}$

$$Q_3(x) = 1 + 2\left(x - \frac{\pi}{4}\right) + \frac{4}{2!}\left(x - \frac{\pi}{4}\right)^2 + \frac{16}{3!}\left(x - \frac{\pi}{4}\right)^3$$

$$= 1 + 2\left(x - \frac{\pi}{4}\right) + 2\left(x - \frac{\pi}{4}\right)^2 + \frac{8}{3}\left(x - \frac{\pi}{4}\right)^3$$

22. $f(x) = \dfrac{1}{x^2 + 1}$

$$f'(x) = \frac{-2x}{(x^2 + 1)^2}$$

$$f''(x) = \frac{2(3x^2 - 1)}{(x^2 + 1)^3}$$

$$f'''(x) = \frac{24x(1 - x^2)}{(x^2 + 1)^4}$$

$$f^{(4)}(x) = \frac{24(5x^4 - 10x^2 + 1)}{(x^2 + 1)^5}$$

(a) $n = 2, c = 0$

$$P_2(x) = 1 + 0x + \frac{-2}{2!}x^2 = 1 - x^2$$

(b) $n = 4, c = 0$

$$P_4(x) = 1 + 0x + \frac{-2}{2!}x^2 + \frac{0}{3!}x^3 + \frac{24}{4!}x^4 = 1 - x^2 + x^4$$

(c) $n = 4, c = 1$

$$Q_4(x) = \frac{1}{2} + \left(-\frac{1}{2}\right)(x - 1) + \frac{1/2}{2!}(x - 1)^2 + \frac{0}{3!}(x - 1)^3 + \frac{-3}{4!}(x - 1)^4 = \frac{1}{2} - \frac{1}{2}(x - 1) + \frac{1}{4}(x - 1)^2 - \frac{1}{8}(x - 1)^4$$

23. $f(x) = \sin x$

$P_1(x) = x$

$P_3(x) = x - \frac{1}{6}x^3$

$P_5(x) = x - \frac{1}{6}x^3 + \frac{1}{120}x^5$

$P_7(x) = x - \frac{1}{6}x^3 + \frac{1}{120}x^5 - \frac{1}{5040}x^7$

(a)

x	0.00	0.25	0.50	0.75	1.00
$\sin x$	0.0000	0.2474	0.4794	0.6816	0.8415
$P_1(x)$	0.0000	0.2500	0.5000	0.7500	1.0000
$P_3(x)$	0.0000	0.2474	0.4792	0.6797	0.8333
$P_5(x)$	0.0000	0.2474	0.4794	0.6817	0.8417
$P_7(x)$	0.0000	0.2474	0.4794	0.6816	0.8415

(b)

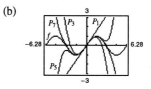

(c) As the distance increases, the accuracy decreases

24. $f(x) = \ln x$

$P_1(x) = x - 1$

$P_4(x) = (x - 1) - \frac{1}{2}(x - 1)^2 + \frac{1}{3}(x - 1)^3 - \frac{1}{4}(x - 1)^4$

(a)

x	1.00	1.25	1.50	1.75	2.00
$\ln x$	0.0000	0.2231	0.4055	0.5596	0.6931
$P_1(x)$	0.0000	0.2500	0.5000	0.7500	1.0000
$P_4(x)$	0.0000	0.2230	0.4010	0.5303	0.5833

(b)

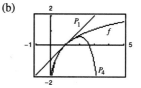

(c) As the distance increases, the accuracy decreases.

25. $f(x) = \arcsin x$

(a) $P_3(x) = x + \dfrac{x^3}{6}$

(b)

x	-0.75	-0.50	-0.25	0	0.25	0.50	0.75
$f(x)$	-0.848	-0.524	-0.253	0	0.253	0.524	0.848
$P_3(x)$	-0.820	-0.521	-0.253	0	0.253	0.521	0.820

(c)

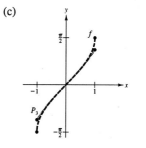

26. (a) $f(x) = \arctan x$

$P_3(x) = x - \dfrac{x^3}{3}$

(b)

x	-0.75	-0.50	-0.25	0	0.25	0.50	0.75
$f(x)$	-0.6435	-0.4636	-0.2450	0	0.2450	0.4636	0.6435
$P_3(x)$	-0.6094	-0.4583	-0.2448	0	0.2448	0.4583	0.6094

(c)

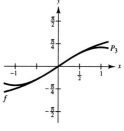

27. $f(x) = \cos x$

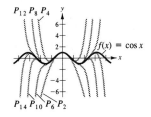

28. $f(x) = \arctan x$

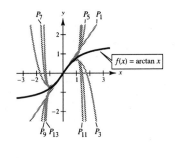

29. $f(x) = e^{-x} \approx 1 - x + \dfrac{x^2}{2} - \dfrac{x^3}{6}$

$f\left(\dfrac{1}{2}\right) \approx 0.6042$

30. $f(x) = x^2 e^{-x} \approx x^2 - x^3 + \dfrac{1}{2}x^4$

$f\left(\dfrac{1}{4}\right) \approx 0.0488$

31. $f(x) = \ln x \approx (x - 1) - \frac{1}{2}(x - 1)^2 + \frac{1}{3}(x - 1)^3 - \frac{1}{4}(x - 1)^4$

$f(1.2) \approx 0.1823$

32. $f(x) = x^2 \cos x \approx -\pi^2 - 2\pi(x - \pi) + \left(\dfrac{\pi^2 - 2}{2}\right)(x - \pi)^2$

$f\left(\dfrac{7\pi}{8}\right) \approx -6.7954$

33. $f(x) = \cos x; f^{(5)}(x) = -\sin x \Longrightarrow$ Max on $[0, 0.3]$ is 1.

$R_4(x) \leq \dfrac{1}{5!}(0.3)^5 = 2.025 \times 10^{-5}$

34. $f(x) = e^x; f^{(6)}(x) = e^x \Longrightarrow$ Max on $[0, 1]$ is e^1.

$R_5(x) \leq \dfrac{e^1}{6!}(1)^6 \approx 0.00378 = 3.78 \times 10^{-3}$

35. $f(x) = \arcsin x; f^{(4)}(x) = \dfrac{x(6x^2 + 9)}{(1 - x^2)^{7/2}} \Longrightarrow$ Max on $[0, 0.4]$ is $f^{(4)}(0.4) \approx 7.3340$.

$R_3(x) \leq \dfrac{7.3340}{4!}(0.4)^4 \approx 0.00782 = 7.82 \times 10^{-3}$

36. $f(x) = \arctan x; f^{(4)}(x) = \dfrac{24x(1 - x^2)}{(x^2 + 1)^4} \Longrightarrow$ Max on $[0, 0.5]$ is $f^{(4)}(0.5) \approx 3.6864$.

$R_3(x) \leq \dfrac{47.4074}{4!}(0.5)^4 \approx 0.0096$

37. $g(x) = \sin x$

$g^{(n+1)}(x) \leq 1$ for all x

$R_n(x) \leq \dfrac{1}{(n + 1)!}(0.3)^{n+1} < 0.001$

By trial and error, $n = 3$.

38. $f(x) = e^x$

$f^{(n+1)}(x) = e^x \Longrightarrow$ Max on $[0, 0.75]$ is $e^{0.75} \approx 2.1170$.

$R_n \leq \dfrac{2.1170}{(n + 1)!}(0.75)^{n+1} < 0.001$

By trial and error, $n = 5$.

39. $f(x) = \ln(x + 1)$

$$f^{(n+1)}(x) = \frac{(-1)^{n+1}n!}{(x + 1)^{n+1}} \Longrightarrow \text{Max on } [0, 0.5] \text{ is } n!.$$

$$R_n \le \frac{n!}{(n + 1)!}(0.5)^{n+1} = \frac{(0.5)^{n+1}}{n + 1} < 0.0001$$

By trial and error, $n = 9$. (See Example 9.) Using 9 terms, $\ln(1.5) \approx 0.4055$.

40. $f(x) = \cos(\pi x^2)$

$$g(x) = \cos x = 1 - \frac{x^2}{2!} + \frac{x^4}{4!} - \frac{x^6}{6!} + \cdots$$

$$f(x) = g(\pi x^2)$$

$$= 1 - \frac{(\pi x^2)^2}{2!} + \frac{(\pi x^2)^4}{4!} - \frac{(\pi x^2)^6}{6!} + \cdots$$

$$= 1 - \frac{\pi^2 x^4}{2!} + \frac{\pi^4 x^8}{4!} - \frac{\pi^6 x^{12}}{6!} + \cdots$$

$$f(0.6) = 1 - \frac{\pi^2}{2!}(0.6)^4 + \frac{\pi^4}{4!}(0.6)^8 - \frac{\pi^6}{6!}(0.6)^{12} + \cdots$$

Since this is an alternating series,

$$R_n \le a_{n+1} = \frac{\pi^{2n}}{(2n)!}(0.6)^{4n} < 0.0001.$$

By trial and error, $n = 4$. Using 4 terms $f(0.6) \approx 0.4257$.

41. $f(x) = e^x \approx 1 + x + \dfrac{x^2}{2} + \dfrac{x^3}{6}, \ x < 0$

$$R_3(x) = \frac{e^z}{4!}x^4 < 0.001$$

$$e^z x^4 < 0.024$$

$$xe^{z/4} < 0.3936$$

$$x < \frac{0.3936}{e^{z/4}} < 0.3936, \ z < 0$$

$$-0.3936 < x < 0$$

42. $f(x) = \sin x \approx x - \dfrac{x^3}{3!}$

$$|R_3(x)| = \left| \frac{\sin z}{4!}x^4 \right| \le \frac{|x^4|}{4!} < 0.001$$

$$x^4 < 0.024$$

$$|x| < 0.3936$$

$$-0.3936 < x < 0.3936$$

43. (a) $f(x) = e^x$

$$P_4(x) = 1 + x + \frac{1}{2}x^2 + \frac{1}{6}x^3 + \frac{1}{24}x^4$$

$$g(x) = xe^x$$

$$Q_5(x) = x + x^2 + \frac{1}{2}x^3 + \frac{1}{6}x^4 + \frac{1}{24}x^5$$

$$Q_5(x) = x P_4(x)$$

(b) $f(x) = \sin x$

$$P_5(x) = x - \frac{x^3}{3!} + \frac{x^5}{5!}$$

$$g(x) = x \sin x$$

$$Q_6(x) = x P_5(x) = x^2 - \frac{x^4}{3!} + \frac{x^6}{5!}$$

(c) $g(x) = \dfrac{\sin x}{x} = \dfrac{1}{x}P_5(x) = 1 - \dfrac{x^2}{3!} + \dfrac{x^4}{5!}$

44. (a) $P_5(x) = x - \dfrac{x^3}{3!} + \dfrac{x^5}{5!}$ for $f(x) = \sin x$

$$P_5'(x) = 1 - \frac{x^2}{2!} + \frac{x^4}{4!}$$

This is the Maclaurin polynomial of degree 4 for $g(x) = \cos x$.

(b) $Q_6(x) = 1 - \dfrac{x^2}{2} + \dfrac{x^4}{4!} - \dfrac{x^6}{6!}$ for $\cos x$

$$Q_6'(x) = -x + \frac{x^3}{3!} - \frac{x^5}{5!} = -P_5(x)$$

(c) $R(x) = 1 + x + \dfrac{x^2}{2!} + \dfrac{x^3}{3!} + \dfrac{x^4}{4!}$

$$R'(x) = 1 + x + \frac{x^2}{2!} + \frac{x^3}{3!}$$

The first four terms are the same!

45. Let f be an odd function and P_n be the nth Maclaurin polynomial for f. Since f is odd, f' is even:

$$f'(-x) = \lim_{h\to 0}\frac{f(-x+h)-f(-x)}{h} = \lim_{h\to 0}\frac{-f(x-h)+f(x)}{h} = \lim_{h\to 0}\frac{f(x+(-h))-f(x)}{-h} = f'(x).$$

Similarly, f'' is odd, f''' is even, etc. Therefore, $f, f'', f^{(4)}$, etc. are all odd functions, which implies that $f(0) = f''(0) = \ldots = 0$. Hence, in the formula

$$P_n(x) = f(0) + f'(0)x + \frac{f''(0)x^2}{2!} + \cdots \text{ all the coefficients of the even power of } x \text{ are zero.}$$

46. Let f be an even function and P_n be the nth Maclaurin polynomial for f. Since f is even, f' is odd, f'' is even, f''' is odd, etc. (see Exercise 45). All of the odd derivatives of f are odd and thus, all of the odd powers of x will have coefficients of zero. P_n will only have terms with even powers of x.

47. Let $P_n(x) = a_0 + a_1(x-c) + a_2(x-c)^2 + \cdots + a_n(x-c)^n$ where $a_i = \frac{f^{(i)}(c)}{i!}$.

$$P_n(c) = a_0 = f(c)$$

For $1 \le k \le n$, $P_n^{(k)}(c) = a_n k! = \left(\frac{f^{(k)}(c)}{k!}\right)k! = f^{(k)}(c)$.

48. As you move away from $x = c$, the Taylor Polynomial becomes less and less accurate.

Section 8.8   Power Series

1. $\sum_{n=0}^{\infty}(-1)^n\frac{x^n}{n+1}$

$L = \lim_{n\to\infty}\left|\frac{u_{n+1}}{u_n}\right| = \lim_{n\to\infty}\left|\frac{(-1)^{n+1}x^{n+1}}{n+2}\cdot\frac{n+1}{(-1)^n x^n}\right|$

$= \lim_{n\to\infty}\left|\frac{n+1}{n+2}\right||x| = |x|$

$|x| < 1 \Rightarrow R = 1$

2. $\sum_{n=0}^{\infty}(4x)^n$

$L = \lim_{n\to\infty}\left|\frac{u_{n+1}}{u_n}\right| = \lim_{n\to\infty}\left|\frac{(4x)^{n+1}}{(4x)^n}\right| = 4|x|$

$4|x| < 1 \Rightarrow R = \frac{1}{4}$

3. $\sum_{n=1}^{\infty}\frac{(2x)^n}{n^2}$

$L = \lim_{n\to\infty}\left|\frac{u_{n+1}}{u_n}\right| = \lim_{n\to\infty}\left|\frac{(2x)^{n+1}}{(n+1)^2}\cdot\frac{n^2}{(2x)^n}\right|$

$= \lim_{n\to\infty}\left|\frac{2n^2 x}{(n+1)^2}\right| = 2|x|$

$2|x| < 1 \Rightarrow R = \frac{1}{2}$

4. $\sum_{n=0}^{\infty}\frac{(-1)^n x^n}{2^n}$

$L = \lim_{n\to\infty}\left|\frac{u_{n+1}}{u_n}\right| = \lim_{n\to\infty}\left|\frac{(-1)^{n+1}x^{n+1}}{2^{n+1}}\cdot\frac{2^n}{(-1)^n x^n}\right|$

$= \frac{1}{2}|x|$

$\frac{1}{2}|x| < 1 \Rightarrow R = 2$

5. $\sum_{n=0}^{\infty}\frac{(2x)^n}{n!}$

$L = \lim_{n\to\infty}\left|\frac{u_{n+1}}{u_n}\right| = \lim_{n\to\infty}\left|\frac{(2x)^{n+1}}{(n+1)!}\cdot\frac{n!}{(2x)^n}\right|$

$= \lim_{n\to\infty}\left|\frac{2x}{n+1}\right| = 0$

Thus, the series converges for all x. R is infinite.

$R = \infty$

6. $\sum_{n=0}^{\infty}\frac{(2n)!x^n}{n!}$

$L = \lim_{n\to\infty}\left|\frac{u_{n+1}}{u_n}\right| = \lim_{n\to\infty}\left|\frac{(2n+2)!x^{n+1}}{(n+1)!}\cdot\frac{n!}{(2n)!x^n}\right|$

$= \lim_{n\to\infty}\left|\frac{(2n+2)(2n+1)(2n)!}{(n+1)n!}\cdot\frac{n!}{(2n)!}\right||x|$

$= \lim_{n\to\infty}\left|\frac{(2n+2)(2n+1)}{n+1}\right||x| = \infty$

The series only converges at $x = 0$. $R = 0$

7. $\displaystyle\sum_{n=0}^{\infty} \left(\frac{x}{2}\right)^n$

Since the series is geometric, it converges only if $|x/2| < 1$ or $-2 < x < 2$.

8. $\displaystyle\sum_{n=0}^{\infty} \left(\frac{x}{k}\right)^n$

Since the series is geometric, it converges only if $|x/k| < 1$ or $-k < x < k$.

9. $\displaystyle\sum_{n=1}^{\infty} \frac{(-1)^n x^n}{n}$

$$\lim_{n \to \infty} \left|\frac{u_{n+1}}{u_n}\right| = \lim_{n \to \infty} \left|\frac{(-1)^{n+1}x^{n+1}}{n+1} \cdot \frac{n}{(-1)^n x^n}\right|$$

$$= \lim_{n \to \infty} \left|\frac{nx}{n+1}\right| = |x|$$

Interval: $-1 < x < 1$

When $x = 1$, the alternating series $\displaystyle\sum_{n=1}^{\infty} \frac{(-1)^n}{n}$ converges.

When $x = -1$, the p-series $\displaystyle\sum_{n=1}^{\infty} \frac{1}{n}$ diverges.

Therefore, the interval of convergence is $-1 < x \leq 1$.

10. $\displaystyle\sum_{n=0}^{\infty} (-1)^{n+1} n x^n$

$$\lim_{n \to \infty} \left|\frac{u_{n+1}}{u_n}\right| = \lim_{n \to \infty} \left|\frac{(-1)^{n+2}(n+1)x^{n+1}}{(-1)^{n+1}nx^n}\right|$$

$$= \lim_{n \to \infty} \left|\frac{(n+1)x}{n}\right| = |x|$$

Interval: $-1 < x < 1$

When $x = 1$, the alternating series $\displaystyle\sum_{n=0}^{\infty} (-1)^{n+1} n$ diverges.

When $x = -1$, the series $\displaystyle\sum_{n=0}^{\infty} (-n)$ diverges.

Therefore, the interval of convergence is $-1 < x < 1$.

11. $\displaystyle\sum_{n=0}^{\infty} \frac{x^n}{n!}$

$$\lim_{n \to \infty} \left|\frac{u_{n+1}}{u_n}\right| = \lim_{n \to \infty} \left|\frac{x^{n+1}}{(n+1)!} \cdot \frac{n!}{x^n}\right|$$

$$= \lim_{n \to \infty} \left|\frac{x}{n+1}\right| = 0$$

The series converges for all x. Therefore, the interval of convergence is $-\infty < x < \infty$.

12. $\displaystyle\sum_{n=0}^{\infty} \frac{(3x)^n}{(2n)!}$

$$\lim_{n \to \infty} \left|\frac{u_{n+1}}{u_n}\right| = \lim_{n \to \infty} \left|\frac{(3x)^{n+1}}{(2n+1)!} \cdot \frac{(2n)!}{(3x)^n}\right|$$

$$= \lim_{n \to \infty} \left|\frac{3x}{(2n+2)(2n+1)}\right| = 0$$

Therefore, the interval of convergence is $-\infty < x < \infty$.

13. $\displaystyle\sum_{n=0}^{\infty} (2n)! \left(\frac{x}{2}\right)^n$

$$\lim_{n \to \infty} \left|\frac{u_{n+1}}{u_n}\right| = \lim_{n \to \infty} \left|\frac{(2n+2)!x^{n+1}}{2^{n+1}} \cdot \frac{2^n}{(2n)!x^n}\right| = \lim_{n \to \infty} \left|\frac{(2n+2)(2n+1)x}{2}\right| = \infty$$

Therefore, the series converges only for $x = 0$.

14. $\displaystyle\sum_{n=0}^{\infty} \frac{(-1)^n x^n}{(n+1)(n+2)}$

$$\lim_{n \to \infty} \left|\frac{u_{n+1}}{u_n}\right| = \lim_{n \to \infty} \left|\frac{(-1)^{n+1}x^{n+1}}{(n+2)(n+3)} \cdot \frac{(n+1)(n+2)}{(-1)^n x^n}\right| = \lim_{n \to \infty} \left|\frac{(n+1)x}{n+3}\right| = |x|$$

Interval: $-1 < x < 1$

When $x = 1$, the alternating series $\displaystyle\sum_{n=0}^{\infty} \frac{(-1)^n}{(n+1)(n+2)}$ converges.

When $x = -1$, the series $\displaystyle\sum_{n=0}^{\infty} \frac{1}{(n+1)(n+2)}$ converges by limit comparison to $\displaystyle\sum_{n=1}^{\infty} \frac{1}{n^2}$.

Therefore, the interval of convergence is $-1 \leq x \leq 1$.

15. $\displaystyle\sum_{n=1}^{\infty} \frac{(-1)^{n+1} x^n}{4^n}$

Since the series is geometric, it converges only if $|x/4| < 1$ or $-4 < x < 4$.

16. $\displaystyle\sum_{n=0}^{\infty} \frac{(-1)^n n!(x-4)^n}{3^n}$

$$\lim_{n\to\infty}\left|\frac{u_{n+1}}{u_n}\right| = \lim_{n\to\infty}\left|\frac{(-1)^{n+1}(n+1)!(x-4)^{n+1}}{3^{n+1}} \cdot \frac{3^n}{(-1)^n n!(x-4)^n}\right| = \lim_{n\to\infty}\left|\frac{(n+1)(x-4)}{3}\right| = \infty$$

$R = 0$

Center: $x = 4$

Therefore, the series converges only for $x = 4$.

17. $\displaystyle\sum_{n=1}^{\infty} \frac{(-1)^{n+1}(x-5)^n}{n5^n}$

$$\lim_{n\to\infty}\left|\frac{u_{n+1}}{u_n}\right| = \lim_{n\to\infty}\left|\frac{(-1)^{n+2}(x-5)^{n+1}}{(n+1)5^{n+1}} \cdot \frac{n5^n}{(-1)^{n+1}(x-5)^n}\right| = \lim_{n\to\infty}\left|\frac{n(x-5)}{5(n+1)}\right| = \frac{1}{5}|x-5|$$

$R = 5$

Center: $x = 5$

Interval: $-5 < x - 5 < 5$ or $0 < x < 10$

When $x = 0$, the p-series $\displaystyle\sum_{n=1}^{\infty} \frac{-1}{n}$ diverges.

When $x = 10$, the alternating series $\displaystyle\sum_{n=1}^{\infty} \frac{(-1)^{n+1}}{n}$ converges.

Therefore, the interval of convergence is $0 < x \le 10$.

18. $\displaystyle\sum_{n=0}^{\infty} \frac{(x-2)^{n+1}}{(n+1)3^{n+1}}$

$$\lim_{n\to\infty}\left|\frac{u_{n+1}}{u_n}\right| = \lim_{n\to\infty}\left|\frac{(x-2)^{n+2}}{(n+2)3^{n+2}} \cdot \frac{(n+1)3^{n+1}}{(x-2)^{n+1}}\right| = \lim_{n\to\infty}\left|\frac{(n+1)(x-2)}{3(n+2)}\right| = \frac{1}{3}|x-2|$$

$R = 3$

Center: $x = 2$

Interval: $-3 < x - 2 < 3$ or $-1 < x < 5$

When $x = -1$, the alternating series $\displaystyle\sum_{n=0}^{\infty} \frac{(-1)^{n+1}}{n+1}$ converges.

When $x = 5$, the series $\displaystyle\sum_{n=0}^{\infty} \frac{1}{n+1}$ diverges by the integral test.

Therefore, the interval of convergence is $-1 \le x < 5$.

19. $\displaystyle\sum_{n=0}^{\infty} \frac{(-1)^{n+1}(x-1)^{n+1}}{n+1}$

$$\lim_{n\to\infty}\left|\frac{u_{n+1}}{u_n}\right| = \lim_{n\to\infty}\left|\frac{(-1)^{n+2}(x-1)^{n+2}}{n+2} \cdot \frac{n+1}{(-1)^{n+1}(x-1)^{n+1}}\right| = \lim_{n\to\infty}\left|\frac{(n+1)(x-1)}{n+2}\right| = |x-1|$$

$R = 1$

Center: $x = 1$

Interval: $-1 < x - 1 < 1$ or $0 < x < 2$

When $x = 0$, the series $\displaystyle\sum_{n=0}^{\infty} \frac{1}{n+1}$ diverges by the integral test.

When $x = 2$, the alternating series $\displaystyle\sum_{n=0}^{\infty} \frac{(-1)^{n+1}}{n+1}$ converges.

Therefore, the interval of convergence is $0 < x \le 2$.

20. $\displaystyle\sum_{n=1}^{\infty} \frac{(-1)^{n+1}(x-c)^n}{nc^n}$

$$\lim_{n\to\infty} \left| \frac{u_{n+1}}{u_n} \right| = \lim_{n\to\infty} \left| \frac{(-1)^{n+2}(x-c)^{n+1}}{(n+1)c^{n+1}} \cdot \frac{nc^n}{(-1)^{n+1}(x-c)^n} \right| = \lim_{n\to\infty} \left| \frac{n(x-c)}{c(n+1)} \right| = \frac{1}{c}|x-c|$$

$R = c$

Center: $x = c$

Interval: $-c < x - c < c$ or $0 < x < 2c$

When $x = 0$, the p-series $\displaystyle\sum_{n=1}^{\infty} \frac{-1}{n}$ diverges.

When $x = 2c$, the alternating series $\displaystyle\sum_{n=1}^{\infty} \frac{(-1)^{n+1}}{n}$ converges. Therefore, the interval of convergence is $0 < x \le 2c$.

21. $\displaystyle\sum_{n=1}^{\infty} \frac{(x-c)^{n-1}}{c^{n-1}}$

$$\lim_{n\to\infty} \left| \frac{u_{n+1}}{u_n} \right| = \lim_{n\to\infty} \left| \frac{(x-c)^n}{c^n} \cdot \frac{c^{n-1}}{(x-c)^{n-1}} \right| = \frac{1}{c}|x-c|$$

$R = c$

Center: $x = c$

Interval: $-c < x - c < c$ or $0 < x < 2c$

When $x = 0$, the series $\displaystyle\sum_{n=1}^{\infty} (-1)^{n-1}$ diverges.

When $x = 2c$, the series $\displaystyle\sum_{n=1}^{\infty} 1$ diverges.

Therefore, the interval of convergence is $0 < x < 2c$.

22. $\displaystyle\sum_{n=1}^{\infty} \frac{(-1)^{n+1}x^{2n-1}}{2n-1}$

$$\lim_{n\to\infty} \left| \frac{u_{n+1}}{u_n} \right| = \lim_{n\to\infty} \left| \frac{(-1)^{n+2}x^{2n+1}}{2n+1} \cdot \frac{2n-1}{(-1)^{n+1}x^{2n-1}} \right|$$

$$= \lim_{n\to\infty} \left| \frac{(2n-1)x^2}{2n+1} \right| = |x^2|$$

Interval: $-1 < x < 1$

When $x = -1$, the alternating series $\displaystyle\sum_{n=1}^{\infty} \frac{(-1)^n}{2n-1}$ converges.

When $x = 1$, the alternating series $\displaystyle\sum_{n=1}^{\infty} \frac{(-1)^{n+1}}{2n-1}$ converges.

Therefore, the interval of convergence is $-1 \le x \le 1$.

23. $\displaystyle\sum_{n=1}^{\infty} \frac{n}{n+1}(-2x)^{n-1}$

$$\lim_{n\to\infty} \left| \frac{u_{n+1}}{u_n} \right| = \lim_{n\to\infty} \left| \frac{(n+1)(-2x)^n}{n+2} \cdot \frac{n+1}{n(-2x)^{n-1}} \right|$$

$$= \lim_{n\to\infty} \left| \frac{(-2x)(n+1)^2}{n(n+2)} \right| = 2|x|$$

$R = \dfrac{1}{2}$

Interval: $-\dfrac{1}{2} < x < \dfrac{1}{2}$

When $x = -\dfrac{1}{2}$, the series $\displaystyle\sum_{n=1}^{\infty} \frac{n}{n+1}$
diverges by the nth Term Test.

When $x = \dfrac{1}{2}$, the alternating series $\displaystyle\sum_{n=1}^{\infty} \frac{(-1)^{n-1}n}{n+1}$ diverges.

Therefore, the interval of convergence is $-\dfrac{1}{2} < x < \dfrac{1}{2}$.

24. $\displaystyle\sum_{n=0}^{\infty} \frac{(-1)^n x^{2n}}{n!}$

$$\lim_{n\to\infty} \left| \frac{u_{n+1}}{u_n} \right| = \lim_{n\to\infty} \left| \frac{(-1)^{n+1}x^{2n+2}}{(n+1)!} \cdot \frac{n!}{(-1)^n x^{2n}} \right|$$

$$= \lim_{n\to\infty} \left| \frac{x^2}{n+1} \right| = 0$$

Therefore, the interval of convergence is $-\infty < x < \infty$.

25. $\displaystyle\sum_{n=0}^{\infty} \frac{x^{2n+1}}{(2n+1)!}$

$$\lim_{n\to\infty} \left| \frac{u_{n+1}}{u_n} \right| = \lim_{n\to\infty} \left| \frac{x^{2n+3}}{(2n+3)!} \cdot \frac{(2n+1)!}{x^{2n+1}} \right|$$

$$= \lim_{n\to\infty} \left| \frac{x^2}{(2n+2)(2n+3)} \right| = 0$$

Therefore, the interval of convergence is $-\infty < x < \infty$.

26. $\displaystyle\sum_{n=1}^{\infty} \frac{n!x^n}{(2n)!}$

$$\lim_{n\to\infty} \left| \frac{u_{n+1}}{u_n} \right| = \lim_{n\to\infty} \left| \frac{(n+1)!x^{n+1}}{(2n+2)!} \cdot \frac{(2n)!}{n!x^n} \right|$$

$$= \lim_{n\to\infty} \left| \frac{(n+1)x}{(2n+2)(2n+1)} \right| = 0$$

Therefore, the interval of convergence is $-\infty < x < \infty$.

27. $\displaystyle\sum_{n=1}^{\infty} \frac{k(k+1)\cdots(k+n-1)x^n}{n!}$

$$\lim_{n\to\infty} \left| \frac{u_{n+1}}{u_n} \right| = \lim_{n\to\infty} \left| \frac{k(k+1)\cdots(k+n-1)(k+n)x^{n+1}}{(n+1)!} \cdot \frac{n!}{k(k+1)\cdots(k+n-1)x^n} \right| = \lim_{n\to\infty} \left| \frac{(k+n)x}{n+1} \right| = |x|$$

$R = 1$

When $x = \pm 1$, the series diverges and the interval of convergence is $-1 < x < 1$.

$$\left[\frac{k(k+1)\cdots(k+n-1)}{1\cdot 2\cdots n} \geq 1 \right]$$

28. $\displaystyle\sum_{n=1}^{\infty} \frac{2\cdot 4\cdot 6\cdots(2n)}{3\cdot 5\cdot 7\cdots(2n+1)}(x^{2n+1})$

$$\lim_{n\to\infty} \left| \frac{u_{n+1}}{u_n} \right| = \lim_{n\to\infty} \left| \frac{2\cdot 4\cdots(2n)(2n+2)x^{2n+3}}{3\cdot 5\cdot 7\cdots(2n+1)(2n+3)} \cdot \frac{3\cdot 5\cdots(2n+1)}{2\cdot 4\cdots(2n)x^{2n+1}} \right| = \lim_{n\to\infty} \left| \frac{(2n+2)x^2}{(2n+3)} \right| = |x^2|$$

$R = 1$

When $x = \pm 1$, the series diverges by comparing it to

$$\sum_{n=1}^{\infty} \frac{1}{2n+1}$$

which diverges. Therefore, the interval of convergence is $-1 < x < 1$.

29. $\displaystyle\sum_{n=1}^{\infty} \frac{(-1)^{n+1}3\cdot 7\cdot 11\cdots(4n-1)(x-3)^n}{4^n}$

$$\lim_{n\to\infty} \left| \frac{u_{n+1}}{u_n} \right| = \lim_{n\to\infty} \left| \frac{(-1)^{n+2}\cdot 3\cdot 7\cdot 11\cdots(4n-1)(4n+3)(x-3)^{n+1}}{4^{n+1}} \cdot \frac{4^n}{(-1)^{n+1}\cdot 3\cdot 7\cdot 11\cdots(4n-1)(x-3)^n} \right|$$

$$= \lim_{n\to\infty} \left| \frac{(4n+3)(x-3)}{4} \right| = \infty$$

$R = 0$

Center: $x = 3$

Therefore, the series converges only for $x = 3$.

30. $\displaystyle\sum_{n=1}^{\infty} \frac{n!(x-c)^n}{1\cdot 3\cdot 5\cdots(2n-1)}$

$$\lim_{n\to\infty} \left| \frac{u_{n+1}}{u_n} \right| = \lim_{n\to\infty} \left| \frac{(n+1)!(x-c)^{n+1}}{1\cdot 3\cdot 5\cdots(2n-1)(2n+1)} \cdot \frac{1\cdot 3\cdot 5\cdots(2n-1)}{n!(x-c)} \right| = \lim_{n\to\infty} \left| \frac{(n+1)(x-c)}{2n+1} \right| = \frac{1}{2}|x-c|$$

$R = 2$

Interval: $-2 < x - c < 2$ or $c - 2 < x < c + 2$

The series diverges at the endpoints. Therefore, the interval of convergence is $c - 2 < x < c + 2$.

$$\left[\frac{n!(c+2-c)^n}{1\cdot 3\cdot 5\cdots(2n-1)} = \frac{n!2^2}{1\cdot 3\cdot 5\cdots(2n-1)} = \frac{2\cdot 4\cdot 6\cdots(2n)}{1\cdot 3\cdot 5\cdots(2n-1)} > 1 \right]$$

31. (a) $f(x) = \sum_{n=0}^{\infty} \left(\frac{x}{2}\right)^n, \, -2 < x < 2$ (Geometric)

(b) $f'(x) = \sum_{n=1}^{\infty} \left(\frac{n}{2}\right)\left(\frac{x}{2}\right)^{n-1}, \, -2 < x < 2$

(c) $f''(x) = \sum_{n=2}^{\infty} \left(\frac{n}{2}\right)\left(\frac{n-1}{2}\right)\left(\frac{x}{2}\right)^{n-2}, \, -2 < x < 2$

(d) $\int f(x)\,dx = \sum_{n=0}^{\infty} \frac{2}{n+1}\left(\frac{x}{2}\right)^{n+1}, \, -2 \le x < 2$

32. (a) $f(x) = \sum_{n=1}^{\infty} \frac{(-1)^{n+1}(x-5)^n}{n5^n}, \, 0 < x \le 10$

(b) $f'(x) = \sum_{n=1}^{\infty} \frac{(-1)^{n+1}(x-5)^{n-1}}{5^n}, \, 0 < x < 10$

(c) $f''(x) = \sum_{n=2}^{\infty} \frac{(-1)^{n+1}(n-1)(x-5)^{n-2}}{5^n}, \, 0 < x < 10$

(d) $\int f(x)\,dx = \sum_{n=1}^{\infty} \frac{(-1)^{n+1}(x-5)^{n+1}}{n(n+1)5^n}, \, 0 \le x \le 10$

33. (a) $f(x) = \sum_{n=0}^{\infty} \frac{(-1)^{n+1}(x-1)^{n+1}}{n+1}, \, 0 < x \le 2$

(b) $f'(x) = \sum_{n=0}^{\infty} (-1)^{n+1}(x-1)^n, \, 0 < x < 2$

(c) $f''(x) = \sum_{n=1}^{\infty} (-1)^{n+1}n(x-1)^{n-1}, \, 0 < x < 2$

(d) $\int f(x)\,dx = \sum_{n=1}^{\infty} \frac{(-1)^{n+1}(x-1)^{n+2}}{(n+1)(n+2)}, \, 0 \le x \le 2$

34. (a) $f(x) = \sum_{n=1}^{\infty} \frac{(-1)^{n+1}(x-1)^n}{n}, \, 0 < x \le 2$

(b) $f'(x) = \sum_{n=1}^{\infty} (-1)^{n+1}(x-1)^{n-1}, \, 0 < x < 2$

(c) $f''(x) = \sum_{n=2}^{\infty} (-1)^{n+1}(n-1)(x-1)^{n-2}, \, 0 < x < 2$

(d) $\int f(x)\,dx = \sum_{n=1}^{\infty} \frac{(-1)^{n+1}(x-1)^{n+1}}{n(n+1)}, \, 0 \le x \le 2$

35. $g(1) = \sum_{n=0}^{\infty} \left(\frac{1}{3}\right)^n = 1 + \frac{1}{3} + \frac{1}{9} + \cdots$

$S_1 = 1, S_2 = 1.33$. Matches (c)

36. $g(2) = \sum_{n=0}^{\infty} \left(\frac{2}{3}\right)^n = 1 + \frac{2}{3} + \frac{4}{9} + \cdots$

$S_1 = 1, S_2 = 1.67$. Matches (a)

37. $g(3.1) = \sum_{n=0}^{\infty} \left(\frac{3.1}{3}\right)^n$ diverges. Matches (b)

38. $g(-2) = \sum_{n=0}^{\infty} \left(-\frac{2}{3}\right)^n$ alternating. Matches (d)

39. (a) $f(x) = \sum_{n=0}^{\infty} \frac{(-1)^n x^{2n+1}}{(2n+1)!}, \, -\infty < x < \infty$ (See Exercise 25)

$g(x) = \sum_{n=0}^{\infty} \frac{(-1)^n x^{2n}}{(2n)!}, \, -\infty < x < \infty$

(b) $f'(x) = \sum_{n=0}^{\infty} \frac{(-1)^n x^{2n}}{(2n)!} = g(x)$

(c) $g'(x) = \sum_{n=1}^{\infty} \frac{(-1)^n x^{2n-1}}{(2n-1)!} = \sum_{n=0}^{\infty} \frac{(-1)^{n+1}x^{2n+1}}{(2n+1)!} = -\sum_{n=0}^{\infty} \frac{(-1)^n x^{2n+1}}{(2n+1)!} = -f(x)$

(d) $f(x) = \sin x$ and $g(x) = \cos x$

40. (a) $f(x) = \sum_{n=0}^{\infty} \frac{x^n}{n!}, \, -\infty < x < \infty$ (See Exercise 11)

(b) $f'(x) = \sum_{n=1}^{\infty} \frac{nx^{n-1}}{n!} = \sum_{n=1}^{\infty} \frac{x^{n-1}}{(n-1)!} = \sum_{n=0}^{\infty} \frac{x^n}{n!} = f(x)$

(c) $f(x) = \sum_{n=1}^{\infty} \frac{x^n}{n!} = 1 + x + \frac{x^2}{2!} + \frac{x^3}{3!} + \frac{x^4}{4!} + \cdots$

$f(0) = 1$

(d) $f(x) = e^x$

41.
$$y = \sum_{n=0}^{\infty} \frac{x^{2n}}{2^n n!}$$

$$y' = \sum_{n=1}^{\infty} \frac{2nx^{2n-1}}{2^n n!}$$

$$y'' = \sum_{n=1}^{\infty} \frac{2n(2n-1)x^{2n-2}}{2^n n!}$$

$$y'' - xy' - y = \sum_{n=1}^{\infty} \frac{2n(2n-1)x^{2n-2}}{2^n n!} - \sum_{n=1}^{\infty} \frac{2nx^{2n}}{2^n n!} - \sum_{n=0}^{\infty} \frac{x^{2n}}{2^n n!} = \sum_{n=1}^{\infty} \frac{2n(2n-1)x^{2n-2}}{2^n n!} - \sum_{n=0}^{\infty} \frac{(2n+1)x^{2n}}{2^n n!}$$

$$= \sum_{n=0}^{\infty} \left[\frac{(2n+2)(2n+1)x^{2n}}{2^{n+1}(n+1)!} - \frac{(2n+1)x^{2n}}{2^n n!} \cdot \frac{2(n+1)}{2(n+1)} \right]$$

$$= \sum_{n=0}^{\infty} \frac{2(n+1)x^{2n}\left[(2n+1) - (2n+1)\right]}{2^{n+1}(n+1)!} = 0$$

42.
$$y = 1 + \sum_{n=1}^{\infty} \frac{(-1)^n x^{4n}}{2^{2n} n! \cdot 3 \cdot 7 \cdot 11 \cdots (4n-1)}$$

$$y' = \sum_{n=1}^{\infty} \frac{(-1)^n 4nx^{4n-1}}{2^{2n} n! \cdot 3 \cdot 7 \cdot 11 \cdots (4n-1)}$$

$$y'' = \sum_{n=1}^{\infty} \frac{(-1)^n 4n(4n-1)x^{4n-2}}{2^{2n} n! \cdot 3 \cdot 7 \cdot 11 \cdots (4n-1)} = -x^2 + \sum_{n=2}^{\infty} \frac{(-1)^n 4nx^{4n-2}}{2^{2n} n! \cdot 3 \cdot 7 \cdot 11 \cdots (4n-5)}$$

$$y'' + x^2 y = -x^2 + \sum_{n=2}^{\infty} \frac{(-1)^n 4nx^{4n-2}}{2^{2n} n! \cdot 3 \cdot 7 \cdot 11 \cdots (4n-5)} + \sum_{n=1}^{\infty} \frac{(-1)^n x^{4n+2}}{2^{2n} n! \cdot 3 \cdot 7 \cdot 11 \cdots (4n-1)} + x^2$$

$$= \sum_{n=1}^{\infty} \frac{(-1)^{n+1} 4(n+1)x^{4n+2}}{2^{2n+2}(n+1)! \cdot 3 \cdot 7 \cdot 11 \cdots (4n-1)} - \sum_{n=1}^{\infty} \frac{(-1)^{n+1} x^{4n+2}}{2^{2n} n! \cdot 3 \cdot 7 \cdot 11 \cdots (4n-1)} \frac{2^2(n+1)}{2^2(n+1)} = 0$$

43. $J_0(x) = \sum_{k=0}^{\infty} \frac{(-1)^k x^{2k}}{2^{2k}(k!)^2}$

(a) $\displaystyle\lim_{k \to \infty} \left| \frac{u_{k+1}}{u_k} \right| = \lim_{k \to \infty} \left| \frac{(-1)^{k+1} x^{2k+2}}{2^{2k+2}[(k+1)!]^2} \cdot \frac{2^{2k}(k!)^2}{(-1)^k x^{2k}} \right| = \lim_{k \to \infty} \left| \frac{(-1)x^2}{2^2(k+1)^2} \right| = 0$

Therefore, the interval of convergence is $-\infty < x < \infty$.

(b)
$$J_0 = \sum_{k=0}^{\infty} (-1)^k \frac{x^{2k}}{4^k(k!)^2}$$

$$J_0' = \sum_{k=1}^{\infty} (-1)^k \frac{2kx^{2k-1}}{4^k(k!)^2} = \sum_{k=0}^{\infty} (-1)^{k+1} \frac{(2k+2)x^{2k+1}}{4^{k+1}[(k+1)!]^2}$$

$$J_0'' = \sum_{k=1}^{\infty} (-1)^k \frac{2k(2k-1)x^{2k-2}}{4^k(k!)^2} = \sum_{k=0}^{\infty} (-1)^{k+1} \frac{(2k+2)(2k+1)x^{2k}}{4^{k+1}[(k+1)!]^2}$$

$$x^2 J_0'' + x J_0' + x^2 J_0 = \sum_{k=0}^{\infty} (-1)^{k+1} \frac{2(2k+1)x^{2k+2}}{4^{k+1}(k+1)!k!} + \sum_{k=0}^{\infty} (-1)^{k+1} \frac{2x^{2k+2}}{4^{k+1}(k+1)!k!} + \sum_{k=0}^{\infty} (-1)^k \frac{x^{2k+2}}{4^k(k!)^2}$$

$$= \sum_{k=0}^{\infty} \frac{(-1)^k x^{2k+2}}{4^k(k!)^2} \left[(-1)\frac{2(2k+1)}{4(k+1)} + (-1)\frac{2}{4(k+1)} + 1 \right]$$

$$= \sum_{k=0}^{\infty} \frac{(-1)^k x^{2k+2}}{4^k(k!)^2} \left[\frac{-4k-2}{4k+4} - \frac{2}{4k+4} + \frac{4k+4}{4k+4} \right] = 0$$

—CONTINUED—

43. —CONTINUED—

(c) $P_6(x) = 1 - \dfrac{x^2}{4} + \dfrac{x^4}{64} - \dfrac{x^6}{2304}$

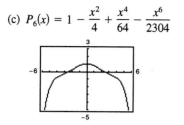

(d) $\displaystyle\int_0^1 J_0\,dx = \int_0^1 \sum_{k=0}^{\infty} \frac{(-1)^k x^{2k}}{4^k (k!)^2}\,dx$

$\displaystyle = \left[\sum_{k=0}^{\infty} \frac{(-1)^k x^{2k+1}}{4^k (k!)^2 (2k+1)} \right]_0^1$

$\displaystyle = \sum_{k=0}^{\infty} \frac{(-1)^k}{4^k (k!)^2 (2k+1)}$

$\displaystyle = 1 - \frac{1}{12} + \frac{1}{320} \approx 0.92$

(exact integral is 0.9197304101)

44. $\displaystyle J_1(x) = x \sum_{k=0}^{\infty} \frac{(-1)^k x^{2k}}{2^{2k+1} k!(k+1)!} = \sum_{k=0}^{\infty} \frac{(-1)^k x^{2k+1}}{2^{2k+1} k!(k+1)!}$

(a) $\displaystyle \lim_{k\to\infty} \left| \frac{u_{k+1}}{u_k} \right| = \lim_{k\to\infty} \left| \frac{(-1)^{k+1} x^{2k+3}}{2^{2k+3}(k+1)!(k+2)!} \cdot \frac{2^{2k+1} k!(k+1)!}{(-1)^k x^{2k+1}} \right| = \lim_{k\to\infty} \left| \frac{(-1)x^2}{2^2 (k+2)(k+1)} \right| = 0$

Therefore, the interval of convergence is $-\infty < x < \infty$.

(b) $\displaystyle J_1(x) = \sum_{k=0}^{\infty} \frac{(-1)^k x^{2k+1}}{2^{2k+1} k!(k+1!)}$

$\displaystyle J_1{}'(x) = \sum_{k=0}^{\infty} \frac{(-1)^k (2k+1)x^{2k}}{2^{2k+1} k!(k+1)!}$

$\displaystyle J_1{}''(x) = \sum_{k=1}^{\infty} \frac{(-1)^k (2k+1)(2k)x^{2k-1}}{2^{2k+1} k!(k+1)!}$

$\displaystyle x^2 J_1{}'' + x J_1{}' + (x^2 - 1)J_1 = \sum_{k=1}^{\infty} \frac{(-1)^k (2k+1)(2k)x^{2k+1}}{2^{2k+1} k!(k+1)!} + \sum_{k=0}^{\infty} \frac{(-1)^k(2k+1)x^{2k+1}}{2^{2k+1} k!(k+1)!}$

$\displaystyle \qquad + \sum_{k=0}^{\infty} \frac{(-1)^k x^{2k+3}}{2^{2k+1} k!(k+1)!} - \sum_{k=0}^{\infty} \frac{(-1)^k x^{2k+1}}{2^{2k+1} k!(k+1)!}$

$\displaystyle = \left[\sum_{k=1}^{\infty} \frac{(-1)^k(2k+1)(2k)x^{2k+1}}{2^{2k+1} k!(k+1)!} + \frac{x}{2} + \sum_{k=1}^{\infty} \frac{(-1)^k(2k+1)x^{2k+1}}{2^{2k+1} k!(k+1)!} \right.$

$\displaystyle \qquad \left. - \frac{x}{2} - \sum_{k=1}^{\infty} \frac{(-1)^k x^{2k+1}}{2^{2k+1} k!(k+1)!} \right] + \sum_{k=0}^{\infty} \frac{(-1)^k x^{2k+3}}{2^{2k+1} k!(k+1)!}$

$\displaystyle = \sum_{k=1}^{\infty} \frac{(-1)^k x^{2k+1}[(2k+1)(2k) + (2k+1) - 1]}{2^{2k+1} k!(k+1)!} + \sum_{k=0}^{\infty} \frac{(-1)^k x^{2k+3}}{2^{2k+1} k!(k+1)!}$

$\displaystyle = \sum_{k=1}^{\infty} \frac{(-1)^k x^{2k+1} 4k(k+1)}{2^{2k+1} k!(k+1)!} + \sum_{k=0}^{\infty} \frac{(-1)^k x^{2k+3}}{2^{2k+1} k!(k+1)!}$

$\displaystyle = \sum_{k=1}^{\infty} \frac{(-1)^k x^{2k+1}}{2^{2k-1}(k-1)!k!} + \sum_{k=0}^{\infty} \frac{(-1)^k x^{2k+3}}{2^{2k+1} k!(k+1)!}$

$\displaystyle = \sum_{k=0}^{\infty} \frac{(-1)^{k+1} x^{2k+3}}{2^{2k+1} k!(k+1)!} + \sum_{k=0}^{\infty} \frac{(-1)^k x^{2k+3}}{2^{2k+1} k!(k+1)!}$

$\displaystyle = \sum_{k=0}^{\infty} \frac{(-1)^k x^{2k+3}[(-1) + 1]}{2^{2k+1} k!(k+1)!} = 0$

(c) $P_7(x) = \dfrac{x}{2} - \dfrac{1}{16}x^3 + \dfrac{1}{384}x^5 - \dfrac{1}{18,432}x^7$

(d) $\displaystyle J_0{}'(x) = \sum_{k=0}^{\infty} \frac{(-1)^{k+1} 2(k+1)x^{2k+1}}{2^{2k+2}(k+1)!(k+1)!} = \sum_{k=0}^{\infty} \frac{(-1)^{k+1} x^{2k+1}}{2^{2k+1} k!(k+1)!}$

$\displaystyle -J_1(x) = -\sum_{k=0}^{\infty} \frac{(-1)^k x^{2k+1}}{2^{2k+1} k!(k+1)!} = \sum_{k=0}^{\infty} \frac{(-1)^{k+1} x^{2k+1}}{2^{2k+1} k!(k+1)!}$

Note: $J_0{}'(x) = -J_1(x)$

45. $f(x) = \displaystyle\sum_{n=0}^{\infty} (-1)^n \frac{x^{2n}}{(2n)!} = \cos x$

(See Exercise 39.)

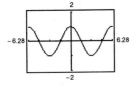
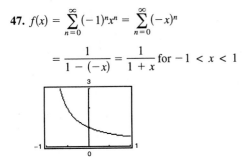

46. $f(x) = \displaystyle\sum_{n=0}^{\infty} (-1)^n \frac{x^{2n+1}}{(2n+1)!} = \sin x$

(See Exercise 39.)

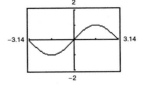

47. $f(x) = \displaystyle\sum_{n=0}^{\infty} (-1)^n x^n = \sum_{n=0}^{\infty} (-x)^n$

$$= \frac{1}{1-(-x)} = \frac{1}{1+x} \text{ for } -1 < x < 1$$

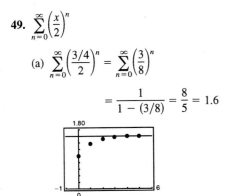

48. $f(x) = \displaystyle\sum_{n=0}^{\infty} (-1)^n \frac{x^{2n+1}}{2n+1} = \arctan x, \ -1 \leq x \leq 1$

(See Exercise 28 in Section 8.7.)

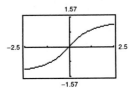

49. $\displaystyle\sum_{n=0}^{\infty} \left(\frac{x}{2}\right)^n$

(a) $\displaystyle\sum_{n=0}^{\infty} \left(\frac{3/4}{2}\right)^n = \sum_{n=0}^{\infty} \left(\frac{3}{8}\right)^n$

$$= \frac{1}{1-(3/8)} = \frac{8}{5} = 1.6$$

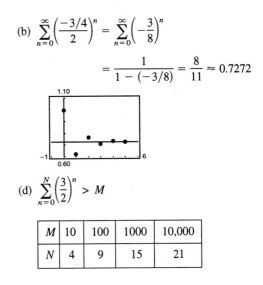

(b) $\displaystyle\sum_{n=0}^{\infty} \left(\frac{-3/4}{2}\right)^n = \sum_{n=0}^{\infty} \left(-\frac{3}{8}\right)^n$

$$= \frac{1}{1-(-3/8)} = \frac{8}{11} \approx 0.7272$$

(c) The alternating series converges more rapidly. The partial sums of the series of positive terms approach the sum from below. The partial sums of the alternating series alternate sides of the horizontal line representing the sum.

(d) $\displaystyle\sum_{n=0}^{N} \left(\frac{3}{2}\right)^n > M$

M	10	100	1000	10,000
N	4	9	15	21

50. $\displaystyle\sum_{n=0}^{\infty} \frac{x^{2n+1}}{(2n+1)!} = \sum_{n=1}^{\infty} \frac{x^{2n-1}}{(2n-1)!}$

Replace n with $n-1$.

51. False;

$$\sum_{n=0}^{\infty} \frac{(-1)^n x^n}{n2^n}$$

converges for $x = 2$ but diverges for $x = -2$.

52. True; if

$$\sum_{n=0}^{\infty} a_n x^n$$

converges for $x = 2$, then we know that it must converge on $(-2, 2]$.

53. True; the radius of convergence is $R = 1$ for both series.

54. True

$$\int_0^1 f(x)\, dx = \int_0^1 \left(\sum_{n=0}^{\infty} a_n x^n \right) dx = \left[\sum_{n=0}^{\infty} \frac{a_n x^{n+1}}{n+1} \right]_0^1$$

$$= \sum_{n=0}^{\infty} \frac{a_n}{n+1}$$

Section 8.9   Representation of Functions by Power Series

1. (a) $\dfrac{1}{2 - x} = \dfrac{1/2}{1 - (x/2)} = \dfrac{a}{1 - r}$

$$= \sum_{n=0}^{\infty} \frac{1}{2}\left(\frac{x}{2}\right)^n = \sum_{n=0}^{\infty} \frac{x^n}{2^{n+1}}$$

This series converges on $(-2, 2)$.

(b)
$$
\begin{array}{r}
\frac{1}{2} + \frac{x}{4} + \frac{x^2}{8} + \frac{x^3}{16} + \cdots \\
2 - x \overline{)\, 1 } \\
\underline{1 - \frac{x}{2}} \\
\frac{x}{2} \\
\underline{\frac{x}{2} - \frac{x^2}{4}} \\
\frac{x^2}{4} \\
\underline{\frac{x^2}{4} - \frac{x^3}{8}} \\
\frac{x^3}{8} \\
\underline{\frac{x^3}{8} - \frac{x^4}{16}} \\
\vdots
\end{array}
$$

2. (a) $\dfrac{3}{4 - x} = \dfrac{3/4}{1 - (x/4)} = \dfrac{a}{1 - r}$

$$= \sum_{n=0}^{\infty} \frac{3}{4}\left(\frac{x}{4}\right)^n = \sum_{n=0}^{\infty} \frac{3x^n}{4^{n+1}}$$

This series converges on $(-4, 4)$.

(b)
$$
\begin{array}{r}
\frac{3}{4} + \frac{3x}{16} + \frac{3x^2}{64} + \frac{3x^3}{256} + \cdots \\
4 - x \overline{)\, 3 } \\
\underline{3 - \frac{3x}{4}} \\
\frac{3x}{4} \\
\underline{\frac{3x}{4} - \frac{3x^2}{16}} \\
\frac{3x^2}{16} \\
\underline{\frac{3x^2}{16} - \frac{3x^3}{64}} \\
\frac{3x^2}{64} \\
\underline{\frac{3x^3}{64} - \frac{3x^4}{256}} \\
\vdots
\end{array}
$$

3. (a) $\dfrac{1}{2 + x} = \dfrac{1/2}{1 - (-x/2)} = \dfrac{a}{1 - r}$

$$= \sum_{n=0}^{\infty} \frac{1}{2}\left(-\frac{x}{2}\right)^n = \sum_{n=0}^{\infty} \frac{(-1)^n x^n}{2^{n+1}}$$

This series converges on $(-2, 2)$.

(b)
$$
\begin{array}{r}
\frac{1}{2} - \frac{x}{4} + \frac{x^2}{8} - \frac{x^3}{16} + \cdots \\
2 + x \overline{)\, 1 } \\
\underline{1 + \frac{x}{2}} \\
-\frac{x}{2} \\
\underline{-\frac{x}{2} - \frac{x^2}{4}} \\
\frac{x^2}{4} \\
\underline{\frac{x^2}{4} + \frac{x^3}{8}} \\
-\frac{x^3}{8} \\
\underline{-\frac{x^3}{8} - \frac{x^4}{16}} \\
\vdots
\end{array}
$$

4. (a) $\dfrac{1}{1 + x} = \dfrac{1}{1 - (-x)} = \dfrac{a}{1 - r}$

$$= \sum_{n=0}^{\infty} (-x)^n = \sum_{n=0}^{\infty} (-1)^n x^n$$

This series converges on $(-1, 1)$.

(b)
$$
\begin{array}{r}
1 - x + x^2 - x^3 + \cdots \\
1 + x \overline{)\, 1 } \\
\underline{1 + x} \\
-x \\
\underline{-x - x^2} \\
x^2 \\
\underline{x^2 + x^3} \\
-x^3 \\
\underline{-x^3 - x^4} \\
\vdots
\end{array}
$$

5. Writing $f(x)$ in the form $a/(1 - r)$, we have

$$\frac{1}{2 - x} = \frac{1}{-3 - (x - 5)} = \frac{-1/3}{1 + (1/3)(x - 5)}$$

which implies that $a = -1/3$ and $r = (-1/3)(x - 5)$.

Therefore, the power series for $f(x)$ is given by

$$\frac{1}{2 - x} = \sum_{n=0}^{\infty} ar^n = \sum_{n=0}^{\infty} -\frac{1}{3}\left[-\frac{1}{3}(x - 5)\right]^n$$

$$= \sum_{n=0}^{\infty} \frac{(x - 5)^n}{(-3)^{n+1}}, |x - 5| < 3 \text{ or } 2 < x < 8.$$

6. Writing $f(x)$ in the form $a/(1 - r)$, we have

$$\frac{3}{4 - x} = \frac{3}{6 - (x + 2)} = \frac{1/2}{1 - (1/6)(x + 2)} = \frac{a}{1 - r}$$

which implies that $a = 1/2$ and $r = (1/6)(x + 2)$.

Therefore, the power series for $f(x)$ is given by

$$\frac{3}{4 - x} = \sum_{n=0}^{\infty} ar^n = \sum_{n=0}^{\infty} \frac{1}{2}\left[\frac{1}{6}(x + 2)\right]^n$$

$$= \sum_{n=0}^{\infty} \frac{(x + 2)^n}{2 \cdot 6^n}, |x + 2| < 6 \text{ or } -8 < x < 4.$$

7. Writing $f(x)$ in the form $a/(1 - r)$, we have

$$\frac{3}{2x - 1} = \frac{-3}{1 - 2x} = \frac{a}{1 - r}$$

which implies that $a = -3$ and $r = 2x$.

Therefore, the power series for $f(x)$ is given by

$$\frac{3}{2x - 1} = \sum_{n=0}^{\infty} ar^n = \sum_{n=0}^{\infty} (-3)(2x)^n$$

$$= -3 \sum_{n=0}^{\infty} (2x)^n, |2x| < 1 \text{ or } -\frac{1}{2} < x < \frac{1}{2}.$$

8. Writing $f(x)$ in the form $a/(1 - r)$, we have

$$\frac{3}{2x - 1} = \frac{3}{3 + 2(x - 2)} = \frac{1}{1 + (2/3)(x - 2)} = \frac{a}{1 - r}$$

which implies that $a = 1$ and $r = (-2/3)(x - 2)$.
Therefore, the power series for $f(x)$ is given by

$$\frac{3}{2x - 1} = \sum_{n=0}^{\infty} ar^n = \sum_{n=0}^{\infty}\left[-\frac{2}{3}(x - 2)\right]^n,$$

$$= \sum_{n=0}^{\infty} \frac{(-2)^n(x - 2)^n}{3^n},$$

$$|x - 2| < \frac{3}{2} \text{ or } \frac{1}{2} < x < \frac{7}{2}.$$

9. Writing $f(x)$ in the form $a/(1 - r)$, we have

$$\frac{1}{2x - 5} = \frac{-1}{11 - 2(x + 3)}$$

$$= \frac{-1/11}{1 - (2/11)(x + 3)} = \frac{a}{1 - r}$$

which implies that $a = -1/11$ and $r = (2/11)(x + 3)$.
Therefore, the power series for $f(x)$ is given by

$$\frac{1}{2x - 5} = \sum_{n=0}^{\infty} ar^n = \sum_{n=0}^{\infty}\left(-\frac{1}{11}\right)\left[\frac{2}{11}(x + 3)\right]^n$$

$$= -\sum_{n=0}^{\infty} \frac{2^n(x + 3)^n}{11^{n+1}},$$

$$|x + 3| < \frac{11}{2} \text{ or } -\frac{17}{2} < x < \frac{5}{2}.$$

10. Writing $f(x)$ in the form $a/(1 - r)$, we have

$$\frac{1}{2x - 5} = \frac{1}{-5 + 2x} = \frac{-1/5}{1 - (2/5)x} = \frac{a}{1 - r}$$

which implies that $a = -1/5$ and $r = (2/5)x$. Therefore, the power series for $f(x)$ is given by

$$\frac{1}{2x - 5} = \sum_{n=0}^{\infty} ar^n = \sum_{n=0}^{\infty}\left(-\frac{1}{5}\right)\left(\frac{2}{5}x\right)^n = -\sum_{n=0}^{\infty} \frac{2^n x^n}{5^{n+1}},$$

$$|x| < \frac{5}{2} \text{ or } -\frac{5}{2} < x < \frac{5}{2}.$$

11. Writing $f(x)$ in the form $a/(1 - r)$, we have

$$\frac{3}{x + 2} = \frac{3}{2 + x} = \frac{3/2}{1 + (1/2)x} = \frac{a}{1 - r}$$

which implies that $a = 3/2$ and $r = (-1/2)x$. Therefore, the power series for $f(x)$ is given by

$$\frac{3}{x + 2} = \sum_{n=0}^{\infty} ar^n = \sum_{n=0}^{\infty} \frac{3}{2}\left(-\frac{1}{2}x\right)^n$$

$$= 3 \sum_{n=0}^{\infty} \frac{(-1)^n x^n}{2^{n+1}} = \frac{3}{2} \sum_{n=0}^{\infty} \left(-\frac{x}{2}\right)^n,$$

$$|x| < 2 \text{ or } -2 < x < 2.$$

12. Writing $f(x)$ in the form $a/(1 - r)$, we have

$$\frac{4}{3x + 2} = \frac{4}{8 + 3(x - 2)} = \frac{1/2}{1 + (3/8)(x - 2)} = \frac{a}{1 - r}$$

which implies that $a = 1/2$ and $r = (-3/8)(x - 2)$.
Therefore, the power series for $f(x)$ is given by

$$\frac{4}{3x + 2} = \sum_{n=0}^{\infty} ar^n = \sum_{n=0}^{\infty} \frac{1}{2}\left[-\frac{3}{8}(x - 2)\right]^n$$

$$= \frac{1}{2} \sum_{n=0}^{\infty} \frac{(-3)^n(x - 2)^n}{8^n},$$

$$|x - 2| < \frac{8}{3} \text{ or } -\frac{2}{3} < x < \frac{14}{3}.$$

13. $\dfrac{3x}{x^2 + x - 2} = \dfrac{2}{x + 2} + \dfrac{1}{x - 1} = \dfrac{2}{2 + x} + \dfrac{1}{-1 + x} = \dfrac{1}{1 + (1/2)x} + \dfrac{-1}{1 - x}$

Writing $f(x)$ as a sum of two geometric series, we have

$$\frac{3x}{x^2 + x - 2} = \sum_{n=0}^{\infty}\left(-\frac{1}{2}x\right)^n + \sum_{n=0}^{\infty}(-1)(x)^n = \sum_{n=0}^{\infty}\left[\frac{1}{(-2)^n} - 1\right]x^n.$$

The interval of convergence is $-1 < x < 1$ since

$$\lim_{n\to\infty}\left|\frac{u_{n+1}}{u_n}\right| = \lim_{n\to\infty}\left|\frac{(1 - (-2)^{n+1})x^{n+1}}{(-2)^{n+1}} \cdot \frac{(-2)^n}{(1 - (-2)^n)x^n}\right| = \lim_{n\to\infty}\left|\frac{(1 - (-2)^{n+1})x}{-2 - (-2)^{n+1}}\right| = |x|.$$

14. $\dfrac{4x - 7}{2x^2 + 3x - 2} = \dfrac{3}{x + 2} - \dfrac{2}{2x - 1} = \dfrac{3}{2 + x} - \dfrac{2}{-1 + 2x} = \dfrac{3/2}{1 + (1/2)x} + \dfrac{2}{1 - 2x}$

Writing $f(x)$ as a sum of two geometric series, we have

$$\frac{4x - 7}{2x^2 + 3x - 2} = \sum_{n=0}^{\infty}\left(\frac{3}{2}\right)\left(-\frac{1}{2}x\right)^n + \sum_{n=0}^{\infty}2(2x)^n = \sum_{n=0}^{\infty}\left[\frac{3(-1)^n}{2^{n+1}} + 2^{n+1}\right]x^n, \ |x| < \frac{1}{2} \text{ or } -\frac{1}{2} < x < \frac{1}{2}.$$

15. $\dfrac{2}{1 - x^2} = \dfrac{1}{1 - x} + \dfrac{1}{1 + x}$

Writing $f(x)$ as a sum of two geometric series, we have

$$\frac{2}{1 - x^2} = \sum_{n=0}^{\infty}x^n + \sum_{n=0}^{\infty}(-x)^n = \sum_{n=0}^{\infty}(1 + (-1)^n)x^n = \sum_{n=0}^{\infty}2x^{2n}.$$

The interval of convergence is $|x^2| < 1$ or $-1 < x < 1$ since $\displaystyle\lim_{n\to\infty}\left|\frac{u_{n+1}}{u_n}\right| = \lim_{n\to\infty}\left|\frac{2x^{2n+2}}{2x^2}\right| = |x^2|.$

16. First finding the power series for $4/(4 + x)$, we have

$$\frac{1}{1 + (1/4)x} = \sum_{n=0}^{\infty}\left(-\frac{1}{4}x\right)^n = \sum_{n=0}^{\infty}\frac{(-1)^n x^n}{4^n}$$

Now replace x with x^2.

$$\frac{4}{4 + x^2} = \sum_{n=0}^{\infty}\frac{(-1)^n x^{2n}}{4^n}.$$

The interval of convergence is $|x^2| < 4$ or $-2 < x < 2$ since

$$\lim_{n\to\infty}\left|\frac{u_{n+1}}{u_n}\right| = \lim_{n\to\infty}\left|\frac{(-1)^{n+1}x^{2n+2}}{4^{n+1}} \cdot \frac{4^n}{(-1)^n x^{2n}}\right| = \left|-\frac{x^2}{4}\right| = \frac{|x^2|}{4}.$$

17. $\dfrac{1}{1 + x} = \displaystyle\sum_{n=0}^{\infty}(-1)^n x^n$

$\dfrac{1}{1 - x} = \displaystyle\sum_{n=0}^{\infty}(-1)^n(-x)^n = \sum_{n=0}^{\infty}(-1)^{2n}x^n = \sum_{n=0}^{\infty}x^n$

$h(x) = \dfrac{-2}{x^2 - 1} = \dfrac{1}{1 + x} + \dfrac{1}{1 - x} = \displaystyle\sum_{n=0}^{\infty}(-1)^n x^n + \sum_{n=0}^{\infty}x^n = \sum_{n=0}^{\infty}[(-1)^n + 1]x^n$

$= 2 + 0x + 2x^2 + 0x^3 + 2x^4 + 0x^5 + 2x^6 + \cdots = \displaystyle\sum_{n=0}^{\infty}2x^{2n}, \ -1 < x < 1$ (See Exercise 15.)

18. $\dfrac{1}{1 + x} = \displaystyle\sum_{n=0}^{\infty} (-1)^n x^n$

$\dfrac{1}{1 - x} = \displaystyle\sum_{n=0}^{\infty} x^n$

$h(x) = \dfrac{2x}{x^2 - 1} = \dfrac{1}{1 + x} - \dfrac{1}{1 - x} = \displaystyle\sum_{n=0}^{\infty} (-1)^n x^n - \sum_{n=0}^{\infty} x^n = \sum_{n=0}^{\infty} [(-1)^n - 1] x^n$

$= 0 - 2x + 0x^2 - 2x^3 + 0x^4 - 2x^5 + \cdots = \displaystyle\sum_{n=0}^{\infty} -2x^{2n+1}, \; -1 < x < 1$

19. By taking the first derivative, we have $\dfrac{d}{dx}\left[\dfrac{1}{x + 1}\right] = \dfrac{-1}{(x + 1)^2}.$ Therefore,

$\dfrac{-1}{(x + 1)^2} = \dfrac{d}{dx}\left[\displaystyle\sum_{n=0}^{\infty} (-1)^n x^n\right] = \sum_{n=1}^{\infty} (-1)^n n x^{n-1}$

$= \displaystyle\sum_{n=0}^{\infty} (-1)^{n+1}(n + 1)x^n, \; -1 < x < 1.$

20. By taking the second derivative, we have $\dfrac{d^2}{dx^2}\left[\dfrac{1}{x + 1}\right] = \dfrac{2}{(x + 1)^3}.$ Therefore,

$\dfrac{2}{(x + 1)^3} = \dfrac{d^2}{dx^2}\left[\displaystyle\sum_{n=0}^{\infty} (-1)^n x^n\right]$

$= \dfrac{d}{dx}\left[\displaystyle\sum_{n=1}^{\infty} (-1)^n n x^{n-1}\right] = \sum_{n=2}^{\infty} (-1)^n n(n - 1)x^{n-2} = \sum_{n=0}^{\infty} (-1)^n (n + 2)(n + 1)x^n, \; -1 < x < 1.$

21. By integrating, we have $\displaystyle\int \dfrac{1}{x + 1}\, dx = \ln(x + 1).$ Therefore,

$\ln(x + 1) = \displaystyle\int \left[\sum_{n=0}^{\infty} (-1)^n x^n\right] dx = C + \sum_{n=0}^{\infty} \dfrac{(-1)^n x^{n+1}}{n + 1}, \; -1 < x \le 1.$

To solve for C, let $x = 0$ and conclude that $C = 0$. Therefore,

$\ln(x + 1) = \displaystyle\sum_{n=0}^{\infty} \dfrac{(-1)^n x^{n+1}}{n + 1}, \; -1 < x \le 1.$

22. By integrating, we have

$\displaystyle\int \dfrac{1}{1 + x}\, dx = \ln(1 + x) + C_1 \text{ and } \int \dfrac{1}{1 - x}\, dx = -\ln(1 - x) + C_2.$

$f(x) = \ln(1 - x^2) = \ln(1 + x) - [-\ln(1 - x)].$ Therefore,

$\ln(1 - x^2) = \displaystyle\int \dfrac{1}{1 + x}\, dx - \int \dfrac{1}{1 - x}\, dx$

$= \displaystyle\int \left[\sum_{n=0}^{\infty} (-1)^n x^n\right] dx - \int \left[\sum_{n=0}^{\infty} x^n\right] dx = \left[C_1 + \sum_{n=0}^{\infty} \dfrac{(-1)^n x^{n+1}}{n + 1}\right] - \left[C_2 + \sum_{n=0}^{\infty} \dfrac{x^{n+1}}{n + 1}\right]$

$= C + \displaystyle\sum_{n=0}^{\infty} \dfrac{[(-1)^n - 1]x^{n+1}}{n + 1} = C + \sum_{n=0}^{\infty} \dfrac{-2x^{2n+2}}{2n + 2} = C + \sum_{n=0}^{\infty} \dfrac{(-1)x^{2n+2}}{n + 1}$

To solve for C, let $x = 0$ and conclude that $C = 0$. Therefore,

$\ln(1 - x^2) = -\displaystyle\sum_{n=0}^{\infty} \dfrac{x^{2n+2}}{n + 1}, \; -1 < x < 1$

23. $\dfrac{1}{x^2+1} = \sum_{n=0}^{\infty} (-1)^n (x^2)^n = \sum_{n=0}^{\infty} (-1)^n x^{2n}, \ -1 < x < 1$

24. $\dfrac{2x}{x^2+1} = 2x \sum_{n=0}^{\infty} (-1)^n x^{2n}$ (See Exercise 23.)

$$= \sum_{n=0}^{\infty} (-1)^n 2x^{2n+1}$$

Since $\dfrac{d}{dx}\left(\ln(x^2+1)\right) = \dfrac{2x}{x^2+1}$, we have

$$\ln(x^2+1) = \int\left[\sum_{n=0}^{\infty} (-1)^n 2x^{2n+1}\right] dx = C + \sum_{n=0}^{\infty} \frac{(-1)^n x^{2n+2}}{n+1}, \ -1 \le x \le 1.$$

To solve for C, let $x = 0$ and conclude that $C = 0$. Therefore,

$$\ln(x^2+1) = \sum_{n=0}^{\infty} \frac{(-1)^n x^{2n+2}}{n+1}, \ -1 \le x \le 1.$$

25. Since, $\dfrac{1}{x+1} = \sum_{n=0}^{\infty} (-1)^n x^n$, we have $\dfrac{1}{4x^2+1} = \sum_{n=0}^{\infty} (-1)^n (4x^2)^n = \sum_{n=0}^{\infty} (-1)^n 4^n x^{2n} = \sum_{n=0}^{\infty} (-1)^n (2x)^{2n}, \ -\dfrac{1}{2} < x < \dfrac{1}{2}.$

26. Since $\displaystyle\int \dfrac{1}{4x^2+1}\, dx = \dfrac{1}{2}\arctan(2x)$, we can use the result of Exercise 25 to obtain

$$\arctan(2x) = 2\int \frac{1}{4x^2+1}\, dx = 2\int\left[\sum_{n=0}^{\infty} (-1)^n 4^n x^{2n}\right] dx = C + 2\sum_{n=0}^{\infty} \frac{(-1)^n 4^n x^{2n+1}}{2n+1}, \ -\frac{1}{2} < x \le \frac{1}{2}.$$

To solve for C, let $x = 0$ and conclude that $C = 0$. Therefore,

$$\arctan(2x) = 2\sum_{n=0}^{\infty} \frac{(-1)^n 4^n x^{2n+1}}{2n+1}, \ -\frac{1}{2} < x \le \frac{1}{2}.$$

27. $x - \dfrac{x^2}{2} \le \ln(x+1) \le x - \dfrac{x^2}{2} + \dfrac{x^3}{3}$

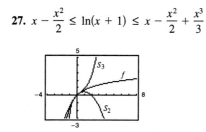

x	0.0	0.2	0.4	0.6	0.8	1.0
$x - \dfrac{x^2}{2}$	0.000	0.180	0.320	0.420	0.480	0.500
$\ln(x+1)$	0.000	0.180	0.336	0.470	0.588	0.693
$x - \dfrac{x^2}{2} + \dfrac{x^3}{3}$	0.000	0.183	0.341	0.492	0.651	0.833

28. $x - \dfrac{x^2}{2} + \dfrac{x^3}{3} - \dfrac{x^4}{4} \le \ln(x+1)$

$$\le x - \frac{x^2}{2} + \frac{x^3}{3} - \frac{x^4}{4} + \frac{x^5}{5}$$

x	0.0	0.2	0.4	0.6	0.8	1.0
$x - \dfrac{x^2}{2} + \dfrac{x^3}{3} - \dfrac{x^4}{4}$	0.0	0.18227	0.33493	0.45960	0.54827	0.58333
$\ln(x+1)$	0.0	0.18232	0.33647	0.47000	0.58779	0.69315
$x - \dfrac{x^2}{2} + \dfrac{x^3}{3} - \dfrac{x^4}{4} + \dfrac{x^5}{5}$	0.0	0.18233	0.33698	0.47515	0.61380	0.78333

In Exercise 35-38, $\arctan x = \displaystyle\sum_{n=0}^{\infty} (-1)^n \dfrac{x^{2n+1}}{2n+1}$.

29. $g(x) = x$, line, Matches (c)

30. $g(x) = x - \dfrac{x^3}{3}$, cubic with 3 zeros.

Matches (d)

31. $g(x) = x - \dfrac{x^3}{3} + \dfrac{x^5}{5}$, Matches (a)

32. $g(x) = x - \dfrac{x^3}{3} + \dfrac{x^5}{5} - \dfrac{x^7}{7}$,

Matches (b)

33. $f(x) = \arctan x$ is an odd function (symmetric to the origin)

34. The approximations of degree 3, 7, 11, . . . $(4n - 1, n = 1, 2, . . .)$ have relative extrema.

In Exercises 35-38, $\arctan x = \displaystyle\sum_{n=0}^{\infty} (-1)^n \dfrac{x^{2n+1}}{2n+1}.$

35. $\arctan \dfrac{1}{4} = \displaystyle\sum_{n=0}^{\infty} (-1)^n \dfrac{(1/4)^{2n+1}}{2n+1} = \displaystyle\sum_{n=0}^{\infty} \dfrac{(-1)^n}{(2n+1)4^{2n+1}} = \dfrac{1}{4} - \dfrac{1}{192} + \dfrac{1}{5120} + \cdots$

Since $\dfrac{1}{5120} < 0.001$, we can approximate the series by its first two terms: $\arctan \dfrac{1}{4} \approx \dfrac{1}{4} - \dfrac{1}{192} \approx 0.245.$

36.
$$\arctan x^2 = \sum_{n=0}^{\infty} (-1)^n \frac{x^{4n+2}}{2n+1}$$

$$\int \arctan x^2 \, dx = \sum_{n=0}^{\infty} (-1)^n \frac{x^{4n+3}}{(4n+3)(2n+1)} + C, C = 0$$

$$\int_0^{3/4} \arctan x^2 \, dx = \sum_{n=0}^{\infty} (-1)^n \frac{(3/4)^{4n+3}}{(4n+3)(2n+1)}$$

$$= \sum_{n=0}^{\infty} (-1)^n \frac{3^{4n+3}}{(4n+3)(2n+1)4^{4n+3}}$$

$$= \frac{27}{192} - \frac{2187}{344,064} + \frac{177,147}{230,686,720}$$

Since $177,147/230,686,720 < 0.001$, we can approximate the series by its first two terms: 0.13427

37.
$$\frac{\arctan x^2}{x} = \sum_{n=0}^{\infty} (-1)^n \frac{x^{4n+1}}{2n+1}$$

$$\int \frac{\arctan x^2}{x} \, dx = \sum_{n=0}^{\infty} (-1)^n \frac{x^{4n+2}}{(4n+2)(2n+1)}$$

$$\int_0^{1/2} \frac{\arctan x^2}{x} \, dx = \sum_{n=0}^{\infty} (-1)^n \frac{1}{(4n+2)(2n+1)2^{4n+2}} = \frac{1}{8} - \frac{1}{1152} + \cdots$$

Since $\dfrac{1}{1152} < 0.001$, we can approximate the series by its first term: $\displaystyle\int_0^{1/2} \frac{\arctan x^2}{x} \, dx \approx 0.125$

38.
$$x^2 \arctan x = \sum_{n=0}^{\infty} (-1)^n \frac{x^{2n+3}}{2n+1}$$

$$\int x^2 \arctan x \, dx = \sum_{n=0}^{\infty} (-1)^n \frac{x^{2n+4}}{(2n+4)(2n+1)}$$

$$\int_0^{1/2} x^2 \arctan x \, dx = \sum_{n=0}^{\infty} (-1)^n \frac{1}{(2n+4)(2n+1)2^{2n+4}} = \frac{1}{64} - \frac{1}{1152} + \cdots$$

Since $\dfrac{1}{1152} < 0.001$, we can approximate the series by its first term: $\displaystyle\int_0^{1/2} x^2 \arctan x \, dx \approx 0.015625.$

In Exercises 39-42, $\dfrac{1}{1-x} = \displaystyle\sum_{n=0}^{\infty} x^n.$

39. $\dfrac{1}{(1-x)^2} = \dfrac{d}{dx}\left[\dfrac{1}{1-x}\right] = \dfrac{d}{dx}\left[\displaystyle\sum_{n=0}^{\infty} x^n\right]$

$\qquad = \displaystyle\sum_{n=1}^{\infty} nx^{n-1}, \; -1 < x < 1$

40. Replace n with $n+1$.

$$\sum_{n=1}^{\infty} nx^{n-1} = \sum_{n=0}^{\infty} (n+1)x^n$$

41. $P(n) = \left(\dfrac{1}{2}\right)^n$

$\quad E(n) = \displaystyle\sum_{n=1}^{\infty} nP(n) = \sum_{n=1}^{\infty} n\left(\dfrac{1}{2}\right)^n = \dfrac{1}{2}\sum_{n=1}^{\infty} n\left(\dfrac{1}{2}\right)^{n-1}$

$\qquad\quad = \dfrac{1}{2}\dfrac{1}{[1-(1/2)]^2} = 2 \quad \text{(Exercise 39)}$

Since the probability of obtaining a head on a single toss is $\frac{1}{2}$, it is expected that, on average, a head will be obtained in two tosses.

42. (a) $\dfrac{1}{3}\displaystyle\sum_{n=1}^{\infty} n\left(\dfrac{2}{3}\right)^n = \dfrac{2}{9}\sum_{n=1}^{\infty} n\left(\dfrac{2}{3}\right)^{n-1} = \dfrac{2}{9}\dfrac{1}{[1-(2/3)]^2} = 2$

$\quad$ (b) $\dfrac{1}{10}\displaystyle\sum_{n=1}^{\infty} n\left(\dfrac{9}{10}\right)^n = \dfrac{9}{100}\sum_{n=1}^{\infty} n\left(\dfrac{9}{10}\right)^{n-1}$

$\qquad\qquad = \dfrac{9}{100}\cdot\dfrac{1}{[1-(9/10)]^2} = 9$

43. Let $\arctan x + \arctan y = \theta.$ Then,

$$\tan(\arctan x + \arctan y) = \tan\theta$$

$$\dfrac{\tan(\arctan x) + \tan(\arctan y)}{1 - \tan(\arctan x)\tan(\arctan y)} = \tan\theta$$

$$\dfrac{x+y}{1-xy} = \tan\theta$$

$$\arctan\left(\dfrac{x+y}{1-xy}\right) = \theta.$$

Therefore, $\arctan x + \arctan y = \arctan\left(\dfrac{x+y}{1-xy}\right)$ for $xy \neq 1.$

44. (a) From Exercise 43, we have

$$\arctan\dfrac{120}{119} - \arctan\dfrac{1}{239} = \arctan\dfrac{120}{119} + \arctan\left(-\dfrac{1}{239}\right)$$

$$= \arctan\left[\dfrac{(120/119)+(-1/239)}{1-(120/119)(-1/239)}\right] = \arctan\left(\dfrac{28,561}{28,561}\right) = \arctan 1 = \dfrac{\pi}{4}$$

$\quad$ (b) $2\arctan\dfrac{1}{5} = \arctan\dfrac{1}{5} + \arctan\dfrac{1}{5} = \arctan\left[\dfrac{2(1/5)}{1-(1/5)^2}\right] = \arctan\dfrac{10}{24} = \arctan\dfrac{5}{12}$

$\qquad 4\arctan\dfrac{1}{5} = 2\arctan\dfrac{1}{5} + 2\arctan\dfrac{1}{5} = \arctan\dfrac{5}{12} + \arctan\dfrac{5}{12} = \arctan\left[\dfrac{2(5/12)}{1-(5/12)^2}\right] = \arctan\dfrac{120}{119}$

$\qquad 4\arctan\dfrac{1}{5} - \arctan\dfrac{1}{239} = \arctan\dfrac{120}{119} - \arctan\dfrac{1}{239} = \dfrac{\pi}{4} \text{ (see part (a).)}$

45. (a) $2\arctan\dfrac{1}{2} = \arctan\dfrac{1}{2} + \arctan\dfrac{1}{2} = \arctan\left[\dfrac{2(1/2)}{1-(1/2)^2}\right] = \arctan\dfrac{4}{3}$

$\qquad 2\arctan\dfrac{1}{2} - \arctan\dfrac{1}{7} = \arctan\dfrac{4}{3} + \arctan\left(-\dfrac{1}{7}\right) = \arctan\left[\dfrac{(4/3)-(1/7)}{1+(4/3)(1/7)}\right] = \arctan\dfrac{25}{25} = \arctan 1 = \dfrac{\pi}{4}$

$\quad$ (b) $\pi = 8\arctan\dfrac{1}{2} - 4\arctan\dfrac{1}{7} \approx 8\left[\dfrac{1}{2} - \dfrac{(0.5)^3}{3} + \dfrac{(0.5)^5}{5} - \dfrac{(0.5)^7}{7}\right] - 4\left[\dfrac{1}{7} - \dfrac{(1/7)^3}{3} + \dfrac{(1/7)^5}{5} - \dfrac{(1/7)^7}{7}\right] \approx 3.14$

46. (a) $2 \arctan \frac{2}{3} = \arctan \frac{2}{3} + \arctan \frac{2}{3} = \arctan \left[\frac{2(2/3)}{1 - (2/3)^2} \right] = \arctan \frac{12}{5}$

$2 \arctan \frac{2}{3} - \arctan \frac{7}{17} = \arctan \frac{12}{5} + \arctan \left(-\frac{7}{17} \right) = \arctan \frac{(12/5) - (7/17)}{1 + (12/5)(7/17)} = \arctan \frac{169}{169} = \arctan 1 = \frac{\pi}{4}$

(b) $\pi = 8 \arctan \frac{2}{3} - 4 \arctan \frac{7}{17}$

$\approx 8 \left[\frac{2}{3} - \frac{(2/3)^3}{3} + \frac{(2/3)^5}{5} - \frac{(2/3)^7}{7} + \frac{(2/3)^9}{9} - \frac{(2/3)^{11}}{11} \right]$

$-4 \left[\frac{7}{17} - \frac{(7/17)^3}{3} + \frac{(7/17)^5}{5} - \frac{(7/17)^7}{7} + \frac{(7/17)^9}{9} - \frac{(7/17)^{11}}{11} \right] \approx 3.14$

47. From Exercise 21, we have

$\ln(x + 1) = \sum_{n=0}^{\infty} \frac{(-1)^n x^{n+1}}{n + 1} = \sum_{n=1}^{\infty} \frac{(-1)^{n-1} x^n}{n}$

$= \sum_{n=1}^{\infty} \frac{(-1)^{n+1} x^n}{n}.$

Thus, $\sum_{n=1}^{\infty} (-1)^{n+1} \frac{1}{2^n n} = \sum_{n=1}^{\infty} \frac{(-1)^{n+1} (1/2)^n}{n}$

$= \ln \left(\frac{1}{2} + 1 \right) = \ln \frac{3}{2} \approx 0.4055$

48. From Exercise 47, we have

$\sum_{n=1}^{\infty} (-1)^{n+1} \frac{1}{3^n n} = \sum_{n=1}^{\infty} \frac{(-1)^{n+1} (1/3)^n}{n}$

$= \ln \left(\frac{1}{3} + 1 \right) = \ln \frac{4}{3} \approx 0.2877.$

49. From Exercise 47, we have

$\sum_{n=1}^{\infty} (-1)^{n+1} \frac{2^n}{5^n n} = \sum_{n=1}^{\infty} \frac{(-1)^{n+1} (2/5)^n}{n}$

$= \ln \left(\frac{2}{5} + 1 \right) = \ln \frac{7}{5} \approx 0.3365.$

50. From Example 5, we have $\arctan x = \sum_{n=0}^{\infty} (-1)^n \frac{x^{2n+1}}{2n + 1}.$

$\sum_{n=0}^{\infty} (-1)^n \frac{1}{2n + 1} = \sum_{n=0}^{\infty} (-1)^n \frac{(1)^{2n+1}}{2n + 1}$

$= \arctan 1 = \frac{\pi}{4} \approx 0.7854$

51. From Exercise 50, we have

$\sum_{n=0}^{\infty} (-1)^n \frac{1}{2^{2n+1}(2n + 1)} = \sum_{n=0}^{\infty} (-1)^n \frac{(1/2)^{2n+1}}{2n + 1}$

$= \arctan \frac{1}{2} \approx 0.4636.$

52. From Exercise 50, we have

$\sum_{n=1}^{\infty} (-1)^{n+1} \frac{1}{3^{2n-1}(2n - 1)} = \sum_{n=0}^{\infty} (-1)^n \frac{1}{3^{2n+1}(2n + 1)}$

$= \sum_{n=0}^{\infty} (-1)^n \frac{(1/3)^{2n+1}}{2n + 1}$

$= \arctan \frac{1}{3} \approx 0.3218.$

53. The series in Exercise 50 converges to its sum at a slower rate because its terms approach 0 at a much slower rate.

54. From Example 5, we have $\arctan x = \sum_{n=0}^{\infty} (-1)^n \frac{x^{2n+1}}{2n + 1}.$

$\sum_{n=0}^{\infty} \frac{(-1)^n}{3^n (2n + 1)} = \sum_{n=0}^{\infty} \frac{(-1)^n}{(\sqrt{3})^{2n} (2n + 1)} \cdot \frac{\sqrt{3}}{\sqrt{3}}$

$= \sqrt{3} \sum_{n=0}^{\infty} \frac{(-1)^n (1/\sqrt{3})^{2n+1}}{2n + 1}$

$= \sqrt{3} \arctan \frac{1}{\sqrt{3}}$

$= \sqrt{3} \left(\frac{\pi}{6} \right) = \frac{\pi}{2\sqrt{3}}$

55. $f(x) = \sum_{n=1}^{\infty} (-1)^{n+1} \frac{(x - 1)^n}{n}, \quad 0 < x \leq 2$

$f(0.5) = \sum_{n=1}^{\infty} (-1)^{n+1} \frac{(-0.5)^n}{n} = \sum_{n=1}^{\infty} -\frac{(1/2)^n}{n}$

$\sum_{n=1}^{\infty} -\frac{(1/2)^n}{n} = -0.6931$

Section 8.10   Taylor and Maclaurin Series

1. For $c = 0$, we have:

$$f(x) = e^{2x}$$

$$f^{(n)}(x) = 2^n e^{2x} \implies f^{(n)}(0) = 2^n$$

$$e^{2x} = 1 + 2x + \frac{4x^2}{2!} + \frac{8x^3}{3!} + \frac{16x^4}{4!} + \ldots = \sum_{n=0}^{\infty} \frac{(2x)^n}{n!}$$

2. For $c = 0$, we have:

$$f(x) = e^{-2x}$$

$$f^{(n)}(x) = (-2)^n e^{-2x} \implies f^{(n)}(0) = (-2)^n$$

$$e^{-2x} = 1 - 2x + \frac{4x^2}{2!} - \frac{8x^3}{3!} + \frac{16x^4}{4!} - \cdots = \sum_{n=0}^{\infty} \frac{(-2x)^n}{n!}$$

3. For $c = \pi/4$, we have:

$$f(x) = \cos(x) \qquad f\left(\frac{\pi}{4}\right) = \frac{\sqrt{2}}{2}$$

$$f'(x) = -\sin(x) \qquad f'\left(\frac{\pi}{4}\right) = -\frac{\sqrt{2}}{2}$$

$$f''(x) = -\cos(x) \qquad f''\left(\frac{\pi}{4}\right) = -\frac{\sqrt{2}}{2}$$

$$f'''(x) = \sin(x) \qquad f'''\left(\frac{\pi}{4}\right) = \frac{\sqrt{2}}{2}$$

$$f^{(4)}(x) = \cos(x) \qquad f^{(4)}\left(\frac{\pi}{4}\right) = \frac{\sqrt{2}}{2}$$

and so on. Therefore we have

$$\cos x = \sum_{n=0}^{\infty} \frac{f^{(n)}(\pi/4)[x - (\pi/4)]^n}{n!}$$

$$= \frac{\sqrt{2}}{2}\left[1 - \left(x - \frac{\pi}{4}\right) - \frac{[x - (\pi/4)]^2}{2!} + \frac{[x - (\pi/4)]^3}{3!} + \frac{[x - (\pi/4)]^4}{4!} - \cdots \right]$$

$$= \frac{\sqrt{2}}{2} \sum_{n=0}^{\infty} \frac{(-1)^{n(n+1)/2}[x - (\pi/4)]^n}{n!}.$$

[**Note:** $(-1)^{n(n+1)/2} = 1, -1, -1, 1, 1, -1, -1, 1, 1, \ldots$]

4. For $c = \pi/4$, we have:

$$f(x) = \sin x \qquad f\left(\frac{\pi}{4}\right) = \frac{\sqrt{2}}{2}$$

$$f'(x) = \cos x \qquad f'\left(\frac{\pi}{4}\right) = \frac{\sqrt{2}}{2}$$

$$f''(x) = -\sin x \qquad f''\left(\frac{\pi}{4}\right) = -\frac{\sqrt{2}}{2}$$

$$f'''(x) = -\cos x \qquad f'''\left(\frac{\pi}{4}\right) = -\frac{\sqrt{2}}{2}$$

$$f^{(4)}(x) = \sin x \qquad f^{(4)}\left(\frac{\pi}{4}\right) = \frac{\sqrt{2}}{2}$$

and so on. Therefore we have:

$$\sin x = \sum_{n=0}^{\infty} \frac{f^{(n)}(\pi/4)[x - (\pi/4)]^n}{n!}$$

$$= \frac{\sqrt{2}}{2}\left[1 + \left(x - \frac{\pi}{4}\right) - \frac{[x - (\pi/4)]^2}{2!} - \frac{[x - (\pi/4)]^3}{3!} + \frac{[x - (\pi/4)]^4}{4!} + \cdots\right]$$

$$= \frac{\sqrt{2}}{2}\left\{\sum_{n=0}^{\infty} \frac{(-1)^{n(n+1)/2}[x - (\pi/4)]^{n+1}}{(n+1)!} + 1\right\}$$

5. For $c = 1$, we have,

$$f(x) = \ln x \qquad f(1) = 0$$

$$f'(x) = \frac{1}{x} \qquad f'(1) = 1$$

$$f''(x) = -\frac{1}{x^2} \qquad f''(1) = -1$$

$$f'''(x) = \frac{2}{x^3} \qquad f'''(1) = 2$$

$$f^{(4)}(x) = -\frac{6}{x^4} \qquad f^{(4)}(1) = -6$$

$$f^{(5)}(x) = \frac{24}{x^5} \qquad f^{(5)}(1) = 24$$

and so on. Therefore, we have:

$$\ln x = \sum_{n=0}^{\infty} \frac{f^{(n)}(1)(x - 1)^n}{n!}$$

$$= 0 + (x - 1) - \frac{(x - 1)^2}{2!} + \frac{2(x - 1)^3}{3!} - \frac{6(x - 1)^4}{4!} + \frac{24(x - 1)^5}{5!} - \cdots$$

$$= (x - 1) - \frac{(x - 1)^2}{2} + \frac{(x - 1)^3}{3} - \frac{(x - 1)^4}{4} + \frac{(x - 1)^5}{5} - \cdots = \sum_{n=0}^{\infty} (-1)^n \frac{(x - 1)^{n+1}}{n + 1}$$

6. For $c = 1$, we have:

$$f(x) = e^x$$

$$f^{(n)}(x) = e^x \Rightarrow f^{(n)}(1) = e$$

$$e^x = \sum_{n=0}^{\infty} \frac{f^{(n)}(1)(x - 1)^n}{n!} = e\left[1 + (x - 1) + \frac{(x - 1)^2}{2!} + \frac{(x - 1)^3}{3!} + \frac{(x - 1)^4}{4!} + \cdots\right] = e\sum_{n=0}^{\infty} \frac{(x - 1)^n}{n!}$$

7. For $c = 0$, we have:

$$f(x) = \sin 2x \qquad\qquad f(0) = 0$$
$$f'(x) = 2 \cos 2x \qquad\qquad f'(0) = 2$$
$$f''(x) = -4 \sin 2x \qquad\qquad f''(0) = 0$$
$$f'''(x) = -8 \cos 2x \qquad\qquad f'''(0) = -8$$
$$f^{(4)}(x) = 16 \sin 2x \qquad\qquad f^{(4)}(0) = 0$$
$$f^{(5)}(x) = 32 \cos 2x \qquad\qquad f^{(5)}(0) = 32$$
$$f^{(6)}(x) = -64 \sin 2x \qquad\qquad f^{(6)}(0) = 0$$
$$f^{(7)}(x) = -128 \cos 2x \qquad\qquad f^{(7)}(0) = -128$$

and so on. Therefore, we have:

$$\sin 2x = \sum_{n=0}^{\infty} \frac{f^{(n)}(0)x^n}{n!} = 0 + 2x + \frac{0x^2}{2!} - \frac{8x^3}{3!} + \frac{0x^4}{4!} + \frac{32x^5}{5!} + \frac{0x^6}{6!} - \frac{128x^7}{7!} + \cdots$$

$$= 2x - \frac{8x^3}{3!} + \frac{32x^5}{5!} - \frac{128x^7}{7!} + \cdots = \sum_{n=0}^{\infty} \frac{(-1)^n (2x)^{2n+1}}{(2n+1)!}$$

8. For $c = 0$, we have:

$$f(x) = \ln(x^2 + 1) \qquad\qquad f(0) = 0$$
$$f'(x) = \frac{2x}{x^2 + 1} \qquad\qquad f'(0) = 0$$
$$f''(x) = \frac{2 - 2x^2}{(x^2 + 1)^2} \qquad\qquad f''(0) = 2$$
$$f'''(x) = \frac{4x(x^2 - 3)}{(x^2 + 1)^3} \qquad\qquad f'''(0) = 0$$
$$f^{(4)}(x) = \frac{12(-x^4 + 6x^2 - 1)}{(x^2 + 1)^4} \qquad\qquad f^{(4)}(0) = -12$$
$$f^{(5)}(x) = \frac{48x(x^4 - 10x^2 + 5)}{(x^2 + 1)^5} \qquad\qquad f^{(5)}(0) = 0$$
$$f^{(6)}(x) = \frac{-240(5x^6 - 15x^4 + 15x^2 - 1)}{(x^2 + 1)^6} \qquad f^{(6)}(0) = 240$$

and so on. Therefore, we have:

$$\ln(x^2 + 1) = \sum_{n=0}^{\infty} \frac{f^{(n)}(0)x^n}{n!} = 0 + 0x + \frac{2x^2}{2!} + \frac{0x^3}{3!} - \frac{12x^4}{4!} + \frac{0x^5}{5!} + \frac{240x^6}{6!} + \cdots$$

$$= x^2 - \frac{x^4}{2} + \frac{x^6}{3} - \cdots = \sum_{n=0}^{\infty} \frac{(-1)^n x^{2n+2}}{n + 1}$$

9. For $c = 0$, we have:

$$f(x) = \sec(x) \qquad\qquad f(0) = 1$$
$$f'(x) = \sec(x)\tan(x) \qquad\qquad f'(0) = 0$$
$$f''(x) = \sec^3(x) + \sec(x)\tan^2(x) \qquad\qquad f''(0) = 1$$
$$f'''(x) = 5 \sec^3(x)\tan(x) + \sec(x)\tan^3(x) \qquad\qquad f'''(0) = 0$$
$$f^{(4)}(x) = 5 \sec^5(x) + 18 \sec^3(x)\tan^2(x) + \sec(x)\tan^4(x) \qquad f^{(4)}(0) = 5$$

$$\sec(x) = \sum_{n=0}^{\infty} \frac{f^{(n)}(0)x^n}{n!} = 1 + \frac{x^2}{2!} + \frac{5x^4}{4!} + \cdots$$

10. For $c = 0$, we have;

$$f(x) = \tan(x) \qquad\qquad\qquad f(0) = 0$$

$$f'(x) = \sec^2(x) \qquad\qquad\qquad f'(0) = 1$$

$$f''(x) = 2\sec^2(x)\tan(x) \qquad\qquad\qquad f''(0) = 0$$

$$f'''(x) = 2[\sec^4(x) + 2\sec^2(x)\tan^2(x)] \qquad\qquad f'''(0) = 2$$

$$f^{(4)}(x) = 8[\sec^4(x)\tan(x) + \sec^2(x)\tan^3(x)] \qquad f^{(4)}(0) = 0$$

$$f^{(5)}(x) = 8[2\sec^6(x) + 11\sec^4(x)\tan^2(x) + 2\sec^2(x)\tan^4(x)] \qquad f^{(5)}(0) = 16$$

$$\tan(x) = \sum_{n=0}^{\infty} \frac{f^{(n)}(0)x^n}{n!} = x + \frac{2x^3}{3!} + \frac{16x^5}{5!} + \cdots = x + \frac{x^3}{3} + \frac{2}{15}x^5 + \cdots$$

11. Since $(1 + x)^{-k} = 1 - kx + \dfrac{k(k+1)x^2}{2!} - \dfrac{k(k+1)(k+2)x^3}{3!} + \cdots$, we have

$$(1 + x)^{-2} = 1 - 2x + \frac{2(3)x^2}{2!} - \frac{2(3)(4)x^3}{3!} + \frac{2(3)(4)(5)x^4}{4!} - \cdots = 1 - 2x + 3x^2 - 4x^3 + 5x^4 - \cdots$$

$$= \sum_{n=0}^{\infty} (-1)^n (n+1)x^n.$$

12. Since $(1 + x)^{-k} = 1 - kx + \dfrac{k(k+1)x^2}{2!} - \dfrac{k(k+1)(k+2)x^3}{3!} + \cdots$, we have

$$\left[1 + (-x)\right]^{-1/2} = 1 + \left(\frac{1}{2}\right)x + \frac{(1/2)(3/2)x^2}{2!} + \frac{(1/2)(3/2)(5/2)x^3}{3!} + \cdots$$

$$= 1 + \frac{x}{2} + \frac{(1)(3)x^2}{2^2 2!} + \frac{(1)(3)(5)x^3}{2^3 3!} + \cdots$$

$$= 1 + \sum_{n=1}^{\infty} \frac{1 \cdot 3 \cdot 5 \cdots (2n-1)x^n}{2^n n!}$$

13. $\dfrac{1}{\sqrt{4 + x^2}} = \left(\dfrac{1}{2}\right)\left[1 + \left(\dfrac{x}{2}\right)^2\right]^{-1/2}$ and since $(1 + x)^{-1/2} = 1 + \displaystyle\sum_{n=1}^{\infty} \frac{(-1)^n 1 \cdot 3 \cdot 5 \cdots (2n-1)x^n}{2^n n!}$, we have

$$\frac{1}{\sqrt{4 + x^2}} = \frac{1}{2}\left[1 + \sum_{n=1}^{\infty} \frac{(-1)^n 1 \cdot 3 \cdot 5 \cdots (2n-1)(x/2)^{2n}}{2^n n!}\right] = \frac{1}{2} + \sum_{n=1}^{\infty} \frac{(-1)^n 1 \cdot 3 \cdot 5 \cdots (2n-1)x^{2n}}{2^{3n+1} n!}.$$

14. Since $(1 + x)^k = 1 + kx + \dfrac{k(k-1)x^2}{2!} + \dfrac{k(k-1)(k-2)x^3}{3!} + \cdots$, we have

$$(1 + x)^{1/2} = 1 + \left(\frac{1}{2}\right)x + \frac{(1/2)(-1/2)x^2}{2!} + \frac{(1/2)(-1/2)(-3/2)x^3}{3!} + \cdots$$

$$= 1 + \frac{x}{2} - \frac{x^2}{2^2 2!} + \frac{1 \cdot 3x^3}{2^3 3!} - \frac{1 \cdot 3 \cdot 5x^4}{2^4 5!} + \cdots$$

$$= 1 + \frac{x}{2} + \sum_{n=2}^{\infty} \frac{(-1)^{n+1} 1 \cdot 3 \cdot 5 \cdots (2n-3)x^n}{2^n n!}.$$

15. Since $(1 + x)^{1/2} = 1 + \dfrac{x}{2} + \displaystyle\sum_{n=2}^{\infty} \frac{(-1)^{n+1} 1 \cdot 3 \cdot 5 \cdots (2n-3)x^n}{2^n n!}$ (Exercise 14)

we have $(1 + x^2)^{1/2} = 1 + \dfrac{x^2}{2} + \displaystyle\sum_{n=2}^{\infty} \frac{(-1)^{n+1} 1 \cdot 3 \cdot 5 \cdots (2n-3)x^{2n}}{2^n n!}.$

16. Since $(1 + x)^{1/2} = 1 + \dfrac{x}{2} + \displaystyle\sum_{n=2}^{\infty} \dfrac{(-1)^{n+1} 1 \cdot 3 \cdot 5 \cdots (2n-3)x^n}{2^n n!}$ (Exercise 14)

we have $(1 + x^3)^{1/2} = 1 + \dfrac{x^3}{2} + \displaystyle\sum_{n=2}^{\infty} \dfrac{(-1)^{n+1} 1 \cdot 3 \cdot 5 \cdots (2n-3)x^{3n}}{2^n n!}$.

17. $e^x = \displaystyle\sum_{n=0}^{\infty} \dfrac{x^n}{n!} = 1 + x + \dfrac{x^2}{2!} + \dfrac{x^3}{3!} + \dfrac{x^4}{4!} + \dfrac{x^5}{5!} + \cdots$

$e^{x^2/2} = \displaystyle\sum_{n=0}^{\infty} \dfrac{(x^2/2)^n}{n!} = \sum_{n=0}^{\infty} \dfrac{x^{2n}}{2^n n!} = 1 + \dfrac{x^2}{2} + \dfrac{x^4}{2^2 2!} + \dfrac{x^6}{2^3 3!} + \dfrac{x^8}{2^4 4!} + \cdots$

18. $e^x = \displaystyle\sum_{n=0}^{\infty} \dfrac{x^n}{n!} = 1 + x + \dfrac{x^2}{2!} + \dfrac{x^3}{3!} + \dfrac{x^4}{4!} + \dfrac{x^5}{5!} + \cdots$

$e^{-3x} = \displaystyle\sum_{n=0}^{\infty} \dfrac{(-3x)^n}{n!} = \sum_{n=0}^{\infty} \dfrac{(-1)^n 3^n x^n}{n!} = 1 - 3x + \dfrac{9x^2}{2!} - \dfrac{27x^3}{3!} + \dfrac{81x^4}{4!} - \dfrac{243x^5}{5!} + \cdots$

19. $\sin x = \displaystyle\sum_{n=0}^{\infty} \dfrac{(-1)^n x^{2n+1}}{(2n+1)!} = x - \dfrac{x^3}{3!} + \dfrac{x^5}{5!} - \dfrac{x^7}{7!} + \cdots$

$\sin 2x = \displaystyle\sum_{n=0}^{\infty} \dfrac{(-1)^n (2x)^{2n+1}}{(2n+1)!} = \sum_{n=0}^{\infty} \dfrac{(-1)^n 2^{2n+1} x^{2n+1}}{(2n+1)!} = 2x - \dfrac{8x^3}{3!} + \dfrac{32x^5}{5!} - \dfrac{128x^7}{7!} + \cdots$

20. $\cos x = \displaystyle\sum_{n=0}^{\infty} \dfrac{(-1)^n x^{2n}}{(2n)!} = 1 - \dfrac{x^2}{2!} + \dfrac{x^4}{4!} - \dfrac{x^6}{6!} + \cdots$

$x \cos x = x \displaystyle\sum_{n=0}^{\infty} \dfrac{(-1)^n x^{2n}}{(2n)!} = \sum_{n=0}^{\infty} \dfrac{(-1)^n x^{2n+1}}{(2n)!} = x - \dfrac{x^3}{2!} + \dfrac{x^5}{4!} - \dfrac{x^7}{6!} + \cdots$

21. $\cos x = \displaystyle\sum_{n=0}^{\infty} \dfrac{(-1)^n x^{2n}}{(2n)!} = 1 - \dfrac{x^2}{2!} + \dfrac{x^4}{4!} - \cdots$

$\cos x^{3/2} = \displaystyle\sum_{n=0}^{\infty} \dfrac{(-1)^n (x^{3/2})^{2n}}{(2n)!} = \sum_{n=0}^{\infty} \dfrac{(-1)^n x^{3n}}{(2n)!} = 1 - \dfrac{x^3}{2!} + \dfrac{x^6}{4!} - \cdots$

22. $\cos x = \displaystyle\sum_{n=0}^{\infty} \dfrac{(-1)^n x^{2n}}{(2n)!} = 1 - \dfrac{x^2}{2!} + \dfrac{x^4}{4!} - \cdots$

$\cos 3x = \displaystyle\sum_{n=0}^{\infty} \dfrac{(-1)^n (3x)^{2n}}{(2n)!} = \sum_{n=0}^{\infty} \dfrac{(-1)^n 3^{2n} x^{2n}}{(2n)!}$

23. $\dfrac{\sin x}{x} = \dfrac{1}{x} \displaystyle\sum_{n=0}^{\infty} \dfrac{(-1)^n x^{2n+1}}{(2n+1)!} = \sum_{n=0}^{\infty} \dfrac{(-1)^n x^{2n}}{(2n+1)!}$

$= 1 - \dfrac{x^2}{3!} + \dfrac{x^4}{5!} - \dfrac{x^6}{7!} + \dfrac{x^8}{9!} - \cdots$

24. $\dfrac{\arcsin x}{x} = \dfrac{1}{x} \displaystyle\sum_{n=0}^{\infty} \dfrac{(2n)! x^{2n+1}}{(2^n n!)^2 (2n+1)} = \sum_{n=0}^{\infty} \dfrac{(2n)! x^{2n}}{(2^n n!)^2 (2n+1)}$

$= 1 + \dfrac{x^2}{2 \cdot 3} + \dfrac{1 \cdot 3 x^4}{2 \cdot 4 \cdot 5} + \dfrac{1 \cdot 3 \cdot 5 x^6}{2 \cdot 4 \cdot 6 \cdot 7} + \cdots$

25. $e^x = 1 + x + \dfrac{x^2}{2!} + \dfrac{x^3}{3!} + \dfrac{x^4}{4!} + \dfrac{x^5}{5!} + \cdots$

$e^{-x} = 1 - x + \dfrac{x^2}{2!} - \dfrac{x^3}{3!} + \dfrac{x^4}{4!} - \dfrac{x^5}{5!} + \cdots$

$e^x - e^{-x} = 2x + \dfrac{2x^3}{3!} + \dfrac{2x^5}{5!} + \dfrac{2x^7}{7!} + \cdots$

$\sinh(x) = \dfrac{1}{2}(e^x - e^{-x}) = x + \dfrac{x^3}{3!} + \dfrac{x^5}{5!} + \dfrac{x^7}{7!} + \cdots = \displaystyle\sum_{n=0}^{\infty} \dfrac{x^{2n+1}}{(2n+1)!}$

26. $e^x = 1 + x + \dfrac{x^2}{2!} + \dfrac{x^3}{3!} + \dfrac{x^4}{4!} + \dfrac{x^5}{5!} + \cdots$

$e^{-x} = 1 - x + \dfrac{x^2}{2!} - \dfrac{x^3}{3!} + \dfrac{x^4}{4!} - \dfrac{x^5}{5!} + \cdots$

$e^x + e^{-x} = 2 + \dfrac{2x^2}{2!} + \dfrac{2x^4}{4!} + \dfrac{2x^6}{6!} + \cdots$

$\cosh(x) = \dfrac{1}{2}(e^x + e^{-x}) = 1 + \dfrac{x^2}{2!} + \dfrac{x^4}{4!} + \dfrac{x^6}{6!} + \cdots$

$= \displaystyle\sum_{n=0}^{\infty} \dfrac{x^{2n}}{(2n)!}$

27. $\cos^2(x) = \dfrac{1}{2}[1 + \cos(2x)]$

$= \dfrac{1}{2}\left[1 + 1 - \dfrac{(2x)^2}{2!} + \dfrac{(2x)^4}{4!} - \dfrac{(2x)^6}{6!} - \cdots \right]$

$= \dfrac{1}{2}\left[1 + \displaystyle\sum_{n=0}^{\infty} \dfrac{(-1)^n (2x)^{2n}}{(2n)!} \right]$

28. The formula for the binomial series gives $(1 + x)^{-1/2} = 1 + \displaystyle\sum_{n=1}^{\infty} \dfrac{(-1)^n \, 1 \cdot 3 \cdot 5 \ldots (2n-1)x^n}{2^n n!}$, which implies that

$(1 + x^2)^{-1/2} = 1 + \displaystyle\sum_{n=1}^{\infty} \dfrac{(-1)^n \, 1 \cdot 3 \cdot 5 \ldots (2n-1)x^{2n}}{2^n n!}$

$\ln\left(x + \sqrt{x^2 + 1}\right) = \displaystyle\int \dfrac{1}{\sqrt{x^2+1}} \, dx$

$= x + \displaystyle\sum_{n=1}^{\infty} \dfrac{(-1)^n 1 \cdot 3 \cdot 5 \ldots (2n-1)^{2n+1}}{2^n (2n+1)n!}$

$= x - \dfrac{x^3}{2 \cdot 3} + \dfrac{1 \cdot 3x^5}{2 \cdot 4 \cdot 5} - \dfrac{1 \cdot 3 \cdot 5x^7}{2 \cdot 4 \cdot 6 \cdot 7} + \cdots.$

29. $e^{ix} = 1 + ix + \dfrac{(ix)^2}{2!} + \dfrac{(ix)^3}{3!} + \dfrac{(ix)^4}{4!} + \cdots = 1 + ix - \dfrac{x^2}{2!} - \dfrac{ix^3}{3!} + \dfrac{x^4}{4!} + \dfrac{ix^5}{5!} - \dfrac{x^6}{6!} - \cdots$

$e^{-ix} = 1 - ix + \dfrac{(-ix)^2}{2!} + \dfrac{(-ix)^3}{3!} + \dfrac{(-ix)^4}{4!} + \cdots = 1 - ix - \dfrac{x^2}{2!} + \dfrac{ix^3}{3!} + \dfrac{x^4}{4!} - \dfrac{ix^5}{5!} - \dfrac{x^6}{6!} + \cdots$

$e^{ix} - e^{-ix} = 2ix - \dfrac{2ix^3}{3!} + \dfrac{2ix^5}{5!} - \dfrac{2ix^7}{7!} + \cdots$

$\dfrac{e^{ix} - e^{-ix}}{2i} = x - \dfrac{x^3}{3!} + \dfrac{x^5}{5!} - \dfrac{x^7}{7!} + \cdots = \displaystyle\sum_{n=0}^{\infty} \dfrac{(-1)^n x^{2n+1}}{(2n+1)!} = \sin(x)$

30. $e^{ix} + e^{-ix} = 2 - \dfrac{2x^2}{2!} + \dfrac{2x^4}{4!} - \dfrac{2x^6}{6!} + \cdots$ (See Exercise 29.)

$\dfrac{e^{ix} + e^{-ix}}{2} = 1 - \dfrac{x^2}{2!} + \dfrac{x^4}{4!} - \dfrac{x^6}{6!} + \cdots = \displaystyle\sum_{n=0}^{\infty} \dfrac{(-1)^n x^{2n}}{(2n)!} = \cos(x)$

31. $f(x) = e^x \sin x$

$= \left(1 + x + \dfrac{x^2}{2} + \dfrac{x^3}{6} + \dfrac{x^4}{24} + \cdots \right)\left(x - \dfrac{x^3}{6} + \dfrac{x^5}{120} - \cdots \right)$

$= x + x^2 + \left(\dfrac{x^3}{2} - \dfrac{x^3}{6} \right) + \left(\dfrac{x^4}{6} - \dfrac{x^4}{6} \right) + \left(\dfrac{x^5}{120} - \dfrac{x^5}{12} + \dfrac{x^5}{24} \right) + \cdots = x + x^2 + \dfrac{x^3}{3} - \dfrac{x^5}{30} + \cdots$

32. $g(x) = e^x \cos x$

$= \left(1 + x + \dfrac{x^2}{2} + \dfrac{x^4}{6} + \dfrac{x^4}{24} + \cdots \right)\left(1 - \dfrac{x^2}{2} + \dfrac{x^4}{24} - \cdots \right)$

$= 1 + x + \left(\dfrac{x^2}{2} - \dfrac{x^2}{2} \right) + \left(\dfrac{x^3}{6} - \dfrac{x^3}{2} \right) + \left(\dfrac{x^4}{24} - \dfrac{x^4}{4} + \dfrac{x^4}{24} \right) + \cdots = 1 + x - \dfrac{x^3}{3} - \dfrac{x^4}{6} + \cdots$

33. $h(x) = \cos x \ln(1 + x)$

$$= \left(1 - \frac{x^2}{2} + \frac{x^4}{24} - \cdots\right)\left(x - \frac{x^2}{2} + \frac{x^3}{3} - \frac{x^4}{4} + \frac{x^5}{5} - \cdots\right)$$

$$= x - \frac{x^2}{2} + \left(\frac{x^3}{3} - \frac{x^3}{2}\right) + \left(\frac{x^4}{4} - \frac{x^4}{4}\right) + \left(\frac{x^5}{5} - \frac{x^5}{6} + \frac{x^5}{24}\right) + \cdots$$

$$= x - \frac{x^2}{2} - \frac{x^3}{6} + \frac{3x^5}{40} + \cdots$$

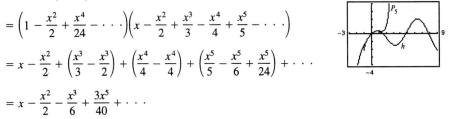

34. $f(x) = e^x \ln(1 + x)$

$$= \left(1 + x + \frac{x^2}{2} + \frac{x^3}{6} + \frac{x^4}{24} + \cdots\right)\left(x - \frac{x^2}{2} + \frac{x^3}{3} - \frac{x^4}{4} + \frac{x^5}{5} - \cdots\right)$$

$$= x + \left(x^2 - \frac{x^2}{2}\right) + \left(\frac{x^3}{3} - \frac{x^3}{2} + \frac{x^3}{2}\right) + \left(-\frac{x^4}{4} + \frac{x^4}{3} - \frac{x^4}{4} + \frac{x^4}{6}\right) + \left(\frac{x^5}{5} - \frac{x^5}{4} + \frac{x^5}{6} - \frac{x^5}{12} + \frac{x^5}{24}\right) + \cdots$$

$$= x + \frac{x^2}{2} + \frac{x^3}{3} + \frac{3x^5}{40} + \cdots$$

35. $g(x) = \dfrac{\sin x}{1 + x}$. Divide the series for $\sin x$ by $(1 + x)$.

$$
\begin{array}{r}
x - x^2 + \dfrac{5x^2}{6} - \dfrac{5x^4}{6} + \\[2mm]
1 + x \overline{\smash{)}\, x + 0x^2 - \dfrac{x^3}{6} + 0x^4 + \dfrac{x^5}{120} + \cdots} \\[2mm]
\underline{x + x^2} \\[1mm]
-x^2 - \dfrac{x^3}{6} \\[1mm]
\underline{-x^2 - x^3} \\[1mm]
\dfrac{5x^3}{6} + 0x^4 \\[1mm]
\underline{\dfrac{5x^3}{6} + \dfrac{5x^4}{6}} \\[1mm]
-\dfrac{5x^4}{6} + \dfrac{x^5}{120} \\[1mm]
\underline{-\dfrac{5x^4}{6} - \dfrac{5x^5}{6}} \\[1mm]
\vdots
\end{array}
$$

$$g(x) = x - x^2 + \frac{5x^3}{6} - \frac{5x^4}{6} + \cdots$$

36. $f(x) = \dfrac{e^x}{1 + x}$. Divide the series for e^x by $(1 + x)$.

$$
\begin{array}{r}
1 + \dfrac{x^2}{2} - \dfrac{x^3}{3} + \dfrac{3x^4}{8} + \cdots \\[2mm]
1 + x \overline{\smash{)}\, 1 + x + \dfrac{x^2}{2} + \dfrac{x^3}{6} + \dfrac{x^4}{24} + \dfrac{x^5}{120} + \cdots} \\[2mm]
\underline{1 + x} \\[1mm]
0 + \dfrac{x^2}{2} + \dfrac{x^3}{6} \\[1mm]
\underline{\dfrac{x^2}{2} + \dfrac{x^3}{2}} \\[1mm]
-\dfrac{x^3}{3} + \dfrac{x^4}{24} \\[1mm]
\underline{-\dfrac{x^3}{3} - \dfrac{x^4}{3}} \\[1mm]
\dfrac{3x^4}{8} + \dfrac{x^5}{120} \\[1mm]
\underline{\dfrac{3x^4}{8} + \dfrac{3x^5}{8}} \\[1mm]
\vdots
\end{array}
$$

$$f(x) = 1 + \frac{x^2}{2} - \frac{x^3}{3} + \frac{3x^4}{8} - \cdots$$

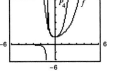

37. $y = x^2 - \dfrac{x^4}{3!} = x\left(x - \dfrac{x^3}{3!}\right) \approx x \sin x.$

Matches (a)

38. $y = x - \dfrac{x^3}{2!} + \dfrac{x^5}{4!} = x\left(1 - \dfrac{x^2}{2!} + \dfrac{x^4}{4!}\right) \approx x \cos x.$

Matches (b)

39. $y = x + x^2 + \dfrac{x^3}{2!} = x\left(1 + x + \dfrac{x^2}{2!}\right) \approx xe^x.$

Matches (c)

40. $y = x^2 - x^3 + x^4 = x^2(1 - x + x^2) \approx x^2\left(\dfrac{1}{1 + x}\right).$

Matches (d)

41. $\displaystyle\int_0^x (e^{-t^2} - 1)dt = \int_0^x \left[\left(\sum_{n=0}^{\infty} \frac{(-1)^n t^{2n}}{n!}\right) - 1\right] dt$

$\displaystyle = \int_0^x \left[\sum_{n=0}^{\infty} \frac{(-1)^{n+1} t^{2n+2}}{(n+1)!}\right] dt = \left[\sum_{n=0}^{\infty} \frac{(-1)^{n+1} t^{2n+3}}{(2n+3)(n+1)!}\right]_0^x = \sum_{n=0}^{\infty} \frac{(-1)^{n+1} x^{2n+3}}{(2n+3)(n+1)!}$

42. $\displaystyle\int_0^x \sqrt{1 + t^3}\, dt = \int_0^x \left[1 + \frac{t^3}{2} + \sum_{n=2}^{\infty} \frac{(-1)^{n-1} 1 \cdot 3 \cdot 5 \cdots (2n-3) t^{3n}}{2^n n!}\right] dt$

$\displaystyle = \left[t + \frac{t^4}{8} + \sum_{n=2}^{\infty} \frac{(-1)^{n-1} 1 \cdot 3 \cdot 5 \cdots (2n-3) t^{3n+1}}{(3n+1)2^n n!}\right]_0^x$

$\displaystyle = x + \frac{x^4}{8} + \sum_{n=2}^{\infty} \frac{(-1)^{n-1} 1 \cdot 3 \cdot 5 \cdots (2n-3) x^{3n+1}}{(3n+1)2^n n!}$

43. Since $\ln x = \displaystyle\sum_{n=0}^{\infty} \frac{(-1)^n (x-1)^{n+1}}{n+1} = (x-1) - \frac{(x-1)^2}{2} + \frac{(x-1)^3}{3} - \frac{(x-1)^4}{4} + \cdots$

we have $\ln 2 = 1 - \dfrac{1}{2} + \dfrac{1}{3} - \dfrac{1}{4} + \cdots = \displaystyle\sum_{n=1}^{\infty} (-1)^{n+1} \frac{1}{n} \approx 0.6931.$ (10,001 terms)

44. Since $\sin(x) = \displaystyle\sum_{n=0}^{\infty} \frac{(-1)^n x^{2n+1}}{(2n+1)!} = x - \frac{x^3}{3!} + \frac{x^5}{5!} - \frac{x^7}{7!} + \cdots,$ we have

$\sin(1) = \displaystyle\sum_{n=0}^{\infty} \frac{(-1)^n}{(2n+1)!} = 1 - \frac{1}{3!} + \frac{1}{5!} - \frac{1}{7!} + \cdots \approx 0.8415.$ (4 terms)

45. Since $e^x = \displaystyle\sum_{n=0}^{\infty} \frac{x^n}{n!} = 1 + x + \frac{x^2}{2!} + \frac{x^3}{3!} + \cdots,$

we have $e^2 = 1 + 2 + \dfrac{2^2}{2!} + \dfrac{2^3}{3!} + \cdots = \displaystyle\sum_{n=0}^{\infty} \frac{2^n}{n!} \approx 7.3891.$ (12 terms)

46. Since $e^x = \displaystyle\sum_{n=0}^{\infty} \frac{x^n}{n!} = 1 + x + \frac{x^2}{2!} + \frac{x^3}{3!} + \frac{x^4}{4!} + \frac{x^5}{5!} + \cdots,$ we have $e^{-1} = 1 - 1 + \dfrac{1}{2!} - \dfrac{1}{3!} + \dfrac{1}{4!} - \dfrac{1}{5!} + \cdots$

and $\dfrac{e-1}{e} = 1 - e^{-1} = 1 - \dfrac{1}{2!} + \dfrac{1}{3!} - \dfrac{1}{4!} + \dfrac{1}{5!} - \dfrac{1}{7!} + \cdots = \displaystyle\sum_{n=1}^{\infty} \frac{(-1)^{n-1}}{n!} \approx 0.6321.$ (6 terms)

47. Since

$$\cos x = \sum_{n=0}^{\infty} \frac{(-1)^n x^{2n}}{(2n)!} = 1 - \frac{x^2}{2!} + \frac{x^4}{4!} - \frac{x^6}{6!} + \frac{x^8}{8!} - \cdots$$

$$1 - \cos x = \frac{x^2}{2!} - \frac{x^4}{4!} + \frac{x^6}{6!} - \frac{x^8}{8!} + \cdots = \sum_{n=0}^{\infty} \frac{(-1)^n x^{2n+2}}{(2n+2)!}$$

$$\frac{1 - \cos}{x} = \frac{x}{2!} - \frac{x^3}{4!} + \frac{x^5}{6!} - \frac{x^7}{8!} + \cdots = \sum_{n=0}^{\infty} \frac{(-1)^n x^{2n+1}}{(2n+2)!}$$

we have $\displaystyle\lim_{x\to 0} \frac{1 - \cos x}{x} = \lim_{x\to 0} \sum_{n=0}^{\infty} \frac{(-1) x^{2n+1}}{(2n+2)!} = 0.$

48. Since

$$\sin x = \sum_{n=0}^{\infty} \frac{(-1)^n x^{2n+1}}{(2n+1)!} = x - \frac{x^3}{3!} + \frac{x^5}{5!} - \frac{x^7}{7!} + \cdots$$

$$\frac{\sin x}{x} = 1 - \frac{x^2}{3!} + \frac{x^4}{5!} - \frac{x^6}{7!} + \cdots = \sum_{n=0}^{\infty} \frac{(-1)^n x^{2n}}{(2n+1)!}$$

we have $\displaystyle\lim_{x \to 0} \frac{\sin x}{x} = \lim_{x \to 0} \sum_{n=0}^{\infty} \frac{(-1)^n x^{2n}}{(2n+1)!} = 1.$

49. $\displaystyle\int_0^1 \frac{\sin x}{x}\,dx = \int_0^1 \left[\sum_{n=0}^{\infty} \frac{(-1)^n x^{2n}}{(2n+1)!}\right] dx = \left[\sum_{n=0}^{\infty} \frac{(-1)^n x^{2n+1}}{(2n+1)(2n+1)!}\right]_0^1 = \sum_{n=0}^{\infty} \frac{(-1)^n}{(2n+1)(2n+1)!}$

Since $1/(7 \cdot 7!) < 0.0001$, we have

$$\int_0^1 \frac{\sin x}{x}\,dx = 1 - \frac{1}{3 \cdot 3!} + \frac{1}{5 \cdot 5!} - \cdots \approx 0.9461.$$

Note: We are using $\displaystyle\lim_{x \to 0^+} \frac{\sin x}{x} = 1.$

50. $\displaystyle\int_0^{1/2} \frac{\arctan x}{x}\,dx = \int_0^{1/2} \left(1 - \frac{x^2}{3} + \frac{x^4}{5} - \frac{x^6}{7} + \cdots\right) dx = \left[x - \frac{x^3}{3^2} + \frac{x^5}{5^2} - \frac{x^7}{7^2} + \cdots\right]_0^{1/2}$

Since $1/(9^2 2^9) < 0.0001$, we have

$$\int_0^{1/2} \frac{\arctan x}{x}\,dx \approx \left(\frac{1}{2} - \frac{1}{3^2 2^3} + \frac{1}{5^2 2^5} - \frac{1}{7^2 2^7} + \frac{1}{9^2 2^9}\right) \approx 0.4872.$$

Note: We are using $\displaystyle\lim_{x \to 0^+} \frac{\arctan x}{x} = 1.$

51. $\displaystyle\int_0^{\pi/2} \sqrt{x}\cos x\,dx = \int_0^{\pi/2} \left[\sum_{n=0}^{\infty} \frac{(-1)^n x^{(4n+1)/2}}{(2n)!}\right] dx = \left[\sum_{n=0}^{\infty} \frac{(-1)^n x^{(4n+3)/2}}{\left(\frac{4n+3}{2}\right)(2n)!}\right]_0^{\pi/2} = \left[\sum_{n=0}^{\infty} \frac{(-1)^n 2x^{(4n+3)/2}}{(4n+3)(2n)!}\right]_0^{\pi/2}$

Since $(\pi/2)^{19/2}/766{,}080 < 0.0001$, we have

$$\int_0^1 \sqrt{x}\cos x\,dx = 2\left[\frac{(\pi/2)^{3/2}}{3} - \frac{(\pi/2)^{7/2}}{14} + \frac{(\pi/2)^{11/2}}{264} - \frac{(\pi/2)^{15/2}}{10{,}800} + \frac{(\pi/2)^{19/2}}{766{,}080}\right] \approx 0.7040.$$

52. $\displaystyle\int_{0.5}^1 \cos\sqrt{x}\,dx = \int_{0.5}^1 \left(1 - \frac{x}{2!} + \frac{x^2}{4!} - \frac{x^3}{6!} + \frac{x^4}{8!} - \cdots\right) dx = \left[x - \frac{x^2}{2(2!)} + \frac{x^3}{3(4!)} - \frac{x^4}{4(6!)} + \frac{x^5}{5(8!)} - \cdots\right]_{0.5}^1$

Since $\frac{1}{210{,}600}(1 - 0.5^5) < 0.0001$, we have

$$\int_{0.5}^1 \cos\sqrt{x}\,dx \approx \left[(1 - 0.5) - \frac{1}{4}(1 - 0.5^2) + \frac{1}{72}(1 - 0.5^3) - \frac{1}{2880}(1 - 0.5^4) + \frac{1}{201{,}600}(1 - 0.5)^5\right] \approx 0.3243.$$

53. $\displaystyle\int_{0.1}^{0.3} \sqrt{1 + x^3}\,dx = \int_{0.1}^{0.3} \left(1 + \frac{x^3}{2} - \frac{x^6}{8} + \frac{x^9}{16} - \frac{5x^{12}}{128} + \cdots\right) dx = \left[x + \frac{x^4}{8} - \frac{x^7}{56} + \frac{x^{10}}{160} - \frac{5x^{13}}{1664} + \cdots\right]_{0.1}^{0.3}$

Since $\frac{1}{56}(0.3^7 - 0.1^7) < 0.0001$, we have

$$\int_{0.1}^{0.3} \sqrt{1 + x^3}\,dx = \left[(0.3 - 0.1) + \frac{1}{8}(0.3^4 - 0.1^4) - \frac{1}{56}(0.3^7 - 0.1^7)\right] \approx 0.2010.$$

54. $\displaystyle\int_0^{1/2} \frac{\ln(x + 1)}{x}\, dx = \int_0^{1/2}\left(1 - \frac{x}{2} + \frac{x^2}{3} - \frac{x^3}{4} + \cdots\right) dx = \left[x - \frac{x^2}{2^2} + \frac{x^3}{3^2} - \frac{x^4}{4^2} + \cdots\right]_0^{1/2}$

Since $1/(8^2 2^8) < 0.0001$, we have

$$\int_0^{1/2} \frac{\ln(x + 1)}{x}\, dx \approx \left(\frac{1}{2} - \frac{1}{2^2 2^2} + \frac{1}{3^2 2^3} - \frac{1}{4^2 2^4} + \frac{1}{5^2 2^5} - \frac{1}{6^2 2^6} + \frac{1}{7^2 2^7} - \frac{1}{8^2 2^8}\right) \approx 0.4484.$$

Note: We are using $\displaystyle\lim_{x \to 0^+} \frac{\ln(x + 1)}{x} = 1$.

55. From Exercise 17, we have

$$\frac{1}{\sqrt{2\pi}} \int_0^1 e^{-x^2/2}\, dx = \frac{1}{\sqrt{2\pi}} \int_0^1 \sum_{n=0}^{\infty} \frac{(-1)^n x^{2n}}{2^n n!}\, dx = \frac{1}{\sqrt{2\pi}}\left[\sum_{n=0}^{\infty} \frac{(-1)^n x^{2n+1}}{2^n n!(2n + 1)}\right]_0^1$$

$$= \frac{1}{\sqrt{2\pi}} \sum_{n=0}^{\infty} \frac{(-1)^n}{2^n n!(2n + 1)}$$

$$\approx \frac{1}{\sqrt{2\pi}}\left[1 - \frac{1}{2 \cdot 1 \cdot 3} + \frac{1}{2^2 \cdot 2! \cdot 5} - \frac{1}{2^3 \cdot 3! \cdot 7}\right] \approx 0.3414.$$

56. From Exercise 17, we have

$$\frac{1}{\sqrt{2\pi}} \int_1^2 e^{-x^2/2}\, dx = \frac{1}{\sqrt{2\pi}} \int_1^2 \sum_{n=0}^{\infty} \frac{(-1)^n x^{2n}}{2^n n!}\, dx = \frac{1}{\sqrt{2\pi}}\left[\sum_{n=0}^{\infty} \frac{(-1)^n x^{2n+1}}{2^n n!(2n + 1)}\right]_1^2$$

$$= \frac{1}{\sqrt{2\pi}} \sum_{n=0}^{\infty} \frac{(-1)^n (2^{n+1} - 1)}{2^n n!(2n + 1)}$$

$$\approx \frac{1}{\sqrt{2\pi}}\left[1 - \frac{7}{2 \cdot 1 \cdot 3} + \frac{31}{2^2 \cdot 2! \cdot 5} - \frac{127}{2^3 \cdot 3! \cdot 7} + \frac{511}{2^4 \cdot 4! \cdot 9} - \frac{2047}{2^5 \cdot 5! \cdot 11}\right.$$

$$\left.+ \frac{8191}{2^6 \cdot 6! \cdot 13} - \frac{32,767}{2^7 \cdot 7! \cdot 15} + \frac{131,071}{2^8 \cdot 8! \cdot 17} - \frac{524,287}{2^9 \cdot 9! \cdot 19}\right] \approx 0.1359.$$

57. $f(x) = x\cos 2x = \displaystyle\sum_{n=0}^{\infty} \frac{(-1)^n 4^n x^{2n+1}}{(2n)!}$

$P_5(x) = x - 2x^3 + \dfrac{2x^5}{3}$

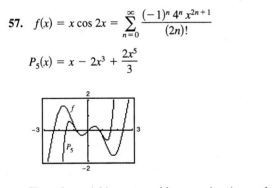

The polynomial is a reasonable approximation on the interval $\left[-\frac{3}{4}, \frac{3}{4}\right]$.

58. $f(x) = \sin\dfrac{x}{2}\ln(1 + x)$

$P_5(x) = \dfrac{x^2}{2} - \dfrac{x^3}{4} + \dfrac{7x^4}{48} - \dfrac{11x^5}{96}$

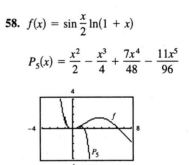

The polynomial is a reasonable approximation on the interval $(-0.60, 0.73)$.

59. $f(x) = \sqrt{x}\ln x,\ c = 1$

$P_5(x) = (x - 1) - \dfrac{(x - 1)^3}{24} + \dfrac{(x - 1)^4}{24} - \dfrac{71(x - 1)^5}{1920}$

The polynomial is a reasonable approximation on the interval $\left[\frac{1}{4}, 2\right]$.

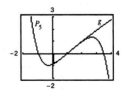

60. $f(x) = \sqrt[3]{x} \cdot \arctan x, c = 1$

$$P_5(x) \approx 0.7854 + 0.7618(x - 1) - 0.3412\left[\frac{(x-1)^2}{2!}\right] - 0.0424\left[\frac{(x-1)^3}{3!}\right]$$

$$+ 1.3025\left[\frac{(x-1)^4}{4!}\right] - 5.5913\left[\frac{(x-1)^5}{5!}\right]$$

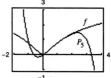

The polynomial is a reasonable approximation on the interval $(0.48, 1.75)$.

61. $y = \left(\tan\theta - \dfrac{g}{kv_0\cos\theta}\right)x - \dfrac{g}{k^2}\ln\left(1 - \dfrac{kx}{v_0\cos\theta}\right)$

$$= (\tan\theta)x - \frac{gx}{kv_0\cos\theta} - \frac{g}{k^2}\left[-\frac{kx}{v_0\cos\theta} - \frac{1}{2}\left(\frac{kx}{v_0\cos\theta}\right)^2 - \frac{1}{3}\left(\frac{kx}{v_0\cos\theta}\right)^3 - \frac{1}{4}\left(\frac{kx}{v_o\cos\theta}\right)^4 - \cdots\right]$$

$$= (\tan\theta)x - \frac{gx}{kv_0\cos\theta} + \frac{gx}{kv_0\cos\theta} + \frac{gx^2}{2v_0^2\cos^2\theta} + \frac{gkx^3}{3v_0^3\cos^3\theta} + \frac{gk^2x^4}{4v_0^4\cos^4\theta} + \cdots\right]$$

$$= (\tan\theta)x + \frac{gx^2}{2v_0^2\cos^2\theta} + \frac{kgx^3}{3v_0^3\cos^3\theta} + \frac{k^2gx^4}{4v_0^4\cos^4\theta} + \cdots$$

62. $\theta = 60°, v_0 = 64, k = \dfrac{1}{16}, g = -32$

$$y = \sqrt{3}x - \frac{32x^2}{2(64)^2(1/2)^2} - \frac{(1/16)(32)x^3}{3(64)^3(1/2)^3} - \frac{(1/16)^2(32)x^4}{4(64)^4(1/2)^4} - \cdots$$

$$= \sqrt{3}x - 32\left[\frac{2^2x^2}{2(64)^2} + \frac{2^3x^3}{3(64)^3 16} + \frac{2^4x^4}{4(64)^4(16)^2} + \cdots\right]$$

$$= \sqrt{3}x - 32\sum_{n=2}^{\infty}\frac{2^nx^n}{n(64)^n(16)^{n-2}} = \sqrt{3}x - 32\sum_{n=2}^{\infty}\frac{x^n}{n(32)^n(16)^{n-2}}$$

63. $f(x) = \begin{cases} e^{-1/x^2}, & x \neq 0 \\ 0, & x = 0 \end{cases}$

(a)

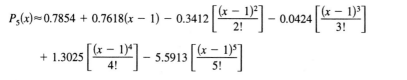

(b) $f'(0) = \lim\limits_{x\to 0}\dfrac{f(x) - f(0)}{x - 0} = \lim\limits_{x\to 0}\dfrac{e^{-1/x^2} - 0}{x}$

Let $y = \lim\limits_{x\to 0}\dfrac{e^{-1/x^2}}{x}$. Then

$$\ln y = \lim_{x\to 0}\ln\left(\frac{e^{-1/x^2}}{x}\right) = \lim_{x\to 0^+}\left[-\frac{1}{x^2} - \ln x\right] = \lim_{x\to 0^+}\left[\frac{-1 - x^2\ln x}{x^2}\right] = -\infty.$$

Thus, $y = e^{-\infty} = 0$ and we have $f'(0) = 0$.

(c) $\sum\limits_{n=0}^{\infty}\dfrac{f^{(n)}(0)}{n!}x^n = f(0) + \dfrac{f'(0)x}{1!} + \dfrac{f''(0)x^2}{2!} + \cdots = 0 \neq f(x)$

This series converges to f at $x = 0$ only.

64. (a) $f(x) = \dfrac{\ln(x^2 + 1)}{x^2}$.

(b)

From Exercise 8, you obtain

$$P_8 = \frac{1}{x^2} \sum_{n=0}^{\infty} \frac{(-1)^n x^{2n+2}}{n+1} = \sum_{n=0}^{\infty} \frac{(-1)^n x^{2n}}{n+1}$$

(c) $F(x) = \displaystyle\int_0^x \frac{\ln(t^2 + 1)}{t^2}\, dt$

$G(x) = \displaystyle\int_0^x P_8(t)\, dt$

x	0.25	0.50	0.75	1.00	1.50	2.00
$F(x)$	0.2475	0.4810	0.6920	0.8776	1.1798	1.4096
$G(x)$	0.2475	0.4810	0.6920	0.8805	5.3064	652.21

The curves are nearly identical for $0 < x < 1$. Hence, the integrals nearly agree on that interval.

65. By the Ratio Test:

$$\lim_{n \to \infty} \left| \frac{x^{n+1}}{(n+1)!} \cdot \frac{n!}{x^n} \right| = \lim_{n \to \infty} \frac{|x|}{n+1} = 0 \text{ which shows that } \sum_{n=0}^{\infty} \frac{x^n}{n!} \text{ converges for all } x.$$

66. Assume $e = p/q$ is rational. Let $N > q$ and form the following.

$$e - \left[1 + 1 + \frac{1}{2!} + \cdots + \frac{1}{N!} \right] = \frac{1}{(N+1)!} + \frac{1}{(N+2)!} + \cdots$$

Set $a = N!\left[e - \left(1 + 1 + \ldots + \frac{1}{N!} \right) \right]$, a positive integer. But,

$$a = N!\left[\frac{1}{(N+1)!} + \frac{1}{(N+2)!} + \cdots \right] = \frac{1}{N+1} + \frac{1}{(N+1)(N+2)} + \cdots < \frac{1}{N+1} + \frac{1}{(N+1)^2} + \cdots$$

$$= \frac{1}{N+1}\left[1 + \frac{1}{N+1} + \frac{1}{(N+1)^2} + \cdots \right] = \frac{1}{N+1}\left[\frac{1}{1 - \left(\dfrac{1}{N+1}\right)} \right] = \frac{1}{N}, \text{ a contradiction.}$$

Review Exercises for Chapter 8

1. $a_n = \dfrac{1}{n!}$

2. $a_n = \dfrac{n}{n^2 + 1}$

3. $a_n = 4 + \dfrac{2}{n}$: $6, 5, 4.67, \ldots$

Matches (a)

4. $a_n = 4 - \dfrac{n}{2}$: $3.5, 3, \ldots$

Matches (c)

5. $a_n = 10(0.3)^{n-1}$: $10, 3, \ldots$

Matches (d)

6. $a_n = 6\left(-\dfrac{2}{3}\right)^{n-1}$: $6, -4, \ldots$

Matches (b)

7. $a_n = \dfrac{5n + 2}{n}$

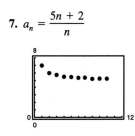

The sequence seems to converge to 5.

$$\lim_{n \to \infty} a_n = \lim_{n \to \infty} \frac{5n + 2}{n} = \lim_{n \to \infty} \left(5 + \frac{2}{n}\right) = 5$$

8. $a_n = \sin \dfrac{n\pi}{2}$

The sequence seems to diverge (oscillates).

$$\sin \frac{n\pi}{2}: \ 1, 0, -1, 0, 1, 0, \ldots$$

9. $\displaystyle\lim_{n \to \infty} \frac{n + 1}{n^2} = 0$

Converges

10. $\displaystyle\lim_{n \to \infty} \frac{1}{\sqrt{n}} = 0$

Converges

11. $\displaystyle\lim_{n \to \infty} \frac{n^3}{n^2 + 1} = \infty$

Diverges

12. $\displaystyle\lim_{n \to \infty} \frac{n}{\ln(n)} = \lim_{n \to \infty} \frac{1}{1/n} = \infty$

Diverges

13. $\displaystyle\lim_{n \to \infty} \left(\sqrt{n + 1} - \sqrt{n}\right) = \lim_{n \to \infty} \left(\sqrt{n + 1} - \sqrt{n}\right)\frac{\sqrt{n + 1} + \sqrt{n}}{\sqrt{n + 1} + \sqrt{n}} = \lim_{n \to \infty} \frac{1}{\sqrt{n + 1} + \sqrt{n}} = 0$ ⠀ Converges

14. $\displaystyle\lim_{n \to \infty} \left(1 + \frac{1}{2n}\right)^n = \lim_{k \to \infty} \left[\left(1 + \frac{1}{k}\right)^k\right]^{1/2} = e^{1/2}$

Converges; $k = 2n$

15. $\displaystyle\lim_{n \to \infty} \frac{\sin(n)}{\sqrt{n}} = 0$

Converges

16. Let ⠀ $y = (b^n + c^n)^{1/n}$

$$\ln y = \frac{\ln(b^n + c^n)}{n}$$

$$\lim_{n \to \infty} \ln y = \lim_{n \to \infty} \frac{1}{b^n + c^n}(b^n \ln b + c^n \ln c)$$

Assume $b \geq c$ and note that the terms

$$\frac{b^n \ln b + c^n \ln c}{b^n + c^n} = \frac{b^n \ln b}{b^n + c^n} + \frac{c^n \ln c}{b^n + c^n}$$

converge as $n \to \infty$. Hence a_n converges.

17. $A_n = 5000\left(1 + \dfrac{0.05}{4}\right)^n = 5000(1.0125)^n$

⠀⠀ $n = 1, 2, 3$

⠀ (a) $A_1 = 5062.50$ ⠀⠀⠀ $A_5 \approx 5320.41$

⠀⠀⠀ $A_2 \approx 5125.78$ ⠀⠀⠀ $A_6 \approx 5386.92$

⠀⠀⠀ $A_3 \approx 5189.85$ ⠀⠀⠀ $A_7 \approx 5454.25$

⠀⠀⠀ $A_4 \approx 5254.73$ ⠀⠀⠀ $A_8 \approx 5522.43$

⠀ (b) $A_{40} \approx 8218.10$

18. (a) $V_n = 120{,}000(0.70)^n, \ n = 1, 2, 3, 4, 5$

⠀ (b) $V_5 = 120{,}000(0.70)^5 = \$20{,}168.40$

19. (a)

k	5	10	15	20	25
S_k	13.2	113.3	873.8	6448.5	50,500.3

⠀ (c) The series diverges $\left(\text{geometric } r = \frac{3}{2} > 1\right)$

⠀ (b)

(graph)

20. (a)

k	5	10	15	20	25
S_k	0.3917	0.3228	0.3627	0.3344	0.3564

(b)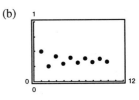

(c) The series converges by the Alternating Series Test.

21. (a)

k	5	10	15	20	25
S_k	0.4597	0.4597	0.4597	0.4597	0.4597

(b)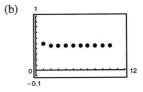

(c) The series converges by the Alternating Series Test.

22. (a)

k	5	10	15	20	25
S_k	0.8333	0.9091	0.9375	0.9524	0.9615

(b)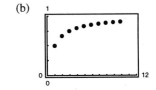

(c) The series converges, by the limit comparison test with $\sum \dfrac{1}{n^2}$.

23. $\displaystyle\sum_{n=0}^{\infty}\left(\frac{2}{3}\right)^n$

Geometric series with $a = 1$ and $r = \frac{2}{3}$.

$$S = \frac{a}{1 - r} = \frac{1}{1 - (2/3)} = \frac{1}{1/3} = 3$$

24. $\displaystyle\sum_{n=0}^{\infty}\frac{2^{n+2}}{3^n} = 4\sum_{n=0}^{\infty}\left(\frac{2}{3}\right)^n = 4(3) = 12$

See Exercise 23.

25. $\displaystyle\sum_{n=0}^{\infty}\left(\frac{1}{2^n} - \frac{1}{3^n}\right) = \sum_{n=0}^{\infty}\left(\frac{1}{2}\right)^n - \sum_{n=0}^{\infty}\left(\frac{1}{3}\right)^n = \frac{1}{1 - (1/2)} - \frac{1}{1 - (1/3)} = 2 - \frac{3}{2} = \frac{1}{2}$

26. $\displaystyle\sum_{n=0}^{\infty}\left[\left(\frac{2}{3}\right)^n - \frac{1}{(n+1)(n+2)}\right] = \sum_{n=0}^{\infty}\left(\frac{2}{3}\right)^n - \sum_{n=0}^{\infty}\left(\frac{1}{n+1} - \frac{1}{n+2}\right)$

$$= \frac{1}{1 - (2/3)} - \left[\left(1 - \frac{1}{2}\right) + \left(\frac{1}{2} - \frac{1}{3}\right) + \left(\frac{1}{3} - \frac{1}{4}\right) + \cdots\right] = 3 - 1 = 2$$

27. $0.\overline{09} = 0.09 + 0.0009 + 0.000009 + \cdots = 0.09(1 + 0.01 + 0.0001 + \cdots) = \displaystyle\sum_{n=0}^{\infty}(0.09)(0.01)^n = \frac{0.09}{1 - 0.01} = \frac{1}{11}$

28. $0.\overline{923076} = 0.923076[1 + 0.000001 + (0.000001)^2 + \cdots]$

$$= \sum_{n=0}^{\infty}(0.923076)(0.000001)^n = \frac{0.923076}{1 - 0.000001} = \frac{923,076}{999,999} = \frac{12(76,923)}{13(76,923)} = \frac{12}{13}$$

29. $D_1 = 8$

$D_2 = 0.7(8) + 0.7(8) = 16(0.7)$

$\vdots$

$D = 8 + 16(0.7) + 16(0.7)^2 + \cdots + 16(0.7)^n + \cdots$

$\quad = -8 + \displaystyle\sum_{n=0}^{\infty}16(0.7)^n = -8 + \frac{16}{1 - 0.7} = 45\frac{1}{3}$ meters

30. $S = \displaystyle\sum_{n=0}^{39}32{,}000(1.055)^n = \frac{32{,}000(1 - 1.055^{40})}{1 - 1.055}$

$\quad \approx \$4{,}371{,}379.65$

31. See Exercise 76 in Section 8.2.

$$A = \frac{P(e^{rt} - 1)}{e^{r/12} - 1}$$

$$= \frac{200(e^{(0.06)(2)} - 1)}{e^{0.06/12} - 1}$$

$$\approx \$5087.14$$

32. See Exercise 76 in Section 8.2.

$$A = P\left(\frac{12}{r}\right)\left[\left(1 + \frac{r}{12}\right)^{12t} - 1\right]$$

$$= 100\left(\frac{12}{0.065}\right)\left[\left(1 + \frac{0.065}{12}\right)^{120} - 1\right]$$

$$\approx \$16,840.32$$

33. (a) Ratio Test: $\displaystyle\lim_{n\to\infty}\left|\frac{a_{n+1}}{a_n}\right| = \lim_{n\to\infty}\frac{(n+1)(3/5)^{n+1}}{n(3/5)^n}$

$$= \lim_{n\to\infty}\left(\frac{n+1}{n}\right)\left(\frac{3}{5}\right) = \frac{3}{5} < 1$$

Converges

(b)

x	5	10	15	20	25
S_n	2.8752	3.6366	3.7377	3.7488	3.7499

(c)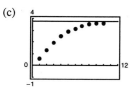

(d) The sum is approximately 3.75.

34. (a) The series converges by the Alternating Series Test.

(b)

x	5	10	15	20	25
S_n	0.0871	0.0669	0.0734	0.0702	0.0721

(c)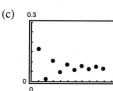

(d) The sum is approximately 0.0714.

35. $\displaystyle\sum_{n=1}^{\infty}\frac{2^n}{n^3}$

$$\lim_{n\to\infty}\left|\frac{a_{n+1}}{a_n}\right| = \lim_{n\to\infty}\left|\frac{2^{n+1}}{(n+1)^3}\cdot\frac{n^3}{2^n}\right|$$

$$= \lim_{n\to\infty}\frac{2n^3}{(n+1)^3} = 2$$

Therefore, by the Ratio Test, the series diverges.

36. $\displaystyle\sum_{n=1}^{\infty}\frac{1}{\sqrt[4]{n^3}} = \sum_{n=1}^{\infty}\frac{1}{n^{3/4}}$

Divergent p-series, $p = \frac{3}{4} < 1$

37. $\displaystyle\sum_{n=1}^{\infty}\frac{1}{\sqrt{n^3 + 2n}}$

$$\lim_{n\to\infty}\frac{1/\sqrt{n^3 + 2n}}{1/(n^{3/2})} = \lim_{n\to\infty}\frac{n^{3/2}}{\sqrt{n^3 + 2n}} = 1$$

By a limit comparison test with the convergent p-series

$$\sum_{n=1}^{\infty}\frac{1}{n^{3/2}},$$ the series converges.

38. $\displaystyle\sum_{n=1}^{\infty}\frac{n+1}{n(n+2)}$

$$\lim_{n\to\infty}\frac{(n+1)/n(n+2)}{1/n} = \lim_{n\to\infty}\frac{n+1}{n+2} = 1$$

By a limit comparison test with $\displaystyle\sum_{n=1}^{\infty}\frac{1}{n}$, the series diverges.

39. $\displaystyle\sum_{n=1}^{\infty} \frac{n}{e^{n^2}}$

$$\lim_{n\to\infty}\left|\frac{a_{n+1}}{a_n}\right| = \lim_{n\to\infty}\left|\frac{n+1}{e^{(n+1)^2}}\cdot\frac{e^{n^2}}{n}\right|$$

$$= \lim_{n\to\infty}\left|\frac{e^{n^2}(n+1)}{e^{n^2+2n+1}n}\right|$$

$$= \lim_{n\to\infty}\left(\frac{1}{e^{2n+1}}\right)\left(\frac{n+1}{n}\right)$$

$$= (0)(1) = 0 < 1$$

By the Ratio Test, the series converges.

41. $\displaystyle\sum_{n=1}^{\infty} \frac{(-1)^n n}{\ln(n)}$

$$\lim_{n\to\infty}\frac{n}{\ln(n)} = \infty \neq 0$$

The series diverges by the nth term test.

43. $\displaystyle\sum_{n=1}^{\infty}\left(\frac{1}{n^2}-\frac{1}{n}\right) = \sum_{n=1}^{\infty}\frac{1}{n^2} - \sum_{n=1}^{\infty}\frac{1}{n}$

Since the second series is a divergent p-series while the first series is a convergent p-series, the difference diverges.

45. $\displaystyle\sum_{n=1}^{\infty} \frac{1\cdot3\cdot5\cdots(2n-1)}{2\cdot4\cdot6\cdots(2n)}$

$$a_n = \frac{1\cdot3\cdot5\cdots(2n-1)}{2\cdot4\cdot6\cdots(2n)} = \left(\frac{3}{2}\cdot\frac{5}{4}\cdots\frac{2n-1}{2n-2}\right)\frac{1}{2n} > \frac{1}{2n}$$

Since $\displaystyle\sum_{n=1}^{\infty}\frac{1}{2n} = \frac{1}{2}\sum_{n=1}^{\infty}\frac{1}{n}$ diverges (harmonic series), so does the original series.

46. $\displaystyle\sum_{n=1}^{\infty} \frac{1\cdot3\cdot5\cdots(2n-1)}{2\cdot5\cdot8\cdots(3n-1)}$

$$\lim_{n\to\infty}\left|\frac{a_{n+1}}{a_n}\right| = \lim_{n\to\infty}\left|\frac{1\cdot3\cdots(2n-1)(2n+1)}{2\cdot5\cdots(3n-1)(3n+2)}\cdot\frac{2\cdot5\cdots(3n-1)}{1\cdot3\cdots(2n-1)}\right| = \lim_{n\to\infty}\frac{2n+1}{3n+2} = \frac{2}{3}$$

By the Ratio Test, the series converges.

47. (a) $\displaystyle\int_N^{\infty}\frac{1}{x^2}\,dx = \left[-\frac{1}{x}\right]_N^{\infty} = \frac{1}{N}$

N	5	10	20	30	40
$\displaystyle\sum_{n=1}^{N}\frac{1}{n^2}$	1.4636	1.5498	1.5962	1.6122	1.6202
$\displaystyle\int_N^{\infty}\frac{1}{x^2}\,dx$	0.2000	0.1000	0.0500	0.0333	0.0250

—CONTINUED—

40. $\displaystyle\sum_{n=1}^{\infty} \frac{n!}{e^n}$

$$\lim_{n\to\infty}\left|\frac{a_{n+1}}{a_n}\right| = \lim_{n\to\infty}\left|\frac{(n+1)!}{e^{n+1}}\cdot\frac{e^n}{n!}\right|$$

$$= \lim_{n\to\infty}\frac{n+1}{e} = \infty$$

By the Ratio Test, the series diverges.

42. $\displaystyle\sum_{n=1}^{\infty} \frac{(-1)^n\sqrt{n}}{n+1}$

$$a_{n+1} = \frac{\sqrt{n+1}}{n+2} \leq \frac{\sqrt{n}}{n+1} = a_n$$

$$\lim_{n\to\infty}\frac{\sqrt{n}}{n+1} = 0$$

By the Alternating Series Test, the series converges.

44. $\displaystyle\sum_{n=1}^{\infty}\left(\frac{1}{n^2}-\frac{1}{2^n}\right) = \sum_{n=1}^{\infty}\frac{1}{n^2} - \sum_{n=1}^{\infty}\frac{1}{2^n}$

The first series is a convergent p-series and the second series is a convergent geometric series. Therefore, their difference converges.

47. —CONTINUED—

(b) $\displaystyle\int_N^\infty \frac{1}{x^5}\,dx = \left[-\frac{1}{4x^4}\right]_N^\infty = \frac{1}{4N^4}$

N	5	10	20	30	40
$\displaystyle\sum_{n=1}^{N} \frac{1}{n^5}$	1.0367	1.0369	1.0369	1.0369	1.0369
$\displaystyle\int_N^\infty \frac{1}{x^5}\,dx$	0.0004	0.0000	0.0000	0.0000	0.0000

The series in part (b) converges more rapidly. The integral values represent the remainders of the partial sums.

48. No. Let $a_n = \dfrac{3937.5}{n^2}$, then $a_{75} = 0.7$. The series $\displaystyle\sum_{n=1}^{\infty} \frac{3937.5}{n^2}$ is a convergent p-series.

49. $\displaystyle\sum_{n=0}^{\infty} \left(\frac{x}{10}\right)^n$

Geometric series which converges only if $|x/10| < 1$ or $-10 < x < 10$.

50. $\displaystyle\sum_{n=0}^{\infty} (2x)^n$

Geometric series which converges only if $|2x| < 1$ or $-\frac{1}{2} < x < \frac{1}{2}$.

51. $\displaystyle\sum_{n=0}^{\infty} \frac{(-1)^n(x-2)^n}{(n+1)^2}$

$\displaystyle\lim_{n\to\infty}\left|\frac{u_{n+1}}{u_n}\right| = \lim_{n\to\infty}\left|\frac{(-1)^{n+1}(x-2)^{n+1}}{(n+2)^2}\cdot\frac{(n+1)^2}{(-1)^n(x-2)^n}\right|$

$\qquad\qquad = |x-2|$

$R = 1$

Center: 2

Since the series converges when $x = 1$ and when $x = 3$, the interval of convergence is $1 \le x \le 3$.

52. $\displaystyle\sum_{n=1}^{\infty} \frac{3^n(x-2)^n}{n}$

$\displaystyle\lim_{n\to\infty}\left|\frac{u_{n+1}}{u_n}\right| = \lim_{n\to\infty}\left|\frac{3^{n+1}(x-2)^{n+1}}{n+1}\cdot\frac{n}{3^n(x-2)^n}\right|$

$\qquad\qquad = 3|x-2|$

$R = \dfrac{1}{3}$

Center: 2

Since the series converges at $\frac{5}{3}$ and diverges at $\frac{7}{3}$, the interval of convergence is $\frac{5}{3} \le x < \frac{7}{3}$.

53. $\displaystyle\sum_{n=0}^{\infty} n!(x-2)^n$

$\displaystyle\lim_{n\to\infty}\left|\frac{u_{n+1}}{u_n}\right| = \lim_{n\to\infty}\left|\frac{(n+1)!(x-2)^{n+1}}{n!(x-2)^n}\right| = \infty$

which implies that the series converges only at the center $x = 2$.

54. $\displaystyle\sum_{n=0}^{\infty} \frac{(x-2)^n}{2^n} = \sum_{n=0}^{\infty}\left(\frac{x-2}{2}\right)^n$

Geometric series which converges only if

$\left|\dfrac{x-2}{2}\right| < 1 \quad\text{or}\quad 0 < x < 4.$

55. $f(x) = \sin(x)$

$f'(x) = \cos(x)$

$f''(x) = -\sin(x)$

$f'''(x) = -\cos(x), \cdots$

$\displaystyle\sin(x) = \sum_{n=0}^{\infty} \frac{f^{(n)}(x)[x-(3\pi/4)]^n}{n!}$

$\displaystyle\quad = \frac{\sqrt{2}}{2} - \frac{\sqrt{2}}{2}\left(x - \frac{3\pi}{4}\right) - \frac{\sqrt{2}}{2\cdot 2!}\left(x - \frac{3\pi}{4}\right)^2 + \cdots = \frac{\sqrt{2}}{2}\sum_{n=0}^{\infty}\frac{(-1)^{n(n+1)/2}[x-(3\pi/4)]^n}{n!}$

56. $f(x) = \cos x$

$f'(x) = -\sin x$

$f''(x) = -\cos x$

$f'''(x) = \sin x$

$$\cos x = \sum_{n=0}^{\infty} \frac{f^{(n)}(-\pi/4)[x + (\pi/4)]^n}{n!}$$

$$= \frac{\sqrt{2}}{2} + \frac{\sqrt{2}}{2}\left(x + \frac{\pi}{4}\right) - \frac{\sqrt{2}}{2 \cdot 2!}\left(x + \frac{\pi}{4}\right)^2 - \frac{\sqrt{2}}{2 \cdot 3!}\left(x + \frac{\pi}{4}\right)^3 + \frac{\sqrt{2}}{2 \cdot 4!}\left(x + \frac{\pi}{4}\right)^4 + \cdots$$

$$= \frac{\sqrt{2}}{2}\left[1 + \left(x + \frac{\pi}{4}\right) + \sum_{n=1}^{\infty} \frac{(-1)^{[n(n+1)]/2}[x + (\pi/4)]^{n+1}}{(n+1)!}\right]$$

57. $3^x = (e^{\ln(3)})^x = e^{x \ln(3)}$ and since $e^x = \sum_{n=0}^{\infty} \frac{x^n}{n!}$, we have

$$3^x = \sum_{n=0}^{\infty} \frac{(x \ln 3)^n}{n!} = 1 + x \ln 3 + \frac{x^2 \ln^2 3}{2!} + \frac{x^3 \ln^3 3}{3!} + \frac{x^4 \ln^4 3}{4!} + \cdots.$$

58. $f(x) = \csc(x)$

$f'(x) = -\csc(x)\cot(x)$

$f''(x) = \csc^3(x) + \csc(x)\cot^2(x)$

$f'''(x) = -5\csc^3(x)\cot(x) - \csc(x)\cot^3(x)$

$f^{(4)}(x) = 5\csc^5(x) + 15\csc^3(x)\cot^2(x) + \csc(x)\cot^4(x)$

$$\csc(x) = \sum_{n=0}^{\infty} \frac{f^{(n)}(\pi/2)[x - (\pi/2)]^n}{n!}$$

$$= 1 + \frac{1}{2!}\left(x - \frac{\pi}{2}\right)^2 + \frac{5}{4!}\left(x - \frac{\pi}{2}\right)^4 + \cdots$$

59. $f(x) = \dfrac{1}{x}$

$f'(x) = -\dfrac{1}{x^2}$

$f''(x) = \dfrac{2}{x^3}$

$f'''(x) = -\dfrac{6}{x^4}, \cdots$

$$\frac{1}{x} = \sum_{n=0}^{\infty} \frac{f^{(n)}(-1)(x + 1)^n}{n!}$$

$$= \sum_{n=0}^{\infty} \frac{-n!(x + 1)^n}{n!} = -\sum_{n=0}^{\infty} (x + 1)^n$$

60. $f(x) = x^{1/2}$

$f'(x) = \dfrac{1}{2}x^{-1/2}$

$f''(x) = -\left(\dfrac{1}{2}\right)\left(\dfrac{1}{2}\right)x^{-3/2}$

$f'''(x) = \left(\dfrac{1}{2}\right)\left(\dfrac{1}{2}\right)\left(\dfrac{3}{2}\right)x^{-5/2}$

$f^{(4)}(x) = -\left(\dfrac{1}{2}\right)\left(\dfrac{1}{2}\right)\left(\dfrac{3}{2}\right)\left(\dfrac{5}{2}\right)x^{-7/2}, \cdots$

$$\sqrt{x} = \sum_{n=0}^{\infty} \frac{f^{(n)}(4)(x - 4)^n}{n!}$$

$$= 2 + \frac{(x - 4)}{2^2} - \frac{(x - 4)^2}{2^5 2!} + \frac{1 \cdot 3(x - 4)^3}{2^8 3!} - \frac{1 \cdot 3 \cdot 5(x - 4)^4}{2^{11} 4!} + \cdots$$

$$= 2 + \frac{(x - 4)}{2^2} + \sum_{n=2}^{\infty} \frac{(-1)^{n+1} 1 \cdot 3 \cdot 5 \cdots (2n - 3)(x - 4)^n}{2^{3n-1} n!}$$

61. $g(x) = \dfrac{2}{3-x} = \dfrac{2/3}{1-(x/3)} = \displaystyle\sum_{n=0}^{\infty} \dfrac{2}{3}\left(\dfrac{x}{3}\right)^n = \displaystyle\sum_{n=0}^{\infty} \dfrac{2x^n}{3^{n+1}}$

62.
$$h(x) = (1+x)^{-3}$$
$$h'(x) = -3(1+x)^{-4}$$
$$h''(x) = 12(1+x)^{-5}$$
$$h'''(x) = -60(1+x)^{-6}$$
$$h^{(4)}(x) = 360(1+x)^{-7}$$
$$h^{(5)}(x) = -2520(1+x)^{-8}$$

$$\dfrac{1}{(1+x)^3} = 1 - 3x + \dfrac{12x^2}{2!} - \dfrac{60x^3}{3!} + \dfrac{360x^4}{4!} - \dfrac{2520x^5}{5!} + \cdots = \sum_{n=0}^{\infty} \dfrac{(-1)^n(n+2)!\,x^n}{2n!} = \sum_{n=0}^{\infty} \dfrac{(-1)^n(n+2)(n+1)x}{2}$$

63. $\ln x = \displaystyle\sum_{n=1}^{\infty} (-1)^{n+1} \dfrac{(x-1)^n}{n}, \quad 0 < x \le 2$

$\ln\left(\dfrac{5}{4}\right) = \displaystyle\sum_{n=1}^{\infty} (-1)^{n+1}\left(\dfrac{(5/4)-1}{n}\right)^n$

$\qquad = \displaystyle\sum_{n=1}^{\infty} (-1)^{n+1} \dfrac{1}{4^n n} \approx 0.2231$

64. $\ln x = \displaystyle\sum_{n=1}^{\infty} (-1)^{n+1} \dfrac{(x-1)^n}{n}, \qquad 0 < x \le 2$

$\ln\left(\dfrac{6}{5}\right) = \displaystyle\sum_{n=1}^{\infty} (-1)^{n+1}\left(\dfrac{(6/5)-1}{n}\right)^n$

$\qquad = \displaystyle\sum_{n=1}^{\infty} (-1)^{n+1} \dfrac{1}{5^n n} \approx 0.1823$

65. $e^x = \displaystyle\sum_{n=0}^{\infty} \dfrac{x^n}{n!}, \quad -\infty < x < \infty$

$e^{1/2} = \displaystyle\sum_{n=0}^{\infty} \dfrac{1}{2^n n!} \approx 1.6487$

66. $e^x = \displaystyle\sum_{n=0}^{\infty} \dfrac{x^n}{n!}, \quad -\infty < x < \infty$

$e^{2/3} = \displaystyle\sum_{n=0}^{\infty} \dfrac{(2/3)^n}{n!} = \displaystyle\sum_{n=0}^{\infty} \dfrac{2^n}{3^n n!} \approx 1.9477$

67. $\cos x = \displaystyle\sum_{n=0}^{\infty} (-1)^n \dfrac{x^{2n}}{(2n)!}, \quad -\infty < x < \infty$

$\cos\left(\dfrac{2}{3}\right) = \displaystyle\sum_{n=0}^{\infty} (-1)^n \dfrac{2^{2n}}{3^{2n}(2n)!} \approx 0.7859$

68. $\sin x = \displaystyle\sum_{n=0}^{\infty} (-1)^n \dfrac{x^{2n+1}}{(2n+1)!}, \quad -\infty < x < \infty$

$\sin\left(\dfrac{1}{3}\right) = \displaystyle\sum_{n=0}^{\infty} (-1)^n \dfrac{1}{3^{2n+1}(2n+1)!} \approx 0.3272$

69. The series for Exercise 37 converges very slowly because the terms approach 0 at a slow rate.

70. $e^x = \displaystyle\sum_{n=0}^{\infty} \dfrac{x^n}{n!} = 1 + x + \dfrac{x^2}{2!} + \cdots$

$xe^x = \displaystyle\sum_{n=0}^{\infty} \dfrac{x^{n+1}}{n!} = x + x^2 + \dfrac{x^3}{2!} + \cdots$

$\displaystyle\int_0^1 xe^x\,dx = \left[xe^x - e^x \right]_0^1 = (e-e) - (0-1) = 1$

$\displaystyle\int_0^1 \sum_{n=0}^{\infty} \dfrac{x^{n+1}}{n!}\,dx = \sum_{n=0}^{\infty} \left[\dfrac{x^{n+2}}{(n+2)n!} \right]_0^1 = \sum_{n=0}^{\infty} \dfrac{1}{(n+2)n!} = 1$

71. (a) $f(x) = e^{2x}$ $f(0) = 1$

     $f'(x) = 2e^{2x}$ $f'(0) = 2$

     $f''(x) = 4e^{2x}$ $f''(0) = 4$

     $f'''(x) = 8e^{2x}$ $f'''(0) = 8$

     $e^{2x} = 1 + 2x + \dfrac{4x^2}{2!} + \dfrac{8x^3}{3!} + \cdots$

        $= 1 + 2x + 2x^2 + \dfrac{4}{3}x^3 + \cdots$

(b) $e^x = \displaystyle\sum_{n=0}^{\infty} \dfrac{x^n}{n!}$

     $e^{2x} = \displaystyle\sum_{n=0}^{\infty} \dfrac{(2x)^n}{n!} = 1 + 2x + \dfrac{4x^2}{2!} + \dfrac{8x^3}{3!} + \cdots$

        $= 1 + 2x + 2x^2 + \dfrac{4}{3}x^3 + \cdots$

(c) $e^{2x} = e^x \cdot e^x = \left(1 + x + \dfrac{x^2}{2} + \dfrac{x^3}{6} + \cdots\right)\left(1 + x + \dfrac{x^2}{2} + \dfrac{x^3}{6} + \cdots\right)$

     $= 1 + (x + x) + \left(x^2 + \dfrac{x^2}{2} + \dfrac{x^2}{2}\right) + \left(\dfrac{x^3}{6} + \dfrac{x^3}{6} + \dfrac{x^3}{2} + \dfrac{x^3}{2}\right) + \cdots = 1 + 2x + 2x^2 + \dfrac{4}{3}x^3 + \cdots$

72. (a) $f(x) = \sin 2x$ $f(0) = 0$

      $f'(x) = 2\cos 2x$ $f'(0) = 2$

      $f''(x) = -4\sin 2x$ $f''(0) = 0$

      $f'''(x) = -8\cos 2x$ $f'''(0) = -8$

      $f^{(4)}(x) = 16\sin 2x$ $f^{(4)}(0) = 0$

      $f^{(5)}(x) = 32\cos 2x$ $f^{(5)}(0) = 32$

      $f^{(6)}(x) = -64\sin 2x$ $f^{(6)}(0) = 0$

      $f^{(7)}(x) = -128\cos 2x$ $f^{(7)}(0) = -128$

      $\sin 2x = 0 + 2x + \dfrac{0x^2}{2!} - \dfrac{8x^3}{3!} + \dfrac{0x^4}{4!} + \dfrac{32x^5}{5!} + \dfrac{0x^6}{6!} - \dfrac{128x^7}{7!} + \cdots = 2x - \dfrac{4}{3}x^3 + \dfrac{4}{15}x^5 - \dfrac{8}{315}x^7 + \cdots$

(b) $\sin x = \displaystyle\sum_{n=0}^{\infty} \dfrac{(-1)^n x^{2n+1}}{(2n+1)!}$

     $\sin 2x = \displaystyle\sum_{n=0}^{\infty} \dfrac{(-1)^n (2x)^{2n+1}}{(2n+1)!} = 2x - \dfrac{(2x)^3}{3!} + \dfrac{(2x)^5}{5!} - \dfrac{(2x)^7}{7!} + \cdots$

        $= 2x - \dfrac{8x^3}{6} + \dfrac{32x^5}{120} - \dfrac{128x^7}{5040} + \cdots = 2x - \dfrac{4}{3}x^3 + \dfrac{4}{15}x^5 - \dfrac{8}{315}x^7 + \cdots$

(c) $\sin 2x = 2\sin x \cos x$

     $= 2\left(x - \dfrac{x^3}{6} + \dfrac{x^5}{120} - \dfrac{x^7}{5040} + \cdots\right)\left(1 - \dfrac{x^2}{2} + \dfrac{x^4}{24} - \dfrac{x^6}{720} + \cdots\right)$

     $= 2\left[x + \left(-\dfrac{x^3}{2} - \dfrac{x^3}{6}\right) + \left(\dfrac{x^5}{24} + \dfrac{x^5}{12} + \dfrac{x^5}{120}\right) + \left(-\dfrac{x^7}{720} - \dfrac{x^7}{144} - \dfrac{x^7}{240} - \dfrac{x^7}{5040}\right) + \cdots\right]$

     $= 2\left[x - \dfrac{2x^3}{3} + \dfrac{2x^5}{15} - \dfrac{4x^7}{315} + \cdots\right] = 2x - \dfrac{4}{3}x^3 + \dfrac{4}{15}x^5 - \dfrac{8}{315}x^7 + \cdots$

73. $1 + \dfrac{2}{3}x + \dfrac{4}{9}x^2 + \dfrac{8}{27}x^3 + \cdots = \displaystyle\sum_{n=0}^{\infty}\left(\dfrac{2x}{3}\right)^n = \dfrac{1}{1 - (2x/3)} = \dfrac{3}{3 - 2x}, \quad -\dfrac{3}{2} < x < \dfrac{3}{2}$

74. $8 - 2(x-3) + \dfrac{1}{2}(x-3)^2 - \dfrac{1}{8}(x-3)^3 + \cdots = \displaystyle\sum_{n=0}^{\infty} 8\left[\dfrac{-(x-3)}{4}\right]^n = \dfrac{8}{1 - [-(x-3)/4]}$

                          $= \dfrac{32}{4 + (x-3)} = \dfrac{32}{1 + x}, \quad -1 < x < 7$

75.
$$\sin t = \sum_{n=0}^{\infty} \frac{(-1)^n t^{2n+1}}{(2n+1)!}$$

$$\frac{\sin t}{t} = \sum_{n=0}^{\infty} \frac{(-1)^n t^{2n}}{(2n+1)!}$$

$$\int_0^x \frac{\sin t}{t}\, dt = \left[\sum_{n=0}^{\infty} \frac{(-1)^n t^{2n+1}}{(2n+1)(2n+1)!} \right]_0^x = \sum_{n=0}^{\infty} \frac{(-1)^n x^{2n+1}}{(2n+1)(2n+1)!}$$

76.
$$\cos t = \sum_{n=0}^{\infty} \frac{(-1)^n t^{2n}}{(2n)!}$$

$$\cos \frac{\sqrt{t}}{2} = \sum_{n=0}^{\infty} \frac{(-1)^n t^n}{2^{2n}(2n)!}$$

$$\int_0^x \cos \frac{\sqrt{t}}{2}\, dt = \left[\sum_{n=0}^{\infty} \frac{(-1)^n t^{n+1}}{2^{2n}(2n)!(n+1)} \right]_0^x = \sum_{n=0}^{\infty} \frac{(-1)^n x^{n+1}}{2^{2n}(2n)!(n+1)}$$

77.
$$\frac{1}{1+t} = \sum_{n=0}^{\infty} (-1)^n t^n$$

$$\ln(1+t) = \int \frac{1}{1+t}\, dt = \sum_{n=0}^{\infty} \frac{(-1)^n t^{n+1}}{n+1}$$

$$\frac{\ln(t+1)}{t} = \sum_{n=0}^{\infty} \frac{(-1)^n t^n}{n+1}$$

$$\int_0^x \frac{\ln(t+1)}{t}\, dt = \left[\sum_{n=0}^{\infty} \frac{(-1)^n t^{n+1}}{(n+1)^2} \right]_0^x = \sum_{n=0}^{\infty} \frac{(-1)^n x^{n+1}}{(n+1)^2}$$

78.
$$e^t = \sum_{n=0}^{\infty} \frac{t^n}{n!}$$

$$e^t - 1 = \sum_{n=1}^{\infty} \frac{t^n}{n!}$$

$$\frac{e^t - 1}{t} = \sum_{n=1}^{\infty} \frac{t^{n-1}}{n!}$$

$$\int_0^x \frac{e^t - 1}{t}\, dt = \left[\sum_{n=1}^{\infty} \frac{t^n}{n \cdot n!} \right]_0^x = \sum_{n=1}^{\infty} \frac{x^n}{n \cdot n!}$$

79.
$$\arctan x = x - \frac{x^3}{3} + \frac{x^5}{5} - \frac{x^7}{7} + \frac{x^9}{9} - \cdots$$

$$\frac{\arctan x}{\sqrt{x}} = \sqrt{x} - \frac{x^{5/2}}{3} + \frac{x^{9/2}}{5} - \frac{x^{13/2}}{7} + \frac{x^{17/2}}{9} - \cdots$$

$$\lim_{x \to 0} \frac{\arctan x}{\sqrt{x}} = 0$$

By L'Hôpital's Rule, $\displaystyle \lim_{x \to 0} \frac{\arctan x}{\sqrt{x}} = \lim_{x \to 0} \frac{\left(\dfrac{1}{1+x^2}\right)}{\left(\dfrac{1}{2\sqrt{x}}\right)} = \lim_{x \to 0} \frac{2\sqrt{x}}{1+x^2} = 0.$

80.
$$\arcsin x = x + \frac{x^3}{2 \cdot 3} + \frac{1 \cdot 3 x^5}{2 \cdot 4 \cdot 5} + \frac{1 \cdot 3 \cdot 5 x^7}{2 \cdot 4 \cdot 6 \cdot 7} + \cdots$$

$$\frac{\arcsin x}{x} = 1 + \frac{x^2}{2 \cdot 3} + \frac{1 \cdot 3 x^4}{2 \cdot 4 \cdot 5} + \frac{1 \cdot 3 \cdot 5 x^6}{2 \cdot 4 \cdot 6 \cdot 7} + \cdots$$

$$\lim_{x \to 0} \frac{\arcsin x}{x} = 1$$

By L'Hôpital's Rule, $\displaystyle \lim_{x \to 0} \frac{\arcsin x}{x} = \lim_{x \to 0} \frac{\left(\dfrac{1}{\sqrt{1-x^2}}\right)}{1} = 1.$

81.
$$y = \sum_{n=0}^{\infty} (-1)^n \frac{x^{2n}}{4^n(n!)^2}$$

$$y' = \sum_{n=1}^{\infty} \frac{(-1)^n(2n)x^{2n-1}}{4^n(n!)^2} = \sum_{n=0}^{\infty} \frac{(-1)^{n+1}(2n+2)x^{2n+1}}{4^{n+1}[(n+1)!]^2}$$

$$y'' = \sum_{n=0}^{\infty} \frac{(-1)^{n+1}(2n+2)(2n+1)x^{2n}}{4^{n+1}[(n+1)!]^2}$$

$$x^2y'' + xy' + x^2y = \sum_{n=0}^{\infty} \frac{(-1)^{n+1}(2n+2)(2n+1)x^{2n+2}}{4^{n+1}[(n+1)!]^2} + \sum_{n=0}^{\infty} \frac{(-1)^{n+1}(2n+2)x^{2n+2}}{4^{n+1}[(n+1)!]^2} + \sum_{n=0}^{\infty} (-1)^n \frac{x^{2n+1}}{4^n(n!)^2}$$

$$= \sum_{n=0}^{\infty} \left[(-1)^{n+1} \frac{(2n+2)(2n+1)}{4^{n+1}[(n+1)!]^2} + \frac{(-1)^{n+1}(2n+2)}{4^{n+1}[(n+1)!]^2} + \frac{(-1)^n}{4^n(n!)^2} \right] x^{2n+2}$$

$$= \sum_{n=0}^{\infty} \left[\frac{(-1)^{n+1}(2n+2)(2n+1+1)}{4^{n+1}[(n+1)!]^2} + (-1)^n \frac{1}{4^n(n!)^2} \right] x^{2n+2}$$

$$= \sum_{n=0}^{\infty} \left[\frac{(-1)^{n+1}4(n+1)^2}{4^{n+1}[(n+1)!]^2} + (-1)^n \frac{1}{4^n(n!)^2} \right] x^{2n+2}$$

$$= \sum_{n=0}^{\infty} \left[\frac{(-1)^{n+1}1}{4^n(n!)^2} + (-1)^n \frac{1}{4^n(n!)^2} \right] x^{2n+2} = 0$$

82.
$$y = \sum_{n=0}^{\infty} \frac{(-3)^n x^{2n}}{2^n n!}$$

$$y' = \sum_{n=1}^{\infty} \frac{(-3)^n(2n)x^{2n-1}}{2^n n!} = \sum_{n=0}^{\infty} \frac{(-3)^{n+1}(2n+2)x^{2n+1}}{2^{n+1}(n+1)!}$$

$$y'' = \sum_{n=0}^{\infty} \frac{(-3)^{n+1}(2n+2)(2n+1)x^{2n}}{2^{n+1}(n+1)!}$$

$$y'' + 3xy' + 3y = \sum_{n=0}^{\infty} \frac{(-3)^{n+1}(2n+2)(2n+1)x^{2n}}{2^{n+1}(n+1)!} + \sum_{n=0}^{\infty} \frac{(-1)^{n+1}3^{n+2}(2n+2)x^{2n+2}}{2^{n+1}(n+1)!} + \sum_{n=0}^{\infty} \frac{(-1)^n 3^{n+1}x^{2n}}{2^n n!}$$

$$= \sum_{n=0}^{\infty} \frac{(-1)^{n+1}3^{n+1}(2n+2)x^{2n}}{2^n n!} + \sum_{n=0}^{\infty} \frac{(-1)^{n+1}3^{n+2}x^{2n+2}}{2^n n!} + \sum_{n=0}^{\infty} \frac{(-1)^n 3^{n+1}x^{2n}}{2^n n!}$$

$$= \sum_{n=0}^{\infty} \frac{(-1)^n 3^{n+1}x^{2n}}{2^n n!} [-(2n+1)+1] + \sum_{n=0}^{\infty} \frac{(-1)^{n+1}3^{n+2}x^{2n+2}}{2^n n!}$$

$$= \sum_{n=0}^{\infty} \frac{(-1)^n 3^{n+1}x^{2n}}{2^n n!} (-2n) + \sum_{n=0}^{\infty} \frac{(-1)^{n+1}3^{n+2}x^{2n+2}}{2^n n!}$$

$$= \sum_{n=1}^{\infty} \frac{(-1)^{n+1}3^{n+1}x^{2n}}{2^n n!} (2n) + \sum_{n=1}^{\infty} \frac{(-1)^n 3^{n+1}x^{2n}}{2^{n-1}(n-1)!} \cdot \frac{2n}{2n}$$

$$= \sum_{n=1}^{\infty} \frac{(-1)^n 3^{n+1}x^{2n}}{2^n n!} [-2n + 2n] = 0$$

83. $\sin(95°) = \sin\left(\frac{95\pi}{180}\right) \approx \frac{95\pi}{180} - \frac{(95\pi)^3}{180^3 3!} + \frac{(95\pi)^5}{180^5 5!} - \frac{(95\pi)^7}{180^7 7!} + \frac{(95\pi)^9}{180^9 9!} \approx 0.996$

84. $\cos(0.75) \approx 1 - \frac{(0.75)^2}{2!} + \frac{(0.75)^4}{4!} - \frac{(0.75)^6}{6!} \approx 0.7317$

85. $\ln(1.75) \approx (0.75) - \frac{(0.75)^2}{2} + \frac{(0.75)^3}{3} - \frac{(0.75)^4}{4} + \frac{(0.75)^5}{5} - \frac{(0.75)^6}{6} + \cdots + \frac{(0.75)^{15}}{15} \approx 0.560$

86. $e^{-0.25} \approx 1 - 0.25 + \frac{(0.25)^2}{2!} - \frac{(0.25)^3}{3!} + \frac{(0.25)^4}{4!} \approx 0.779$

87. $f(x) = \cos x, \ c = 0$

$$R_n(x) = \frac{f^{(n+1)}(z)}{(n+1)!} x^{n+1}$$

$$\left| f^{(n+1)}(z) \right| \le 1 \implies R_n(x) \le \frac{x^{n+1}}{(n+1)!}$$

(a) $R_n(x) \le \dfrac{(0.5)^{n+1}}{(n+1)!} < 0.001$

This inequality is true for $n = 4$.

(c) $R_n(x) \le \dfrac{(0.5)^{n+1}}{(n+1)!} < 0.0001$

This inequality is true for $n = 5$.

(b) $R_n(x) \le \dfrac{(1)^{n+1}}{(n+1)!} < 0.001$

This inequality is true for $n = 6$.

(d) $R_n(x) \le \dfrac{2^{n+1}}{(n+1)!} < 0.0001$

This inequality is true for $n = 10$.

88. $f(x) = \cos x$

$$P_4(x) = 1 - \frac{x^2}{2!} + \frac{x^4}{4!}$$

$$P_6(x) = 1 - \frac{x^2}{2!} + \frac{x^4}{4!} - \frac{x^6}{6!}$$

$$P_{10}(x) = 1 - \frac{x^2}{2!} + \frac{x^4}{4!} - \frac{x^6}{6!} + \frac{x^8}{8!} - \frac{x^{10}}{10!}$$

CHAPTER 9
Conics, Parametric Equations, and Polar Coordinates

CHAPTER 9
Conics, Parametric Equations, and Polar Coordinates

Section 9.1   Conics and Calculus

Solutions to Exercises

1. $y^2 = 4x$

Vertex: $(0, 0)$

$p = 1 > 0$

Opens to the right
Matches graph (h).

2. $x^2 = 8y$

Vertex: $(0, 0)$

$p = 2 > 0$

Opens upward
Matches graph (a).

3. $(x + 3)^2 = -2(y - 2)$

Vertex: $(-3, 2)$

$p = -\frac{1}{2} < 0$

Opens downward
Matches graph (e).

4. $\dfrac{(x - 2)^2}{16} + \dfrac{(y + 1)^2}{4} = 1$

Center: $(2, -1)$
Ellipse
Matches (b)

5. $\dfrac{x^2}{9} + \dfrac{y^2}{4} = 1$

Center: $(0, 0)$
Ellipse
Matches (f)

6. $\dfrac{x^2}{9} + \dfrac{y^2}{9} = 1$

Circle radius 3.
Matches (g)

7. $\dfrac{y^2}{16} - \dfrac{x^2}{1} = 1$

Hyperbola

Center: $(0, 0)$

Vertical transverse axis.
Matches (c)

8. $\dfrac{(x - 2)^2}{9} - \dfrac{y^2}{4} = 1$

Hyperbola

Center: $(-2, 0)$

Horizontal transverse axis.
Matches (d)

9. $y^2 = -6x = 4\left(-\frac{3}{2}\right)x$

Vertex: $(0, 0)$

Focus: $\left(-\frac{3}{2}, 0\right)$

Directrix: $x = \frac{3}{2}$

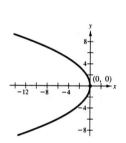

10. $x^2 + 8y = 0$

$x^2 = 4(-2)y$

Vertex: $(0, 0)$

Focus: $(0, -2)$

Directrix: $y = 2$

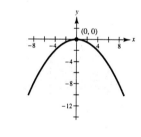

11. $(x + 3) + (y - 2)^2 = 0$

$(y - 2)^2 = 4\left(-\frac{1}{4}\right)(x + 3)$

Vertex: $(-3, 2)$

Focus: $(-3.25, 2)$

Directrix: $x = -2.75$

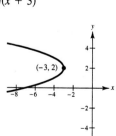

12. $(x - 1)^2 + 8(y + 2) = 0$

$(x - 1)^2 = 4(-2)(y + 2)$

Vertex: $(1, -2)$

Focus: $(1, -4)$

Directrix: $y = 0$

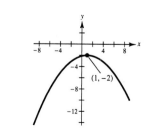

13. $y^2 - 4y - 4x = 0$

$\qquad y^2 - 4y + 4 = 4x + 4$

$\qquad\quad (y - 2)^2 = 4(1)(x + 1)$

Vertex: $(-1, 2)$

Focus: $(0, 2)$

Directrix: $x = -2$

14. $y^2 + 6y + 8x + 25 = 0$

$\qquad y^2 + 6y + 9 = -8x - 25 + 9$

$\qquad\quad (y + 3)^2 = 4(-2)(x + 2)$

Vertex: $(-2, -3)$

Focus: $(-4, -3)$

Directrix: $x = 0$

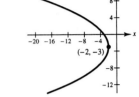

15. $x^2 + 4x + 4y - 4 = 0$

$\qquad x^2 + 4x + 4 = -4y + 4 + 4$

$\qquad\quad (x + 2)^2 = 4(-1)(y - 2)$

Vertex: $(-2, 2)$

Focus: $(-2, 1)$

Directrix: $y = 3$

16. $y^2 + 4y + 8x - 12 = 0$

$\qquad y^2 + 4y + 4 = -8x + 12 + 4$

$\qquad\quad (y + 2)^2 = 4(-2)(x - 2)$

Vertex: $(2, -2)$

Focus: $(0, -2)$

Directrix: $x = 4$

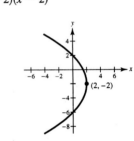

17. $y^2 + x + y = 0$

$\qquad y^2 + y + \frac{1}{4} = -x + \frac{1}{4}$

$\qquad\left(y + \frac{1}{2}\right)^2 = 4\left(-\frac{1}{4}\right)\left(x - \frac{1}{4}\right)$

Vertex: $\left(\frac{1}{4}, -\frac{1}{2}\right)$

Focus: $\left(0, -\frac{1}{2}\right)$

Directrix: $x = \frac{1}{2}$

18. $\qquad y = -\frac{1}{6}(x^2 + 4x - 2)$

$\qquad -6y + 6 = (x + 2)^2$

$\qquad (x + 2)^2 = 4\left(-\frac{3}{2}\right)(y - 1)$

Vertex: $(-2, 1)$

Focus: $\left(-2, -\frac{1}{2}\right)$

Directrix: $y = \frac{5}{2}$

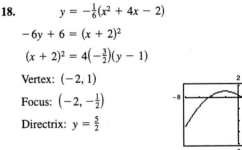

19. $y^2 - 4x - 4 = 0$

$\qquad\quad y^2 = 4x + 4$

$\qquad\quad\ = 4(1)(x + 1)$

Vertex: $(-1, 0)$

Focus: $(0, 0)$

Directrix: $x = -2$

20. $x^2 - 2x + 8y + 9 = 0$

$\qquad x^2 - 2x + 1 = -8y - 9 + 1$

$\qquad (x - 1)^2 = 4(-2)(y + 1)$

Vertex: $(1, -1)$

Focus: $(1, -3)$

Directrix: $y = 1$

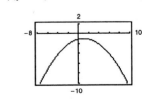

21. $\qquad (y - 2)^2 = 4(-2)(x - 3)$

$\quad y^2 - 4y + 8x - 20 = 0$

22. $\qquad (x + 1)^2 = 4(-2)(y - 2)$

$\quad x^2 + 2x + 8y - 15 = 0$

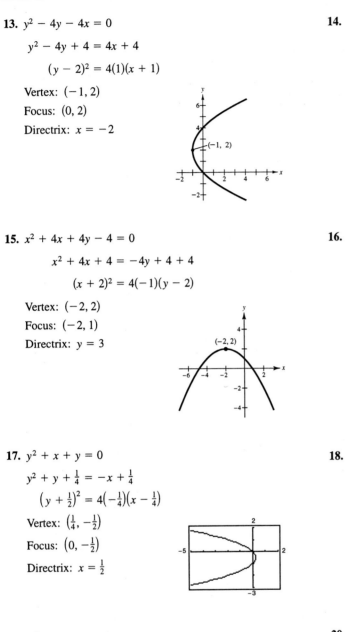
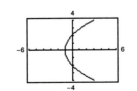

23.
$$(x - h)^2 = 4p(y - k)$$
$$x^2 = 4(6)(y - 4)$$
$$x^2 - 24y + 96 = 0$$

24. Vertex: $(0, 2)$
$$(y - 2)^2 = 4(2)(x - 0)$$
$$y^2 - 8x - 4y + 4 = 0$$

25.
$$y = 4 - x^2$$
$$x^2 + y - 4 = 0$$

26.
$$y = 4 - (x - 2)^2 = 4x - x^2$$
$$x^2 - 4x + y = 0$$

27. Since the axis of the parabola is vertical, the form of the equation is $y = ax^2 + bx + c$. Now, substituting the values of the given coordinates into this equation, we obtain

$$3 = c, 4 = 9a + 3b + c, 11 = 16a + 4b + c.$$

Solving this system, we have $a = \frac{5}{3}, b = -\frac{14}{3}, c = 3$. Therefore,

$$y = \frac{5}{3}x^2 - \frac{14}{3}x + 3 \text{ or } 5x^2 - 14x - 3y + 9 = 0.$$

28. From Example 2: $4p = 8$ or $p = 2$

Vertex: $(4, 0)$
$$(x - 4)^2 = 8(y - 0)$$
$$x^2 - 8x - 8y + 16 = 0$$

29. Assume that the vertex is at the origin.

$$x^2 = 4py$$
$$(3)^2 = 4p(1)$$
$$\frac{9}{4} = p$$

The pipe is located $\frac{9}{4}$ meters from the vertex.

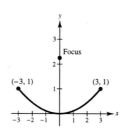

30. Assume that the vertex is at the origin.

(a)
$$x^2 = 4py$$
$$8^2 = 4p\left(\frac{3}{100}\right)$$
$$\frac{1600}{3} = p$$
$$x^2 = 4\left(\frac{1600}{3}\right)y = \frac{6400}{3}y$$

(b) The deflection is 1 cm when

$$y = \frac{2}{100} \implies x = \pm\sqrt{\frac{128}{3}} \approx \pm 6.53 \text{ meters.}$$

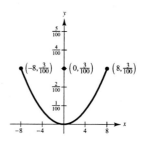

31. $y = ax^2$

$y' = 2ax$

The equation of the tangent line is

$$y - ax_0^2 = 2ax_0(x - x_0) \text{ or } y = 2ax_0 x - ax_0^2.$$

Let $y = 0$. Then:

$$-ax_0^2 = 2ax_0 x - 2ax_0^2$$
$$ax_0^2 = 2ax_0 x$$

Therefore, $\frac{x_0}{2} = x$ is the x-intercept.

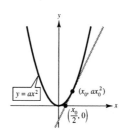

32. (a) Without loss of generality, place the coordinate system so that the equation of the parabola is $x^2 = 4py$ and, hence,

$$y' = \left(\frac{1}{2p}\right)x.$$

Therefore, for distinct tangent lines, the slopes are unequal and the lines intersect.

(b) $x^2 - 4x - 4y = 0$

$$2x - 4 - 4\frac{dy}{dx} = 0$$

$$\frac{dy}{dx} = \frac{1}{2}x - 1$$

At $(0, 0)$, the slope is -1: $y = -x$. At $(6, 3)$, the slope is 2: $y = 2x - 9$. Solving for x,

$$-x = 2x - 9$$

$$-3x = -9$$

$$x = 3$$

$$y = -3.$$

Point of intersection: $(3, -3)$

33. (a) Consider the parabola $x^2 = 4py$. Let m_0 be the slope of the one tangent line at (x_1, y_1) and therefore, $-1/m_0$ is the slope of the second at (x_2, y_2). From the derivative given in Exercise 32 we have:

$$m_0 = \frac{1}{2p}x_1 \text{ or } x_1 = 2pm_0$$

$$\frac{-1}{m_0} = \frac{1}{2p}x_2 \text{ or } x_2 = \frac{-2p}{m_0}$$

Substituting these values of x into the equation $x^2 = 4py$, we have the coordinates of the points of tangency $(2pm_0, pm_0^2)$ and $(-2p/m_0, p/m_0^2)$ and the equations of the tangent lines are

$$(y - pm_0^2) = m_0(x - 2pm_0) \quad \text{and} \quad \left(y - \frac{p}{m_0^2}\right) = \frac{-1}{m_0}\left(x + \frac{2p}{m_0}\right).$$

The point of intersection of these lines is

$$\left(\frac{p(m_0^2 - 1)}{m_0}, -p\right) \text{ and is on the directrix, } y = -p.$$

(b) $x^2 - 4x - 4y + 8 = 0$

$$(x - 2)^2 = 4(y - 1). \text{ Vertex } (2, 1)$$

$$2x - 4 - 4\frac{dy}{dx} = 0$$

$$\frac{dy}{dx} = \frac{1}{2}x - 1$$

At $(-2, 5)$, $dy/dx = -2$. At $\left(3, \frac{5}{4}\right)$, $dy/dx = \frac{1}{2}$.

Tangent line at $(-2, 5)$: $y - 5 = -2(x + 2) \implies 2x + y - 1 = 0.$

Tangent line at $\left(3, \frac{5}{4}\right)$: $y - \frac{5}{4} = \frac{1}{2}(x - 3) \implies 2x - 4y - 1 = 0.$

Since $m_1 m_2 = (-2)\left(\frac{1}{2}\right) = -1$, the lines are perpendicular.

Point of intersection: $-2x + 1 = \frac{1}{2}x - \frac{1}{4}$

$$-\frac{5}{2}x = -\frac{5}{4}$$

$$x = \frac{1}{2}$$

$$y = 0$$

Directrix: $y = 0$ and the point of intersection $\left(\frac{1}{2}, 0\right)$ lies on this line.

34. The focus of $x^2 = 8y = 4(2)y$ is $(0, 2)$. The distance from a point on the parabola, $(x, x^2/8)$, and the focus, $(0, 2)$, is

$$d = \sqrt{(x - 0)^2 + \left(\frac{x^2}{8} - 2\right)^2}.$$

Since d is minimized when d^2 is minimized, it is sufficient to minimize the function

$$f(x) = x^2 + \left(\frac{x^2}{8} - 2\right)^2.$$

$$f'(x) = 2x + 2\left(\frac{x^2}{8} - 2\right)\left(\frac{x}{4}\right) = \frac{x^3}{16} + x.$$

$f'(x) = 0$ implies that

$$\frac{x^3}{16} + x = x\left(\frac{x^2}{16} + 1\right) = 0 \implies x = 0.$$

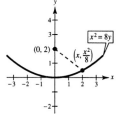

This is a minimum by the First Derivative Test. Hence, the closest point to the focus is the vertex, $(0, 0)$.

35. $y = x - x^2$

$$\frac{dy}{dx} = 1 - 2x$$

At (x_1, y_1) on the mountain, $m = 1 - 2x_1$. Also, $m = \dfrac{y_1 - 1}{x_1 + 1}$.

$$\frac{y_1 - 1}{x_1 + 1} = 1 - 2x_1$$

$$(x_1 - x_1^2) - 1 = (1 - 2x_1)(x_1 + 1)$$

$$-x_1^2 + x_1 - 1 = -2x_1^2 - x_1 + 1$$

$$x_1^2 + 2x_1 - 2 = 0$$

$$x_1 = \frac{-2 \pm \sqrt{2^2 - 4(1)(-2)}}{2(1)} = \frac{-2 \pm 2\sqrt{3}}{2} = -1 \pm \sqrt{3}$$

Choosing the positive value for x_1, we have $x_1 = -1 + \sqrt{3}$.

$$m = 1 - 2\left(-1 + \sqrt{3}\right) = 3 - 2\sqrt{3}$$

$$m = \frac{0 - 1}{x_0 + 1} = -\frac{1}{x_0 + 1}$$

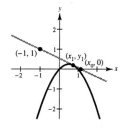

Thus, $-\dfrac{1}{x_0 + 1} = 3 - 2\sqrt{3}$

$$\frac{-1}{3 - 2\sqrt{3}} = x_0 + 1$$

$$\frac{3 + 2\sqrt{3}}{3} - 1 = x_0$$

$$\frac{2\sqrt{3}}{3} = x_0.$$

The closest the receiver can be to the hill is $\left(2\sqrt{3}/3\right) - 1 \approx 0.155$.

36. (a) $C = 0.07658t^2 - 7.64323t + 212.53412$

(c) Consumption was minimum in 1993.

37. Parabola

Vertex: $(0, 4)$

$$x^2 = 4p(y - 4)$$

$$4^2 = 4p(0 - 4)$$

$$p = -1$$

$$x^2 = -4(y - 4)$$

$$y = 4 - \frac{x^2}{4}$$

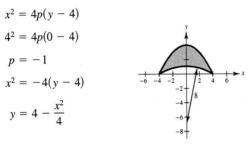

Circle

Center: $(0, k)$

Radius: 8

$$x^2 + (y - k)^2 = 64$$

$$4^2 + (0 - k)^2 = 64$$

$$k^2 = 48$$

$$k = -4\sqrt{3} \quad \text{(Center is on the negative } y\text{-axis.)}$$

$$x^2 + (y + 4\sqrt{3})^2 = 64$$

$$y = -4\sqrt{3} \pm \sqrt{64 - x^2}$$

Since the y value is positive when $x = 0$, we have $y = -4\sqrt{3} + \sqrt{64 - x^2}$.

$$A = 2\int_0^4 \left[\left(4 - \frac{x^2}{4}\right) - \left(-4\sqrt{3} + \sqrt{64 - x^2}\right)\right] dx$$

$$= 2\left[4x - \frac{x^3}{12} + 4\sqrt{3}x - \frac{1}{2}\left(x\sqrt{64 - x^2} + 64 \arcsin \frac{x}{8}\right)\right]_0^4$$

$$= 2\left[16 - \frac{64}{12} + 16\sqrt{3} - 2\sqrt{48} - 32 \arcsin \frac{1}{2}\right]$$

$$= \frac{16\left(4 + 3\sqrt{3} - 2\pi\right)}{3} \approx 15.536 \text{ square feet}$$

38.

$$x = \frac{1}{4}y^2$$

$$x' = \frac{1}{2}y$$

$$1 + (x')^2 = 1 + \frac{y^2}{4}$$

$$s = \int_0^4 \sqrt{1 + \left(\frac{y^2}{4}\right)} \, dy = \frac{1}{2}\int_0^4 \sqrt{4 + y^2} \, dy = \frac{1}{4}\left[y\sqrt{4 + y^2} + 4 \ln\left|y + \sqrt{4 + y^2}\right|\right]_0^4$$

$$= \frac{1}{4}\left[4\sqrt{20} + 4 \ln\left|4 + \sqrt{20}\right| - 4 \ln 2\right] = 2\sqrt{5} + \ln\left(2 + \sqrt{5}\right) \approx 5.916$$

39. (a) Assume that $y = ax^2$.

$$20 = a(60)^2 \implies a = \frac{2}{360} = \frac{1}{180} \implies y = \frac{1}{180}x^2$$

(b) $f(x) = \frac{1}{180}x^2, \, f'(x) = \frac{1}{90}x$

$$S = 2\int_0^{60} \sqrt{1 + \left(\frac{1}{90}x\right)^2} \, dx = \frac{2}{90}\int_0^{60} \sqrt{90^2 + x^2} \, dx$$

$$= \frac{2}{90}\frac{1}{2}\left[x\sqrt{90^2 + x^2} + 90^2 \ln\left|x + \sqrt{90^2 + x^2}\right|\right]_0^{60} \quad \text{(formula 26)}$$

$$= \frac{1}{90}\left[60\sqrt{11,700} + 90^2 \ln\left(60 + \sqrt{11,700}\right) - 90^2 \ln 90\right]$$

$$= \frac{1}{90}\left[1800\sqrt{13} + 90^2 \ln\left(60 + 30\sqrt{13}\right) - 90^2 \ln 90\right]$$

$$= 20\sqrt{13} + 90 \ln\left(\frac{60 + 30\sqrt{13}}{90}\right)$$

$$= 10\left[2\sqrt{13} + 9 \ln\left(\frac{2 + \sqrt{13}}{3}\right)\right] \approx 128.4 \text{ m}$$

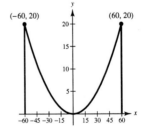

40. $x^2 = 20y$

$$y = \frac{x^2}{20}$$

$$y' = \frac{x}{10}$$

$$S = 2\pi \int_0^r x \sqrt{1 + \left(\frac{x}{10}\right)^2} \, dx \;=\; 2\pi \int_0^r \frac{x\sqrt{100 + x^2}}{10} \, dx$$

$$= \left[\frac{\pi}{10} \cdot \frac{2}{3}(100 + x^2)^{3/2} \right]_0^r = \frac{\pi}{15}\left[(100 + r^2)^{3/2} - 1000\right]$$

41. $x^2 + 4y^2 = 4$

$$\frac{x^2}{4} + \frac{y^2}{1} = 1$$

$a^2 = 4,\ b^2 = 1,\ c^2 = 3$

Center: $(0, 0)$

Foci: $\left(\pm\sqrt{3}, 0\right)$

Vertices: $(\pm 2, 0)$

$$e = \frac{\sqrt{3}}{2}$$

42. $5x^2 + 7y^2 = 70$

$$\frac{x^2}{14} + \frac{y^2}{10} = 1$$

$a^2 = 14,\ b^2 = 10,\ c^2 = 4$

Center: $(0, 0)$

Foci: $(\pm 2, 0)$

Vertices: $\left(\pm\sqrt{14}, 0\right)$

$$e = \frac{2}{\sqrt{14}} = \frac{\sqrt{14}}{7}$$

43. $\dfrac{(x - 1)^2}{9} + \dfrac{(y - 5)^2}{25} = 1$

$a^2 = 25,\ b^2 = 9,\ c^2 = 16$

Center: $(1, 5)$

Foci: $(1, 9),\ (1, 1)$

Vertices: $(1, 10),\ (1, 0)$

$$e = \frac{4}{5}$$

44. $\dfrac{(x + 2)^2}{1} + \dfrac{(y + 4)^2}{1/4} = 1$

$a^2 = 1,\ b^2 = \dfrac{1}{4},\ c^2 = \dfrac{3}{4}$

Center: $(-2, -4)$

Foci: $\left(-2 \pm \dfrac{\sqrt{3}}{2}, -4\right)$

Vertices: $(-1, -4),\ (-3, -4)$

$$e = \frac{\sqrt{3}}{2}$$

45. $\quad 9x^2 + 4y^2 + 36x - 24y + 36 = 0$

$$9(x^2 + 4x + 4) + 4(y^2 - 6y + 9) = -36 + 36 + 36$$

$$= 36$$

$$\frac{(x + 2)^2}{4} + \frac{(y - 3)^2}{9} = 1$$

$a^2 = 9,\ b^2 = 4,\ c^2 = 5$

Center: $(-2, 3)$

Foci: $\left(-2, 3 \pm \sqrt{5}\right)$

Vertices: $(-2, 6),\ (-2, 0)$

$$e = \frac{\sqrt{5}}{3}$$

46. $\quad 16x^2 + 25y^2 - 32x + 50y + 31 = 0$

$$16(x^2 - 2x + 1) + 25(y^2 + 2y + 1) = -31 + 16 + 25$$

$$= 10$$

$$\frac{(x - 1)^2}{5/8} + \frac{(y + 1)^2}{2/5} = 1$$

$a^2 = \dfrac{5}{8},\ b^2 = \dfrac{2}{5},\ c^2 = \dfrac{9}{40}$

Center: $(1, -1)$

Foci: $\left(1 \pm \dfrac{3\sqrt{10}}{20}, -1\right)$

Vertices: $\left(1 \pm \dfrac{\sqrt{10}}{4}, -1\right)$

$$e = \frac{3}{5}$$

47. $12x^2 + 20y^2 - 12x + 40y - 37 = 0$

$$12\left(x^2 - x + \frac{1}{4}\right) + 20(y^2 + 2y + 1) = 37 + 3 + 20$$

$$= 60$$

$$\frac{[x - (1/2)]^2}{5} + \frac{(y + 1)^2}{3} = 1$$

$a^2 = 5, b^2 = 3, c^2 = 2$

Center: $\left(\frac{1}{2}, -1\right)$

Foci: $\left(\frac{1}{2} \pm \sqrt{2}, -1\right)$

Vertices: $\left(\frac{1}{2} \pm \sqrt{5}, -1\right)$

Solve for y:

$$20(y^2 + 2y + 1) = -12x^2 + 12x + 37 + 20$$

$$(y + 1)^2 = \frac{57 + 12x - 12x^2}{20}$$

$$y = -1 \pm \sqrt{\frac{57 + 12x - 12x^2}{20}}$$

(Graph each of these separately.)

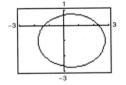

48. $36x^2 + 9y^2 + 48x - 36y + 43 = 0$

$$36\left(x^2 + \frac{4}{3}x + \frac{4}{9}\right) + 9(y^2 - 4y + 4) = -43 + 16 + 36$$

$$= 9$$

$$\frac{[x + (2/3)]^2}{1/4} + \frac{(y - 2)^2}{1} = 1$$

$a^2 = 1, b^2 = \frac{1}{4}, c^2 = \frac{3}{4}$

Center: $\left(-\frac{2}{3}, 2\right)$

Foci: $\left(-\frac{2}{3}, 2 \pm \frac{\sqrt{3}}{2}\right)$

Vertices: $\left(-\frac{2}{3}, 3\right), \left(-\frac{2}{3}, 1\right)$

Solve for y:

$$9(y^2 - 4y + 4) = -36x^2 - 48x - 43 + 36$$

$$(y - 2)^2 = \frac{-(36x^2 + 48x + 7)}{9}$$

$$y = 2 \pm \frac{1}{3}\sqrt{-(36x^2 + 48x + 7)}$$

(Graph each of these separately.)

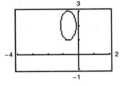

49. $x^2 + 2y^2 - 3x + 4y + 0.25 = 0$

$$\left(x^2 - 3x + \frac{9}{4}\right) + 2(y^2 + 2y + 1) = -\frac{1}{4} + \frac{9}{4} + 2 = 4$$

$$\frac{[x - (3/2)]^2}{4} + \frac{(y + 1)^2}{2} = 1$$

$a^2 = 4, b^2 = 2, c^2 = 2$

Center: $\left(\frac{3}{2}, -1\right)$

Foci: $\left(\frac{3}{2} \pm \sqrt{2}, -1\right)$

Vertices: $\left(-\frac{1}{2}, -1\right), \left(\frac{7}{2}, -1\right)$

Solve for y: $2(y^2 + 2y + 1) = -x^2 + 3x - \frac{1}{4} + 2$

$$(y + 1)^2 = \frac{1}{2}\left(\frac{7}{4} + 3x - x^2\right)$$

$$y = -1 \pm \sqrt{\frac{7 + 12x - 4x^2}{8}}$$

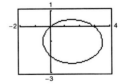

(Graph each of these separately.)

50. $2x^2 + y^2 + 4.8x - 6.4y + 3.12 = 0$

$50x^2 + 25y^2 + 120x - 160y + 78 = 0$

$50\left(x^2 + \frac{12}{5}x + \frac{36}{25}\right) + 25\left(y^2 - \frac{32}{5}y + \frac{256}{25}\right) = -78 + 72 + 256 = 250$

$\dfrac{[x + (6/5)]^2}{5} + \dfrac{[y - (16/5)]^2}{10} = 1$

$a^2 = 10, b^2 = 5, c^2 = 5$

Center: $\left(-\dfrac{6}{5}, \dfrac{16}{5}\right)$

Foci: $\left(-\dfrac{6}{5}, \dfrac{16}{5} \pm \sqrt{5}\right)$

Vertices: $\left(-\dfrac{6}{5}, \dfrac{16}{5} \pm \sqrt{10}\right)$

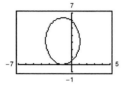

Solve for y: $(y^2 - 6.4y + 10.24) = -2x^2 - 4.8x - 3.12 + 10.24$

$(y - 3.2)^2 = 7.12 - 4x - 2x^2$

$y = 3.2 \pm \sqrt{7.12 - 4x - 2x^2}$ (Graph each of these separately.)

51. Center: $(0, 0)$
Focus: $(2, 0)$
Vertex: $(3, 0)$
Horizontal major axis

$a = 3, c = 2 \implies b = \sqrt{5}$

$\dfrac{x^2}{9} + \dfrac{y^2}{5} = 1$

52. Vertices: $(0, 2), (4, 2)$

Eccentricity: $\dfrac{1}{2}$

Horizontal major axis
Center: $(2, 2)$

$a = 2, c = 1 \implies b = \sqrt{3}$

$\dfrac{(x - 2)^2}{4} + \dfrac{(y - 2)^2}{3} = 1$

53. Vertices: $(3, 1), (3, 9)$
Minor axis length: 6
Vertical major axis
Center: $(3, 5)$

$a = 4, b = 3$

$\dfrac{(x - 3)^2}{9} + \dfrac{(y - 5)^2}{16} = 1$

54. Foci: $(0, \pm 5)$
Major axis length: 14
Vertical major axis
Center: $(0, 0)$

$c = 5, a = 7 \implies b = \sqrt{24}$

$\dfrac{x^2}{24} + \dfrac{y^2}{49} = 1$

55. Center: $(0, 0)$
Horizontal major axis
Points on ellipse: $(3, 1), (4, 0)$

Since the major axis is horizontal,

$$\left(\dfrac{x^2}{a^2}\right) + \left(\dfrac{y^2}{b^2}\right) = 1.$$

Substituting the values of the coordinates of the given points into this equation, we have

$$\left(\dfrac{9}{a^2}\right) + \left(\dfrac{1}{b^2}\right) = 1, \text{ and } \dfrac{16}{a^2} = 1.$$

The solution to this system is $a^2 = 16, b^2 = 16/7$.
Therefore,

$$\dfrac{x^2}{16} + \dfrac{y^2}{16/7} = 1, \dfrac{x^2}{16} + \dfrac{7y^2}{16} = 1.$$

56. Center: $(1, 2)$
Vertical major axis
Points on ellipse: $(1, 6), (3, 2)$

From the sketch, we can see that $h = 1, k = 2, a = 4,$
$b = 2$

$$\dfrac{(x - 1)^2}{4} + \dfrac{(y - 2)^2}{16} = 1.$$

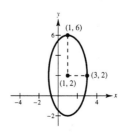

57.

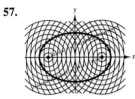

58. (a) At the vertices we notice that the string is horizontal and has a length of $2a$.

(b) The thumbtacks are located at the foci and the length of string is the constant sum of the distances from the foci.

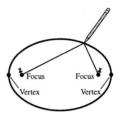

59. $a = \dfrac{5}{2}, b = 2, c = \sqrt{\left(\dfrac{5}{2}\right)^2 - (2)^2} = \dfrac{3}{2}$

The tacks should be placed 1.5 feet from the center. The string should be $2a = 5$ feet long.

60. $e = \dfrac{c}{a}$

$0.0167 = \dfrac{c}{14,957,000}$

$c \approx 249,782$

Least distance: $a - c = 14,707,218$ km
Greatest distance: $a + c = 15,206,782$ km

61. $e = \dfrac{c}{a}$

$A + P = 2a$

$a = \dfrac{A + P}{2}$

$c = a - P = \dfrac{A + P}{2} - P = \dfrac{A - P}{2}$

$e = \dfrac{c}{a} = \dfrac{(A - P)/2}{(A + P)/2} = \dfrac{A - P}{A + P}$

62. $e = \dfrac{A - P}{A + P}$

$= \dfrac{(122,000 + 4000) - (119 + 4000)}{(122,000 + 4000) + (119 + 4000)}$

$= \dfrac{121,881}{130,119} \approx 0.9367$

63. $e = \dfrac{A - P}{A + P}$

$= \dfrac{35.34au - 0.59au}{35.34au + 0.59au} \approx 0.9672$

64. $\dfrac{x^2}{a^2} + \dfrac{y^2}{b^2} = 1$

$\dfrac{x^2}{a^2} + \dfrac{y^2}{a^2(b^2/a^2)} = 1$

$\dfrac{x^2}{a^2} + \dfrac{y^2}{a^2(a^2 - c^2)/a^2} = 1$

$\dfrac{x^2}{a^2} + \dfrac{y^2}{a^2(1 - e^2)} = 1$

As $e \to 0, 1 - e^2 \to 1$ and we have

$\dfrac{x^2}{a^2} + \dfrac{y^2}{a^2} = 1$

or the circle $x^2 + y^2 = a^2$.

65. $16x^2 + 9y^2 + 96x + 36y + 36 = 0$

$32x + 18yy' + 96 + 36y' = 0$

$y'(18y + 36) = -(32x + 96)$

$y' = \dfrac{-(32x + 96)}{18y + 36}$

$y' = 0$ when $x = -3$. y' is undefined when $y = -2$.

At $x = -3, y = 2$ or -6.

Endpoints of major axis: $(-3, 2), (-3, -6)$

At $y = -2, x = 0$ or -6.

Endpoints of minor axis: $(0, -2), (-6, -2)$

Note: Equation of ellipse is $\dfrac{(x + 3)^2}{9} + \dfrac{(y + 2)^2}{16} = 1$

66. $9x^2 + 4y^2 + 36x - 24y + 36 = 0$

$18x + 8yy' + 36 - 24y' = 0$

$(8y - 24)y' = -(18x + 36)$

$y' = \dfrac{-(18x + 36)}{8y - 24}$

$y' = 0$ when $x = -2$. y' undefined when $y = 3$.

At $x = -2$, $y = 0$ or 6.

Endpoints of major axis: $(-2, 0), (-2, 6)$

At $y = 3$, $x = 0$ or -4.

Endpoints of minor axis: $(0, 3), (-4, 3)$

Note: Equation of ellipse is $\dfrac{(x + 2)^2}{4} + \dfrac{(y - 3)^2}{9} = 1$

67. $\dfrac{x^2}{10^2} + \dfrac{y^2}{5^2} = 1$

$\dfrac{2x}{10^2} + \dfrac{2yy'}{5^2} = 0$

$y' = \dfrac{-5^2 x}{10^2 y} = \dfrac{-x}{4y}$

At $(-8, 3)$: $y' = \dfrac{8}{12} = \dfrac{2}{3}$

The equation of the tangent line is $y - 3 = \frac{2}{3}(x + 8)$. It will cross the y-axis when $x = 0$ and $y = \frac{2}{3}(8) + 3 = \frac{25}{3}$.

68. $\dfrac{x^2}{(4.5)^2} + \dfrac{y^2}{(2.5)^2} = 1$

$x^2 = (4.5)^2 \left[1 - \dfrac{y^2}{(2.5)^2} \right]$

$x = \pm \dfrac{9}{5} \sqrt{(2.5)^2 - y^2}$

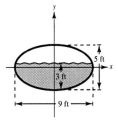

$V = $ (Area of bottom)(Length) $+$ (Area of top)(Length)

$V = \left[\dfrac{\pi(4.5)(2.5)}{2} \right](16) + 16 \displaystyle\int_0^{0.5} \dfrac{9}{5} \sqrt{(2.5)^2 - y^2} \, dy$ (Recall: Area of ellipse is πab.)

$= 90\pi + \dfrac{144}{5} \cdot \dfrac{1}{2} \left[y\sqrt{(2.5)^2 - y^2} + (2.5)^2 \arcsin \dfrac{y}{2.5} \right]_0^{0.5} = 90\pi + \dfrac{72}{5} \left[0.5\sqrt{6} + (2.5)^2 \arcsin \dfrac{1}{5} \right] \approx 318.5 \text{ ft}^3$

69. (a) $A = 4 \displaystyle\int_0^2 \dfrac{1}{2}\sqrt{4 - x^2} \, dx = \left[x\sqrt{4 - x^2} + 4\arcsin\left(\dfrac{x}{2}\right) \right]_0^2 = 2\pi$ [or, $A = \pi ab = \pi(2)(1) = 2\pi$]

(b) **Disc:** $V = 2\pi \displaystyle\int_0^2 \dfrac{1}{4}(4 - x^2) \, dx = \dfrac{1}{2}\pi \left[4x - \dfrac{1}{3}x^3 \right]_0^2 = \dfrac{8\pi}{3}$

$y = \dfrac{1}{2}\sqrt{4 - x^2}$

$y' = \dfrac{-x}{2\sqrt{4 - x^2}}$

$\sqrt{1 + (y')^2} = \sqrt{1 + \dfrac{x^2}{16 - 4x^2}} = \sqrt{\dfrac{16 - 3x^2}{4y}}$

$S = 2(2\pi) \displaystyle\int_0^2 y\left(\dfrac{\sqrt{16 - 3x^2}}{4y} \right) dx = \dfrac{\pi}{2\sqrt{3}} \left[\sqrt{3}x\sqrt{16 - 3x^2} + 16 \arcsin\left(\dfrac{\sqrt{3}x}{4} \right) \right]_0^2 = \dfrac{2\pi}{9}(9 + 4\sqrt{3}\pi) \approx 21.48$

(c) **Shell:** $V = 2\pi \displaystyle\int_0^2 x\sqrt{4 - x^2} \, dx = -\pi \displaystyle\int_0^2 -2x(4 - x^2)^{1/2} \, dx = -\dfrac{2\pi}{3}\left[(4 - x^2)^{3/2} \right]_0^2 = \dfrac{16\pi}{3}$

$x = 2\sqrt{1 - y^2}$

$x' = \dfrac{-2y}{\sqrt{1 - y^2}}$

$\sqrt{1 + (x')^2} = \sqrt{1 + \dfrac{4y^2}{1 - y^2}} = \dfrac{\sqrt{1 + 3y^2}}{\sqrt{1 - y^2}}$

$S = 2(2\pi) \displaystyle\int_0^1 2\sqrt{1 - y^2} \, \dfrac{\sqrt{1 + 3y^2}}{\sqrt{1 - y^2}} \, dy = 8\pi \displaystyle\int_0^1 \sqrt{1 + 3y^2} \, dy$

$= \dfrac{8\pi}{2\sqrt{3}} \left[\sqrt{3}y\sqrt{1 + 3y^2} + \ln\left| \sqrt{3}y + \sqrt{1 + 3y^2} \right| \right]_0^1 = \dfrac{4\pi}{3}\left| 6 + \sqrt{3}\ln(2 + \sqrt{3}) \right| \approx 34.69$

70. (a) $A = 4\int_0^4 \frac{3}{4}\sqrt{16 - x^2}\, dx = \frac{3}{2}\left[x\sqrt{16 - x^2} + 16\arcsin\frac{x}{4}\right]_0^4 = 12\pi$

(b) **Disc:** $V = 2\pi\int_0^4 \frac{9}{16}(16 - x^2)\, dx = \frac{9\pi}{8}\left[\left(16x - \frac{1}{3}x^3\right)\right]_0^4 = 48\pi$

$$y = \frac{3}{4}\sqrt{16 - x^2}$$

$$y' = \frac{-3x}{4\sqrt{16 - x^2}}$$

$$\sqrt{1 + (y')^2} = \sqrt{1 + \frac{9x^2}{16(16 - x^2)}}$$

$$S = 2(2\pi)\int_0^4 \frac{3}{4}\sqrt{16 - x^2}\sqrt{\frac{16(16 - x^2) + 9x^2}{16(16 - x^2)}}\, dx = 4\pi\int_0^4 \frac{3}{4}\sqrt{16 - x^2}\frac{\sqrt{256 - 7x^2}}{4\sqrt{16 - x^2}}\, dx = \frac{3\pi}{4}\int_0^4 \sqrt{256 - 7x^2}\, dx$$

$$= \frac{3\pi}{8\sqrt{7}}\left[\sqrt{7}x\sqrt{256 - 7x^2} + 256\arcsin\frac{\sqrt{7}x}{16}\right]_0^4 = \frac{3\pi}{8\sqrt{7}}\left(48\sqrt{7} + 256\arcsin\frac{\sqrt{7}}{4}\right) \approx 138.93$$

(c) **Shell:** $V = 4\pi\int_0^4 x\left[\frac{3}{4}\sqrt{16 - x^2}\right]dx = 3\pi\left[\left(-\frac{1}{2}\right)\left(\frac{2}{3}\right)(16 - x^2)^{3/2}\right]_0^4 = 64\pi$

$$x = \frac{4}{3}\sqrt{9 - y^2}$$

$$x' = \frac{-4y}{3\sqrt{9 - y^2}}$$

$$\sqrt{1 + (x')^2} = \sqrt{1 + \frac{16y^2}{9(9 - y^2)}}$$

$$S = 2(2\pi)\int_0^3 \frac{4}{3}\sqrt{9 - y^2}\sqrt{\frac{9(9 - y^2) + 16y^2}{9(9 - y^2)}}\, dy = 4\pi\int_0^3 \frac{4}{9}\sqrt{81 + 7y^2}\, dy$$

$$= \frac{16}{9}\left(\frac{\pi}{2\sqrt{7}}\right)\left[\sqrt{7}y\sqrt{81 + 7y^2} + 81\ln\left|\sqrt{7}y + \sqrt{81 + 7y^2}\right|\right]_0^3$$

$$= \frac{8\pi}{9\sqrt{7}}\left[3\sqrt{7}(12) + 81\ln\left(3\sqrt{7} + 12\right) - 81\ln 9\right] \approx 168.53$$

71. From Example 5 we have

$$C = 4a\int_0^{\pi/2} \sqrt{1 - e^2\sin^2\theta}\, d\theta.$$

For $(x^2/9) + (y^2/16) = 1$ we have

$$a = 4, b = 3, c = \sqrt{7}, e = \frac{\sqrt{7}}{4}, C = 4(4)\int_0^{\pi/2}\sqrt{1 - \left(\frac{7}{16}\right)\sin^2\theta}\, d\theta.$$

Applying Simpson's Rule with $n = 12$, or the integration capability of a graphing utility, produces $C \approx 22.10$.

72. (a) $\frac{x^2}{a^2} + \frac{y^2}{b^2} = 1$

(b) Slope of line through $(-c, 0)$ and (x_0, y_0): $m_1 = \frac{y_0}{x_0 + c}$

$$\frac{2x}{a^2} + \frac{2yy'}{b^2} = 0$$

Slope of line through $(c, 0)$ and (x_0, y_0): $m_2 = \frac{y_0}{x_0 - c}$

$$y' = -\frac{xb^2}{ya^2}$$

At P, $y' = -\frac{b^2}{a^2}\cdot\frac{x_0}{y_0} = m.$

—CONTINUED—

72. **—CONTINUED—**

(c) $\tan \alpha = \dfrac{m_2 - m}{1 + m_2 m} = \dfrac{\dfrac{y_0}{x_0 - c} - \left(-\dfrac{b^2 x_0}{a^2 y_0}\right)}{1 + \left(\dfrac{y_0}{x_0 - c}\right)\left(-\dfrac{b^2 x_0}{a^2 y_0}\right)} = \dfrac{a^2 y_0{}^2 + b^2 x_0(x_0 - c)}{a^2 y_0(x_0 - c) - b^2 x_0 y_0}$

$\qquad = \dfrac{a^2 y_0{}^2 + b^2 x_0{}^2 - b^2 x_0 c}{x_0 y_0(a^2 - b^2) - a^2 y_0 c} = \dfrac{a^2 b^2 - b^2 x_0 c}{x_0 y_0 c^2 - a^2 y_0 c} = \dfrac{b^2(a^2 - x_0 c)}{y_0 c(x_0 c - a^2)} = -\dfrac{b^2}{y_0 c}$

$\qquad \alpha = \arctan\left(-\dfrac{b^2}{y_0 c}\right) = -\arctan\left(\dfrac{b^2}{y_0 c}\right)$

$\quad \tan \beta = \dfrac{m_1 - m}{1 + m_1 m} = \dfrac{\dfrac{y_0}{x_0 + c} - \left(-\dfrac{b^2 x_0}{a^2 y_0}\right)}{1 + \left(\dfrac{y_0}{x_0 + c}\right)\left(-\dfrac{b^2 x_0}{a^2 y_0}\right)} = \dfrac{a^2 y_0{}^2 + b^2 x_0(x_0 + c)}{a^2 y_0(x_0 + c) - b^2 x_0 y_0}$

$\qquad = \dfrac{a^2 y_0{}^2 + b^2 x_0{}^2 + b^2 x_0 c}{a^2 x_0 y_0 + a^2 c y_0 - b^2 x_0 y_0} = \dfrac{a^2 b^2 + b^2 x_0 c}{x_0 y_0(a^2 - b^2) + a^2 c y_0} = \dfrac{b^2(a^2 + x_0 c)}{y_0 c(x_0 c + a^2)} = \dfrac{b^2}{y_0 c}$

$\qquad \beta = \arctan\left(\dfrac{b^2}{y_0 c}\right)$

Since $|\alpha| = |\beta|$, the tangent line to an ellipse at a point P makes equal angles with the lines through P and the foci.

73. Area circle $= \pi r^2 = 100\pi$

Area ellipse $= \pi a b = \pi a(10)$

$\qquad 2(100\pi) = 10\pi a \implies a = 20$

Hence, the length of the major axis is $2a = 40$.

74. (a) $e = \dfrac{c}{a} = \dfrac{\sqrt{a^2 + b^2}}{a} \implies (ea)^2 - a^2 = b^2$. Hence,

$\qquad \dfrac{(x - h)^2}{a^2} + \dfrac{(y - k)^2}{b^2} = 1$

$\qquad \dfrac{(x - h)^2}{a^2} + \dfrac{(y - k)^2}{a^2(1 - e^2)} = 1.$

(b) $\dfrac{(x - 2)^2}{4} + \dfrac{(y - 3)^2}{4(1 - e^2)} = 1$

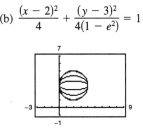

(c) As e approaches 0, the ellipse approaches a circle.

75. $\dfrac{y^2}{1} - \dfrac{x^2}{4} = 1$

$a = 1, b = 2, c = \sqrt{5}$

Center: $(0, 0)$

Vertices: $(0, \pm 1)$

Foci: $\left(0, \pm\sqrt{5}\right)$

Asymptotes: $y = \pm\dfrac{1}{2}x$

76. $\dfrac{x^2}{36} - \dfrac{y^2}{4} = 1$

$a = 6, b = 2, c = 2\sqrt{10}$

Center: $(0, 0)$

Vertices: $(\pm 6, 0)$

Foci: $\left(\pm 2\sqrt{10}, 0\right)$

Asymptotes: $y = \pm\dfrac{1}{3}x$

77. $\dfrac{(x - 1)^2}{4} - \dfrac{(y + 2)^2}{1} = 1$

$a = 2, b = 1, c = \sqrt{5}$

Center: $(1, -2)$

Vertices: $(-1, -2), (3, -2)$

Foci: $\left(1 \pm \sqrt{5}, -2\right)$

Asymptotes: $y = -2 \pm \dfrac{1}{2}(x - 1)$

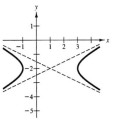

78. $\dfrac{(y+1)^2}{12^2} - \dfrac{(x-4)^2}{5^2} = 1$

$a = 12, b = 5, c = \sqrt{a^2 + b^2} = 13$

Center: $(4, -1)$

Vertices: $(4, 11), (4, -13)$

Foci: $(4, -14), (4, 12)$

Asymptotes: $y = -1 \pm \dfrac{12}{5}(x-4)$

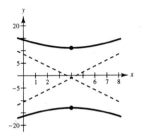

79. $9x^2 - y^2 - 36x - 6y + 18 = 0$

$9(x^2 - 4x + 4) - (y^2 + 6y + 9) = -18 + 36 - 9$

$$\dfrac{(x-2)^2}{1} - \dfrac{(y+3)^2}{9} = 1$$

$a = 1, b = 3, c = \sqrt{10}$

Center: $(2, -3)$

Vertices: $(1, -3), (3, -3)$

Foci: $\left(2 \pm \sqrt{10}, -3\right)$

Asymptotes: $y = -3 \pm 3(x-2)$

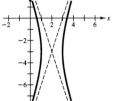

80. $y^2 - 9x^2 + 36x - 72 = 0$

$y^2 - 9(x^2 - 4x + 4) = 72 - 36 = 36$

$$\dfrac{y^2}{36} - \dfrac{(x-2)^2}{4} = 1$$

$a = 6, b = 2, c = \sqrt{a^2 + b^2} = 2\sqrt{10}$

Center: $(2, 0)$

Vertices: $(2, 6), (2, -6)$

Foci: $\left(2, 2\sqrt{10}\right), \left(2, -2\sqrt{10}\right)$

Asymptotes: $y = \pm 3(x-2)$

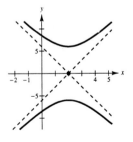

81. $x^2 - 9y^2 + 2x - 54y - 80 = 0$

$(x^2 + 2x + 1) - 9(y^2 + 6y + 9) = 80 + 1 - 81 = 0$

$(x+1)^2 - 9(y+3)^2 = 0$

$$y + 3 = \pm\dfrac{1}{3}(x+1)$$

Degenerate hyperbola is two lines intersecting at $(-1, -3)$.

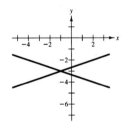

82. $9(x^2 + 6x + 9) - 4(y^2 - 2y + 1) = -78 + 81 - 4 = -1$

$9(x+3)^2 - 4(y-1)^2 = -1$

$$\dfrac{(y-1)^2}{1/4} - \dfrac{(x+3)^2}{1/9} = 1$$

$a = \dfrac{1}{2}, b = \dfrac{1}{3}, c = \dfrac{\sqrt{13}}{6}$

Center: $(-3, 1)$

Vertices: $\left(-3, \dfrac{1}{2}\right), \left(-3, \dfrac{3}{2}\right)$

Foci: $\left(-3, 1 \pm \dfrac{1}{6}\sqrt{13}\right)$

Asymptotes: $y = 1 \pm \dfrac{3}{2}(x+3)$

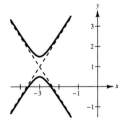

83. $9y^2 - x^2 + 2x + 54y + 62 = 0$

$9(y^2 + 6y + 9) - (x^2 - 2x + 1) = -62 - 1 + 81 = 18$

$$\frac{(y + 3)^2}{2} - \frac{(x - 1)^2}{18} = 1$$

$a = \sqrt{2}, b = 3\sqrt{2}, c = 2\sqrt{5}$

Center: $(1, -3)$

Vertices: $\left(1, -3 \pm \sqrt{2}\right)$

Foci: $\left(1, -3 \pm 2\sqrt{5}\right)$

Solve for y:

$$9(y^2 + 6y + 9) = x^2 - 2x - 62 + 81$$

$$(y + 3)^2 = \frac{x^2 - 2x + 19}{9}$$

$$y = -3 \pm \frac{1}{3}\sqrt{x^2 - 2x + 19}$$

(Graph each curve separately.)

84. $9x^2 - y^2 + 54x + 10y + 55 = 0$

$9(x^2 + 6x + 9) - (y^2 - 10y + 25) = -55 + 81 - 25$

$$= 1$$

$$\frac{(x + 3)^2}{1/9} - \frac{(y - 5)^2}{1} = 1$$

$a = \frac{1}{3}, b = 1, c = \frac{\sqrt{10}}{3}$

Center: $(-3, 5)$

Vertices: $\left(-3 \pm \frac{1}{3}, 5\right)$

Foci: $\left(-3 \pm \frac{\sqrt{10}}{3}, 5\right)$

Solve for y:

$$y^2 - 10y + 25 = 9x^2 + 54x + 55 + 25$$

$$(y - 5)^2 = 9x^2 + 54x + 80$$

$$y = 5 \pm \sqrt{9x^2 + 54x + 80}$$

(Graph each curve separately.)

85. $3x^2 - 2y^2 - 6x - 12y - 27 = 0$

$3(x^2 - 2x + 1) - 2(y^2 + 6y + 9) = 27 + 3 - 18 = 12$

$$\frac{(x - 1)^2}{4} - \frac{(y + 3)^2}{6} = 1$$

$a = 2, b = \sqrt{6}, c = \sqrt{10}$

Center: $(1, -3)$

Vertices: $(-1, -3), (3, -3)$

Foci: $\left(1 \pm \sqrt{10}, -3\right)$

Solve for y:

$$2(y^2 + 6y + 9) = 3x^2 - 6x - 27 + 18$$

$$(y + 3)^2 = \frac{3x^2 - 6x - 9}{2}$$

$$y = -3 \pm \sqrt{\frac{3(x^2 - 2x - 3)}{2}}$$

(Graph each curve separately.)

86. $3y^2 - x^2 + 6x - 12y = 0$

$3(y^2 - 4y + 4) - (x^2 - 6x + 9) = 0 + 12 - 9 = 3$

$$\frac{(y - 2)^2}{1} - \frac{(x - 3)^2}{3} = 1$$

$a = 1, b = \sqrt{3}, c = 2$

Center: $(3, 2)$

Vertices: $(3, 1), (3, 3)$

Foci: $(3, 0), (3, 4)$

Solve for y:

$$3(y^2 - 4y + 4) = x^2 - 6x + 12$$

$$(y - 2)^2 = \frac{x^2 - 6x + 12}{3}$$

$$y = 2 \pm \sqrt{\frac{x^2 - 6x + 12}{3}}$$

(Graph each curve separately.)

87. Vertices: $(\pm 1, 0)$

Asymptotes: $y = \pm 3x$

Horizontal transverse axis

Center: $(0, 0)$

$a = 1, \pm\frac{b}{a} = \pm\frac{b}{1} = \pm 3 \implies b = 3$

Therefore, $\dfrac{x^2}{1} - \dfrac{y^2}{9} = 1.$

88. Vertices: $(0, \pm 3)$

Asymptotes: $y = \pm 3x$

Vertical transverse axis

$a = 3$

Slopes of asymptotes: $\pm\dfrac{a}{b} = \pm 3$

Thus, $b = 1$. Therefore,

$$\frac{y^2}{9} - \frac{x^2}{1} = 1.$$

89. Vertices: $(2, \pm 3)$

Point on graph: $(0, 5)$

Vertical transverse axis

Center: $(2, 0)$

$a = 3$

Therefore, the equation is of the form

$$\frac{y^2}{9} - \frac{(x-2)^2}{b^2} = 1.$$

Substituting the coordinates of the point $(0, 5)$, we have

$$\frac{25}{9} - \frac{4}{b^2} = 1 \quad \text{or} \quad b^2 = \frac{9}{4}.$$

Therefore, the equation is $\dfrac{y^2}{9} - \dfrac{(x-2)^2}{9/4} = 1.$

90. Vertices: $(2, \pm 3)$

Foci: $(2, \pm 5)$

Vertical transverse axis

Center: $(2, 0)$

$a = 3, c = 5, b^2 = c^2 - a^2 = 16$

Therefore, $\dfrac{y^2}{9} - \dfrac{(x-2)^2}{16} = 1.$

91. Center: $(0, 0)$

Vertex: $(0, 2)$

Focus: $(0, 4)$

Vertical transverse axis

$a = 2, c = 4, b^2 = c^2 - a^2 = 12$

Therefore, $\dfrac{y^2}{4} - \dfrac{x^2}{12} = 1.$

92. Center: $(0, 0)$

Vertex: $(3, 0)$

Focus: $(5, 0)$

Horizontal transverse axis

$a = 3, c = 5, b^2 = c^2 - a^2 = 16$

Therefore, $\dfrac{x^2}{9} - \dfrac{y^2}{16} = 1.$

93. Vertices: $(0, 2), (6, 2)$

Asymptotes: $y = \dfrac{2}{3}x, y = 4 - \dfrac{2}{3}x$

Horizontal transverse axis

Center: $(3, 2)$

$a = 3$

Slopes of asymptotes: $\pm\dfrac{b}{a} = \pm\dfrac{2}{3}$

Thus, $b = 2$. Therefore,

$$\frac{(x-3)^2}{9} - \frac{(y-2)^2}{4} = 1.$$

94. Focus: $(10, 0)$

Asymptotes: $y = \pm\dfrac{3}{4}x$

Horizontal transverse axis

Center: $(0, 0)$ since asymptotes intersect at the origin.

$c = 10$

Slopes of asymptotes: $\pm\dfrac{b}{a} = \pm\dfrac{3}{4}$ and $b = \dfrac{3}{4}a$

$c^2 = a^2 + b^2 = 100$

Solving these equations, we have $a^2 = 64$ and $b^2 = 36$. Therefore, the equation is

$$\frac{x^2}{64} - \frac{y^2}{36} = 1.$$

95. The transverse axis is horizontal since $(2, 2)$ and $(10, 2)$ are the foci (see definition of hyperbola).

Center: $(6, 2)$

$c = 4, 2a = 6, b^2 = c^2 - a^2 = 7$

Therefore, the equation is

$$\frac{(x-6)^2}{9} - \frac{(y-2)^2}{7} = 1.$$

96. $2a = 10 \implies a = 5$

$c = 6 \implies b = \sqrt{11}$

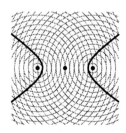

97. The transverse axis is vertical since $(-3, 0)$ and $(-3, 3)$ are the foci.

Center: $\left(-3, \dfrac{3}{2}\right)$

$c = \dfrac{3}{2}, 2a = 2, b^2 = c^2 - a^2 = \dfrac{5}{4}$

Therefore, the equation is

$$\frac{[y - (3/2)]^2}{1} - \frac{(x+3)^2}{5/4} = 1.$$

98. Center: $(0, 0)$

Horizontal transverse axis

Foci: $(\pm c, 0)$

Vertices: $(\pm a, 0)$

The difference of the distances from any point on the hyperbola is constant. At a vertex, this constant difference is

$$(a + c) - (c - a) = 2a.$$

Now, for any point (x, y) on the hyperbola, the difference of the distances between (x, y) and the two foci must also be $2a$.

$$\sqrt{(x - c)^2 + (y - 0)^2} - \sqrt{(x + c)^2 + (y - 0)^2} = 2a$$

$$\sqrt{(x - c)^2 + y^2} = 2a + \sqrt{(x + c)^2 + y^2}$$

$$(x - c)^2 + y^2 = 4a^2 + 4a\sqrt{(x + c)^2 + y^2} + (x + c)^2 + y^2$$

$$-4xc - 4a^2 = 4a\sqrt{(x + c)^2 + y^2}$$

$$-(xc + a^2) = a\sqrt{(x + c)^2 + y^2}$$

$$x^2c^2 + 2a^2cx + a^4 = a^2[x^2 + 2cx + c^2 + y^2]$$

$$x^2(c^2 - a^2) - a^2y^2 = a^2(c^2 - a^2)$$

$$\frac{x^2}{a^2} - \frac{y^2}{c^2 - a^2} = 1$$

Since $a^2 + b^2 = c^2$, we have $(x^2/a^2) - (y^2/b^2) = 1$.

99. Time for sound of bullet hitting target to reach (x, y): $\dfrac{2c}{v_m} + \dfrac{\sqrt{(x - c)^2 + y^2}}{v_s}$

Time for sound of rifle to reach (x, y): $\dfrac{\sqrt{(x + c)^2 + y^2}}{v_s}$

Since the times are the same, we have:

$$\frac{2c}{v_m} + \frac{\sqrt{(x - c)^2 + y^2}}{v_s} = \frac{\sqrt{(x + c)^2 + y^2}}{v_s}$$

$$\frac{4c^2}{v_m^2} + \frac{4c}{v_m v_s}\sqrt{(x - c)^2 + y^2} + \frac{(x - c)^2 + y^2}{v_s^2} = \frac{(x + c)^2 + y^2}{v_s^2}$$

$$\sqrt{(x - c)^2 + y^2} = \frac{v_m^2 x - v_s^2 c}{v_s v_m}$$

$$\left(1 - \frac{v_m^2}{v_s^2}\right)x^2 + y^2 = \left(\frac{v_s^2}{v_m^2} - 1\right)c^2$$

$$\frac{x^2}{c^2 v_s^2/v_m^2} - \frac{y^2}{c^2(v_m^2 - v_s^2)/v_m^2} = 1$$

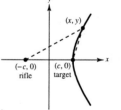

100. $c = 150$, $2a = 0.001(186,000)$, $a = 93$, $b = \sqrt{150^2 - 93^2} = \sqrt{13,851}$

$$\frac{x^2}{93^2} - \frac{y^2}{13,851} = 1$$

When $y = 75$, we have

$$x^2 = 93^2\left(1 + \frac{75^2}{13,851}\right)$$

$$x \approx 110.3 \text{ miles.}$$

101. The point (x, y) lies on the line between $(0, 10)$ and $(10, 0)$. Thus, $y = 10 - x$. The point also lies on the hyperbola $(x^2/36) - (y^2/64) = 1$. Using substitution, we have:

$$\frac{x^2}{36} - \frac{(10 - x)^2}{64} = 1$$

$$16x^2 - 9(10 - x)^2 = 576$$

$$7x^2 + 180x - 1476 = 0$$

$$x = \frac{-180 \pm \sqrt{180^2 - 4(7)(-1476)}}{2(7)} = \frac{-180 \pm 192\sqrt{2}}{14} = \frac{-90 \pm 96\sqrt{2}}{7}$$

Choosing the positive value for x we have:

$$x = \frac{-90 + 96\sqrt{2}}{7} \approx 6.538 \text{ and } y = \frac{160 - 96\sqrt{2}}{7} \approx 3.462$$

102. (a) $\dfrac{x^2}{9} - y^2 = 1, \dfrac{2x}{9} - 2yy' = 0, \dfrac{x}{9y} = y'$

At $x = 6$: $y = \pm\sqrt{3}, y' = \dfrac{\pm 6}{9\sqrt{3}} = \dfrac{\pm 2\sqrt{3}}{9}$

At $\left(6, \sqrt{3}\right)$: $y - \sqrt{3} = \dfrac{2\sqrt{3}}{9}(x - 6)$

or $2x - 3\sqrt{3}y - 3 = 0$

At $\left(6, -\sqrt{3}\right)$: $y + \sqrt{3} = \dfrac{-2\sqrt{3}}{9}(x - 6)$

or $2x + 3\sqrt{3}y - 3 = 0$

(b) From part (a) we know that the slopes of the normal lines must be $\mp 9/\left(2\sqrt{3}\right)$.

At $\left(6, \sqrt{3}\right)$: $y - \sqrt{3} = -\dfrac{9}{2\sqrt{3}}(x - 6)$

or $9x + 2\sqrt{3}y - 60 = 0$

At $\left(6, -\sqrt{3}\right)$: $y + \sqrt{3} = \dfrac{9}{2\sqrt{3}}(x - 6)$

or $9x - 2\sqrt{3}y - 60 = 0$

103. (a) $\dfrac{y^2}{4} - \dfrac{x^2}{2} = 1, y^2 - 2x^2 = 4, 2yy' - 4x = 0,$

$y' = \dfrac{4x}{2y} = \dfrac{2x}{y}$

At $x = 4$: $y = \pm 6, y' = \dfrac{\pm 2(4)}{6} = \pm\dfrac{4}{3}$

At $(4, 6)$: $y - 6 = -\dfrac{4}{3}(x - 4)$ or $4x - 3y + 2 = 0$

At $(4, -6)$: $y + 6 = -\dfrac{4}{3}(x - 4)$ or $4x + 3y + 2 = 0$

(b) From part (a) we know that the slopes of the normal lines must be $\mp 3/4$.

At $(4, 6)$: $y - 6 = -\dfrac{3}{4}(x - 4)$ or $3x + 4y - 36 = 0$

At $(4, -6)$: $y + 6 = \dfrac{3}{4}(x - 4)$ or $3x - 4y - 36 = 0$

104. $\dfrac{x^2}{a^2} - \dfrac{y^2}{b^2} = 1$

$\dfrac{2x}{a^2} - \dfrac{2yy'}{b^2} = 0$ or $y' = \dfrac{b^2x}{a^2y}$

$y - y_0 = \dfrac{b^2x_0}{a^2y_0}(x - x_0)$

$a^2y_0y - a^2y_0^2 = b^2x_0x - b^2x_0^2$

$b^2x_0^2 - a^2y_0^2 = b^2x_0x - a^2y_0y$

$a^2b^2 = b^2x_0x - a^2y_0y$

$\dfrac{x_0x}{a^2} - \dfrac{y_0y}{b^2} = 1$

105.
$$\frac{x^2}{a^2} + \frac{2y^2}{b^2} = 1 \implies \frac{2y^2}{b^2} = 1 - \frac{x^2}{a^2}, \ c^2 = a^2 - b^2$$

$$\frac{x^2}{a^2 - b^2} - \frac{2y^2}{b^2} = 1 \implies \frac{2y^2}{b^2} = \frac{x^2}{a^2 - b^2} - 1$$

$$1 - \frac{x^2}{a^2} = \frac{x^2}{a^2 - b^2} - 1 \implies 2 = x^2\left(\frac{1}{a^2} + \frac{1}{a^2 - b^2}\right)$$

$$x^2 = \frac{2a^2(a^2 - b^2)}{2a^2 - b^2} \implies x = \pm\frac{\sqrt{2}a\sqrt{a^2 - b^2}}{\sqrt{2a^2 - b^2}} = \pm\frac{\sqrt{2}ac}{\sqrt{2a^2 - b^2}}$$

$$\frac{2y^2}{b^2} = 1 - \frac{1}{a^2}\left(\frac{2a^2c^2}{2a^2 - b^2}\right) \implies \frac{2y^2}{b^2} = \frac{b^2}{2a^2 - b^2}$$

$$y^2 = \frac{b^4}{2(2a^2 - b^2)} \implies y = \pm\frac{b^2}{\sqrt{2}\sqrt{2a^2 - b^2}}$$

There are four points of intersection: $\left(\dfrac{\sqrt{2}ac}{\sqrt{2a^2 - b^2}}, \pm\dfrac{b^2}{\sqrt{2}\sqrt{2a^2 - b^2}}\right),$ $\left(-\dfrac{\sqrt{2}ac}{\sqrt{2a^2 - b^2}}, \pm\dfrac{b^2}{\sqrt{2}\sqrt{2a^2 - b^2}}\right)$

$$\frac{x^2}{a^2} + \frac{2y^2}{b^2} = 1 \implies \frac{2x}{a^2} + \frac{4yy'}{b^2} = 0 \implies y'_e = -\frac{b^2 x}{2a^2 y}$$

$$\frac{x^2}{a^2 - b^2} - \frac{2y^2}{b^2} = 1 \implies \frac{2x}{c^2} - \frac{4yy'}{b^2} = 0 \implies y'_h = \frac{b^2 x}{2c^2 y}$$

At $\left(\dfrac{\sqrt{2}ac}{\sqrt{2a^2 - b^2}}, \dfrac{b^2}{\sqrt{2}\sqrt{2a^2 - b^2}}\right)$, the slopes of the tangent lines are:

$$y'_e = \frac{-b^2\left(\dfrac{\sqrt{2}ac}{\sqrt{2a^2 - b^2}}\right)}{2a^2\left(\dfrac{b^2}{\sqrt{2}\sqrt{2a^2 - b^2}}\right)} = -\frac{c}{a} \quad \text{and} \quad y'_h = \frac{b^2\left(\dfrac{\sqrt{2}ac}{\sqrt{2a^2 - b^2}}\right)}{2c^2\left(\dfrac{b^2}{\sqrt{2}\sqrt{2a^2 - b^2}}\right)} = \frac{a}{c}$$

Since the slopes are negative reciprocals, the tangent lines are perpendicular. Similarly, the curves are perpendicular at the other three points of intersection.

106. False. See the definition of a parabola.

107. True

108. True

109. False. The y^4 term should be y^2.

110. False. $y^2 - x^2 + 2x + 2y = 0$ yields two intersecting lines.

111. True

112. True

113.
$$Ax^2 + Cy^2 + Dx + Ey + F = 0 \quad (\text{Assume } A \neq 0 \text{ and } C \neq 0; \text{ see (b) below)}$$

$$A\left(x^2 + \frac{D}{A}x\right) + C\left(y^2 + \frac{E}{C}y\right) = -F$$

$$A\left(x^2 + \frac{D}{A}x + \frac{D^2}{4A^2}\right) + C\left(y^2 + \frac{E}{C}y + \frac{E^2}{4C^2}\right) = -F + \frac{D^2}{4A} + \frac{E^2}{4C} = R$$

$$\frac{\left[x + \left(\dfrac{D}{2A}\right)\right]^2}{C} + \frac{\left[y + \left(\dfrac{E}{2C}\right)\right]^2}{A} = \frac{R}{AC}$$

(a) If $A = C$, we have

$$\left(x + \frac{D}{2A}\right)^2 + \left(y + \frac{E}{2C}\right)^2 = \frac{R}{A}$$

which is the standard equation of a circle.

(b) If $C = 0$, we have

$$A\left(x + \frac{D}{2A}\right)^2 = -F - Ey + \frac{D^2}{4A}.$$

If $A = 0$, we have

$$C\left(y + \frac{E}{2C}\right)^2 = -F - Dx + \frac{E^2}{4C}.$$

These are the equations of parabolas.

—CONTINUED—

113. **—CONTINUED—**

(c) If $AC > 0$, we have

$$\frac{\left[x + \left(\dfrac{D}{2A}\right)\right]^2}{\left|\dfrac{R}{A}\right|} + \frac{\left[y + \left(\dfrac{E}{2C}\right)\right]^2}{\left|\dfrac{R}{C}\right|} = 1$$

which is the equation of an ellipse.

(d) If $AC < 0$, we have

$$\frac{\left[x + \left(\dfrac{D}{2A}\right)\right]^2}{\left|\dfrac{R}{A}\right|} - \frac{\left[y + \left(\dfrac{E}{2C}\right)\right]^2}{\left|\dfrac{R}{C}\right|} = \pm 1$$

which is the equation of a hyperbola.

114. $x^2 + 4y^2 - 6x + 16y + 21 = 0$

$A = 1, C = 4$

$AC = 4 > 0$

Ellipse

115. $4x^2 - y^2 - 4x - 3 = 0$

$A = 4, C = -1$

$AC < 0$

Hyperbola

116. $y^2 - 4y - 4x = 0$

$A = 0, C = 1$

Parabola

117. $25x^2 - 10x - 200y - 119 = 0$

$A = 25, C = 0$

Parabola

118. $4x^2 + 4y^2 - 16y + 15 = 0$

$A = C = 4$

Circle

119. $y^2 - x - 4y - 5 = 0$

$A = 0, C = 1$

Parabola

120. $9x^2 + 9y^2 - 36x + 6y + 34 = 0$

$A = C = 9$

Circle

121. $2x^2 - 2xy = 3y - y^2 - 2xy$

$2x^2 + y^2 - 3y = 0$

$A = 2, C = 1, AC > 0$

Ellipse

122. $3x^2 - 6x + 3 = 6 + 2y^2 + 4y + 2$

$3x^2 - 2y^2 - 6x - 4y - 5 = 0$

$A = 3, C = -2, AC < 0$

Hyperbola

123. $9x^2 + 54x + 81 = 36 - 4(y^2 - 4y + 4)$

$9x^2 + 4y^2 + 54x - 16y + 61 = 0$

$A = 9, C = 4, AC > 0$

Ellipse

Section 9.2   Plane Curves and Parametric Equations

1. $x = \sqrt{t}, \ y = 1 - t$

(a)

t	0	1	2	3	4
x	0	1	$\sqrt{2}$	$\sqrt{3}$	2
y	1	0	-1	-2	-3

(b)

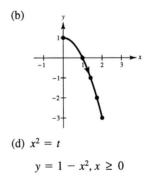

(c)

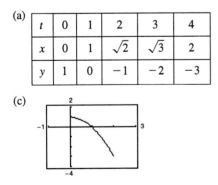

(d) $x^2 = t$

$y = 1 - x^2, x \geq 0$

2. $x = 4 \cos^2 \theta$ $y = 2 \sin \theta$

$0 \leq x \leq 4$ $-2 \leq y \leq 2$

(a)

θ	$-\dfrac{\pi}{2}$	$-\dfrac{\pi}{4}$	0	$\dfrac{\pi}{4}$	$\dfrac{\pi}{2}$
x	0	2	4	2	0
y	-2	$-\sqrt{2}$	0	$\sqrt{2}$	2

(b)

(c)

(d) $\dfrac{x}{4} = \cos^2 \theta$

$\dfrac{y^2}{4} = \sin^2 \theta$

$\dfrac{x}{4} + \dfrac{y^2}{4} = 1$

$x = 4 - y^2,\ -2 \leq y \leq 2$

(e) The graph would be oriented in the opposite direction.

3. $x = 3t - 1$

$y = 2t + 1$

$y = 2\left(\dfrac{x+1}{3}\right) + 1$

$2x - 3y + 5 = 0$

4. $x = 3 - 2t$

$y = 2 + 3t$

$y = 2 + 3\left(\dfrac{3-x}{2}\right)$

$2y + 3x - 13 = 0$

5. $x = t + 1$

$y = t^2$

$y = (x - 1)^2$

6. $x = \sqrt[3]{t}$

$y = 1 - t$

$y = 1 - x^3$

7. $x = t^3$

$y = \tfrac{1}{2}t^2$

$x = t^3$ implies $t = x^{1/3}$

$y = \tfrac{1}{2}x^{2/3}$

8. $x = t^2 + t,\ y = t^2 - t$

Subtracting the second equation from the first, we have

$x - y = 2t$ or $t = \dfrac{x - y}{2}$

$y = \dfrac{(x-y)^2}{4} - \dfrac{x-y}{2}$

t	-2	-1	0	1	2
x	2	0	0	2	6
y	6	2	0	0	2

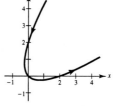

Since the discriminant is

$B^2 - 4AC = (-2)^2 - 4(1)(1) = 0,$

the graph is a rotated parabola.

9. $x = t - 1$

$$y = \frac{t}{t - 1}$$

$$y = \frac{x + 1}{x}$$

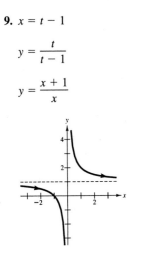

10. $x = 1 + \dfrac{1}{t}$

$$y = t - 1$$

$x = 1 + \dfrac{1}{t}$ implies $t = \dfrac{1}{x - 1}$

$$y = \frac{1}{x - 1} - 1$$

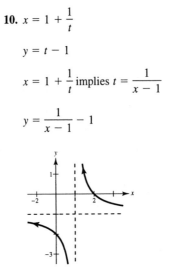

11. $x = 2t$

$$y = |t - 2|$$

$$y = \left| \frac{x}{2} - 2 \right| = \frac{|x - 4|}{2}$$

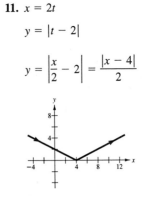

12. $x = |t - 1|$

$$y = t + 2$$

$$x = |(y - 2) - 1| = |y - 3|$$

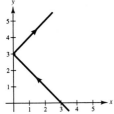

13. $x = \sec \theta$

$$y = \cos \theta$$

$$0 \le \theta < \frac{\pi}{2}, \ \frac{\pi}{2} < \theta \le \pi$$

$$xy = 1$$

$$y = \frac{1}{x}$$

$$|x| \ge 1, \ |y| \le 1$$

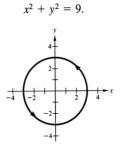

14. $x = \tan^2 \theta$

$$y = \sec^2 \theta$$

$$\sec^2 \theta = \tan^2 \theta + 1$$

$$y = x + 1$$

$$x \ge 0$$

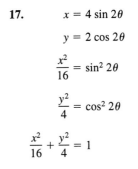

15. $x = 3 \cos \theta, \ y = 3 \sin \theta$

Squaring both equations and adding, we have

$$x^2 + y^2 = 9.$$

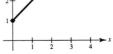

16. $x = \cos \theta$

$$y = 3 \sin \theta$$

$$x^2 = \cos^2 \theta$$

$$y^2 = 9 \sin^2 \theta$$

$$\frac{y^2}{9} = \sin^2 \theta$$

$$x^2 + \frac{y^2}{9} = 1 \quad \text{ellipse}$$

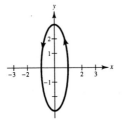

17. $x = 4 \sin 2\theta$

$$y = 2 \cos 2\theta$$

$$\frac{x^2}{16} = \sin^2 2\theta$$

$$\frac{y^2}{4} = \cos^2 2\theta$$

$$\frac{x^2}{16} + \frac{y^2}{4} = 1$$

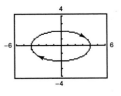

18. $x = \cos \theta$

$y = 2 \sin 2\theta$

$\theta = \arccos x$

$y = 4 \sin \theta \cos \theta$

$y = 4x \sin(\arccos x)$

19. $x = 4 + 2 \cos \theta$

$y = -1 + \sin \theta$

$\dfrac{(x-4)^2}{4} = \cos^2 \theta$

$\dfrac{(y+1)^2}{1} = \sin^2 \theta$

$\dfrac{(x-4)^2}{4} + \dfrac{(y+1)^2}{1} = 1$

20. $x = 4 + 2 \cos \theta$

$y = -1 + 2 \sin \theta$

$(x-4)^2 = 4 \cos^2 \theta$

$(y+1)g2 = 4 \sin^2 \theta$

21. $x = 4 + 2 \cos \theta$

$y = -1 + 4 \sin \theta$

$\dfrac{(x-4)^2}{4} = \cos^2 \theta$

$\dfrac{(y+1)^2}{16} = \sin^2 \theta$

$\dfrac{(x-4)^2}{4} + \dfrac{(y+1)^2}{16} = 1$

22. $x = \sec \theta$

$y = \tan \theta$

$x^2 = \sec^2 \theta$

$y^2 = \tan^2 \theta$

23. $x = 4 \sec \theta$

$y = 3 \tan \theta$

$\dfrac{x^2}{16} = \sec^2 \theta$

$\dfrac{y^2}{9} = \tan^2 \theta$

$\dfrac{x^2}{16} - \dfrac{y^2}{9} = 1$

24. $x = \cos^3 \theta$

$y = \sin^3 \theta$

$x^{2/3} = \cos^2 \theta$

$y^{2/3} = \sin^2 \theta$

25. $x = t^3$

$y = 3 \ln t$

$y = 3 \ln \sqrt[3]{x} = \ln x$

26. $x = \ln 2t$

$y = t^2$

$t = \dfrac{e^x}{2}$

$y = \dfrac{e^{2x}}{r} = \dfrac{1}{4}e^{2x}$

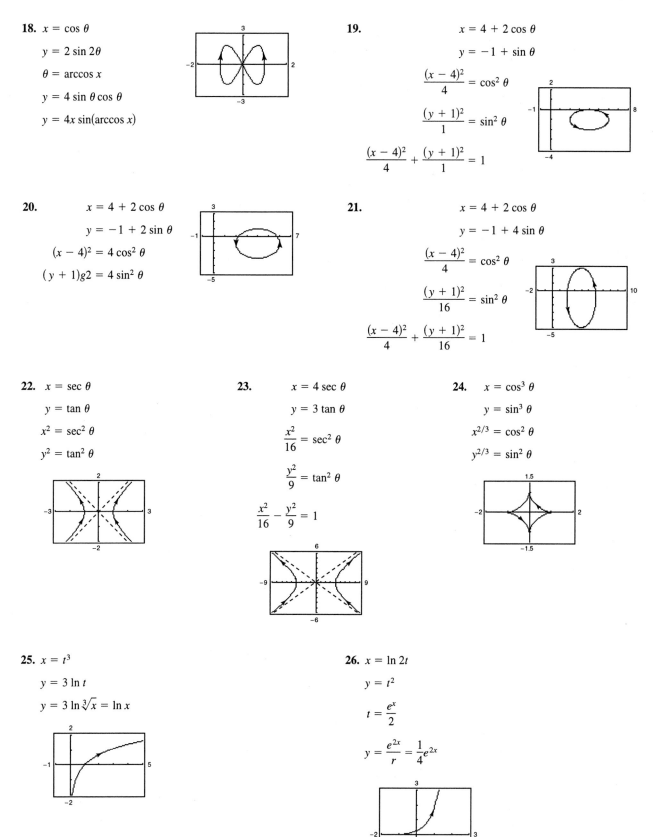

27. $x = e^{-t}$

       $y = e^{3t}$

       $e^t = \dfrac{1}{x}$

       $e^t = \sqrt[3]{y}$

   $\sqrt[3]{y} = \dfrac{1}{x}$

       $y = \dfrac{1}{x^3}$

       $x > 0$

       $y > 0$

28. $x = e^{2t}$

       $y = e^t$

       $y^2 = x$

       $y > 0$

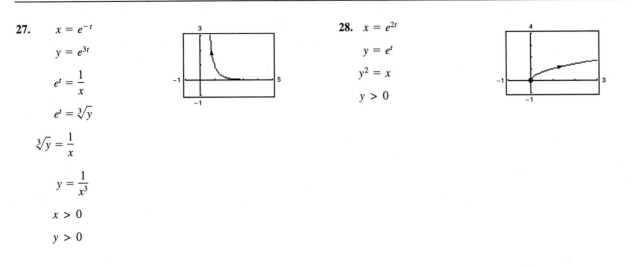

29. By eliminating the parameters in (a) – (d), we get $y = 2x + 1$. They differ from each other in orientation and in restricted domains. These curves are all smooth except for (b).

(a) $x = t$, $y = 2t + 1$

(b) $x = \cos\theta$ $y = 2\cos\theta + 1$

     $-1 \le x \le 1$ $-1 \le y \le 3$

     $\dfrac{dx}{d\theta} = \dfrac{dy}{d\theta} = 0$ when $\theta = 0, \pm\pi, \pm2\pi, \ldots$

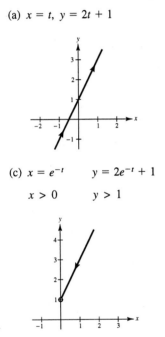

(c) $x = e^{-t}$ $y = 2e^{-t} + 1$

     $x > 0$ $y > 1$

(d) $x = e^t$ $y = 2e^t + 1$

     $x > 0$ $y > 1$

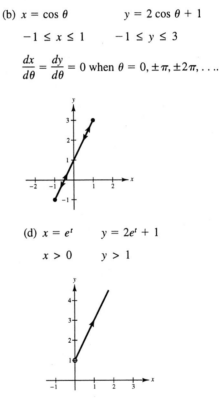

30. By eliminating the parameters in (a) – (d), we get $x^2 + y^2 = 4$. They differ from each other in orientation and in restricted domains. These curves are all smooth.

(a) $x = 2\cos\theta$, $y = 2\sin\theta$

(b) $x = \dfrac{\sqrt{4t^2 - 1}}{|t|} = \sqrt{4 - \dfrac{1}{t^2}}$ $y = \dfrac{1}{t}$

     $x \ge 0,\ x \ne 2$ $y \ne 0$

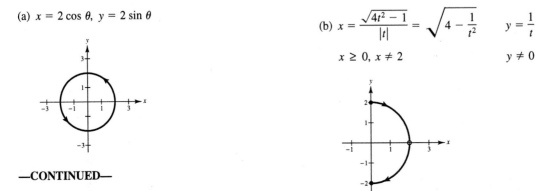

—CONTINUED—

30. —CONTINUED—

(c) $x = \sqrt{t}$ $y = \sqrt{4 - t}$

$\quad$ $x \geq 0$ $\quad$ $y \geq 0$

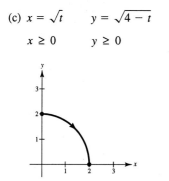

(d) $x = -\sqrt{4 - e^{2t}}$ $y = e^t$

$\quad$ $-2 < x \leq 0$ $\quad$ $y > 0$

31. The curves are identical on $0 < \theta < \pi$. They are both smooth.

32. The orientations are reversed. The graphs are the same. They are both smooth.

33. (a)

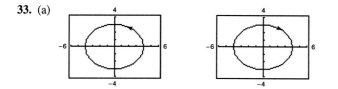

(b) The orientation of the second curve is reversed.

(c) The orientation will be reversed.

(d) Many answers possible. For example, $x = 1 + t$, $y = 1 + 2t$, and $x = 1 - t$, $x = 1 - 2t$.

34. The set of points (x, y) corresponding to the rectangular equation of a set of parametric equations does not show the orientation of the curve nor any restriction on the domain of the original parametric equations.

35. $\quad x = x_1 + t(x_2 - x_1)$

$\quad y = y_1 + t(y_2 - y_1)$

$\dfrac{x - x_1}{x_2 - x_1} = t$

$y = y_1 + \left(\dfrac{x - x_1}{x_2 - x_1}\right)(y_2 - y_1)$

$y - y_1 = \dfrac{y_2 - y_1}{x_2 - x_1}(x - x_1)$

$y - y_1 = m(x - x_1)$

36. $\quad x = h + r \cos \theta$

$\quad y = k + r \sin \theta$

$\cos \theta = \dfrac{x - h}{r}$

$\sin \theta = \dfrac{y - k}{r}$

$\cos^2 \theta + \sin^2 \theta = \dfrac{(x - h)^2}{r^2} + \dfrac{(y - k)^2}{r^2} = 1$

$(x - h)^2 + (y - k)^2 = r^2$

37. $\quad x = h + a \cos \theta$

$\quad y = k + b \sin \theta$

$\dfrac{x - h}{a} = \cos \theta$

$\dfrac{y - k}{b} = \sin \theta$

$\dfrac{(x - h)^2}{a^2} + \dfrac{(y - k)^2}{b^2} = 1$

38. $\quad x = h + a \sec \theta$

$\quad y = k + b \tan \theta$

$\dfrac{x - h}{a} = \sec \theta$

$\dfrac{y - k}{b} = \tan \theta$

$\dfrac{(x - h)^2}{a^2} - \dfrac{(y - k)^2}{b^2} = 1$

39. From Exercise 35 we have

$\quad x = 5t$

$\quad y = -2t.$

Solution not unique

40. From Exercise 35 we have

$\quad x = 1 + 4t$

$\quad y = 4 - 6t.$

Solution not unique

41. From Exercise 36 we have

$\quad x = 2 + 4 \cos \theta$

$\quad y = 1 + 4 \sin \theta.$

Solution not unique

42. From Exercise 36 we have

$$x = -3 + 3 \cos \theta$$

$$y = 1 + 3 \sin \theta.$$

Solution not unique

43. From Exercise 37 we have

$$a = 5, c = 4 \implies b = 3$$

$$x = 5 \cos \theta$$

$$y = 3 \sin \theta.$$

Center: $(0, 0)$
Solution not unique

44. From Exercise 37 we have

$$a = 5, c = 3 \implies b = 4$$

$$x = 4 + 5 \cos$$

$$y = 2 + 4 \sin \theta.$$

Center: $(4, 2)$
Solution not unique

45. From Exercise 38 we have

$$a = 4, c = 5 \implies b = 3$$

$$x = 4 \sec \theta$$

$$y = 3 \tan \theta.$$

Center: $(0, 0)$
Solution not unique

46. From Exercise 38 we have

$$a = 1, c = 2 \implies b = \sqrt{3}$$

$$x = \sqrt{3} \tan \theta$$

$$y = \sec \theta.$$

Center: $(0, 0)$
Solution not unique
The transverse axis is vertical,
therefore, x and y are interchanged.

47. $y = 3x - 2$

Example

$$x = t, \qquad y = 3t - 2$$

$$x = t - 3, \quad y = 3t - 11$$

48. $y = \dfrac{1}{x}$

Example

$$x = t, \qquad y = \frac{1}{t}$$

$$x = \frac{1}{t}, \qquad y = t$$

49. $y = x^3$

Example

$$x = t, \qquad y = t^3$$

$$x = \sqrt[3]{t}, \qquad y = t$$

$$x = \tan t, \qquad y = \tan^3 t$$

50. $y = x^2$

Example

$$x = t, \qquad y = t^2$$

$$x = t^3, \qquad y = t^6$$

51. $x = 2(\theta - \sin \theta)$

$$y = 2(1 - \cos \theta)$$

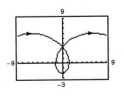

Not smooth at $\theta = 2n\pi$

52. $x = \theta + \sin \theta$

$$y = 1 - \cos \theta$$

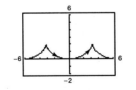

Not smooth at $x = (2n - 1)\pi$

53. $x = \theta - \dfrac{3}{2} \sin \theta$

$$y = 1 - \frac{3}{2} \cos \theta$$

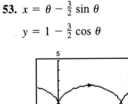

54. $x = 2\theta - 4 \sin \theta$

$$y = 2 - 4 \cos \theta$$

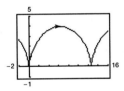

55. $x = 3 \cos^3 \theta$

$$y = 3 \sin^3 \theta$$

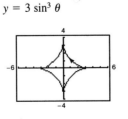

Not smooth at $(x, y) = (\pm 3, 0)$
and $(0, \pm 3)$, or $\theta = \frac{1}{2}n\pi$.

56. $x = 2\theta - \sin \theta$

$$y = 2 - \cos \theta$$

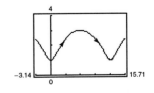

Smooth everywhere

57. $x = 2 \cot \theta$

    $y = 2 \sin^2 \theta$

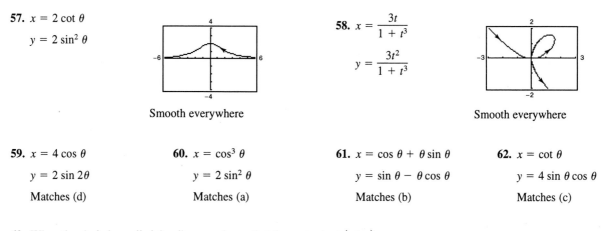

58. $x = \dfrac{3t}{1 + t^3}$

    $y = \dfrac{3t^2}{1 + t^3}$

Smooth everywhere   Smooth everywhere

59. $x = 4 \cos \theta$

    $y = 2 \sin 2\theta$

    Matches (d)

60. $x = \cos^3 \theta$

    $y = 2 \sin^2 \theta$

    Matches (a)

61. $x = \cos \theta + \theta \sin \theta$

    $y = \sin \theta - \theta \cos \theta$

    Matches (b)

62. $x = \cot \theta$

    $y = 4 \sin \theta \cos \theta$

    Matches (c)

63. When the circle has rolled θ radians, we know that the center is at $(a\theta, a)$.

$$\sin \theta = \sin(180° - \theta) = \frac{|AC|}{b} = \frac{|BD|}{b} \quad \text{or} \quad |BD| = b \sin \theta$$

$$\cos \theta = -\cos(180° - \theta) = \frac{|AP|}{-b} \quad \text{or} \quad |AP| = -b \cos \theta$$

Therefore, $x = a\theta - b \sin \theta$ and $y = a - b \cos \theta$.

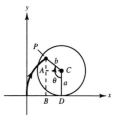

64. Let the circle of radius 1 be centered at C. A is the point of tangency on the line OC. $OA = 2$, $AC = 1$, $OC = 3$. $P = (x, y)$ is the point on the curve being traced out as the angle θ changes $\overset{\frown}{AB} = \overset{\frown}{AP}$. $\overset{\frown}{AB} = 2\theta$ and $\overset{\frown}{AP} = \alpha \implies \alpha = 2\theta$. Form the right triangle $\triangle CDP$. The angle $OCE = (\pi/2) - \theta$ and

$$\angle DCP = \alpha - \left(\frac{\pi}{2} - \theta\right) = \alpha + \theta - \left(\frac{\pi}{2}\right) = 3\theta - \left(\frac{\pi}{2}\right).$$

$$x = OE + Ex = 3\sin\left(\frac{\pi}{2} - \theta\right) + \sin\left(3\theta - \frac{\pi}{2}\right) = 3\cos\theta - \cos 3\theta$$

$$y = EC - CD = 3\sin\theta - \cos\left(3\theta - \frac{\pi}{2}\right) = 3\sin\theta - \sin 3\theta$$

Hence, $x = 3\cos\theta - \cos 3\theta$, $y = 3\sin\theta - \sin 3\theta$.

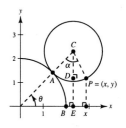

65. True

66. False

$$x = t^2 \implies x \geq 0$$

$$x = t^2 \implies y \geq 0$$

The graph of the parametric equations is only a portion of the line $y = x$.

67. False. Let $x = t^2$ and $y = t$. Then $x = y^2$ and y is not a function of x.

68. False. Let $x = \sin t$ and $y = \cos t$. $x(0) = 0$ and $y(\pi/2) = 0$, but the graph of $x^2 + y^2 = 1$ does not pass through the origin.

69. (a) $100 \text{ mi/hr} = \dfrac{(100)(5280)}{3600} = \dfrac{440}{3} \text{ ft/sec}$

$$x = (v_0 \cos \theta)t = \left(\frac{440}{3} \cos \theta\right)t$$

$$y = h + (v_0 \sin \theta)t - 16t^2$$

$$= 3 + \left(\frac{440}{3} \sin \theta\right)t - 16t^2$$

(b)

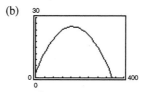

It is not a home run—when $x = 400$, $y \leq 20$.

—CONTINUED—

69. **—CONTINUED—**

(c)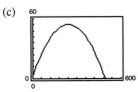

Yes, it's a home run when $x = 400$, $y > 10$.

(d) We need to find the angle θ (and time t) such that

$$x = \left(\frac{440}{3}\cos\theta\right)t = 400$$

$$y = 3 + \left(\frac{440}{3}\sin\theta\right)t - 16t^2 = 10.$$

From the first equation $t = 1200/440 \cos\theta$. Substituting into the second equation,

$$10 = 3 + \left(\frac{440}{3}\sin\theta\right)\left(\frac{1200}{440\cos\theta}\right) - 16\left(\frac{1200}{440\cos\theta}\right)^2$$

$$7 = 400\tan\theta - 16\left(\frac{120}{44}\right)^2\sec^2\theta = 400\tan\theta - 16\left(\frac{120}{44}\right)^2(\tan^2\theta + 1).$$

We now solve the quadratic for $\tan\theta$:

$$16\left(\frac{120}{44}\right)^2\tan^2\theta - 400\tan\theta + 7 + 16\left(\frac{120}{44}\right)^2 = 0$$

$$\tan\theta \approx 0.35185 \implies \theta \approx 19.4°$$

70. (a) $x = (v_0\cos\theta)t$

$y = h + (v_0\sin\theta)t - 16t^2$

$$t = \frac{x}{v_0\cos\theta} \implies y = h + (v_0\sin\theta)\frac{x}{v_0\cos\theta} - 16\left(\frac{x}{v_0\cos\theta}\right)^2$$

$$y = h + (\tan\theta)x - \frac{16\sec^2\theta}{v_0^2}x^2$$

(b) $y = 5 + x - 0.005x^2 = h + (\tan\theta)x - \dfrac{16\sec^2\theta}{v_0^2}x^2$

(c)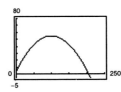

$$h = 5, \tan\theta = 1 \implies \theta = \frac{\pi}{4}, \text{ and}$$

$$0.005 = \frac{16\sec^2(\pi/4)}{v_0^2} = \frac{16}{v_0^2}(2)$$

$$v_0^2 = \frac{32}{0.005} = 6400 \implies v_0 = 80.$$

(d) Maximum height: $y = 55$ (at $x = 100$)

Range: 204.88

Hence, $x = (80\cos(45°))t$

$y = 5 + (80\sin(45°))t - 16t^2.$

71. $x = \dfrac{1 - t^2}{1 + t^2}$ and $y = \dfrac{2t}{1 + t^2}$

The graph is the circle $x^2 + y^2 = 1$ except the point $(-1, 0)$. Thus,

$$\left(\frac{1 - t^2}{1 + t^2}\right)^2 + \left(\frac{2t}{1 + t^2}\right)^2 = 1$$

$$(1 - t^2)^2 + (2t)^2 = (1 + t^2)^2.$$

When $t = 2$: $(-3)^2 + (4)^2 = (5)^2$

When $t = 3$: $(-8)^2 + (6)^2 = (10)^2$

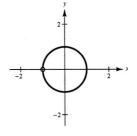

Section 9.3   Parametric Equations and Calculus

1. $x = 2t, \; y = 3t - 1$

$$\frac{dy}{dx} = \frac{dy/dt}{dx/dt} = \frac{3}{2}$$

$$\frac{d^2y}{dx^2} = 0$$

2. $x = \sqrt{t}, \; y = 3t - 1$

$$\frac{dy}{dx} = \frac{3}{1/(2\sqrt{t})} = 6\sqrt{t} = 6 \text{ when } t = 1.$$

$$\frac{d^2y}{dx^2} = \frac{3/\sqrt{t}}{1/(2\sqrt{t})} = 6$$

3. $x = t + 1, \; y = t^2 + 3t$

$$\frac{dy}{dx} = \frac{2t + 3}{1} = 1 \text{ when } t = -1.$$

$$\frac{d^2y}{dx^2} = 2$$

4. $x = t^2 + 3t, \; y = t + 1$

$$\frac{dy}{dx} = \frac{1}{2t + 3} = \frac{1}{3} \text{ when } t = 0.$$

$$\frac{d^2y}{dx^2} = \frac{-2/(2t + 3)^2}{2t + 3}$$

$$= \frac{-2}{(2t + 3)^3} = \frac{-2}{27} \text{ when } t = 0.$$

5. $x = 2\cos\theta, \; y = 2\sin\theta$

$$\frac{dy}{dx} = \frac{2\cos\theta}{-2\sin\theta} = -\cot\theta = -1 \text{ when } \theta = \frac{\pi}{4}.$$

$$\frac{d^2y}{dx^2} = \frac{\csc^2\theta}{-2\sin\theta} = \frac{-\csc^3\theta}{2} = -\sqrt{2} \text{ when } \theta = \frac{\pi}{4}.$$

6. $x = \cos\theta, \; y = 3\sin\theta$

$$\frac{dy}{dx} = \frac{3\cos\theta}{-\sin\theta} = -3\cot\theta \cdot \frac{dy}{dx} \text{ is undefined when } \theta = 0.$$

$$\frac{d^2y}{dx^2} = \frac{3\csc^2\theta}{-\sin\theta} = \frac{-3}{\sin^3\theta} \cdot \frac{d^2y}{dx^2} \text{ is undefined when } \theta = 0.$$

7. $x = 2 + \sec\theta, \; y = 1 + 2\tan\theta$

$$\frac{dy}{dx} = \frac{2\sec^2\theta}{\sec\theta\tan\theta}$$

$$= \frac{2\sec\theta}{\tan\theta} = 2\csc\theta = 4 \text{ when } \theta = \frac{\pi}{6}.$$

$$\frac{d^2y}{dx^2} = \frac{-2\csc\theta\cot\theta}{\sec\theta\tan\theta}$$

$$= -2\cot^3\theta = -6\sqrt{3} \text{ when } \theta = \frac{\pi}{6}.$$

8. $x = \sqrt{t}, \; y = \sqrt{t - 1}$

$$\frac{dy}{dx} = \frac{1/(2\sqrt{t - 1})}{1/(2\sqrt{t})}$$

$$= \frac{\sqrt{t}}{\sqrt{t - 1}} = \sqrt{2} \text{ when } t = 2.$$

$$\frac{d^2y}{dx^2} = \frac{\left[\sqrt{t - 1}/(2\sqrt{t}) - \sqrt{t}(1/2\sqrt{t - 1})\right]/(t - 1)}{1/(2\sqrt{t})}$$

$$= \frac{-1}{(t - 1)^{3/2}} = -1 \text{ when } t = 2.$$

9. $x = \cos^3\theta, \; y = \sin^3\theta$

$$\frac{dy}{dx} = \frac{3\sin^2\theta\cos\theta}{-3\cos^2\theta\sin\theta}$$

$$= -\tan\theta = -1 \text{ when } \theta = \frac{\pi}{4}.$$

$$\frac{d^2y}{dx^2} = \frac{-\sec^2\theta}{-3\cos^2\theta\sin\theta} = \frac{1}{3\cos^4\theta\sin\theta}$$

$$= \frac{\sec^4\theta\csc\theta}{3} = \frac{4\sqrt{2}}{3} \text{ when } \theta = \frac{\pi}{4}.$$

10. $x = \theta - \sin\theta, \; y = 1 - \cos\theta$

$$\frac{dy}{dx} = \frac{\sin\theta}{1 - \cos\theta} = 0 \text{ when } \theta = \pi.$$

$$\frac{d^2y}{dx^2} = \frac{\dfrac{[(1 - \cos\theta)\cos\theta - \sin^2\theta]}{(1 - \cos\theta)^2}}{(1 - \cos\theta)}$$

$$= \frac{-1}{(1 - \cos\theta)^2} = -\frac{1}{4} \text{ when } \theta = \pi.$$

11. $x = 2 \cot \theta,\ y = 2 \sin^2 \theta$

$$\frac{dy}{dx} = \frac{4 \sin \theta \cos \theta}{-2 \csc^2 \theta} = -2 \sin^3 \theta \cos \theta$$

At $\left(-\dfrac{2}{\sqrt{3}}, \dfrac{3}{2}\right)$, $\theta = \dfrac{2\pi}{3}$, and $\dfrac{dy}{dx} = \dfrac{3\sqrt{3}}{8}$.

Tangent line: $\qquad y - \dfrac{3}{2} = \dfrac{3\sqrt{3}}{8}\left(x + \dfrac{2}{\sqrt{3}}\right)$

$$3\sqrt{3}x - 8y + 18 = 0$$

At $(0, 2)$, $\theta = \dfrac{\pi}{2}$, and $\dfrac{dy}{dx} = 0$.

Tangent line: $y - 2 = 0$

At $\left(2\sqrt{3}, \dfrac{1}{2}\right)$, $\theta = \dfrac{\pi}{6}$, and $\dfrac{dy}{dx} = -\dfrac{\sqrt{3}}{8}$.

Tangent line: $\qquad y - \dfrac{1}{2} = -\dfrac{\sqrt{3}}{8}(x - 2\sqrt{3})$

$$\sqrt{3}x + 8y - 10 = 0$$

12. $x = 2 - 3 \cos \theta,\ y = 3 + 2 \sin \theta$

$$\frac{dy}{dx} = \frac{2 \cos \theta}{3 \sin \theta} = \frac{2}{3} \cot \theta$$

At $(-1, 3)$, $\theta = 0$, and $\dfrac{dy}{dx}$ is undefined.

Tangent line: $x = -1$

At $(2, 5)$, $\theta = \dfrac{\pi}{2}$, and $\dfrac{dy}{dx} = 0$.

Tangent line: $y = 5$

At $\left(\dfrac{4 + 3\sqrt{3}}{2}, 2\right)$, $\theta = \dfrac{7\pi}{6}$, and $\dfrac{dy}{dx} = \dfrac{2\sqrt{3}}{3}$.

Tangent line:

$$y - 2 = \frac{2\sqrt{3}}{3}\left(x - \frac{4 + 3\sqrt{3}}{2}\right)$$

$$2\sqrt{3}x - 3y - 4\sqrt{3} - 3 = 0$$

13. $x = 2t,\ y = t^2 - 1,\ t = 2$

(a)

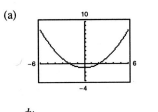

(b) At $t = 2$, $(x, y) = (4, 3)$, and $\dfrac{dx}{dt} = 2$, $\dfrac{dy}{dt} = 4$, $\dfrac{dy}{dx} = 2$

(d)

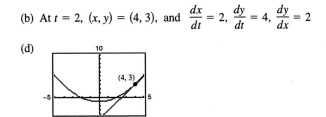

(c) $\dfrac{dy}{dx} = 2$. At $(4, 3)$, $y - 3 = 2(x - 4)$

$$y = 2x - 5$$

14. $x = t - 1,\ y = \dfrac{1}{t} + 1,\ t = 1$

(a)

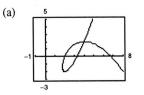

(b) At $t = 1$, $(x, y) = (0, 2)$, and

$$\frac{dx}{dt} = 1,\ \frac{dy}{dt} = -1,\ \frac{dy}{dx} = -1$$

(d)

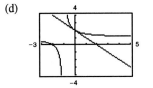

(c) $\dfrac{dy}{dx} = -1$. At $(0, 2)$, $y - 2 = -1(x - 0)$

$$y = -x + 2$$

15. $x = t^2 - t + 2,\ y = t^3 - 3t,\ t = -1$

(a)

(b) At $t = -1$, $(x, y) = (4, 2)$, and

$$\frac{dx}{dt} = -3,\ \frac{dy}{dt} = 0,\ \frac{dy}{dx} = 0$$

—CONTINUED—

15. —CONTINUED—

(c) $\dfrac{dy}{dx} = 0$. At $(4, 2)$, $y - 2 = 0(x - 4)$

$$y = 2$$

(d)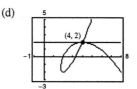

16. $x = 4 \cos \theta$, $y = 3 \sin \theta$, $\theta = \dfrac{3\pi}{4}$

(a)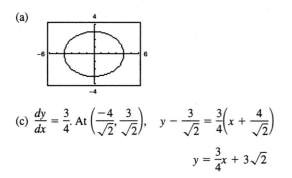

(b) At $\theta = \dfrac{3\pi}{4}$, $(x, y) = \left(\dfrac{-4}{\sqrt{2}}, \dfrac{3}{\sqrt{2}}\right)$, and

$$\dfrac{dx}{dt} = -2\sqrt{2}, \ \dfrac{dy}{dt} = -\dfrac{3\sqrt{2}}{2}, \ \dfrac{dy}{dx} = \dfrac{3}{4}$$

(c) $\dfrac{dy}{dx} = \dfrac{3}{4}$. At $\left(\dfrac{-4}{\sqrt{2}}, \dfrac{3}{\sqrt{2}}\right)$, $\quad y - \dfrac{3}{\sqrt{2}} = \dfrac{3}{4}\left(x + \dfrac{4}{\sqrt{2}}\right)$

$$y = \dfrac{3}{4}x + 3\sqrt{2}$$

(d)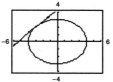

17. $x = \cos \theta + \theta \sin \theta$, $y = \sin \theta - \theta \cos \theta$

Horizontal tangents: $\dfrac{dy}{d\theta} = \theta \sin \theta = 0$ when $\theta = 0, \ \pi, \ 2\pi, \ 3\pi, \ldots$.

Points: $(-1, [2n - 1]\pi)$, $(1, 2n\pi)$ where n is an integer.
Points shown: $(1, 0)$, $(-1, \pi)$, $(1, -2\pi)$

Vertical tangents: $\dfrac{dx}{d\theta} = \theta \cos \theta = 0$ when $\theta = \dfrac{\pi}{2}, \ \dfrac{3\pi}{2}, \ \dfrac{5\pi}{2}, \ldots$.

Points: $\left(\dfrac{(-1)^{n+1}(2n - 1)\pi}{2}, (-1)^{n+1}\right)$

Points shown: $\left(\dfrac{\pi}{2}, 1\right)$, $\left(-\dfrac{3\pi}{2}, -1\right)$, $\left(\dfrac{5\pi}{2}, 1\right)$

18. $x = 2\theta$, $y = 2(1 - \cos \theta)$

Horizontal tangents: $\dfrac{dy}{d\theta} = 2 \sin \theta = 0$ when $\theta = 0, \ \pm\pi, \ \pm 2\pi, \ldots$.

Points: $(4n\pi, 0)$, $(2[2n - 1]\pi, 4)$ where n is an integer.
Points shown: $(0, 0)$, $(2\pi, 4)$, $(4\pi, 0)$

Vertical tangents: $\dfrac{dx}{d\theta} = 2 \neq 0$; none

19. $x = 1 - t$, $y = t^2$

Horizontal tangents: $\dfrac{dy}{dt} = 2t = 0$ when $t = 0$.

Point: $(1, 0)$

Vertical tangents: $\dfrac{dx}{dt} = -1 \neq 0$; none

20. $x = t + 1$, $y = t^2 + 3t$

Horizontal tangents: $\dfrac{dy}{dt} = 2t + 3 = 0$ when $t = -\dfrac{3}{2}$.

Point: $\left(-\dfrac{1}{2}, -\dfrac{9}{4}\right)$

Vertical tangents: $\dfrac{dx}{dt} = 1 \neq 0$; none

21. $x = 1 - t,\ y = t^3 - 3t$

Horizontal tangents: $\dfrac{dy}{dt} = 3t^2 - 3 = 0$ when $t = \pm 1$.

Points: $(0, -2), (2, 2)$

Vertical tangents: $\dfrac{dx}{dt} = -1 \neq 0$; none

22. $x = t^2 - t + 2,\ y = t^3 - 3t$

Horizontal tangents: $\dfrac{dy}{dt} = 3t^2 - 3 = 0$ when $t = \pm 1$.

Points: $(2, -2), (4, 2)$

Vertical tangents: $\dfrac{dx}{dt} = 2t - 1 = 0$ when $t = \dfrac{1}{2}$.

Point: $\left(\dfrac{7}{4}, -\dfrac{11}{8}\right)$

23. $x = 3 \cos \theta,\ y = 3 \sin \theta$

Horizontal tangents: $\dfrac{dy}{d\theta} = 3 \cos \theta = 0$ when $\theta = \dfrac{\pi}{2}, \dfrac{3\pi}{2}$.

Points: $(0, 3), (0, -3)$

Vertical tangents: $\dfrac{dx}{d\theta} = -3 \sin \theta = 0$ when $\theta = 0, \pi$.

Points: $(3, 0), (-3, 0)$

24. $x = \cos \theta,\ y = 2 \sin 2\theta$

Horizontal tangents: $\dfrac{dy}{d\theta} = 4 \cos 2\theta = 0$ when $\theta = \dfrac{\pi}{4}, \dfrac{3\pi}{4}, \dfrac{5\pi}{4}, \dfrac{7\pi}{4}$.

Points: $\left(\dfrac{\sqrt{2}}{2}, 2\right), \left(-\dfrac{\sqrt{2}}{2}, -2\right), \left(-\dfrac{\sqrt{2}}{2}, 2\right), \left(\dfrac{\sqrt{2}}{2}, -2\right)$

Vertical tangents: $\dfrac{dx}{d\theta} = -\sin \theta = 0$ when $\theta = 0, \pi$.

Points: $(1, 0), (-1, 0)$

25. $x = 4 + 2 \cos \theta,\ y = -1 + \sin \theta$

Horizontal tangents: $\dfrac{dy}{d\theta} = \cos \theta = 0$ when $\theta = \dfrac{\pi}{2}, \dfrac{3\pi}{2}$.

Points: $(4, 0), (4, -2)$

Vertical tangents: $\dfrac{dx}{d\theta} = -2 \sin \theta = 0$ when $x = 0, \pi$.

Points: $(6, -1), (2, -1)$

26. $x = 4 \cos^2 \theta,\ y = 2 \sin \theta$

Horizontal tangents: $\dfrac{dy}{d\theta} = 2 \cos \theta = 0$ when $\theta = \dfrac{\pi}{2}, \dfrac{3\pi}{2}$.

Since $dx/d\theta = 0$ at $\pi/2$ and $3\pi/2$, exclude them.

Vertical tangents: $\dfrac{dx}{d\theta} = -8 \cos \theta \sin \theta = 0$ when

$\theta = 0, \pi$.

Point: $(4, 0)$

27. $x = \sec \theta,\ y = \tan \theta$

Horizontal tangents: $\dfrac{dy}{d\theta} = \sec^2 \theta \neq 0$; none

Vertical tangents: $\dfrac{dx}{d\theta} = \sec \theta \tan \theta = 0$ when $x = 0, \pi$.

Points: $(1, 0), (-1, 0)$

28. $x = \cos^2 \theta,\ y = \cos \theta$

Horizontal tangents: $\dfrac{dy}{d\theta} = -\sin \theta = 0$ when $x = 0, \pi$.

Since $dx/d\theta = 0$ at these values, exclude them.

Vertical tangents: $\dfrac{dx}{d\theta} = -2 \cos \theta \sin \theta = 0$ when

$\theta = \dfrac{\pi}{2}, \dfrac{3\pi}{2}$.

(Exclude $0, \pi$.)

Point: $(0, 0)$

29. One possible answer is the graph given by

$$x = t, y = -t.$$

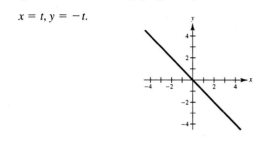

30. One possible answer is the graph given by

$$x = -t, y = -t.$$

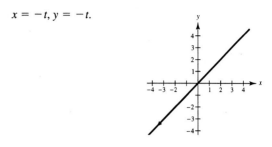

31. $x = e^{-t} \cos t, \; y = e^{-t} \sin t, \; 0 \le t \le \dfrac{\pi}{2}$

$$\frac{dx}{dt} = -e^{-t}(\sin t + \cos t), \; \frac{dy}{dt} = e^{-t}(\cos t - \sin t)$$

$$s = \int_0^{\pi/2} \sqrt{\left(\frac{dx}{dt}\right)^2 + \left(\frac{dy}{dt}\right)^2} \; dt$$

$$= \int_0^{\pi/2} \sqrt{2e^{-2t}} \; dt = -\sqrt{2} \int_0^{\pi/2} e^{-t}(-1) \; dt$$

$$= \left[-\sqrt{2} e^{-t} \right]_0^{\pi/2} = \sqrt{2}(1 - e^{-\pi/2}) \approx 1.12$$

32. $x = t^2, \; y = 4t^3 - 1, \; -1 \le t \le 1$

$$\frac{dx}{dt} = 2t, \; \frac{dy}{dt} = 12t^2$$

$$s = \int_{-1}^{1} \sqrt{4t^2 + 144t^4} \; dt$$

$$= 2 \int_0^1 2t \sqrt{1 + 36t^2} \; dt = \frac{1}{18} \int_0^1 (1 + 36t^2)^{1/2}(72t) \; dt$$

$$= \left[\frac{1}{27}(1 + 36t^2)^{3/2} \right]_0^1 \approx 8.30$$

33. $x = t^2, \; y = 2t, \; 0 \le t \le 2$

$$\frac{dx}{dt} = 2t, \; \frac{dy}{dt} = 2, \; \left(\frac{dx}{dt}\right)^2 + \left(\frac{dy}{dt}\right)^2 = 4t^2 + 4 = 4(t^2 + 1)$$

$$s = 2 \int_0^2 \sqrt{t^2 + 1} \; dt$$

$$= \left[t\sqrt{t^2 + 1} + \ln\left| t + \sqrt{t^2 + 1} \right| \right]_0^2$$

$$= 2\sqrt{5} + \ln\left(2 + \sqrt{5}\right) \approx 5.916$$

34. $x = \arcsin t, \; y = \ln\sqrt{1 - t^2}, \; 0 \le t \le \dfrac{1}{2}$

$$\frac{dx}{dt} = \frac{1}{\sqrt{1 - t^2}}, \; \frac{dy}{dt} = \frac{1}{2}\left(\frac{-2t}{1 - t^2}\right) = \frac{t}{1 - t^2}$$

$$s = \int_0^{1/2} \sqrt{\left(\frac{dx}{dt}\right)^2 + \left(\frac{dy}{dt}\right)^2} \; dt$$

$$= \int_0^{1/2} \sqrt{\frac{1}{(1 - t^2)^2}} \; dt = \int_0^{1/2} \frac{1}{1 - t^2} \; dt$$

$$= \left[-\frac{1}{2} \ln\left| \frac{t - 1}{t + 1} \right| \right]_0^{1/2}$$

$$= -\frac{1}{2} \ln\left(\frac{1}{3}\right) = \frac{1}{2} \ln(3) \approx 0.549$$

35. $x = \sqrt{t}, \; y = 3t - 1, \; \dfrac{dx}{dt} = \dfrac{1}{2\sqrt{t}}, \; \dfrac{dy}{dt} = 3$

$$S = \int_0^1 \sqrt{\frac{1}{4t} + 9} \; dt = \frac{1}{2} \int_0^1 \frac{\sqrt{1 + 36t}}{\sqrt{t}} \; dt$$

$$= \frac{1}{6} \int_0^6 \sqrt{1 + u^2} \; du$$

$$= \frac{1}{12}\left[\ln\left(\sqrt{1 + u^2} + u\right) + u\sqrt{1 + u^2} \right]_0^6$$

$$= \frac{1}{12}\left[\ln\left(\sqrt{37} + 6\right) + 6\sqrt{37} \right] \approx 3.249$$

$$u = 6\sqrt{t}, \; du = \frac{3}{\sqrt{t}} \; dt$$

36. $x = t, \; y = \dfrac{t^5}{10} + \dfrac{1}{6t^3}, \; \dfrac{dx}{dt} = 1, \; \dfrac{dy}{dt} = \dfrac{t^4}{2} - \dfrac{1}{2t^4}$

$$S = \int_1^2 \sqrt{1 + \left(\frac{t^4}{2} - \frac{1}{2t^4}\right)^2} \; dt =$$

$$= \int_1^2 \sqrt{\left(\frac{t^4}{2} + \frac{1}{2t^4}\right)^2} \; dt$$

$$= \int_1^2 \left(\frac{t^4}{2} + \frac{1}{2t^4}\right) \; dt$$

$$= \left[\frac{t^5}{10} - \frac{1}{6t^3} \right]_1^2 = \frac{779}{240}$$

37. $x = a \cos^3 \theta$, $y = a \sin^3 \theta$, $\dfrac{dx}{d\theta} = -3a \cos^2 \theta \sin \theta$,

$\dfrac{dy}{d\theta} = 3a \sin^2 \theta \cos \theta$

$S = 4 \displaystyle\int_0^{\pi/2} \sqrt{9a^2 \cos^4 \theta \sin^2 + 9a^2 \sin^4 \theta \cos^2 \theta} \, d\theta$

$= 12a \displaystyle\int_0^{\pi/2} \sin \theta \cos \theta \sqrt{\cos^2 \theta + \sin^2 \theta} \, d\theta$

$= 6a \displaystyle\int_0^{\pi/2} \sin 2\theta \, d\theta = \left[-3a \cos 2\theta \right]_0^{\pi/2} = 6a$

38. $x = a \cos \theta$, $y = a \sin \theta$, $\dfrac{dx}{d\theta} = -a \sin \theta$, $\dfrac{dy}{d\theta} = a \cos \theta$

$S = 4 \displaystyle\int_0^{\pi/2} \sqrt{a^2 \sin^2 \theta + a^2 \cos^2 \theta} \, d\theta$

$= 4a \displaystyle\int_0^{\pi/2} d\theta = \left[4a\theta \right]_0^{\pi/2} = 2\pi a$

39. $x = a(\theta - \sin \theta)$, $y = a(1 - \cos \theta)$, $\dfrac{dx}{d\theta} = a(1 - \cos \theta)$,

$\dfrac{dy}{d\theta} = a \sin \theta$

$S = 2 \displaystyle\int_0^{\pi} \sqrt{a^2(1 - \cos \theta)^2 + a^2 \sin^2 \theta} \, d\theta$

$= 2\sqrt{2}a \displaystyle\int_0^{\pi} \sqrt{1 - \cos \theta} \, d\theta$

$= 2\sqrt{2}a \displaystyle\int_0^{\pi} \dfrac{\sin \theta}{\sqrt{1 + \cos \theta}} \, d\theta$

$= \left[-4\sqrt{2}a\sqrt{1 + \cos \theta} \right]_0^{\pi} = 8a$

40. $x = \cos \theta + \theta \sin \theta$, $y = \sin \theta - \theta \cos \theta$, $\dfrac{dx}{d\theta} = \theta \cos \theta$

$\dfrac{dy}{d\theta} = \theta \sin \theta$

$S = \displaystyle\int_0^{2\pi} \sqrt{\theta^2 \cos^2 \theta + \theta^2 \sin^2 \theta} \, d\theta$

$= \displaystyle\int_0^{2\pi} \theta \, d\theta = \left[\dfrac{\theta^2}{2} \right]_0^{2\pi} = 2\pi^2$

41. $x = (90 \cos 30°)t$, $y = (90 \sin 30°)t - 16t^2$

(a)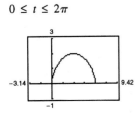

(b) Range: 219.2 ft

(c) $\dfrac{dx}{dt} = 90 \cos 30°$, $\dfrac{dy}{dt} = 90 \sin 30° - 32t$. $y = 0$ for $t = \dfrac{45}{16}$.

$s = \displaystyle\int_0^{45/16} \sqrt{(90 \cos 30°)^2 + (90 \sin 30° - 32t)^2} \, dt$

$= 230.8$ ft

42. $x = 3 \cos \theta$, $y = 4 \sin \theta$

$\dfrac{dx}{d\theta} = -3 \sin \theta$, $\dfrac{dy}{d\theta} = 4 \cos \theta$

$s = \displaystyle\int_0^{2\pi} \sqrt{9 \sin^2 \theta + 16 \cos^2 \theta} \, d\theta \approx 22.1$

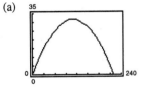

43. (a) $x = t - \sin t$ $x = 2t - \sin(2t)$

$y = 1 - \cos t$ $y = 1 - \cos(2t)$

$0 \le t \le 2\pi$ $0 \le t \le \pi$

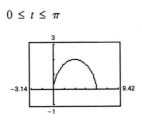

(b) The average speed of the particle on the second path is twice the average speed of a particle on the first path.

(c) $x = \frac{1}{2}t - \sin(\frac{1}{2}t)$

$y = 1 - \cos(\frac{1}{2}t)$

The time required for the particle to traverse the same path is $t = 4\pi$.

44. $x = \dfrac{4t}{1 + t^3}$, $y = \dfrac{4t^2}{1 + t^3}$

(a) $x^3 + y^3 = 4xy$

(b) $\dfrac{dy}{dt} = \dfrac{(1 + t^3)(8t) - 4t^2(3t^2)}{(1 + t^3)^2}$

$\qquad = \dfrac{4t(2 - t^3)}{(1 + t^3)^2} = 0$ when $t = 0$ or $t = \sqrt[3]{2}$.

Points: $(0, 0), \left(\dfrac{4\sqrt[3]{2}}{3}, \dfrac{4\sqrt[3]{4}}{3}\right) \approx (1.6799, 2.1165)$

(c) $s = 2\displaystyle\int_0^1 \sqrt{\left[\dfrac{4(1 - 2t^3)}{(1 + t^3)^2}\right]^2 + \left[\dfrac{4t(2 - t^3)}{(1 + t^3)^2}\right]^2}\, dt = 2\int_0^1 \sqrt{\dfrac{16}{(1 + t^3)^4}[t^8 + 4t^6 - 4t^5 - 4t^3 + 4t^2 + 1]}\, dt$

$\quad = 8\displaystyle\int_0^1 \dfrac{\sqrt{t^8 + 4t^6 - 4t^5 - 4t^3 + 4t^2 + 1}}{(1 + t^3)^2}\, dt \approx 6.557$

45. $x = t$, $y = 2t$, $\dfrac{dx}{dt} = 1$, $\dfrac{dy}{dt} = 2$

(a) $S = 2\pi\displaystyle\int_0^4 2t\sqrt{1 + 4}\, dt = 4\sqrt{5}\pi\int_0^4 t\, dt$

$\quad = \left[2\sqrt{5}\pi t^2\right]_0^4 = 32\pi\sqrt{5}$

(b) $S = 2\pi\displaystyle\int_0^4 t\sqrt{1 + 4}\, dt = 2\sqrt{5}\pi\int_0^4 t\, dt$

$\quad = \left[\sqrt{5}\pi\, t^2\right]_0^4 = 16\pi\sqrt{5}$

46. $x = t$, $y = 4 - 2t$, $\dfrac{dx}{dt} = 1$, $\dfrac{dy}{dt} = -2$

(a) $S = 2\pi\displaystyle\int_0^2 (4 - 2t)\sqrt{1 + 4}\, dt$

$\quad = \left[2\sqrt{5}\pi(4t - t^2)\right]_0^2 = 8\pi\sqrt{5}$

(b) $S = 2\pi\displaystyle\int_0^2 t\sqrt{1 + 4}\, dt = \left[\sqrt{5}\pi t^2\right]_0^2 = 4\pi\sqrt{5}$

47. $x = 4\cos\theta$, $y = 4\sin\theta$, $\dfrac{dx}{d\theta} = -4\sin\theta$, $\dfrac{dy}{d\theta} = 4\cos\theta$

$S = 2\pi\displaystyle\int_0^{\pi/2} 4\cos\theta\sqrt{(-4\sin\theta)^2 + (4\cos\theta)^2}\, d\theta$

$\quad = 32\pi\displaystyle\int_0^{\pi/2} \cos\theta\, d\theta = \left[32\pi\sin\theta\right]_0^{\pi/2} = 32\pi$

48. $x = t^3$, $y = t + 2$, $\dfrac{dx}{dt} = 3t^2$, $\dfrac{dy}{dt} = 1$

$S = 2\pi\displaystyle\int_1^2 t^3\sqrt{9t^4 + 1}\, dt = \left[\dfrac{\pi}{27}(9t^4 + 1)^{3/2}\right]_1^2$

$\quad = \dfrac{\pi}{27}\left(145\sqrt{145} - 10\sqrt{10}\right) \approx 199.48$

49. $x = a\cos^3\theta$, $y = a\sin^3\theta$, $\dfrac{dx}{d\theta} = -3a\cos^2\theta\sin\theta$, $\dfrac{dy}{d\theta} = 3a\sin^2\theta\cos\theta$

$S = 4\pi\displaystyle\int_0^{\pi/2} a\sin^3\theta\sqrt{9a^2\cos^4\theta\sin^2\theta + 9a^2\sin^4\theta\cos^2\theta}\, d\theta = 12a^2\pi\int_0^{\pi/2}\sin^4\theta\cos\theta\, d\theta = \dfrac{12\pi a^2}{5}\left[\sin^5\theta\right]_0^{\pi/2} = \dfrac{12}{5}\pi a^2$

50. $x = a\cos\theta$, $y = b\sin\theta$, $\dfrac{dx}{d\theta} = -a\sin\theta$, $\dfrac{dy}{d\theta} = b\cos\theta$

(a) $S = 4\pi\displaystyle\int_0^{\pi/2} b\sin\theta\sqrt{a^2\sin^2\theta + b^2\cos^2\theta}\, d\theta$

$\quad = 4\pi\displaystyle\int_0^{\pi/2} ab\sin\theta\sqrt{1 - \left(\dfrac{a^2 - b^2}{a^2}\right)\cos^2\theta}\, d\theta = \dfrac{-4ab\pi}{e}\int_0^{\pi/2} (-e\sin\theta)\sqrt{1 - e^2\cos^2\theta}\, d\theta$

$\quad = \dfrac{-2ab\pi}{e}\left[e\cos\theta\sqrt{1 - e^2\cos^2\theta} + \arcsin(e\cos\theta)\right]_0^{\pi/2} = \dfrac{-ab\pi}{e}\left[e\sqrt{1 - e^2} + \arcsin(e)\right]$

$\quad = 2\pi b^2 + \left(\dfrac{2\pi a^2 b}{\sqrt{a^2 - b^2}}\right)\arcsin\left(\dfrac{\sqrt{a^2 - b^2}}{a}\right) = 2\pi b^2 + 2\pi\left(\dfrac{ab}{e}\right)\arcsin(e)$

$\quad \left(e = \dfrac{\sqrt{a^2 - b^2}}{a} = \dfrac{c}{a}\text{: eccentricity}\right)$

—CONTINUED—

50. —CONTINUED—

(b) $S = 4\pi \int_0^{\pi/2} a \cos \theta \sqrt{a^2 \sin^2 \theta + b^2 \cos^2 \theta} \, d\theta$

$$= 4\pi \int_0^{\pi/2} a \cos \theta \sqrt{b^2 + c^2 \sin^2 \theta} \, d\theta = \frac{4a\pi}{c} \int_0^{\pi/2} c \cos \theta \sqrt{b^2 + c^2 \sin^2 \theta} \, d\theta$$

$$= \frac{2a\pi}{c} \left[c \sin \theta \sqrt{b^2 + c^2 \sin^2 \theta} + b^2 \ln \left| c \sin \theta + \sqrt{b^2 + c^2 \sin^2 \theta} \right| \right]_0^{\pi/2}$$

$$= \frac{2a\pi}{c} \left[c \sqrt{b^2 + c^2} + b^2 \ln \left| c + \sqrt{b^2 + c^2} \right| - b^2 \ln b \right]$$

$$= 2\pi a^2 + \frac{2\pi a b^2}{\sqrt{a^2 - b^2}} \ln \left| \frac{a + \sqrt{a^2 - b^2}}{b} \right| = 2\pi a^2 + \left(\frac{\pi b^2}{e} \right) \ln \left| \frac{1 + e}{1 - e} \right|$$

51. $x = r \cos \phi, \ y = r \sin \phi$

$$S = 2\pi \int_0^\theta r \sin \phi \sqrt{r^2 \sin^2 \phi + r^2 \cos^2 \phi} \, d\phi$$

$$= 2\pi r^2 \int_0^\theta \sin \phi \, d\phi$$

$$= \left[-2\pi r^2 \cos \phi \right]_0^\theta$$

$$= 2\pi r^2 (1 - \cos \theta)$$

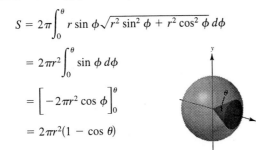

52. Let y be a continuous function of x on $a \le x \le b$. Suppose that $x = f(t), \ y = g(t),$ and $f(t_1) = a, f(t_2) = b$. Then using integration by substitution, $dx = f'(t) \, dt$ and

$$\int_a^b y \, dx = \int_{t_1}^{t_2} g(t) f'(t) \, dt.$$

53. $x = \sqrt{t}, \ y = 4 - t, \ 0 \le t \le 4$

$$A = \int_0^4 (4 - t) \frac{1}{2\sqrt{t}} \, dt = \frac{1}{2} \int_0^4 (4t^{-1/2} - t^{1/2}) \, dt = \left[\frac{1}{2} \left(8\sqrt{t} - \frac{2}{3} t \sqrt{t} \right) \right]_0^4 = \frac{16}{3}$$

$$\bar{x} = \frac{3}{16} \int_0^4 (4 - t) \sqrt{t} \left(\frac{1}{2\sqrt{t}} \right) dt = \frac{3}{32} \int_0^4 (4 - t) \, dt = \left[\frac{3}{32} \left(4t - \frac{t^2}{2} \right) \right]_0^4 = \frac{3}{4}$$

$$\bar{y} = \frac{3}{32} \int_0^4 (4 - t)^2 \frac{1}{2\sqrt{t}} \, dt = \frac{3}{64} \int_0^4 [16t^{-1/2} - 8t^{1/2} + t^{3/2}] \, dt = \frac{3}{64} \left[32\sqrt{t} - \frac{16}{3} t \sqrt{t} + \frac{2}{5} t^2 \sqrt{t} \right]_0^4 = \frac{8}{5}$$

$$(\bar{x}, \bar{y}) = \left(\frac{3}{4}, \frac{8}{5} \right)$$

54. $x = \sqrt{4 - t}, \ y = \sqrt{t}, \ \dfrac{dx}{dt} = -\dfrac{1}{2\sqrt{4 - t}}, \ 0 \le t \le 4$

$$A = \int_4^0 \sqrt{t} \left(-\frac{1}{2\sqrt{4 - t}} \right) dt = \int_0^2 \sqrt{4 - u^2} \, du = \frac{1}{2} \left[u\sqrt{4 - u^2} + 4 \arcsin \frac{u}{2} \right]_0^2 = \pi$$

Let $u = \sqrt{4 - t}$, then $du = -1/(2\sqrt{4 - t}) \, dt$ and $\sqrt{t} = \sqrt{4 - u^2}$.

$$\bar{x} = \frac{1}{\pi} \int_4^0 \sqrt{4 - t} \sqrt{t} \left(-\frac{1}{2\sqrt{4 - t}} \right) dt = -\frac{1}{2\pi} \int_4^0 \sqrt{t} \, dt = \left[-\frac{1}{2\pi} \frac{2}{3} t^{3/2} \right]_4^0 = \frac{8}{3\pi}$$

$$\bar{y} = \frac{1}{2\pi} \int_4^0 (\sqrt{t})^2 \left(-\frac{1}{2\sqrt{4 - t}} \right) dt = -\frac{1}{4\pi} \int_4^0 \frac{t}{\sqrt{4 - t}} \, dt = -\frac{1}{4\pi} \left[\frac{-2(8 + t)}{3} \sqrt{4 - t} \right]_4^0 = \frac{8}{3\pi}$$

$$(\bar{x}, \bar{y}) = \left(\frac{8}{3\pi}, \frac{8}{3\pi} \right)$$

55. $x = 3\cos\theta$, $y = 3\sin\theta$, $\dfrac{dx}{d\theta} = -3\sin\theta$

$$V = 2\pi\int_{\pi/2}^{0}(3\sin\theta)^2(-3\sin\theta)\,d\theta$$

$$= -54\pi\int_{\pi/2}^{0}\sin^3\theta\,d\theta$$

$$= -54\pi\int_{\pi/2}^{0}(1-\cos^2\theta)\sin\theta\,d\theta$$

$$= -54\pi\left[-\cos\theta + \frac{\cos^3\theta}{3}\right]_{\pi/2}^{0} = 36\pi$$

56. $x = \cos\theta$, $y = 3\sin\theta$, $\dfrac{dx}{d\theta} = -\sin\theta$

$$V = 2\pi\int_{\pi/2}^{0}(3\sin\theta)^2(-\sin\theta)\,d\theta$$

$$= -18\pi\int_{\pi/2}^{0}\sin^3\theta\,d\theta$$

$$= -18\pi\left[-\cos\theta + \frac{\cos^3\theta}{3}\right]_{\pi/2}^{0} = 12\pi$$

57. $x = 2\sin^2\theta$

$y = 2\sin^2\theta\tan\theta$

$\dfrac{dx}{d\theta} = 4\sin\theta\cos\theta$

$$A = \int_{0}^{\pi/2}2\sin^2\theta\tan\theta(4\sin\theta\cos\theta)\,d\theta = 8\int_{0}^{\pi/2}\sin^4\theta\,d\theta$$

$$= 8\left[\frac{-\sin^3\theta\cos\theta}{4} - \frac{3}{8}\sin\theta\cos\theta + \frac{3}{8}\theta\right]_{0}^{\pi/2} = \frac{3\pi}{2}$$

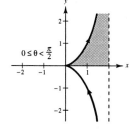

$0 \le \theta < \dfrac{\pi}{2}$

58. $x = 2\cot\theta$, $y = 2\sin^2\theta$, $\dfrac{dx}{d\theta} = -2\csc^2\theta$

$$A = 2\int_{\pi/2}^{0}(2\sin^2\theta)(-2\csc^2\theta)\,d\theta = -8\int_{\pi/2}^{0}d\theta = \left[-8\theta\right]_{\pi/2}^{0} = 4\pi$$

59. πab is area of ellipse (d).

60. $\frac{3}{8}\pi a^2$ is area of asteroid (b).

61. $6\pi a^2$ is area of cardioid (f).

62. $2\pi a^2$ is area of deltoid (c).

63. $\frac{8}{3}ab$ is area of hourglass (a).

64. $2\pi ab$ is area of teardrop (e).

65. (a) $x = \dfrac{1-t^2}{1+t^2}$, $y = \dfrac{2t}{1+t^2}$, $-20 \le t \le 20$

The graph is the circle $x^2 + y^2 = 1$, except the point $(-1, 0)$.

Verify: $x^2 + y^2 = \left(\dfrac{1-t^2}{1+t^2}\right)^2 + \left(\dfrac{2t}{1+t^2}\right)^2 = \dfrac{1-2t^2+t^4+4t^2}{(1+t^2)^2} = \dfrac{(1+t^2)^2}{(1+t^2)^2} = 1$

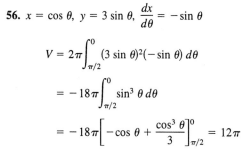

(b) As t increases from -20 to 0, the speed increases, and as t increases from 0 to 20, the speed decreases.

66. (a) $y = -12\ln\left(\dfrac{12-\sqrt{144-x^2}}{x}\right) - \sqrt{144-x^2}$

$0 < x \le 12$

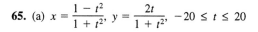

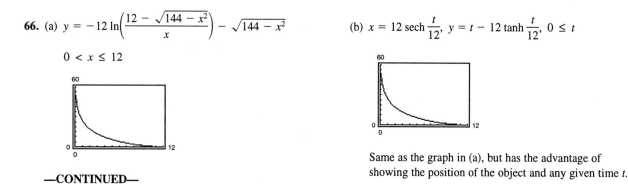

(b) $x = 12\,\text{sech}\dfrac{t}{12}$, $y = t - 12\tanh\dfrac{t}{12}$, $0 \le t$

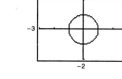

Same as the graph in (a), but has the advantage of showing the position of the object and any given time t.

—CONTINUED—

66. —CONTINUED—

(c) $\dfrac{dy}{dx} = \dfrac{1 - \operatorname{sech}^2(t/12)}{-\operatorname{sech}(t/12)\tan(t/12)} = -\sinh\dfrac{t}{12}$

Tangent line: $y - \left(t_0 - 12\tanh\dfrac{t_0}{12}\right) = -\sinh\dfrac{t_0}{12}\left(x - 12\operatorname{sech}\dfrac{t_0}{12}\right)$

$$y = t_0 - \left(\sinh\dfrac{t_0}{12}\right)x$$

y-intercept: $(0, t_0)$

Distance between $(0, t_0)$ and (x, y): $d = \sqrt{\left(12\operatorname{sech}\dfrac{t_0}{12}\right)^2 + \left(-12\tanh\dfrac{t_0}{12}\right)^2} = 12$

$$d = 12 \text{ for any } t \geq 0.$$

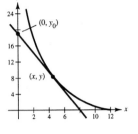

67. (a) The first plane makes an angle of 70° with the positive x-axis, and is 150 miles from P:

$x_1 = \cos 70°(150 - 375t)$

$y_1 = \sin 70°(150 - 375t)$

Similarly for the second plane,

$x_2 = \cos 135°(190 - 450t)$

$\quad = \cos 45°(-190 + 450t)$

$y_2 = \sin 135°(190 - 450t)$

$\quad = \sin 45°(190 - 450t)$

(b) $d = \sqrt{(x_2 - x_1)^2 + (y_2 - y_1)^2}$

$\quad = [[\cos 45(-190 + 450t) - \cos 70(150 - 375t)]^2 + [\sin 45(190 - 450t) - \sin 70(150 - 375t)]^2]^{1/2}$

(c)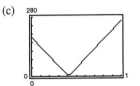

The minimum distance is 7.59 miles when $t = 0.4145$.

68. If $x = g(t)$ and $y = f(t)$ are differentiable on (a, b) and continuous on $[a, b]$ such that $g'(t) \neq 0$ for any t in (a, b), then there exists t_0 in (a, b) such that

$\dfrac{f'(t_0)}{g'(t_0)} = \dfrac{f(b) - f(a)}{g(b) - g(a)}$

$\dfrac{\frac{dy}{dt}(t_0)}{\frac{dx}{dt}(t_0)} = \dfrac{y_2 - y_1}{x_2 - x_1}$

$\dfrac{dy}{dx}(t_0) = \dfrac{y_2 - y_1}{x_2 - x_1}$

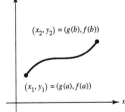

This is just the Mean Value Theorem.

69. False

$$\dfrac{d^2y}{dx^2} = \dfrac{\dfrac{d}{dt}\left[\dfrac{g'(t)}{f'(t)}\right]}{f'(t)} = \dfrac{f'(t)g''(t) - g'(t)f''(t)}{[f'(t)]^3}$$

70. False. Both dx/dt and dy/dt are zero when $t = 0$. By eliminating the parameter, we have $y = x^{2/3}$ which does not have a horizontal tangent at the origin.

Section 9.4   Polar Coordinates and Polar Graphs

1. $\left(4, \dfrac{\pi}{2}\right)$

$x = 4\cos\left(\dfrac{\pi}{2}\right) = 0$

$y = 4\sin\left(\dfrac{\pi}{2}\right) = 4$

$(x, y) = (0, 4)$

2. $\left(-1, \dfrac{5\pi}{4}\right)$

$x = -1\cos\left(\dfrac{5\pi}{4}\right) = \dfrac{\sqrt{2}}{2}$

$y = -1\sin\left(\dfrac{5\pi}{4}\right) = \dfrac{\sqrt{2}}{2}$

$(x, y) = \left(\dfrac{\sqrt{2}}{2}, \dfrac{\sqrt{2}}{2}\right)$

3. $\left(-4, -\dfrac{\pi}{3}\right)$

$x = -4\cos\left(-\dfrac{\pi}{3}\right) = -2$

$y = -4\sin\left(-\dfrac{\pi}{3}\right) = 2\sqrt{3}$

$(x, y) = \left(-2, 2\sqrt{3}\right)$

4. $\left(0, -\dfrac{7\pi}{6}\right)$

$x = 0\cos\left(-\dfrac{7\pi}{6}\right) = 0$

$y = 0\sin\left(-\dfrac{7\pi}{6}\right) = 0$

$(x, y) = (0, 0)$

5. $\left(\sqrt{2}, 2.36\right)$

$x = \sqrt{2}\cos(2.36) \approx -1.004$

$y = \sqrt{2}\sin(2.36) \approx 0.996$

$(x, y) = (-1.004, 0.996)$

6. $(-3, -1.57)$

$x = -3\cos(-1.57) \approx -0.0024$

$y = -3\sin(-1.57) \approx 3$

$(x, y) = (-0.0024, 3)$

7. $(r, \theta) = \left(5, \dfrac{3\pi}{4}\right)$

$(x, y) = (-3.5355, 3.5355)$

8. $(r, \theta) = \left(-2, \dfrac{11\pi}{6}\right)$

$(x, y) = (-1.7321, 1)$

9. $(r, \theta) = (-3.5, 2.5)$

$(x, y) = (2.804, -2.095)$

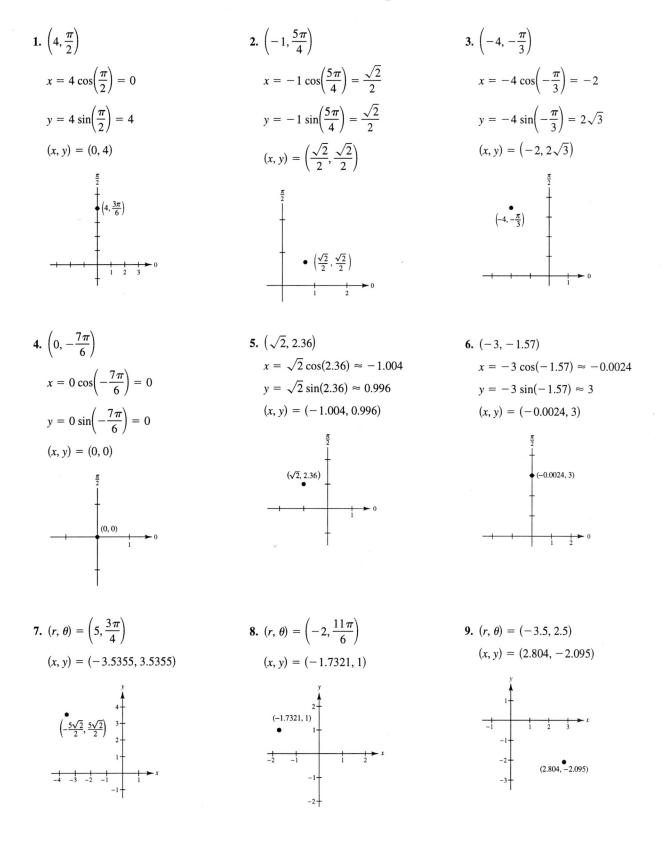

10. $(r, \theta) = (8.25, 1.3)$

$(x, y) = (2.2069, 7.9494)$

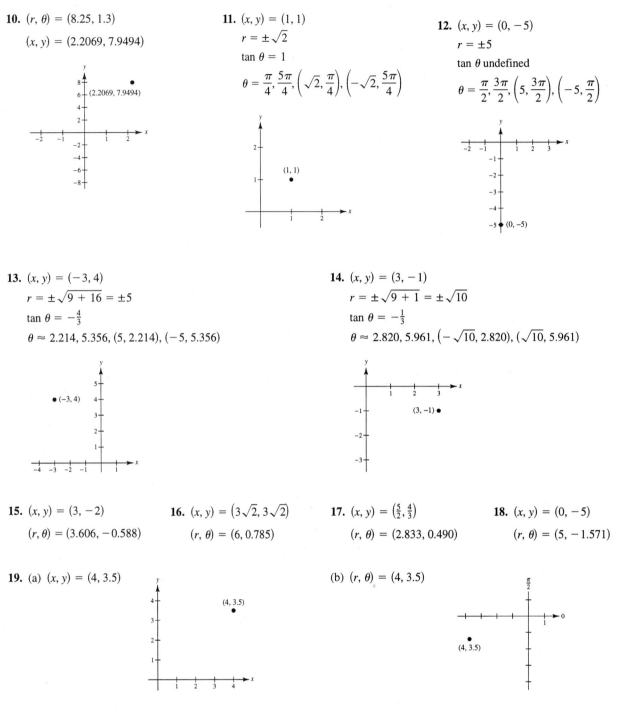

11. $(x, y) = (1, 1)$

$r = \pm\sqrt{2}$

$\tan \theta = 1$

$\theta = \dfrac{\pi}{4}, \dfrac{5\pi}{4}, \left(\sqrt{2}, \dfrac{\pi}{4}\right), \left(-\sqrt{2}, \dfrac{5\pi}{4}\right)$

12. $(x, y) = (0, -5)$

$r = \pm 5$

$\tan \theta$ undefined

$\theta = \dfrac{\pi}{2}, \dfrac{3\pi}{2}, \left(5, \dfrac{3\pi}{2}\right), \left(-5, \dfrac{\pi}{2}\right)$

13. $(x, y) = (-3, 4)$

$r = \pm\sqrt{9 + 16} = \pm 5$

$\tan \theta = -\dfrac{4}{3}$

$\theta \approx 2.214, 5.356, (5, 2.214), (-5, 5.356)$

14. $(x, y) = (3, -1)$

$r = \pm\sqrt{9 + 1} = \pm\sqrt{10}$

$\tan \theta = -\dfrac{1}{3}$

$\theta \approx 2.820, 5.961, \left(-\sqrt{10}, 2.820\right), \left(\sqrt{10}, 5.961\right)$

15. $(x, y) = (3, -2)$

$(r, \theta) = (3.606, -0.588)$

16. $(x, y) = \left(3\sqrt{2}, 3\sqrt{2}\right)$

$(r, \theta) = (6, 0.785)$

17. $(x, y) = \left(\dfrac{5}{2}, \dfrac{4}{3}\right)$

$(r, \theta) = (2.833, 0.490)$

18. $(x, y) = (0, -5)$

$(r, \theta) = (5, -1.571)$

19. (a) $(x, y) = (4, 3.5)$

(b) $(r, \theta) = (4, 3.5)$

20. (a) Moving horizontally, the x-coordinate changes. Moving vertically, the y-coordinate changes.

(b) Both r and θ values change.

(c) In polar mode, horizontal (or vertical) changes result in changes in both r and θ.

21. $x^2 + y^2 = a^2$

$r = a$

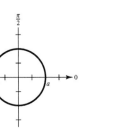

22. $x^2 + y^2 - 2ax = 0$

$r^2 - 2ar\cos\theta = 0$

$r(r - 2a\cos\theta) = 0$

$r = 2a\cos\theta$

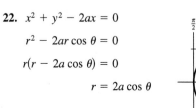
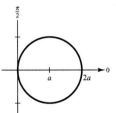

23. $y = 4$

$r \sin \theta = 4$

$r = 4 \csc \theta$

24. $x = 10$

$r \cos \theta = 10$

$r = 10 \sec \theta$

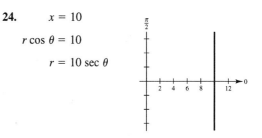

25. $3x - y + 2 = 0$

$3r \cos \theta - r \sin \theta + 2 = 0$

$r(3 \cos \theta - \sin \theta) = -2$

$$r = \frac{-2}{3 \cos \theta - \sin \theta}$$

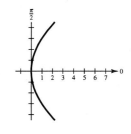

26. $xy = 4$

$(r \cos \theta)(r \sin \theta) = 4$

$r^2 = 4 \sec \theta \csc \theta$

$= 8 \csc 2\theta$

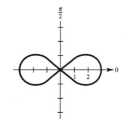

27. $y^2 = 9x$

$r^2 \sin^2 \theta = 9r \cos \theta$

$$r = \frac{9 \cos \theta}{\sin^2 \theta}$$

$r = 9 \csc^2 \theta \cos \theta$

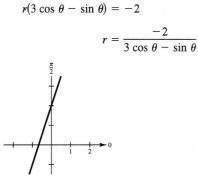

28. $(x^2 + y^2)^2 - 9(x^2 - y^2) = 0$

$(r^2)^2 - 9(r^2 \cos^2 \theta - r^2 \sin^2 \theta) = 0$

$r^2[r^2 - 9(\cos 2\theta)] = 0$

$r^2 = 9 \cos 2\theta$

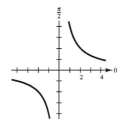

29. $r = 3$

$r^2 = 9$

$x^2 + y^2 = 9$

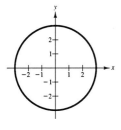

30. $r = -2$

$r^2 = 4$

$x^2 + y^2 = 4$

31. $r = \sin \theta$

$r^2 = r \sin \theta$

$x^2 + y^2 = y$

$x^2 + \left(y - \frac{1}{2}\right)^2 = \frac{1}{4}$

$x^2 + y^2 - y = 0$

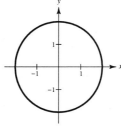

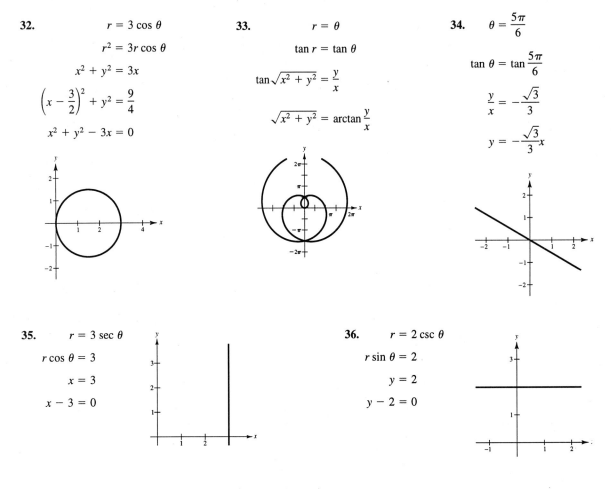

32.
$$r = 3 \cos \theta$$
$$r^2 = 3r \cos \theta$$
$$x^2 + y^2 = 3x$$
$$\left(x - \frac{3}{2}\right)^2 + y^2 = \frac{9}{4}$$
$$x^2 + y^2 - 3x = 0$$

33.
$$r = \theta$$
$$\tan r = \tan \theta$$
$$\tan \sqrt{x^2 + y^2} = \frac{y}{x}$$
$$\sqrt{x^2 + y^2} = \arctan \frac{y}{x}$$

34.
$$\theta = \frac{5\pi}{6}$$
$$\tan \theta = \tan \frac{5\pi}{6}$$
$$\frac{y}{x} = -\frac{\sqrt{3}}{3}$$
$$y = -\frac{\sqrt{3}}{3}x$$

35.
$$r = 3 \sec \theta$$
$$r \cos \theta = 3$$
$$x = 3$$
$$x - 3 = 0$$

36.
$$r = 2 \csc \theta$$
$$r \sin \theta = 2$$
$$y = 2$$
$$y - 2 = 0$$

37.
$$r = 2(h \cos \theta + k \sin \theta)$$
$$r^2 = 2r(h \cos \theta + k \sin \theta)$$
$$r^2 = 2[h(r \cos \theta) + k(r \sin \theta)]$$
$$x^2 + y^2 = 2(hx + ky)$$
$$x^2 + y^2 - 2hx - 2ky = 0$$
$$(x^2 - 2hx + h^2) + (y^2 - 2ky + k^2) = 0 + h^2 + k^2$$
$$(x - h)^2 + (v - k)^2 = h^2 + k^2$$

Radius: $\sqrt{h^2 + k^2}$

Center: (h, k)

38. (a) The rectangular coordinates of (r_1, θ_1) are $(r_1 \cos \theta_1, r_1 \sin \theta_1)$. The rectangular coordinates of (r_2, θ_2) are $(r_2 \cos \theta_2, r_2 \sin \theta_2)$.

$$d^2 = (x_2 - x_1)^2 + (y_2 - y_1)^2$$
$$= (r_2 \cos \theta_2 - r_1 \cos \theta_1)^2 + (r_2 \sin \theta_2 - r_1 \sin \theta_1)^2$$
$$= r_2{}^2 \cos^2 \theta_2 - 2r_1 r_2 \cos \theta_1 \cos \theta_2 + r_1{}^2 \cos^2 \theta_1 + r_2{}^2 \sin^2 \theta_2 - 2r_1 r_2 \sin \theta_1 \sin \theta_2 + r_1{}^2 \sin^2 \theta_1$$
$$= r_2{}^2 (\cos^2 \theta_2 + \sin^2 \theta_2) + r_1{}^2 (\cos^2 \theta_1 + \sin^2 \theta_1) - 2 r_1 r_2 (\cos \theta_1 \cos \theta_2 + \sin \theta_1 \sin \theta_2)$$
$$= r_1{}^2 + r_2{}^2 - 2r_1 r_2 \cos(\theta_1 - \theta_2)$$
$$d = \sqrt{r_1{}^2 + r_2{}^2 - 2r_1 r_2 \cos(\theta_1 - \theta_2)}$$

—CONTINUED—

38. —CONTINUED—

(b) If $\theta_1 = \theta_2$, the points lie on the same line passing through the origin. In this case,

$$d = \sqrt{r_1^2 + r_2^2 - 2r_1r_2 \cos(0)}$$
$$= \sqrt{(r_1 - r_2)^2} = |r_1 - r_2|$$

(c) If $\theta_1 - \theta_2 = 90°$, then $\cos(\theta_1 - \theta_2) = 0$ and
$$d = \sqrt{r_1^2 + r_2^2},$$

the Pythagorean Theorem!

(d) Many answers are possible. For example, consider the two points $(r_1, \theta_1) = (1, 0)$ and $(r_2, \theta_2) = (2, \pi/2)$.

$$d = \sqrt{1 + 2^2 - 2(1)(2) \cos\left(0 - \frac{\pi}{2}\right)} = \sqrt{5}$$

Using $(r_1, \theta_1) = (-1, \pi)$ and $(r_2, \theta_2) = [2, (5\pi/2)]$,

$$d = \sqrt{(-1)^2 + (2)^2 - 2(-1)(2) \cos\left(\pi - \frac{5\pi}{2}\right)}$$
$$= \sqrt{5}.$$

You always obtain the same distance.

39. $\left(4, \dfrac{2\pi}{3}\right), \left(2, \dfrac{\pi}{6}\right)$

$$d = \sqrt{4^2 + 2^2 - 2(4)(2) \cos\left(\frac{2\pi}{3} - \frac{\pi}{6}\right)}$$
$$= \sqrt{20 - 16 \cos \frac{\pi}{2}} = 2\sqrt{5} \approx 4.5$$

40. $\left(10, \dfrac{7\pi}{6}\right), (3, \pi)$

$$d = \sqrt{10^2 + 3^2 - 2(10)(3) \cos\left(\frac{7\pi}{6} - \pi\right)}$$
$$= \sqrt{109 - 60 \cos \frac{\pi}{6}} = \sqrt{109 - 30\sqrt{3}} \approx 7.6$$

41. $(2, 0.5), (7, 1.2)$

$$d = \sqrt{2^2 + 7^2 - 2(2)(7) \cos(0.5 - 1.2)}$$
$$= \sqrt{53 - 28 \cos(-0.7)} \approx 5.6$$

42. $(4, 2.5), (12, 1)$

$$d = \sqrt{4^2 + 12^2 - 2(4)(12) \cos(2.5 - 1)}$$
$$= \sqrt{160 - 96 \cos 1.5} \approx 12.3$$

43. $r = 2 + 3 \sin \theta$

$$\frac{dy}{dx} = \frac{3 \cos \theta \sin \theta + \cos \theta(2 + 3 \sin \theta)}{3 \cos \theta \cos \theta - \sin \theta(2 + 3 \sin \theta)}$$

$$= \frac{2 \cos \theta(3 \sin \theta + 1)}{3 \cos 2\theta - 2 \sin \theta} = \frac{2 \cos \theta(3 \sin \theta + 1)}{6 \cos^2 \theta - 2 \sin \theta - 3}$$

At $\left(5, \dfrac{\pi}{2}\right)$, $\dfrac{dy}{dx} = 0$.

At $(2, \pi)$, $\dfrac{dy}{dx} = -\dfrac{2}{3}$.

At $\left(-1, \dfrac{3\pi}{2}\right)$, $\dfrac{dy}{dx} = 0$.

44. $r = 2(1 - \sin \theta)$

$$\frac{dy}{dx} = \frac{-2 \cos \theta \sin \theta + 2 \cos \theta(1 - \sin \theta)}{-2 \cos \theta \cos \theta - 2 \sin \theta(1 - \sin \theta)}$$

At $(2, 0)$, $\dfrac{dy}{dx} = -1$.

At $\left(3, \dfrac{7\pi}{6}\right)$, $\dfrac{dy}{dx}$ is undefined.

At $\left(4, \dfrac{3\pi}{2}\right)$, $\dfrac{dy}{dx} = 0$.

45. (a), (b) $r = 3(1 - \cos \theta)$

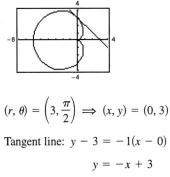

$$(r, \theta) = \left(3, \frac{\pi}{2}\right) \implies (x, y) = (0, 3)$$

Tangent line: $y - 3 = -1(x - 0)$

$$y = -x + 3$$

(c) At $\theta = \dfrac{\pi}{2}, \dfrac{dy}{dx} = -1.0$.

46. (a), (b) $r = 3 - 2 \cos \theta$

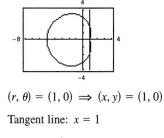

$$(r, \theta) = (1, 0) \implies (x, y) = (1, 0)$$

Tangent line: $x = 1$

(c) At $\theta = 0$, $\dfrac{dy}{dx}$ does not exist (vertical tangent).

47. (a), (b) $r = 3 \sin \theta$

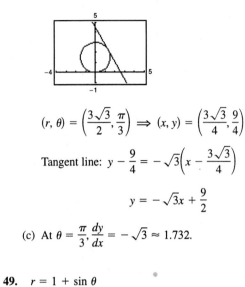

$$(r, \theta) = \left(\frac{3\sqrt{3}}{2}, \frac{\pi}{3}\right) \Rightarrow (x, y) = \left(\frac{3\sqrt{3}}{4}, \frac{9}{4}\right)$$

Tangent line: $y - \dfrac{9}{4} = -\sqrt{3}\left(x - \dfrac{3\sqrt{3}}{4}\right)$

$$y = -\sqrt{3}x + \frac{9}{2}$$

(c) At $\theta = \dfrac{\pi}{3}, \dfrac{dy}{dx} = -\sqrt{3} \approx 1.732$.

48. (a), (b) $r = 4$

at $(r, \theta) = \left(4, \dfrac{\pi}{4}\right) \Rightarrow (x, y) = \left(2\sqrt{2}, 2\sqrt{2}\right)$

Tangent line: $y - 2\sqrt{2} = -1\left(x - 2\sqrt{2}\right)$

$$y = -x + 4\sqrt{2}$$

(c) At $\theta = \dfrac{\pi}{4}, \dfrac{dy}{dx} = -1$.

49. $r = 1 + \sin \theta$

$\dfrac{dy}{d\theta} = (1 + \sin \theta) \cos \theta + \cos \theta \sin \theta$

$\quad = \cos \theta(1 + 2 \sin \theta) = 0$

$\cos \theta = 0, \ \sin \theta = -\dfrac{1}{2}, \ \theta = \dfrac{\pi}{2}, \dfrac{3\pi}{2}, \ \theta = \dfrac{7\pi}{6}, \dfrac{11\pi}{6}$

Horizontal: $\left(2, \dfrac{\pi}{2}\right), \left(\dfrac{1}{2}, \dfrac{7\pi}{6}\right), \left(\dfrac{1}{2}, \dfrac{11\pi}{6}\right)$

$\dfrac{dx}{d\theta} = -(1 + \sin \theta) \sin \theta + \cos^2 \theta$

$\quad = (1 + \sin \theta)(1 - 2 \sin \theta) = 0$

$\sin \theta = -1, \ \sin \theta = \dfrac{1}{2}, \ \theta = \dfrac{3\pi}{2}, \ \theta = \dfrac{\pi}{6}, \dfrac{5\pi}{6}$

Vertical: $\left(\dfrac{3}{2}, \dfrac{\pi}{6}\right), \left(\dfrac{3}{2}, \dfrac{5\pi}{6}\right)$

50. $r = a \sin \theta$

$\dfrac{dy}{d\theta} = a \sin \theta \cos \theta + a \cos \theta \sin \theta$

$\quad = 2a \sin \theta \cos \theta = 0$

$\theta = 0, \ \dfrac{\pi}{2}, \ \pi, \ \dfrac{3\pi}{2}$

$\dfrac{dx}{d\theta} = -a \sin^2 \theta + a \cos^2 \theta = a(1 - 2 \sin^2 \theta) = 0$

$\sin \theta = \pm \dfrac{1}{\sqrt{2}}, \ \theta = \dfrac{\pi}{4}, \dfrac{3\pi}{4}, \dfrac{5\pi}{4}, \dfrac{7\pi}{4}$

Horizontal: $(0, 0), \left(a, \dfrac{\pi}{2}\right)$

Vertical: $\left(\dfrac{a\sqrt{2}}{2}, \dfrac{\pi}{4}\right), \left(\dfrac{a\sqrt{2}}{2}, \dfrac{3\pi}{4}\right)$

51. $r = 2 \csc \theta + 3$

$\dfrac{dy}{d\theta} = (2 \csc \theta + 3) \cos \theta + (-2 \csc \theta \cot \theta) \sin \theta$

$\quad = 3 \cos \theta = 0$

$\theta = \dfrac{\pi}{2}, \dfrac{3\pi}{2}$

Horizontal: $\left(5, \dfrac{\pi}{2}\right), \left(1, \dfrac{3\pi}{2}\right)$

52. $r = a \sin \theta \cos^2 \theta$

$\dfrac{dy}{d\theta} = a \sin \theta \cos^3 \theta + [-2a \sin^2 \theta \cos \theta + a \cos^3 \theta] \sin \theta$

$\quad = 2a[\sin \theta \cos^3 \theta - \sin^3 \theta \cos \theta]$

$\quad = 2a \sin \theta \cos \theta(\cos^2 \theta - \sin^2 \theta) = 0$

$\theta = 0, \ \tan^2 \theta = 1, \ \theta = \dfrac{\pi}{4}, \dfrac{3\pi}{4}$

Horizontal: $\left(\dfrac{\sqrt{2}a}{4}, \dfrac{\pi}{4}\right), \left(\dfrac{\sqrt{2}a}{4}, \dfrac{3\pi}{4}\right), (0, 0)$

53. $r = 4 \sin \theta \cos^2 \theta$

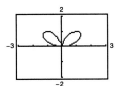

Horizontal tangents:

$(0, 0), (1.4142, 0.7854), (1.4142, 2.3562)$

54. $r = 3 \cos 2\theta \sec \theta$

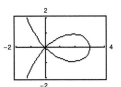

Horizontal tangents: $(2.133, \pm 0.4352)$

55. $r = 2 \csc \theta + 5$

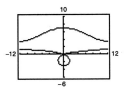

Horizontal tangents: $\left(7, \dfrac{\pi}{2}\right), \left(3, \dfrac{3\pi}{2}\right)$

56. $r = 2 \cos(3\theta - 2)$

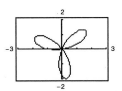

Horizontal tangents:

$(1.894, 0.776), (1.755, 2.594), (1.998, -1.442)$

57.

$$r = 3 \sin \theta$$

$$r^2 = 3r \sin \theta$$

$$x^2 + y^2 = 3y$$

$$x^2 + \left(y - \tfrac{3}{2}\right)^2 = \tfrac{9}{4}$$

Circle $r = \dfrac{3}{2}$

Center: $\left(0, \dfrac{3}{2}\right)$

Tangent at the pole: $\theta = 0$

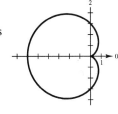

58. $r = 3(1 - \cos \theta)$

Cardioid

Symmetric to polar axis since r is a function of $\cos \theta$.

θ	0	$\dfrac{\pi}{3}$	$\dfrac{\pi}{2}$	$\dfrac{2\pi}{3}$	π
r	0	$\dfrac{3}{2}$	3	$\dfrac{9}{2}$	6

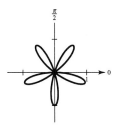

59. $r = 2 \cos(3\theta)$

Rose curve with three petals

Symmetric to the polar axis

Relative extrema: $(2, 0), \left(-2, \dfrac{\pi}{3}\right), \left(2, \dfrac{2\pi}{3}\right)$

θ	0	$\dfrac{\pi}{6}$	$\dfrac{\pi}{4}$	$\dfrac{\pi}{3}$	$\dfrac{\pi}{2}$	$\dfrac{2\pi}{3}$	$\dfrac{5\pi}{6}$	π
r	2	0	$-\sqrt{2}$	-2	0	2	0	-2

Tangents at the pole: $\theta = \dfrac{\pi}{6}, \dfrac{\pi}{2}, \dfrac{5\pi}{6}$

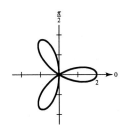

60. $r = -\sin(5\theta)$

Rose curve with five petals

Symmetric to $\theta = \dfrac{\pi}{2}$

Relative extrema occur when

$$\frac{dr}{d\theta} = -5 \cos(5\theta) = 0 \text{ at } \theta = \frac{\pi}{10}, \frac{3\pi}{10}, \frac{5\pi}{10}, \frac{7\pi}{10}, \frac{9\pi}{10}.$$

Tangents at the pole: $\theta = 0, \dfrac{\pi}{5}, \dfrac{2\pi}{5}, \dfrac{3\pi}{5}, \dfrac{4\pi}{5}$

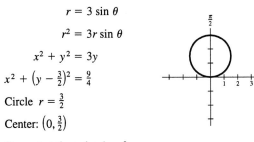

61. $r = 3 \sin 2\theta$

Rose curve with four petals

Symmetric to the polar axis, $\theta = \dfrac{\pi}{2}$, and pole

Relative extrema: $\left(\pm 3, \dfrac{\pi}{4} \right), \left(\pm 3, \dfrac{5\pi}{4} \right)$

Tangents at the pole: $\theta = 0, \dfrac{\pi}{2}$

$(\theta = \pi, \ 3\pi/2$ give the same tangents.)

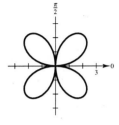

62. $r = 3 \cos 2\theta$

Rose curve with four petals

Symmetric to the polar axis, $\theta = \dfrac{\pi}{2}$, and pole

Relative extrema: $(3, 0), \left(-3, \dfrac{\pi}{2} \right), (3, \pi), \left(-3, \dfrac{3\pi}{2} \right)$

Tangents at the pole: $\theta = \dfrac{\pi}{4}, \dfrac{3\pi}{4}$

$\theta = \dfrac{5\pi}{4}$ and $\dfrac{7\pi}{4}$ given the same tangents.

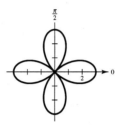

63. $r = 3 - 2 \cos \theta$

Limaçon

Symmetric to polar axis

θ	0	$\dfrac{\pi}{3}$	$\dfrac{\pi}{2}$	$\dfrac{2\pi}{3}$	π
r	1	2	3	4	5

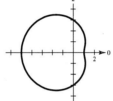

64. $r = 5 - 4 \sin \theta$

Limaçon

Symmetric to $\theta = \dfrac{\pi}{2}$

θ	$-\dfrac{\pi}{2}$	$-\dfrac{\pi}{6}$	0	$\dfrac{\pi}{6}$	$\dfrac{\pi}{2}$
r	9	7	5	3	1

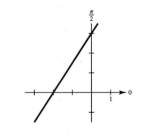

65. $r = 3 \csc \theta$

$r \sin \theta = 3$

$y = 3$

Horizontal line

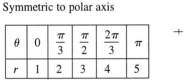

66. $r = \dfrac{6}{2 \sin \theta - 3 \cos \theta}$

$2r \sin \theta - 3r \cos \theta = 6$

$2y - 3x = 6$

Line

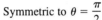

67. $r = 2\theta$

Spiral of Archimedes

Symmetric to $\theta = \dfrac{\pi}{2}$

θ	0	$\dfrac{\pi}{4}$	$\dfrac{\pi}{2}$	$\dfrac{3\pi}{4}$	π	$\dfrac{5\pi}{4}$	$\dfrac{3\pi}{2}$
r	0	$\dfrac{\pi}{2}$	π	$\dfrac{3\pi}{2}$	2π	$\dfrac{5\pi}{2}$	3π

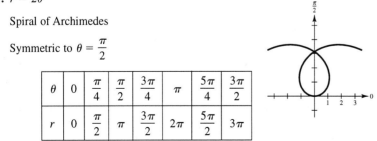

Tangent at the pole: $\theta = 0$

68. $r = \dfrac{1}{\theta}$

Hyperbolic spiral

θ	$\dfrac{\pi}{4}$	$\dfrac{\pi}{2}$	$\dfrac{3\pi}{4}$	π	$\dfrac{5\pi}{4}$	$\dfrac{3\pi}{2}$
r	$\dfrac{4}{\pi}$	$\dfrac{2}{\pi}$	$\dfrac{4}{3\pi}$	$\dfrac{1}{\pi}$	$\dfrac{4}{5\pi}$	$\dfrac{2}{3\pi}$

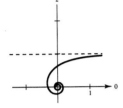

69. $r^2 = 4\cos(2\theta)$

Lemniscate

Symmetric to the polar axis, $\theta = \dfrac{\pi}{2}$, and pole

Relative extrema: $(\pm 2, 0)$

θ	0	$\dfrac{\pi}{6}$	$\dfrac{\pi}{4}$
r	± 2	$\pm\sqrt{2}$	0

Tangents at the pole: $\theta = \dfrac{\pi}{4}, \dfrac{3\pi}{4}$

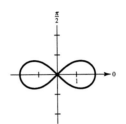

70. $r^2 = 4\sin\theta$

Lemniscate

Symmetric to the polar axis, $\theta = \dfrac{\pi}{2}$, and pole

Relative extrema: $\left(\pm 2, \dfrac{\pi}{2}\right)$

θ	0	$\dfrac{\pi}{6}$	$\dfrac{\pi}{2}$	$\dfrac{5\pi}{6}$	π
r	0	$\pm\sqrt{2}$	± 2	$\pm\sqrt{2}$	0

Tangent at the pole: $\theta = 0$

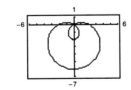

71. $r = 3 - 4\cos\theta$

$0 \le \theta < 2\pi$

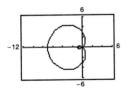

72. $r = 2(1 - 2\sin\theta)$

$0 \le \theta < 2\pi$

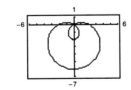

73. $r = 2 + \sin\theta$

$0 \le \theta < 2\pi$

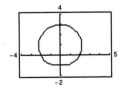

74. $r = 4 + 3\cos\theta$

$0 \le \theta < 2\pi$

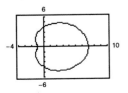

75. $r = \dfrac{2}{1 + \cos\theta}$

Traced out once on $-\pi < \theta < \pi$

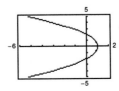

76. $r = \dfrac{2}{4 - 3\sin\theta}$

Traced out once on $0 \le \theta \le 2\pi$

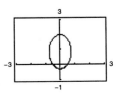

77. $r = 2 \cos\left(\dfrac{3\theta}{2}\right)$

$0 \le \theta < 4\pi$

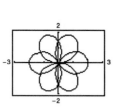

78. $r = 3 \sin\left(\dfrac{5\pi}{2}\right)$

$0 \le \theta < 4\pi$

79. $r^2 = 4 \sin 2\theta$

$0 \le \theta < \dfrac{\pi}{2}$

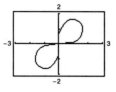

80. $r^2 = \dfrac{1}{\theta}$.

Graph as

$$r_1 = \dfrac{1}{\sqrt{\theta}}, r_2 = -\dfrac{1}{\sqrt{\theta}}.$$

It is traced out once on $[0, \infty)$.

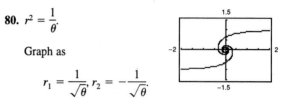

81. Since

$$r = 2 - \sec\theta = 2 - \dfrac{1}{\cos\theta},$$

the graph has polar axis symmetry and the lengths at the pole are

$$\theta = \dfrac{\pi}{3}, \dfrac{-\pi}{3}.$$

Furthermore,

$$r \Rightarrow -\infty \text{ as } \theta \Rightarrow \dfrac{\pi}{2}^-$$

$$r \Rightarrow \infty \text{ as } \theta \Rightarrow -\dfrac{\pi}{2}^+.$$

Also, $r = 2 - \dfrac{1}{\cos\theta} = 2 - \dfrac{r}{r\cos\theta} = 2 - \dfrac{r}{x}$

$$rx = 2x - r$$

$$r = \dfrac{2x}{1 + x}.$$

Thus, $r \Rightarrow \pm\infty$ as $x \Rightarrow -1$.

82. Since

$$r = 2 + \csc\theta = 2 + \dfrac{1}{\sin\theta},$$

the graphs has symmetry with respect to $\theta = \pi/2$. Furthermore,

$$r \Rightarrow \infty \text{ as } \theta \Rightarrow 0^+$$

$$r \Rightarrow \infty \text{ as } \theta \Rightarrow \pi^-$$

Also, $r = 2 + \dfrac{1}{\sin\theta} = 2 + \dfrac{r}{\sin\theta} = 2 + \dfrac{r}{y}$

$$ry = 2y + r$$

$$r = \dfrac{2y}{y - 1}.$$

Thus, $r \Rightarrow \pm\infty$ as $y \Rightarrow 1$.

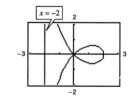

83. $r = \dfrac{2}{\theta}$

Hyperbolic spiral

$r \Rightarrow \infty$ as $\theta \Rightarrow 0$

$$r = \dfrac{2}{\theta} \Rightarrow \theta = \dfrac{2}{r} = \dfrac{2\sin\theta}{r\sin\theta} = \dfrac{2\sin\theta}{y}$$

$$y = \dfrac{2\sin\theta}{\theta}$$

$$\lim_{\theta \to 0} \dfrac{2\sin\theta}{\theta} = \lim_{\theta \to 0} \dfrac{2\cos\theta}{1} = 2$$

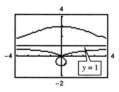

84. $r = 2\cos 2\theta \sec\theta$

Strophoid

$$r \Rightarrow -\infty \text{ as } \theta \Rightarrow \dfrac{\pi}{2}^-$$

$$r \Rightarrow \infty \text{ as } \theta \Rightarrow \dfrac{-\pi}{2}^+$$

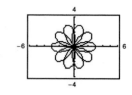

$$r = 2\cos 2\theta \sec\theta = 2(2\cos^2\theta - 1)\sec\theta$$

$$r\cos\theta = 4\cos^2\theta - 2$$

$$x = 4\cos^2\theta - 2$$

$$\lim_{\theta \to \pm\pi/2} (4\cos^2\theta - 2) = -2$$

85. $r = 4 \sin \theta$

(a) $0 \le \theta \le \dfrac{\pi}{2}$

(b) $\dfrac{\pi}{2} \le \theta \le \pi$

(c) $-\dfrac{\pi}{2} \le \theta \le \dfrac{\pi}{2}$

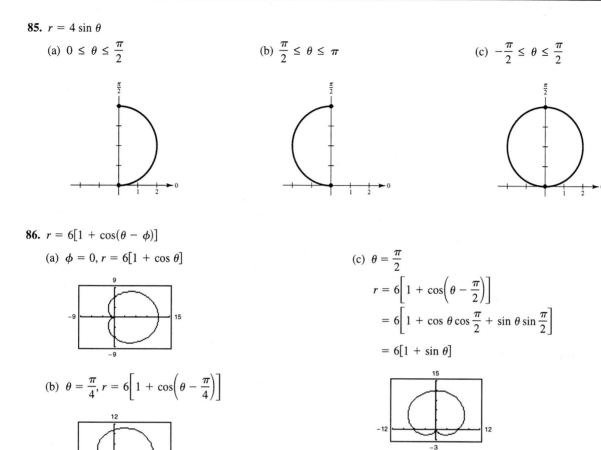

86. $r = 6[1 + \cos(\theta - \phi)]$

(a) $\phi = 0,\ r = 6[1 + \cos \theta]$

(b) $\theta = \dfrac{\pi}{4},\ r = 6\left[1 + \cos\left(\theta - \dfrac{\pi}{4}\right)\right]$

The graph of $r = 6[1 + \cos \theta]$ is rotated through the angle $\pi/4$.

(c) $\theta = \dfrac{\pi}{2}$

$$r = 6\left[1 + \cos\left(\theta - \dfrac{\pi}{2}\right)\right]$$
$$= 6\left[1 + \cos \theta \cos \dfrac{\pi}{2} + \sin \theta \sin \dfrac{\pi}{2}\right]$$
$$= 6[1 + \sin \theta]$$

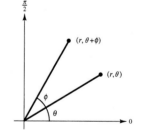

The graph of $r = 6[1 + \cos \theta]$ is rotated through the angle $\pi/2$.

87. Let the curve $r = f(\theta)$ be rotated by ϕ to form the curve $r = g(\theta)$. If (r_1, θ_1) is a point on $r = f(\theta)$, then $(r_1, \theta_1 + \phi)$ is on $r = g(\theta)$. That is,

$$g(\theta_1 + \phi) = r_1 = f(\theta_1).$$

Letting $\theta = \theta_1 + \phi$, or $\theta_1 = \theta - \phi$, we see that

$$g(\theta) = g(\theta_1 + \phi) = f(\theta_1) = f(\theta - \phi).$$

88. (a) $\sin\left(\theta - \dfrac{\pi}{2}\right) = \sin \theta \cos\left(\dfrac{\pi}{2}\right) - \cos \theta \sin\left(\dfrac{\pi}{2}\right)$

$$= -\cos \theta$$

$$r = f\left[\sin\left(\theta - \dfrac{\pi}{2}\right)\right]$$

$$= f(-\cos \theta)$$

(c) $\sin\left(\theta - \dfrac{3\pi}{2}\right) = \sin \theta \cos\left(\dfrac{3\pi}{2}\right) - \cos \theta \sin\left(\dfrac{3\pi}{2}\right)$

$$= \cos \theta$$

$$r = f\left[\sin\left(\theta - \dfrac{3\pi}{2}\right)\right] = f(\cos \theta)$$

(b) $\sin(\theta - \pi) = \sin \theta \cos \pi - \cos \theta \sin \pi$

$$= -\sin \theta$$

$$r = f[\sin(\theta - \pi)]$$

$$= f(-\sin \theta)$$

89. $r = 2 - \sin\theta$

(a) $r = 2 - \sin\left(\theta - \dfrac{\pi}{4}\right) = 2 - \dfrac{\sqrt{2}}{2}(\sin\theta - \cos\theta)$

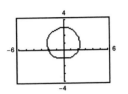

(b) $r = 2 - (-\cos\theta) = 2 + \cos\theta$

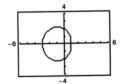

(c) $r = 2 - (-\sin\theta) = 2 + \sin\theta$

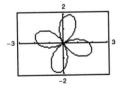

(d) $r = 2 - \cos\theta$

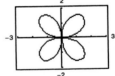

90. $r = 2\sin 2\theta = 4\sin\theta\cos\theta$

(a) $r = 4\sin\left(\theta - \dfrac{\pi}{6}\right)\cos\left(\theta - \dfrac{\pi}{6}\right)$

(b) $r = 4\sin\left(\theta - \dfrac{\pi}{2}\right)\cos\left(\theta - \dfrac{\pi}{2}\right) = -4\sin\theta\cos\theta$

(c) $r = 4\sin\left(\theta - \dfrac{2\pi}{3}\right)\cos\left(\theta - \dfrac{2\pi}{3}\right)$

• (d) $r = 4\sin(\theta - \pi)\cos(\theta - \pi) = 4\sin\theta\cos\theta$

91. (a) $r = 1 - \sin\theta$

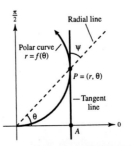

(b) $r = 1 - \sin\left(\theta - \dfrac{\pi}{4}\right)$

Rotate the graph of

$r = 1 - \sin\theta$

through the angle $\pi/4$.

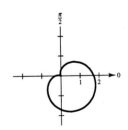

92. By Theorem 9.11, the slope of the tangent line through A and P is

$$\frac{f\cos\theta + f'\sin\theta}{-f\sin\theta + f'\cos\theta}$$

This is equal to

$$\tan(\theta + \psi) = \frac{\tan\theta + \tan\psi}{1 - \tan\theta\tan\psi} = \frac{\sin\theta + \cos\theta\tan\psi}{\cos\theta - \sin\theta\tan\psi}.$$

—CONTINUED—

92. —CONTINUED—

Equating the expressions and cross-multiplying, you obtain

$$(f\cos\theta + f'\sin\theta)(\cos\theta - \sin\theta\tan\psi) = (\sin\theta + \cos\theta\tan\psi)(-f\sin\theta + f'\cos\theta)$$

$$f\cos^2\theta - f\cos\theta\sin\theta\tan\psi + f'\sin\theta\cos\theta - f'\sin^2\theta\tan\psi = -f\sin^2\theta - f\sin\theta\cos\theta\tan\psi + f'\sin\theta\cos\theta + f'\cos^2\theta\tan\psi$$

$$f(\cos^2\theta + \sin^2\theta) = f'\tan\psi(\cos^2\theta + \sin^2\theta)$$

$$\tan\psi = \frac{f}{f'} = \frac{r}{dr/d\theta}.$$

93. $\tan\psi = \dfrac{r}{dr/d\theta} = \dfrac{2(1-\cos\theta)}{2\sin\theta}$

At $\theta = \pi$, $\tan\psi$ is undefined $\implies \psi = \dfrac{\pi}{2}$.

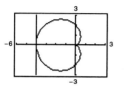

94. $\tan\psi = \dfrac{r}{dr/d\theta} = \dfrac{3(1-\cos\theta)}{3\sin\theta}$

At $\theta = \dfrac{3\pi}{4}$, $\tan\psi = \dfrac{1 + \left(\sqrt{2}/2\right)}{\sqrt{2}} = \dfrac{2 + \sqrt{2}}{\sqrt{2}}$.

$$\psi = \arctan\left(\frac{2+\sqrt{2}}{\sqrt{2}}\right) \approx 1.041(\approx 59.64°)$$

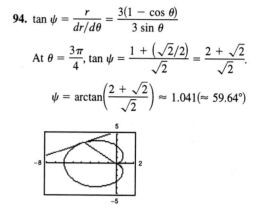

95. $\tan\psi = \dfrac{r}{dr/d\theta} = \dfrac{2\cos 3\theta}{-6\sin 3\theta}$

At $\theta = \dfrac{\pi}{6}$, $\tan\psi = 0 \implies \psi = 0$.

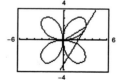

96. $\tan\psi = \dfrac{r}{dr/d\theta} = \dfrac{4\sin 2\theta}{8\cos 2\theta}$

At $\theta = \dfrac{\pi}{6}$, $\tan\psi = \dfrac{\sin(\pi/3)}{2\cos(\pi/3)} = \dfrac{\sqrt{3}}{2}$.

$$\psi = \arctan\left(\frac{\sqrt{3}}{2}\right) \approx 0.7137(\approx 40.89°)$$

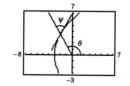

97. $r = \dfrac{6}{1-\cos\theta} = 6(1-\cos\theta)^{-1} \implies \dfrac{dr}{d\theta} = \dfrac{6\sin\theta}{(1-\cos\theta)^2}$

$$\tan\psi = \frac{r}{\dfrac{dr}{d\theta}} = \frac{\dfrac{6}{(1-\cos\theta)}}{\dfrac{6\sin\theta}{(1-\cos\theta)^2}} = \frac{1-\cos\theta}{\sin\theta}$$

At $\theta = \dfrac{2\pi}{3}$, $\tan\psi = \dfrac{1 - \left(-\dfrac{1}{2}\right)}{\dfrac{\sqrt{3}}{2}} = \sqrt{3}$.

$$\psi = \frac{\pi}{3} \, (60°)$$

98. $\tan \psi = \dfrac{r}{dr/d\theta} = \dfrac{5}{0}$ undefined $\implies \psi = \dfrac{\pi}{2}$.

99. The curve is produced over the interval

$$0 \le \theta \le 9\pi.$$

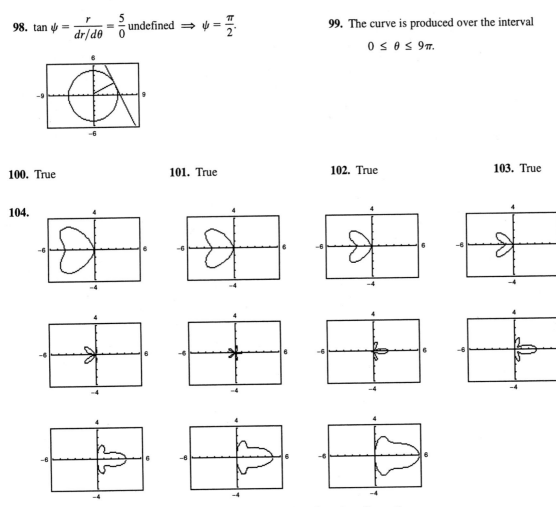

$n = 1, 2, 3, 4, 5$ produce "bells"; $n = -1, -2, -3, -4, -5$ produce "hearts".

100. True   **101.** True   **102.** True   **103.** True

104.

Section 9.5   Area and Arc Length in Polar Coordinates

1. (a) $r = 8 \sin \theta$

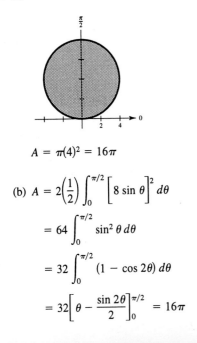

$$A = \pi(4)^2 = 16\pi$$

(b) $A = 2\left(\dfrac{1}{2}\right) \displaystyle\int_0^{\pi/2} \Big[8 \sin \theta \Big]^2 \, d\theta$

$$= 64 \int_0^{\pi/2} \sin^2 \theta \, d\theta$$

$$= 32 \int_0^{\pi/2} (1 - \cos 2\theta) \, d\theta$$

$$= 32 \left[\theta - \dfrac{\sin 2\theta}{2} \right]_0^{\pi/2} = 16\pi$$

2. (a) $r = 3 \cos \theta$

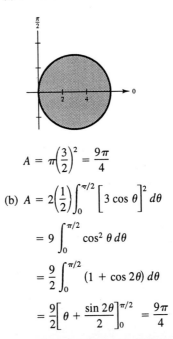

$$A = \pi \left(\dfrac{3}{2}\right)^2 = \dfrac{9\pi}{4}$$

(b) $A = 2\left(\dfrac{1}{2}\right) \displaystyle\int_0^{\pi/2} \Big[3 \cos \theta \Big]^2 \, d\theta$

$$= 9 \int_0^{\pi/2} \cos^2 \theta \, d\theta$$

$$= \dfrac{9}{2} \int_0^{\pi/2} (1 + \cos 2\theta) \, d\theta$$

$$= \dfrac{9}{2} \left[\theta + \dfrac{\sin 2\theta}{2} \right]_0^{\pi/2} = \dfrac{9\pi}{4}$$

3. $A = 2\left[\dfrac{1}{2}\displaystyle\int_0^{\pi/6} (2\cos 3\theta)^2\,d\theta\right] = 2\left[\theta + \dfrac{1}{6}\sin 6\theta\right]_0^{\pi/6} = \dfrac{\pi}{3}$

4. $A = 2\left[\dfrac{1}{2}\displaystyle\int_0^{\pi/4} (4\sin 2\theta)^2\,d\theta\right] = 8\left[\theta - \dfrac{1}{4}\sin 4\theta\right]_0^{\pi/4} = 2\pi$

5. $A = 2\left[\dfrac{1}{2}\displaystyle\int_0^{\pi/4} (\cos 2\theta)^2\,d\theta\right]$

$= \dfrac{1}{2}\left[\theta + \dfrac{1}{4}\sin 4\theta\right]_0^{\pi/4} = \dfrac{\pi}{8}$

6. $A = 2\left[\dfrac{1}{2}\displaystyle\int_0^{\pi/10} (\cos 5\theta)^2\,d\theta\right]$

$= \dfrac{1}{2}\left[\theta + \dfrac{1}{10}\sin(10\,\theta)\right]_0^{\pi/10} = \dfrac{\pi}{20}$

7. $A = 2\left[\dfrac{1}{2}\displaystyle\int_{-\pi/2}^{\pi/2} (1 - \sin\theta)^2\,d\theta\right]$

$= \left[\dfrac{3}{2}\theta + 2\cos\theta - \dfrac{1}{4}\sin 2\theta\right]_{-\pi/2}^{\pi/2} = \dfrac{3\pi}{2}$

8. $A = 2\left[\dfrac{1}{2}\displaystyle\int_0^{\pi/2} (1 - \sin\theta)^2\,d\theta\right]$

$= \left[\dfrac{3}{2}\theta + 2\cos\theta - \dfrac{1}{4}\sin 2\theta\right]_0^{\pi/2} = \dfrac{3\pi - 8}{4}$

9. $A = 2\left[\dfrac{1}{2}\displaystyle\int_{2\pi/3}^{\pi} (1 + 2\cos\theta)^2\,d\theta\right]$

$= \left[3\theta + 4\sin\theta + \sin 2\theta\right]_{2\pi/3}^{\pi} = \dfrac{2\pi - 3\sqrt{3}}{2}$

10. $A = 2\left[\dfrac{1}{2}\displaystyle\int_{-\pi/2}^{\arcsin(-3/4)} (3 + 4\sin\theta)^2\,d\theta\right]$

$= \left[17\theta - 24\cos\theta - 4\sin(2\theta)\right]_{-\pi/2}^{\arcsin(-3/4)} \approx 0.3806$

11. The area inside the outer loop is

$$2\left[\dfrac{1}{2}\int_0^{2\pi/3} (1 + 2\cos\theta)^2\,d\theta\right] = \left[3\theta + 4\sin\theta + \sin 2\theta\right]_0^{2\pi/3} = \dfrac{4\pi + 3\sqrt{3}}{2}.$$

From the result of Exercise 9, the area between the loops is

$$A = \left(\dfrac{4\pi + 3\sqrt{3}}{2}\right) - \left(\dfrac{2\pi - 3\sqrt{3}}{2}\right) = \pi + 3\sqrt{3}.$$

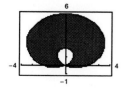

12. Four times the area in Exercise 11, $A = 4\left(\pi + 3\sqrt{3}\right)$. More specifically, we see that the area inside the outer loop is

$$2\left[\dfrac{1}{2}\int_{-\pi/6}^{\pi/2} (2(1 + 2\sin\theta))^2\,d\theta\right] = \int_{-\pi/6}^{\pi/2} (4 + 16\sin\theta + 16\sin^2\theta)\,d\theta = 8\pi + 6\sqrt{3}.$$

The area inside the inner loop is

$$2\,\dfrac{1}{2}\left[\int_{7\pi/6}^{3\pi/2} (2(1 + 2\sin\theta))^2\,d\theta\right] = 4\pi - 6\sqrt{3}.$$

Thus, the area between the loops is $\left(8\pi + 6\sqrt{3}\right) - \left(4\pi - 6\sqrt{3}\right) = 4\pi + 12\sqrt{3}$.

13. $r = 1 + \cos \theta$

$r = 1 - \cos \theta$

Solving simultaneously,

$1 + \cos \theta = 1 - \cos \theta$

$2 \cos \theta = 0$

$\theta = \dfrac{\pi}{2}, \dfrac{3\pi}{2}$.

Replacing r by $-r$ and θ by $\theta + \pi$ in the first equation and solving, $-1 + \cos \theta = 1 - \cos \theta$, $\cos \theta = 1$, $\theta = 0$. Both curves pass through the pole, $(0, \pi)$, and $(0, 0)$, respectively.

Points of intersection: $\left(1, \dfrac{\pi}{2}\right), \left(1, \dfrac{3\pi}{2}\right), (0, 0)$

14. $r = 3(1 + \sin \theta)$

$r = 3(1 - \sin \theta)$

Solving simultaneously,

$3(1 + \sin \theta) = 3(1 - \sin \theta)$

$2 \sin \theta = 0$

$\theta = 0, \pi$.

Replacing r by $-r$ and θ by $\theta + \pi$ in the first equation and solving, $-3(1 - \sin \theta) = 3(1 - \sin \theta)$, $\sin \theta = 1$, $\theta = \pi/2$. Both curves pass through the pole, $(0, 3\pi/2)$, and $(0, \pi/2)$, respectively.

Points of intersection: $(3, 0), (3, \pi), (0, 0)$

15. $r = 1 + \cos \theta$

$r = 1 - \sin \theta$

Solving simultaneously,

$1 + \cos \theta = 1 - \sin \theta$

$\cos \theta = -\sin \theta$

$\tan \theta = -1$

$\theta = \dfrac{3\pi}{4}, \dfrac{7\pi}{4}$.

Replacing r by $-r$ and θ by $\theta + \pi$ in the first equation and solving, $-1 + \cos \theta = 1 - \sin \theta$, $\sin \theta + \cos \theta = 2$, which has no solution. Both curves pass through the pole, $(0, \pi)$, and $(0, \pi/2)$, respectively.

Points of intersection: $\left(\dfrac{2 - \sqrt{2}}{2}, \dfrac{3\pi}{4}\right), \left(\dfrac{2 + \sqrt{2}}{2}, \dfrac{7\pi}{4}\right), (0, 0)$

16. $r = 2 - 3 \cos \theta$

$r = \cos \theta$

Solving simultaneously,

$2 - 3 \cos \theta = \cos \theta$

$\cos \theta = \dfrac{1}{2}$

$\theta = \dfrac{\pi}{3}, \dfrac{5\pi}{3}$.

Both curves pass through the pole, $(0, \arccos 2/3)$, and $(0, \pi/2)$, respectively.

Points of intersection: $\left(\dfrac{1}{2}, \dfrac{\pi}{3}\right), \left(\dfrac{1}{2}, \dfrac{5\pi}{3}\right), (0, 0)$

17. $r = 4 - 5 \sin \theta$

$r = 3 \sin \theta$

Solving simultaneously,

$4 - 5 \sin \theta = 3 \sin \theta$

$\sin \theta = \dfrac{1}{2}$

$\theta = \dfrac{\pi}{6}, \dfrac{5\pi}{6}$.

Both curves pass through the pole, $(0, \arcsin 4/5)$, and $(0, 0)$, respectively.

Points of intersection: $\left(\dfrac{3}{2}, \dfrac{\pi}{6}\right), \left(\dfrac{3}{2}, \dfrac{5\pi}{6}\right), (0, 0)$

18. $r = 1 + \cos \theta$

$r = 3 \cos \theta$

Solving simultaneously,

$1 + \cos \theta = 3 \cos \theta$

$\cos \theta = \dfrac{1}{2}$

$\theta = \dfrac{\pi}{3}, \dfrac{5\pi}{3}$.

Both curves pass through the pole, $(0, \pi)$, and $(0, \pi/2)$, respectively.

Points of intersection: $\left(\dfrac{3}{2}, \dfrac{\pi}{3}\right), \left(\dfrac{3}{2}, \dfrac{5\pi}{3}\right), (0, 0)$

19. $r = \dfrac{\theta}{2}$

$r = 2$

Solving simultaneously, we have

$\theta/2 = 2,\ \theta = 4.$

Points of intersection:

$(2, 4), (-2, -4)$

20. $\theta = \dfrac{\pi}{4}$

$r = 2$

Line of slope 1 passing through the pole and a circle of radius 2 centered at the pole.

Points of intersection:

$$\left(2, \frac{\pi}{4}\right), \left(-2, \frac{\pi}{4}\right)$$

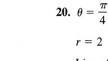

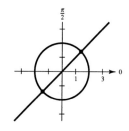

21. $r = 4 \sin 2\theta$

$r = 2$

$r = 4 \sin 2\theta$ is the equation of a rose curve with four petals and is symmetric to the polar axis, $\theta = \pi/2$, and the pole. Also, $r = 2$ is the equation of a circle of radius 2 centered at the pole. Solving simultaneously,

$4 \sin 2\theta = 2$

$2\theta = \dfrac{\pi}{6}, \dfrac{5\pi}{6}$

$\theta = \dfrac{\pi}{12}, \dfrac{5\pi}{12}.$

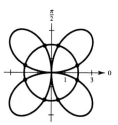

Therefore, the points of intersection for one petal are $(2, \pi/12)$ and $(2, 5\pi/12)$. By symmetry, the other points of intersection are $(2, 7\pi/12)$, $(2, 11\pi/12)$, $(2, 13\pi/12)$, $(2, 17\pi/12)$, $(2, 19\pi/12)$, and $(2, 23\pi/12)$.

22. $r = 3 + \sin \theta$

$r = 2 \csc \theta$

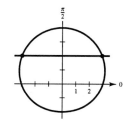

The graph of $r = 3 + \sin \theta$ is a limaçon symmetric to $\theta = \pi/2$, and the graph of $r = 2 \csc \theta$ is the horizontal line $y = 2$. Therefore, there are two points of intersection. Solving simultaneously,

$3 + \sin \theta = 2 \csc \theta$

$\sin^2 \theta + 3 \sin \theta - 2 = 0$

$\sin \theta = \dfrac{-3 \pm \sqrt{17}}{2}$

$\theta = \arcsin\left(\dfrac{\sqrt{17} - 3}{2}\right) \approx 0.596.$

Points of intersection:

$$\left(\frac{\sqrt{17} + 3}{2}, \arcsin\left(\frac{\sqrt{17} - 3}{2}\right)\right),$$

$$\left(\frac{\sqrt{17} + 3}{2}, \pi - \arcsin\left(\frac{\sqrt{17} - 3}{2}\right)\right),$$

$(3.56, 0.596), (3.56, 2.545)$

23. $r = 2 + 3 \cos \theta$

$r = \dfrac{\sec \theta}{2}$

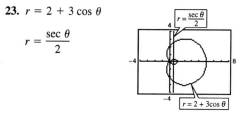

The graph of $r = 2 + 3 \cos \theta$ is a limaçon with an inner loop $(b > a)$ and is symmetric to the polar axis. The graph of $r = (\sec \theta)/2$ is the vertical line $x = 1/2$. Therefore, there are four points of intersection. Solving simultaneously,

$2 + 3 \cos \theta = \dfrac{\sec \theta}{2}$

$6 \cos^2 \theta + 4 \cos \theta - 1 = 0$

$\cos \theta = \dfrac{-2 \pm \sqrt{10}}{6}$

$\theta = \arccos\left(\dfrac{-2 + \sqrt{10}}{6}\right) \approx 1.376$

$\theta = \arccos\left(\dfrac{-2 - \sqrt{10}}{6}\right) \approx 2.6068.$

Points of intersection: $(-0.581, \pm 2.607), (2.581, \pm 1.376)$

24. $r = 3(1 - \cos \theta)$

$$r = \frac{6}{1 - \cos \theta}$$

The graph of $r = 3(1 - \cos \theta)$ is a cardioid with polar axis symmetry. The graph of

$$r = 6/(1 - \cos \theta)$$

is a parabola with focus at the pole, vertex$(3, \pi)$, and polar axis symmetry. Therefore, there are two points of intersection. Solving simultaneously,

$$3(1 - \cos \theta) = \frac{6}{1 - \cos \theta}$$

$$(1 - \cos \theta)^2 = 2$$

$$\cos \theta = 1 \pm \sqrt{2}$$

$$\theta = \arccos\left(1 - \sqrt{2}\right).$$

Points of intersection: $\left(3\sqrt{2}, \arccos\left(1 - \sqrt{2}\right)\right) \approx (4.243, 1.998), \left(3\sqrt{2}, 2\pi - \arccos(1 - \sqrt{2})\right) \approx (4.243, 4.285)$

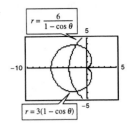

25. $r = \cos \theta$

$r = 2 - 3 \sin \theta$

Points of intersection:

$(0, 0), (0.935, 0.363), (0.535, -1.006)$

The graphs reach the pole at different times (θ values).

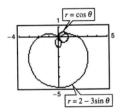

26. $r = 4 \sin \theta$

$r = 2(1 + \sin \theta)$

Points of intersection: $(0, 0), \left(4, \dfrac{\pi}{2}\right)$

The graphs reach the pole at different times (θ values).

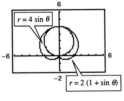

27. From Exercise 21, the points of intersection for one petal are $(2, \pi/12)$ and $(2, 5\pi/12)$. The area within one petal is

$$A = \frac{1}{2}\int_0^{\pi/12} (4 \sin 2\theta)^2 \, d\theta + \frac{1}{2}\int_{\pi/12}^{5\pi/12} (2)^2 \, d\theta + \frac{1}{2}\int_{5\pi/12}^{\pi/2} (4 \sin 2\theta)^2 \, d\theta$$

$$= 16 \int_0^{\pi/12} \sin^2(2\theta) \, d\theta + 2 \int_{\pi/12}^{5\pi/12} d\theta \quad \text{(by symmetry of the petal)}$$

$$= 8\left[\theta - \frac{1}{4}\sin 4\theta\right]_0^{\pi/12} + \left[2\theta\right]_{\pi/12}^{5\pi/12} = \frac{4\pi}{3} - \sqrt{3}.$$

Total area $= 4\left(\dfrac{4\pi}{3} - \sqrt{3}\right) = \dfrac{16\pi}{3} - 4\sqrt{3} = \dfrac{4}{3}(4\pi - 3\sqrt{3})$

28. $A = 4\left[\dfrac{1}{2}\displaystyle\int_0^{\pi/2} 9(1 - \sin \theta)^2 \, d\theta\right]$

$$= 18 \int_0^{\pi/2} (1 - \sin \theta)^2 \, d\theta = \frac{9}{2}(3\pi - 8)$$

(from Exercise 14)

29. $A = 4\left[\dfrac{1}{2}\displaystyle\int_0^{\pi/2} (3 - 2 \sin \theta)^2 \, d\theta\right]$

$$= 2\left[11\theta + 12 \cos \theta - \sin(2\theta)\right]_0^{\pi/2} = 11\pi - 24$$

30. $A = 2\left[\dfrac{1}{2}\displaystyle\int_{\pi/4}^{5\pi/4} (3 - 2\sin\theta)^2\, d\theta\right]$

$= \left[11\theta + 12\cos\theta - \sin(2\theta)\right]_{\pi/4}^{5\pi/4} = 11\pi - 12\sqrt{2}$

31. $A = 2\left[\dfrac{1}{2}\displaystyle\int_{0}^{\pi/6} (4\sin\theta)^2\, d\theta + \dfrac{1}{2}\displaystyle\int_{\pi/6}^{\pi/2} (2)^2\, d\theta\right]$

$= 16\left[\dfrac{1}{2}\theta - \dfrac{1}{4}\sin(2\theta)\right]_{0}^{\pi/6} + \left[4\theta\right]_{\pi/6}^{\pi/2}$

$= \dfrac{8\pi}{3} - 2\sqrt{3} = \dfrac{2}{3}\left(4\pi - 3\sqrt{3}\right)$

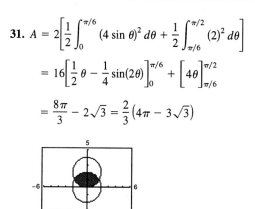

32. $A = 2\left[\dfrac{1}{2}\displaystyle\int_{\pi/6}^{\pi/2} (3\sin\theta)^2\, d\theta - \dfrac{1}{2}\displaystyle\int_{\pi/6}^{\pi/2} (2 - \sin\theta)^2\, d\theta\right]$

$= \displaystyle\int_{\pi/6}^{\pi/2} (-4\cos 2\theta + 4\sin\theta)\, d\theta$

$= \left[-2\sin(2\theta) - 4\cos\theta\right]_{\pi/6}^{\pi/2} = 3\sqrt{3}$

33. $A = 2\left[\dfrac{1}{2}\displaystyle\int_{0}^{\pi} [a(1 + \cos\theta)]^2\, d\theta\right] - \dfrac{a^2\pi}{4}$

$= a^2\left[\dfrac{3}{2}\theta + 2\sin\theta + \dfrac{\sin 2\theta}{4}\right]_{0}^{\pi} - \dfrac{a^2\pi}{4}$

$= \dfrac{3a^2\pi}{2} - \dfrac{a^2\pi}{4} = \dfrac{5a^2\pi}{4}$

34. Area = Area of $r = 2a\cos\theta$ − Area of sector − twice area between $r = 2a\cos\theta$ and the lines

$\theta = \dfrac{\pi}{3}, \theta = \dfrac{\pi}{2}.$

$A = \pi a^2 - \left(\dfrac{\pi}{3}\right)a^2 - 2\left[\dfrac{1}{2}\displaystyle\int_{\pi/3}^{\pi/2} (2a\cos\theta)^2\, d\theta\right]$

$= \dfrac{2\pi a^2}{3} - 2a^2\displaystyle\int_{\pi/3}^{\pi/2} (1 + \cos 2\theta)\, d\theta$

$= \dfrac{2\pi a^2}{3} - 2a^2\left[\theta + \dfrac{\sin 2\theta}{2}\right]_{\pi/3}^{\pi/2}$

$= \dfrac{2\pi a^2}{3} - 2a^2\left[\dfrac{\pi}{2} - \dfrac{\pi}{3} - \dfrac{\sqrt{3}}{4}\right] = \dfrac{2\pi a^2 + 3\sqrt{3}a^2}{6}$

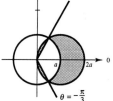

35. $A = \dfrac{\pi a^2}{8} + \dfrac{1}{2}\displaystyle\int_{\pi/2}^{\pi} [a(1 + \cos\theta)]^2\, d\theta$

$= \dfrac{\pi a^2}{8} + \dfrac{a^2}{2}\displaystyle\int_{\pi/2}^{\pi} \left(\dfrac{3}{2} + 2\cos\theta + \dfrac{\cos 2\theta}{2}\right)\, d\theta$

$= \dfrac{\pi a^2}{8} + \dfrac{a^2}{2}\left[\dfrac{3}{2}\theta + 2\sin\theta + \dfrac{\sin 2\theta}{4}\right]_{\pi/2}^{\pi}$

$= \dfrac{\pi a^2}{8} + \dfrac{a^2}{2}\left[\dfrac{3\pi}{2} - \dfrac{3\pi}{4} - 2\right] = \dfrac{a^2}{2}[\pi - 2]$

36. $r = \dfrac{ab}{a \sin \theta + b \cos \theta} \Rightarrow ar \sin \theta + br \cos \theta = ab \Rightarrow ay + bx = ab$

$\dfrac{y}{b} + \dfrac{x}{a} = 1 \quad$ (line)

Assume $a > 0$ and $b > 0$.

$A = \dfrac{1}{2} ab$

37. (a) $r = a \cos^2 \theta$

$r^3 = ar^2 \cos^2 \theta$

$(x^2 + y^2)^{3/2} = ax^2$

(b)

(c) $A = 4\left(\dfrac{1}{2}\right) \displaystyle\int_0^{\pi/2} [(6 \cos^2 \theta)^2 - (4 \cos^2 \theta)^2] \, d\theta \;=\; 40 \int_0^{\pi/2} \cos^4 \theta \, d\theta \;=\; 10 \int_0^{\pi/2} (1 + \cos 2\theta)^2 \, d\theta$

$= 10 \displaystyle\int_0^{\pi/2} \left(1 + 2 \cos 2\theta + \dfrac{1 - \cos 4\theta}{2}\right) d\theta \;=\; 10\left[\dfrac{3}{2}\theta + \sin 2\theta + \dfrac{1}{8}\sin 4\theta\right]_0^{\pi/2} = \dfrac{15\pi}{2}$

38. By symmetry, $A_1 = A_2$ and $A_3 = A_4$.

$A_1 = A_2 = \dfrac{1}{2} \displaystyle\int_{-\pi/3}^{\pi/6} [(2a \cos \theta)^2 - (a)^2] \, d\theta + \dfrac{1}{2} \int_{\pi/6}^{\pi/4} [(2a \cos \theta)^2 - (2a \sin \theta)^2] \, d\theta$

$= \dfrac{a^2}{2} \displaystyle\int_{-\pi/3}^{\pi/6} (4 \cos^2 \theta - 1) \, d\theta + 2a^2 \int_{\pi/6}^{\pi/4} \cos 2\theta \, d\theta$

$= \dfrac{a^2}{2}\Big[\theta + \sin 2\theta\Big]_{-\pi/3}^{\pi/6} + a^2\Big[\sin 2\theta\Big]_{\pi/6}^{\pi/4} = \dfrac{a^2}{2}\left(\dfrac{\pi}{2} + \sqrt{3}\right) + a^2\left(1 - \dfrac{\sqrt{3}}{2}\right) = a^2\left(\dfrac{\pi}{4} + 1\right)$

$A_3 = A_4 = \dfrac{1}{2}\left(\dfrac{\pi}{2}\right) a^2 = \dfrac{\pi a^2}{4}$

$A_5 = \dfrac{1}{2}\left(\dfrac{5\pi}{6}\right) a^2 - 2\left(\dfrac{1}{2}\right) \displaystyle\int_{5\pi/6}^{\pi} (2a \sin \theta)^2 \, d\theta$

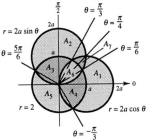

$= \dfrac{5\pi a^2}{12} - 2a^2 \displaystyle\int_{5\pi/6}^{\pi} (1 - \cos 2\theta) \, d\theta$

$= \dfrac{5\pi a^2}{12} - a^2\Big[2\theta - \sin 2\theta\Big]_{5\pi/6}^{\pi} = \dfrac{5\pi a^2}{12} - a^2\left(\dfrac{\pi}{3} - \dfrac{\sqrt{3}}{2}\right) = a^2\left(\dfrac{\pi}{12} + \dfrac{\sqrt{3}}{2}\right)$

$A_6 = 2\left(\dfrac{1}{2}\right) \displaystyle\int_0^{\pi/6} (2a \sin \theta)^2 \, d\theta + 2\left(\dfrac{1}{2}\right) \int_{\pi/6}^{\pi/4} a^2 \, d\theta$

$= 2a^2 \displaystyle\int_0^{\pi/6} (1 - \cos 2\theta) \, d\theta + \Big[a^2\theta\Big]_{\pi/6}^{\pi/4}$

$= a^2\Big[2\theta - \sin 2\theta\Big]_0^{\pi/6} + \dfrac{\pi a^2}{12} = a^2\left(\dfrac{\pi}{3} - \dfrac{\sqrt{3}}{2}\right) + \dfrac{\pi a^2}{12} = a^2\left(\dfrac{5\pi}{12} - \dfrac{\sqrt{3}}{2}\right)$

$A_7 = 2\left(\dfrac{1}{2}\right) \displaystyle\int_{\pi/6}^{\pi/4} [(2a \sin \theta)^2 - (a)^2] \, d\theta$

$= a^2 \displaystyle\int_{\pi/6}^{\pi/4} (4 \sin^2 \theta - 1) \, d\theta = a^2\Big[\theta - \sin 2\theta\Big]_{\pi/6}^{\pi/4} = a^2\left(\dfrac{\pi}{12} - 1 + \dfrac{\sqrt{3}}{2}\right)$

[Note: $A_1 + A_6 + A_7 + A_4 = \pi a^2 = $ area of circle of radius a]

39. $r = a \cos(n\theta)$

For $n = 1$:

$r = a \cos \theta$

$$A = \pi \left(\frac{a}{2}\right)^2 = \frac{\pi a^2}{4}$$

For $n = 2$:

$r = a \cos 2\theta$

$$A = 8\left(\frac{1}{2}\right) \int_0^{\pi/4} (a \cos 2\theta)^2 \, d\theta = \frac{\pi a^2}{2}$$

For $n = 3$:

$r = a \cos 3\theta$

$$A = 6\left(\frac{1}{2}\right) \int_0^{\pi/6} (a \cos 3\theta)^2 \, d\theta = \frac{\pi a^2}{4}$$

For $n = 4$:

$r = a \cos 4\theta$

$$A = 16\left(\frac{1}{2}\right) \int_0^{\pi/8} (a \cos 4\theta)^2 \, d\theta = \frac{\pi a^2}{2}$$

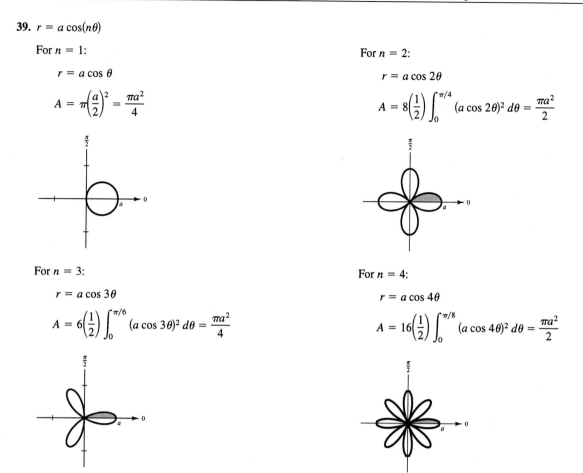

In general, the area of the region enclosed by $r = a \cos(n\theta)$ for $n = 1, 2, 3, \ldots$ is $(\pi a^2)/4$ if n is odd and is $(\pi a^2)/2$ if n is even.

40.
$$r = \sec \theta - 2 \cos \theta, \quad -\frac{\pi}{2} < \theta < \frac{\pi}{2}$$

$$r \cos \theta = 1 - 2 \cos^2 \theta$$

$$x = 1 - 2 \frac{r^2 \cos^2 \theta}{r^2} = 1 - 2\left(\frac{x^2}{x^2 + y^2}\right)$$

$$(x^2 + y^2)x = x^2 + y^2 - 2x^2$$

$$y^2(x - 1) = -x^2 - x^3$$

$$y^2 = \frac{x^2(1 + x)}{1 - x}$$

$$A = 2\left(\frac{1}{2}\right) \int_0^{\pi/4} (\sec \theta - 2 \cos \theta)^2 \, d\theta$$

$$= \int_0^{\pi/4} (\sec^2 \theta - 4 + 4 \cos^2 \theta) \, d\theta = \int_0^{\pi/4} (\sec^2 \theta - 4 + 2(1 + \cos 2\theta)) \, d\theta = \left[\tan \theta - 2\theta + \sin 2\theta\right]_0^{\pi/4} = 2 - \frac{\pi}{2}$$

41. $r = a$

$r' = 0$

$$s = \int_0^{2\pi} \sqrt{a^2 + 0^2} \, d\theta = \left[a\theta\right]_0^{2\pi} = 2\pi a$$

(circumference of circle of radius a)

42. $r = 2a \cos \theta$

$r' = -2a \sin \theta$

$$s = \int_{-\pi/2}^{\pi/2} \sqrt{(2a \cos \theta)^2 + (-2a \sin \theta)^2} \, d\theta$$

$$= \int_{-\pi/2}^{\pi/2} 2a \, d\theta = \left[2\theta\right]_{-\pi/2}^{\pi/2} = 2\pi a$$

43. $r = 1 + \sin\theta$

$r' = \cos\theta$

$s = 2\displaystyle\int_{\pi/2}^{3\pi/2} \sqrt{(1 + \sin\theta)^2 + (\cos\theta)^2}\, d\theta$

$= 2\sqrt{2}\displaystyle\int_{\pi/2}^{3\pi/2} \sqrt{1 + \sin\theta}\, d\theta$

$= 2\sqrt{2}\displaystyle\int_{\pi/2}^{3\pi/2} \dfrac{-\cos\theta}{\sqrt{1 - \sin\theta}}\, d\theta$

$= \left[4\sqrt{2}\,\sqrt{1 - \sin\theta}\,\right]_{\pi/2}^{3\pi/2}$

$= 4\sqrt{2}\left(\sqrt{2} - 0\right) = 8$

44. $r = 5(1 + \cos\theta)$

$r' = -5\sin\theta$

$s = 2\displaystyle\int_{0}^{\pi} \sqrt{[5(1 + \cos\theta)]^2 + (-5\sin\theta)^2}\, d\theta$

$= 10\sqrt{2}\displaystyle\int_{0}^{\pi} \sqrt{1 + \cos\theta}\, d\theta$

$= 10\sqrt{2}\displaystyle\int_{0}^{\pi} \dfrac{\sin\theta}{\sqrt{1 - \cos\theta}}\, d\theta$

$= \left[20\sqrt{2}\,\sqrt{1 - \cos\theta}\,\right]_{0}^{\pi} = 40$

45. $r = 2\theta, 0 \le \theta \le \dfrac{\pi}{2}$

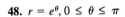

Length ≈ 4.16

46. $r = \sec\theta, 0 \le \theta \le \dfrac{\pi}{3}$

Length $\approx 1.73 \left(\text{exact } \sqrt{3}\right)$

47. $r = \dfrac{1}{\theta}, \pi \le \theta \le 2\pi$

Length ≈ 0.71

48. $r = e^{\theta}, 0 \le \theta \le \pi$

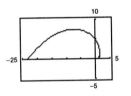

Length ≈ 31.31

49. $r = \sin(3\cos\theta), 0 \le \theta \le \pi$

Length ≈ 4.39

50. $r = 2\sin(2\cos\theta), 0 \le \theta \le \pi$

Length ≈ 7.78

51. $r = 2\cos\theta$

$r' = -2\sin\theta$

$S = 2\pi\displaystyle\int_{0}^{\pi/2} 2\cos\theta\sin\theta\sqrt{4\cos^2\theta + 4\sin^2\theta}\, d\theta$

$= 8\pi\displaystyle\int_{0}^{\pi/2} \sin\theta\cos\theta\, d\theta = \left[4\pi\sin^2\theta\right]_{0}^{\pi/2} = 4\pi$

52. $r = a\cos\theta$

$r' = -a\sin\theta$

$S = 2\pi\displaystyle\int_{0}^{\pi/2} a\cos\theta\,(\cos\theta)\sqrt{a^2\cos^2\theta + a^2\sin^2\theta}\, d\theta$

$= 2\pi a^2\displaystyle\int_{0}^{\pi/2} \cos^2\theta\, d\theta = \pi a^2\displaystyle\int_{0}^{\pi/2} (1 + \cos 2\theta)\, d\theta$

$= \left[\pi a^2\left(\theta + \dfrac{\sin 2\theta}{2}\right)\right]_{0}^{\pi/2} = \dfrac{\pi^2 a^2}{2}$

53. $r = e^{a\theta}$

$r' = ae^{a\theta}$

$S = 2\pi\displaystyle\int_{0}^{\pi/2} e^{a\theta}\cos\theta\sqrt{(e^{a\theta})^2 + (ae^{a\theta})^2}\, d\theta$

$= 2\pi\sqrt{1 + a^2}\displaystyle\int_{0}^{\pi/2} e^{2a\theta}\cos\theta\, d\theta = 2\pi\sqrt{1 + a^2}\left[\dfrac{e^{2a\theta}}{4a^2 + 1}(2a\cos\theta + \sin\theta)\right]_{0}^{\pi/2} = \dfrac{2\pi\sqrt{1 + a^2}}{4a^2 + 1}\left(e^{\pi a} - 2a\right)$

54. $r = a(1 + \cos\theta)$

$r' = -a\sin\theta$

$$S = 2\pi \int_0^\pi a(1 + \cos\theta)\sin\theta\sqrt{a^2(1 + \cos\theta)^2 + a^2\sin^2\theta}\,d\theta = 2\pi a^2 \int_0^\pi \sin\theta(1 + \cos\theta)\sqrt{2 + 2\cos\theta}\,d\theta$$

$$= -2\sqrt{2}\pi a^2 \int_0^\pi (1 + \cos\theta)^{3/2}(-\sin\theta)\,d\theta = -\frac{4\sqrt{2}\pi a^2}{5}\left[(1 + \cos\theta)^{5/2}\right]_0^\pi = \frac{32\pi a^2}{5}$$

55. $r = 4\cos 2\theta$

$r' = -8\sin 2\theta$

$$S = 2\pi \int_0^{\pi/4} 4\cos 2\theta \sin\theta\sqrt{16\cos^2 2\theta + 64\sin^2 2\theta}\,d\theta$$

$$= 32\pi \int_0^{\pi/4} \cos 2\theta \sin\theta\sqrt{\cos^2 2\theta + 4\sin^2 2\theta}\,d\theta \approx 21.87$$

56. $r = \theta$

$r' = 1$

$$S = 2\pi \int_0^\pi \theta \sin\theta\sqrt{\theta^2 + 1}\,d\theta \approx 42.32$$

57. Revolve $r = a$ about the line $r = b\sec\theta$ where $b > a > 0$.

$$f(\theta) = a$$

$$f'(\theta) = 0$$

$$S = 2\pi \int_0^{2\pi} [b - a\cos\theta]\sqrt{a^2 + 0^2}\,d\theta$$

$$= 2\pi a \Big[b\theta - a\sin\theta \Big]_0^{2\pi}$$

$$= 2\pi a(2\pi b) = 4\pi^2 ab$$

58. In parametric form,

$$s = \int_a^b \sqrt{\left(\frac{dx}{dt}\right)^2 + \left(\frac{dy}{dt}\right)^2}\,dt.$$

Using θ instead of t, we have $x = r\cos\theta = f(\theta)\cos\theta$ and $y = r\sin\theta = f(\theta)\sin\theta$. Thus,

$$\frac{dx}{d\theta} = f'(\theta)\cos\theta - f(\theta)\sin\theta \text{ and}$$

$$\frac{dy}{d\theta} = f'(\theta)\sin\theta + f(\theta)\cos\theta.$$

It follows that

$$\left(\frac{dx}{d\theta}\right)^2 + \left(\frac{dy}{d\theta}\right)^2 = [f(\theta)]^2 + [f'(\theta)]^2.$$

Therefore, $s = \int_\alpha^\beta \sqrt{[f(\theta)]^2 + [f'(\theta)]^2}\,d\theta$

59. $r = 8\cos\theta, 0 \le \theta \le \pi$

(a) $A = \dfrac{1}{2}\displaystyle\int_0^\pi r^2\,d\theta = \dfrac{1}{2}\displaystyle\int_0^\pi 64\cos^2\theta\,d\theta = 32\displaystyle\int_0^\pi \dfrac{1 + \cos 2\theta}{2}\,d\theta = 16\left[\theta + \dfrac{\sin 2\theta}{2}\right]_0^\pi = 16\pi$

(Area circle $= \pi r^2 = \pi 4^2 = 16\pi$)

(b)

θ	0.2	0.4	0.6	0.8	1.0	1.2	1.4
A	6.32	12.14	17.06	20.80	23.27	24.60	25.08

(c), (d) For $\frac{1}{4}$ of area $(4\pi \approx 12.57)$: 0.42

For $\frac{1}{2}$ of area $(8\pi \approx 25.13)$: 1.57 $(\pi/2)$

For $\frac{3}{4}$ of area $(12\pi \approx 37.70)$: 2.73

(e) No, it does not depend on the radius.

60. False. The area is given by $\displaystyle\int_0^{\pi/2} \sin^2\theta\,d\theta$.

61. False. $f(\theta) = 1$ and $g(\theta) = -1$ have the same graphs.

62. False. $f(\theta) = 0$ and $g(\theta) = \sin 2\theta$ have only one point of intersection.

63. True. The area enclosed by $r = \sin(n\theta)$ is $\pi/2$ and the area enclosed by $r = \sin[(n + 1)\theta]$ is $\pi/4$.

Section 9.6   Polar Equations of Conics and Kepler's Laws

1. $r = \dfrac{2e}{1 + e \cos \theta}$

   (a) $e = 1, r = \dfrac{2}{1 + \cos \theta}$, parabola

   (b) $e = 0.5, r = \dfrac{1}{1 + 0.5 \cos \theta} = \dfrac{2}{2 + \cos \theta}$, ellipse

   (c) $e = 1.5, r = \dfrac{3}{1 + 1.5 \cos \theta} = \dfrac{6}{2 + 3 \cos \theta}$, hyperbola

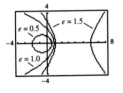

2. $r = \dfrac{2e}{1 - e \cos \theta}$

   (a) $e = 1, r = \dfrac{2}{1 - \cos \theta}$, parabola

   (b) $e = 0.5, r = \dfrac{1}{1 - 0.5 \cos \theta} = \dfrac{2}{2 - \cos \theta}$, ellipse

   (c) $e = 1.5, r = \dfrac{3}{1 - 1.5 \cos \theta} = \dfrac{6}{2 - 3 \cos \theta}$, hyperbola

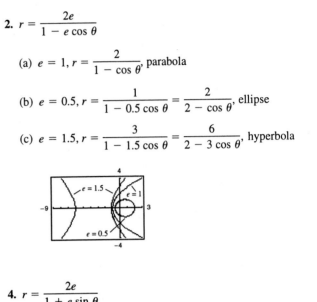

3. $r = \dfrac{2e}{1 - e \sin \theta}$

   (a) $e = 1, r = \dfrac{2}{1 - \sin \theta}$, parabola

   (b) $e = 0.5, r = \dfrac{1}{1 - 0.5 \sin \theta} = \dfrac{2}{2 - \sin \theta}$, ellipse

   (c) $e = 1.5, r = \dfrac{3}{1 - 1.5 \sin \theta} = \dfrac{6}{2 - 3 \sin \theta}$, hyperbola

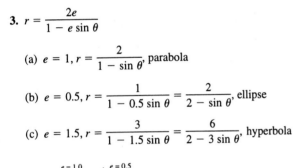

4. $r = \dfrac{2e}{1 + e \sin \theta}$

   (a) $e = 1, r = \dfrac{2}{1 + \sin \theta}$, parabola

   (b) $e = 0.5, r = \dfrac{1}{1 + 0.5 \sin \theta} = \dfrac{2}{2 + \sin \theta}$, ellipse

   (c) $e = 1.5, r = \dfrac{3}{1 + 1.5 \sin \theta} = \dfrac{6}{2 + 3 \sin \theta}$, hyperbola

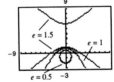

5. $r = \dfrac{4}{1 + e \sin \theta}$

   (a)

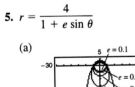

   (b)

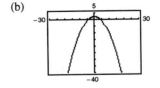

The conic is a parabola.

The conic is an ellipse. As $e \to 1^{-}$, the ellipse becomes more elliptical, and as $e \to 0^{+}$, it becomes more circular.

   (c)

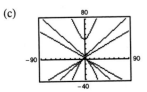

The conic is a hyperbola. As $e \to 1^{+}$, the hyperbolas opens more slowly, and as $e \to \infty$, they open more rapidly.

6. $r = \dfrac{4}{1 - 0.4 \cos \theta}$

(a) Because $e = 0.4 < 1$, the conic is an ellipse with vertical directrix to the left of the pole.

(b) $r = \dfrac{4}{1 + 0.4 \cos \theta}$

The ellipse is shifted to the left. The vertical directrix is to the right of the pole

$$r = \dfrac{4}{1 - 0.4 \sin \theta}.$$

The ellipse has a horizontal directrix below the pole.

(c)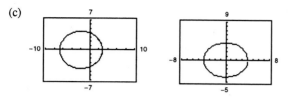

7. Parabola; Matches (c)

8. Ellipse; Matches (f)

9. Hyperbola; Matches (a)

10. Parabola; Matches (e)

11. Ellipse; Matches (b)

12. Hyperbola; Matches (d)

13. $r = \dfrac{-1}{1 - \sin \theta}$

Parabola since $e = 1$

Vertex: $\left(-\dfrac{1}{2}, \dfrac{3\pi}{2}\right)$

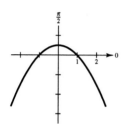

14. $r = \dfrac{6}{1 + \cos \theta}$

Parabola since $e = 1$

Vertex: $(3, 0)$

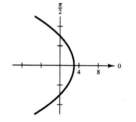

15. $r = \dfrac{6}{2 + \cos \theta}$

$= \dfrac{3}{1 + (1/2) \cos \theta}$

Ellipse since $e = \dfrac{1}{2} < 1$

Vertices: $(2, 0), (6, \pi)$

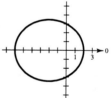

16. $r = \dfrac{3}{3 + 2 \sin \theta}$

$= \dfrac{1}{1 + (2/3) \sin \theta}$

Ellipse since $e = \dfrac{2}{3} < 1$

Vertices: $\left(\dfrac{3}{5}, \dfrac{\pi}{2}\right), \left(3, \dfrac{3\pi}{2}\right)$

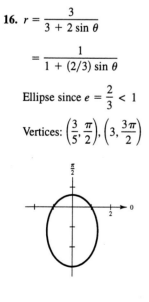

17. $r(2 + \sin \theta) = 4$

$r = \dfrac{4}{2 + \sin \theta}$

$= \dfrac{2}{1 + (1/2) \sin \theta}$

Ellipse since $e = \dfrac{1}{2} < 1$

Vertices: $\left(\dfrac{4}{3}, \dfrac{\pi}{2}\right), \left(4, \dfrac{3\pi}{2}\right)$

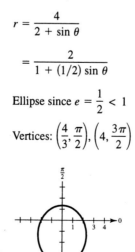

18. $r(3 - 2 \cos \theta) = 6$

$r = \dfrac{6}{3 - 2 \cos \theta}$

$= \dfrac{2}{1 - (2/3) \cos \theta}$

Ellipse since $e = \dfrac{2}{3} < 1$

Vertices: $(6, 0), \left(\dfrac{6}{5}, \pi\right)$

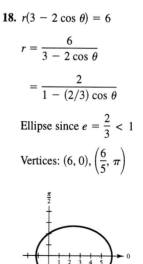

19. $r = \dfrac{5}{-1 + 2\cos\theta} = \dfrac{-5}{1 - 2\cos\theta}$

Hyperbola since $e = 2 > 1$

Vertices: $(5, 0), \left(-\dfrac{5}{3}, \pi\right)$

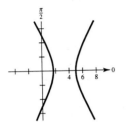

20. $r = \dfrac{-3}{2 + 4\sin\theta} = \dfrac{-3/2}{1 + 2\sin\theta}$

Hyperbola since $e = 2 > 1$

Vertices: $\left(-\dfrac{1}{2}, \dfrac{\pi}{2}\right), \left(\dfrac{3}{2}, \dfrac{3\pi}{2}\right)$

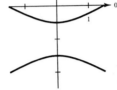

21. $r = \dfrac{3}{2 + 6\sin\theta} = \dfrac{3/2}{1 + 3\sin\theta}$

Hyperbola since $e = 3 > 1$

Vertices: $\left(\dfrac{3}{8}, \dfrac{\pi}{2}\right), \left(-\dfrac{3}{4}, \dfrac{3\pi}{2}\right)$

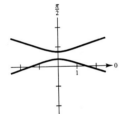

22. $r = \dfrac{4}{1 + 2\cos\theta}$

Hyperbola since $e = 2 > 1$

Vertices: $\left(\dfrac{4}{3}, 0\right), (-4, \pi)$

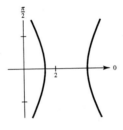

23.

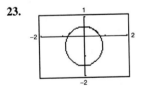

Ellipse

24.

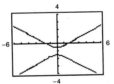

Hyperbola

25.

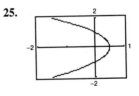

Parabola

26.

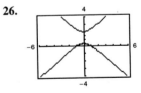

Hyperbola

27. $r = \dfrac{-1}{1 - \sin\left(\theta - \dfrac{\pi}{4}\right)}$

Rotate the graph of

$$r = \dfrac{-1}{1 - \sin\theta}$$

counterclockwise through the angle $\dfrac{\pi}{4}$.

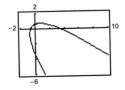

28. $r = \dfrac{6}{1 + \cos\left(\theta - \dfrac{\pi}{3}\right)}$

Rotate the graph of

$$r = \dfrac{6}{1 + \cos\theta}$$

counterclockwise through the angle $\dfrac{\pi}{3}$.

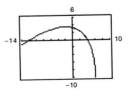

29. $r = \dfrac{6}{2 + \cos\left(\theta + \dfrac{\pi}{6}\right)}$

Rotate the graph of

$$r = \dfrac{6}{2 + \cos\theta}$$

clockwise through the angle $\dfrac{\pi}{6}$.

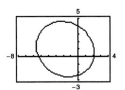

30. $r = \dfrac{-3}{2 + 4\sin\left(\theta + \dfrac{2\pi}{3}\right)}$

Rotate the graph of

$$r = \dfrac{-3}{2 + 4\sin\theta}$$

clockwise through the angle $\dfrac{2\pi}{3}$.

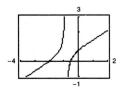

31. Change θ to $\theta + \dfrac{\pi}{4}$: $r = \dfrac{5}{5 + 3\cos\left(\theta + \dfrac{\pi}{4}\right)}$.

32. Change θ to $\theta - \dfrac{\pi}{6}$: $r = \dfrac{2}{1 + \sin\left(\theta - \dfrac{\pi}{6}\right)}$

33. Parabola

$e = 1, x = -1, d = 1$

$$r = \dfrac{ed}{1 - e\cos\theta} = \dfrac{1}{1 - \cos\theta}$$

34. Parabola

$e = 1, y = 1, d = 1$

$$r = \dfrac{ed}{1 + e\sin\theta} = \dfrac{1}{1 + \sin\theta}$$

35. Ellipse

$e = \dfrac{1}{2}, y = 1, d = 1$

$$r = \dfrac{ed}{1 + e\sin\theta}$$

$$= \dfrac{1/2}{1 + (1/2)\sin\theta}$$

$$= \dfrac{1}{2 + \sin\theta}$$

36. Ellipse

$e = \dfrac{3}{4}, y = -2, d = 2$

$$r = \dfrac{ed}{1 - e\sin\theta}$$

$$= \dfrac{2(3/4)}{1 - (3/4)\sin\theta}$$

$$= \dfrac{6}{4 - 3\sin\theta}$$

37. Hyperbola

$e = 2, x = 1, d = 1$

$$r = \dfrac{ed}{1 + e\cos\theta} = \dfrac{2}{1 + 2\cos\theta}$$

38. Hyperbola

$e = \dfrac{3}{2}, x = -1, d = 1$

$$r = \dfrac{ed}{1 - e\cos\theta}$$

$$= \dfrac{3/2}{1 - (3/2)\cos\theta}$$

$$= \dfrac{3}{2 - 3\cos\theta}$$

39. Parabola

Vertex: $\left(1, -\dfrac{\pi}{2}\right)$

$e = 1, d = 2, r = \dfrac{2}{1 - \sin\theta}$

40. Parabola

Vertex: $(5, \pi)$

$e = 1, d = 10$

$$r = \dfrac{ed}{1 - e\cos\theta} = \dfrac{10}{1 - \cos\theta}$$

41. Ellipse

Vertices: $(2, 0), (8, \pi)$

$e = \dfrac{3}{5}, d = \dfrac{16}{3}$

$$r = \dfrac{ed}{1 + e\cos\theta}$$

$$= \dfrac{16/5}{1 + (3/5)\cos\theta}$$

$$= \dfrac{16}{5 + 3\cos\theta}$$

42. Ellipse

Vertices: $\left(2, \frac{\pi}{2}\right), \left(4, \frac{3\pi}{2}\right)$

$e = \frac{1}{3}, d = 8$

$r = \dfrac{ed}{1 + e \sin \theta}$

$= \dfrac{8/3}{1 + (1/3) \sin \theta}$

$= \dfrac{8}{3 + \sin \theta}$

43. Hyperbola

Vertices: $\left(1, \frac{3\pi}{2}\right), \left(9, \frac{3\pi}{2}\right)$

$e = \frac{5}{4}, d = \frac{9}{5}$

$r = \dfrac{ed}{1 - e \sin \theta}$

$= \dfrac{9/4}{1 - (5/4) \sin \theta}$

$= \dfrac{9}{4 - 5 \sin \theta}$

44. Hyperbola

Vertices: $(2, 0), (10, 0)$

$e = \frac{3}{2}, d = \frac{10}{3}$

$r = \dfrac{ed}{1 + e \cos \theta}$

$= \dfrac{5}{1 + (3/2) \cos \theta}$

$= \dfrac{10}{2 + 3 \cos \theta}$

45.

$\dfrac{x^2}{a^2} + \dfrac{y^2}{b^2} = 1$

$x^2 b^2 + y^2 a^2 = a^2 b^2$

$b^2 r^2 \cos^2 \theta + a^2 r^2 \sin^2 \theta = a^2 b^2$

$r^2 [b^2 \cos^2 \theta + a^2 (1 - \cos^2 \theta)] = a^2 b^2$

$r^2 [a^2 + \cos^2 \theta (b^2 - a^2)] = a^2 b^2$

$r^2 = \dfrac{a^2 b^2}{a^2 + (b^2 - a^2) \cos^2 \theta} = \dfrac{a^2 b^2}{a^2 - c^2 \cos^2 \theta}$

$= \dfrac{b^2}{1 - (c/a)^2 \cos^2 \theta} = \dfrac{b^2}{1 - e^2 \cos^2 \theta}$

46.

$\dfrac{x^2}{a^2} - \dfrac{y^2}{b^2} = 1$

$x^2 b^2 - y^2 a^2 = a^2 b^2$

$b^2 r^2 \cos^2 \theta - a^2 r^2 \sin^2 \theta = a^2 b^2$

$r^2 [b^2 \cos^2 \theta - a^2 (1 - \cos^2 \theta)] = a^2 b^2$

$r^2 [-a^2 + \cos^2 \theta (a^2 + b^2)] = a^2 b^2$

$r^2 = \dfrac{a^2 b^2}{-a^2 + c^2 \cos^2 \theta} = \dfrac{b^2}{-1 + (c^2/a^2) \cos^2 \theta}$

$= \dfrac{-b^2}{1 - e^2 \cos^2 \theta}$

47. $a = 5, c = 4, e = \dfrac{4}{5}, b = 3$

$r^2 = \dfrac{9}{1 - (16/25) \cos^2 \theta}$

48. $a = 4, c = 5, b = 3, e = \dfrac{5}{4}$

$r^2 = \dfrac{-9}{1 - (25/16) \cos^2 \theta}$

49. $a = 3, b = 4, c = 5, e = \dfrac{5}{3}$

$r^2 = \dfrac{-16}{1 - (25/9) \cos^2 \theta}$

50. $a = 2, b = 1, c = \sqrt{3}, e = \dfrac{\sqrt{3}}{2}$

$r^2 = \dfrac{1}{1 - (3/4) \cos^2 \theta}$

51. $A = 2 \left[\dfrac{1}{2} \displaystyle\int_0^\pi \left(\dfrac{3}{2 - \cos \theta} \right)^2 d\theta \right] = 9 \displaystyle\int_0^\pi \dfrac{1}{(2 - \cos \theta)^2} d\theta \approx 10.88$

52. $A = 2 \left[\dfrac{1}{2} \displaystyle\int_{-\pi/2}^{\pi/2} \left(\dfrac{2}{3 - 2 \sin \theta} \right)^2 d\theta \right] = 4 \displaystyle\int_{-\pi/2}^{\pi/2} \dfrac{1}{(3 - 2 \sin \theta)^2} d\theta \approx 3.37$

53. Vertices: $(126,000, 0), (4119, \pi)$

$a = \dfrac{126,000 + 4119}{2} = 65,059.5, c = 65,059.5 - 4119 = 60,940.5, e = \dfrac{c}{a} = \dfrac{40,627}{43,373}, d = 4119 \left(\dfrac{84,000}{40,627} \right)$

$r = \dfrac{ed}{1 - e \cos \theta} = \dfrac{4119(84,000/43,373)}{1 - (40,627/43,373) \cos \theta} = \dfrac{345,996,000}{43,373 - 40,627 \cos \theta}$

When $\theta = 60°, r = \dfrac{345,996,000}{23,059.5} \approx 15,004.49$.

Distance between the surface of the earth and the satellite is $r - 4000 = 11,004.49$ miles.

54. (a) $r = \dfrac{ed}{1 - e\cos\theta}$

When $\theta = 0, r = c + a = ea + a = a(1 + e)$.

Therefore,

$$a(1 + e) = \frac{ed}{1 - e}$$

$$a(1 + e)(1 - e) = ed$$

$$a(1 - e^2) = ed.$$

Thus, $r = \dfrac{(1 - e^2)a}{1 - e\cos\theta}$.

(b) The perihelion distance is $a - c = a - ea = a(1 - e)$.

When $\theta = \pi,\ r = \dfrac{(1 - e^2)a}{1 + e} = a(1 - e)$.

The aphelion distance is $a + c = a + ea = a(1 + e)$.

When $\theta = 0,\ r = \dfrac{(1 - e^2)a}{1 - e} = a(1 + e)$.

55. $a = 92.957 \times 10^6$ mi, $e = 0.0167$

$$r = \frac{(1 - e^2)a}{1 - e\cos\theta} = \frac{92{,}931{,}075.2223}{1 - 0.0167\cos\theta}$$

Perihelion distance: $a(1 - e) \approx 91{,}404{,}618$ mi

Aphelion distance: $a(1 + e) \approx 94{,}509{,}382$ mi

56. $a = 1.427 \times 10^9$ km

$e = 0.0543$

$$r = \frac{(1 - e^2)a}{1 - e\cos\theta} = \frac{1.422792505 \times 10^9}{1 - 0.0543\cos\theta}$$

Perihelion distance: $a(1 - e) = 1.3495139 \times 10^9$ km

Aphelion distance: $a(1 + e) = 1.5044861 \times 10^9$ km

57. $a = 5.900 \times 10^9$ km, $e = 0.2481$

$$r = \frac{(1 - e^2)a}{1 - e\cos\theta} \approx \frac{5.537 \times 10^9}{1 - 0.2481\cos\theta}$$

Perihelion distance: $a(1 - e) = 4.436 \times 10^9$ km

Aphelion distance: $a(1 + e) = 7.364 \times 10^9$ km

58. $a = 36.0 \times 10^6$ mi, $e = 0.206$

$$r = \frac{(1 - e^2)a}{1 - e\cos\theta} \approx \frac{34.472 \times 10^6}{1 - 0.206\cos\theta}$$

Perihelion distance: $a(1 - e) = 28.582 \times 10^6$ mi

Aphelion distance: $a(1 + e) = 43.416 \times 10^6$ mi

59. $r = \dfrac{5.537 \times 10^9}{1 - 0.2481\cos\theta}$

(a) $A = \dfrac{1}{2} \displaystyle\int_0^{\pi/9} \left[\dfrac{5.537 \times 10^9}{1 - 0.2481\cos\theta}\right]^2 d\theta \approx 9.341 \times 10^{18}$ km^2

$$248 \left[\frac{\dfrac{1}{2}\displaystyle\int_0^{\pi/9}\left[\dfrac{5.537 \times 10^9}{1 - 0.2481\cos\theta}\right]^2 d\theta}{\dfrac{1}{2}\displaystyle\int_0^{2\pi}\left[\dfrac{5.537 \times 10^9}{1 - 0.2481\cos\theta}\right]^2 d\theta}\right] \approx 21.867 \text{ yr}$$

(b) $\dfrac{1}{2}\displaystyle\int_\pi^{\alpha - \pi}\left[\dfrac{5.537 \times 10^9}{1 - 0.2481\cos\theta}\right]^2 d\theta = 9.341 \times 10^{18}$

$\alpha \approx \pi + 0.8995$ rad

In part (a) the ray swept through a smaller angle to generate the same area since the length of the ray is longer than in part (b).

(c) $r' = \dfrac{(-5.537 \times 10^9)(0.2481\sin\theta)}{(1 - 0.2481\cos\theta)^2}$

$$s = \int_0^{\pi/9} \sqrt{\left(\frac{5.537 \times 10^9}{1 - 0.2481\cos\theta}\right)^2 + \left[\frac{-1.3737297 \times 10^9\sin\theta}{(1 - 0.2481\cos\theta)^2}\right]^2}\, d\theta \approx 2.559 \times 10^9 \text{ km}$$

$\dfrac{2.559 \times 10^9 \text{ km}}{21.867 \text{ yr}} \approx 1.17 \times 10^8$ km/yr

$$s = \int_\pi^{\pi + 0.899} \sqrt{\left(\frac{5.537 \times 10^9}{1 - 0.2481\cos\theta}\right)^2 + \left[\frac{-1.3737297 \times 10^9\sin\theta}{(1 - 0.2481\cos\theta)^2}\right]^2}\, d\theta \approx 4.119 \times 10^9 \text{ km}$$

$\dfrac{4.119 \times 10^9 \text{ km}}{21.867 \text{ yr}} \approx 1.88 \times 10^8$ km/yr

60.
$$r = a \sin \theta + b \cos \theta$$
$$r^2 = ar \sin \theta + br \cos \theta$$
$$x^2 + y^2 = ay + bx$$

$x^2 + y^2 - bx - ay = 0$ represents a circle.

61. $r_1 = \dfrac{ed}{1 + \sin \theta}$ and $r_2 = \dfrac{ed}{1 - \sin \theta}$

Points of intersection: $(ed, 0)$, (ed, π)

$r_1: \dfrac{dy}{dx} = \dfrac{\left(\dfrac{ed}{1 + \sin \theta}\right)(\cos \theta) + \left(\dfrac{-ed \cos \theta}{(1 + \sin \theta)^2}\right)(\sin \theta)}{\left(\dfrac{-ed}{1 + \sin \theta}\right)(\sin \theta) + \left(\dfrac{-ed \cos \theta}{(1 + \sin \theta)^2}\right)(\cos \theta)}$

At $(ed, 0)$, $\dfrac{dy}{dx} = -1$. At (ed, π), $\dfrac{dy}{dx} = 1$.

$r_2: \dfrac{dy}{dx} = \dfrac{\left(\dfrac{ed}{1 - \sin \theta}\right)(\cos \theta) + \left(\dfrac{ed \cos \theta}{(1 - \sin \theta)^2}\right)(\sin \theta)}{\left(\dfrac{-ed}{1 - \sin \theta}\right)(\sin \theta) + \left(\dfrac{ed \cos \theta}{(1 - \sin \theta)^2}\right)(\cos \theta)}$

At $(ed, 0)$, $\dfrac{dy}{dx} = 1$. At (ed, π), $\dfrac{dy}{dx} = -1$.

Therefore, at $(ed, 0)$ we have $m_1 m_2 = (-1)(1) = -1$, and at (ed, π) we have $m_1 m_2 = 1(-1) = -1$. The curves intersect at right angles.

Review Exercises for Chapter 9

1. Matches (d) - ellipse   **2.** Matches (b) - hyperbola   **3.** Matches (a) - parabola   **4.** Matches (c) - hyperbola

5. $16x^2 + 16y^2 - 16x + 24y - 3 = 0$

$\left(x^2 - x + \dfrac{1}{4}\right) + \left(y^2 + \dfrac{3}{2}y + \dfrac{9}{16}\right) = \dfrac{3}{16} + \dfrac{1}{4} + \dfrac{9}{16}$

$\left(x - \dfrac{1}{2}\right)^2 + \left(y + \dfrac{3}{4}\right)^2 = 1$

Circle

Center: $\left(\dfrac{1}{2}, -\dfrac{3}{4}\right)$

Radius: 1

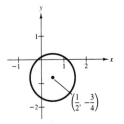

6. $y^2 - 12y - 8x + 20 = 0$

$y^2 - 12y + 36 = 8x - 20 + 36$

$(y - 6)^2 = 4(2)(x + 2)$

Parabola

Vertex: $(-2, 6)$

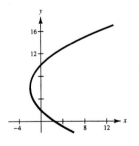

7. $3x^2 - 2y^2 + 24x + 12y + 24 = 0$

$3(x^2 + 8x + 16) - 2(y^2 - 6y + 9) = -24 + 48 - 18$

$$\frac{(x + 4)^2}{2} - \frac{(y - 3)^2}{3} = 1$$

Hyperbola

Center: $(-4, 3)$

Vertices: $\left(-4 \pm \sqrt{2}, 3\right)$

Asymptotes: $y = 3 \pm \sqrt{\frac{3}{2}}(x + 4)$

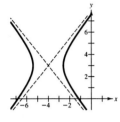

8. $4x^2 + y^2 - 16x + 15 = 0$

$4(x^2 - 4x + 4) + y^2 = -15 + 16$

$$\frac{(x - 2)^2}{1/4} + \frac{y^2}{1} = 1$$

Ellipse

Center: $(2, 0)$

Vertices: $(2, \pm 1)$

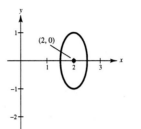

9. $3x^2 + 2y^2 - 12x + 12y + 29 = 0$

$3(x^2 - 4x + 4) + 2(y^2 + 6y + 9) = -29 + 12 + 18$

$$\frac{(x - 2)^2}{1/3} + \frac{(y + 3)^2}{1/2} = 1$$

Ellipse

Center: $(2, -3)$

Vertices: $\left(2, -3 \pm \frac{\sqrt{2}}{2}\right)$

10. $4x^2 - 4y^2 - 4x + 8y - 11 = 0$

$4\left(x^2 - x + \frac{1}{4}\right) - 4(y^2 - 2y + 1) = 11 + 1 - 4$

$$\frac{[x - (1/2)]^2}{2} - \frac{(y - 1)^2}{2} = 1$$

Hyperbola

Center: $\left(\frac{1}{2}, 1\right)$

Vertices: $\left(\frac{1}{2} \pm \sqrt{2}, 1\right)$

Asymptotes: $y = 1 \pm \left(x - \frac{1}{2}\right)$

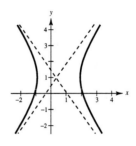

11. Vertex: $(0, 2)$

Directrix: $x = -3$

Parabola opens to the right

$p = 3$

$(y - 2)^2 = 4(3)(x - 0)$

$y^2 - 4y - 12x + 4 = 0$

12. Vertex: $(4, 2)$

Focus: $(4, 0)$

Parabola opens downward

$p = -2$

$(x - 4)^2 = 4(-2)(y - 2)$

$x^2 - 8x + 8y = 0$

13. Vertices: $(-3, 0), (7, 0)$

Foci: $(0, 0), (4, 0)$

Horizontal major axis

Center: $(2, 0)$

$a = 5, c = 2, b = \sqrt{21}$

$$\frac{(x - 2)^2}{25} + \frac{y^2}{21} = 1$$

14. Center: $(0, 0)$

Solution points: $(1, 2), (2, 0)$

Substituting the values of the coordinates of the given points into

$$\left(\frac{x^2}{b^2}\right) + \left(\frac{y^2}{a^2}\right) = 1,$$

we obtain the system

$$\left(\frac{1}{b^2}\right) + \left(\frac{4}{a^2}\right) = 1, \, 4/b^2 = 1.$$

Solving the system, we have

$$a^2 = \frac{16}{3} \text{ and } b^2 = 4, \left(\frac{x^2}{4}\right) + \left(\frac{3y^2}{16}\right) = 1.$$

15. $\dfrac{x^2}{9} + \dfrac{y^2}{4} = 1, a = 3, b = 2, c = \sqrt{5}, e = \dfrac{\sqrt{5}}{3}$

By Example 5 of Section 9.1,

$$C = 12 \int_0^{\pi/2} \sqrt{1 - \left(\frac{5}{9}\right) \sin^2 \theta} \, d\theta \approx 15.87.$$

16. $\dfrac{x^2}{4} + \dfrac{y^2}{25} = 1, a = 5, b = 2, c = \sqrt{21}, e = \dfrac{\sqrt{21}}{5}$

By Example 5 of Section 9.1,

$$C = 20 \int_0^{\pi/2} \sqrt{1 - \frac{21}{25} \sin^2 \theta} \, d\theta \approx 23.01.$$

17. $y = x - 2$ has a slope of 1. The perpendicular slope is -1.

$$y = x^2 - 2x + 2$$

$$\frac{dy}{dx} = 2x - 2 = -1 \text{ when } x = \frac{1}{2} \text{ and } y = \frac{5}{4}.$$

Perpendicular line: $y - \dfrac{5}{4} = -1\left(x - \dfrac{1}{2}\right)$

$$4x + 4y - 7 = 0$$

18. $y = \dfrac{1}{200}x^2$

(a) $x^2 = 200y$

$x^2 = 4(50)y$

Focus: $(0, 50)$

(b) $y = \dfrac{1}{200}x^2$

$$y' = \frac{1}{100}x$$

$$\sqrt{1 + (y')^2} = \sqrt{1 + \frac{x^2}{10,000}}$$

$$S = 2\pi \int_0^{100} x\sqrt{1 + \frac{x^2}{10,000}} \, dx \approx 38,294.49$$

19. (a) $V = (\pi ab)(\text{Length}) = 12\pi(16) = 192\pi \text{ ft}^3$

(b) $F = 2(62.4) \displaystyle\int_{-3}^{3} (3 - y)\frac{4}{3}\sqrt{9 - y^2} \, dy = \frac{8}{3}(62.4)\left[3\int_{-3}^{3} \sqrt{9 - y^2} \, dy - \int_{-3}^{3} y\sqrt{9 - y^2} \, dy\right]$

$$= \frac{8}{3}(62.4)\left[\frac{3}{2}\left(y\sqrt{9 - y^2} + 9 \arcsin \frac{y}{3}\right) + \frac{1}{3}(9 - y^2)^{3/2}\right]_{-3}^{3}$$

$$= \frac{8}{3}(62.4)\left[\frac{3}{2}\left(\frac{9\pi}{2}\right) - \frac{3}{2}\left(-\frac{9\pi}{2}\right)\right] = \frac{8}{3}(62.4)\left(\frac{27\pi}{2}\right) \approx 7057.274$$

—CONTINUED—

19. —CONTINUED—

(c) You want $\frac{3}{4}$ of the total area of 12π covered. Find h so that

$$2\int_0^h \frac{4}{3}\sqrt{9-y^2}\,dy = 3\pi$$

$$\int_0^h \sqrt{9-y^2}\,dy = \frac{9\pi}{8}$$

$$\frac{1}{2}\left[y\sqrt{9-y^2} + 9\arcsin\left(\frac{y}{3}\right)\right]_0^h = \frac{9\pi}{8}$$

$$h\sqrt{9-h^2} + 9\arcsin\left(\frac{h}{3}\right) = \frac{9\pi}{4}.$$

By Newton's Method, $h \approx 1.212$. Therefore, the total height of the water is $1.212 + 3 = 4.212$ ft.

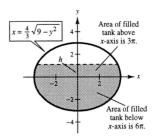

Area of filled tank above x-axis is 3π.

$x = \frac{4}{3}\sqrt{9-y^2}$

Area of filled tank below x-axis is 6π.

(d) Area of ends $= 2(12\pi) = 24\pi$

Area of sides $=$ (Perimeter)(Length)

$$= 16\int_0^{\pi/2}\left(\sqrt{1-\left(\frac{7}{16}\right)\sin^2\theta}\right)d\theta(16) \quad \text{[from Example 5 of Section 9.1]}$$

$$\approx 256\left(\frac{\pi/2}{12}\right)\left[\sqrt{1-\left(\frac{7}{16}\right)\sin^2(0)} + 4\sqrt{1-\left(\frac{7}{16}\right)\sin^2\left(\frac{\pi}{8}\right)} + 2\sqrt{1-\left(\frac{7}{16}\right)\sin^2\left(\frac{\pi}{4}\right)}\right.$$

$$\left. + 4\sqrt{1-\left(\frac{7}{16}\right)\sin^2\left(\frac{3\pi}{8}\right)} + \sqrt{1-\left(\frac{7}{16}\right)\sin^2\left(\frac{\pi}{2}\right)}\right] \approx 353.65$$

Total area $= 24\pi + 353.65 \approx 429.05$

20. (a) $A = 4\int_0^a \frac{b}{a}\sqrt{a^2-x^2}\,dx = \frac{4b}{a}\left(\frac{1}{2}\right)\left[x\sqrt{a^2-x^2} + a^2\arcsin\left(\frac{x}{a}\right)\right]_0^a = \pi ab$

(b) Disc: $V = 2\pi\int_0^b \frac{a^2}{b^2}(b^2-y^2)\,dy = \frac{2\pi a^2}{b^2}\int_0^b (b^2-y^2)\,dy = \frac{2\pi a^2}{b^2}\left[b^2y - \frac{1}{3}y^3\right]_0^b = \frac{4}{3}\pi a^2 b$

$$S = 4\pi\int_0^b \frac{a}{b}\sqrt{b^2-y^2}\left(\frac{\sqrt{b^4+(a^2-b^2)y^2}}{b\sqrt{b^2-y^2}}\right)dy$$

$$= \frac{4\pi a}{b^2}\int_0^b \sqrt{b^4+c^2y^2}\,dy = \frac{2\pi a}{b^2 c}\left[cy\sqrt{b^4+c^2y^2} + b^4\ln\left|cy + \sqrt{b^4+c^2y^2}\right|\right]_0^b$$

$$= \frac{2\pi a}{b^2 c}\left[b^2 c\sqrt{b^2+c^2} + b^4\ln\left|cb + b\sqrt{b^2+c^2}\right| - b^4\ln(b^2)\right]$$

$$= 2\pi a^2 + \frac{\pi ab^2}{c}\ln\left(\frac{c+a}{e}\right)^2 = 2\pi a^2 + \left(\frac{\pi b^2}{e}\right)\ln\left(\frac{1+e}{1-e}\right)$$

(c) Disc: $V = 2\pi\int_0^a \frac{b^2}{a^2}(a^2-x^2)\,dx = \frac{2\pi b^2}{a^2}\int_0^a (a^2-x^2)\,dx = \frac{2\pi b^2}{a^2}\left[a^2x - \frac{1}{3}x^3\right]_0^a = \frac{4}{3}\pi ab^2$

$$S = 2(2\pi)\int_0^a \frac{b}{a}\sqrt{a^2-x^2}\left(\frac{\sqrt{a^4-(a^2-b^2)x^2}}{a\sqrt{a^2-x^2}}\right)dx$$

$$= \frac{4\pi b}{a^2}\int_0^a \sqrt{a^4-c^2x^2}\,dx = \frac{2\pi b}{a^2 c}\left[cx\sqrt{a^4-c^2x^2} + a^4\arcsin\left(\frac{cx}{a^2}\right)\right]_0^a$$

$$= \frac{a\pi b}{a^2 c}\left[a^2 c\sqrt{a^2-c^2} + a^4\arcsin\left(\frac{c}{a}\right)\right] = 2\pi b^2 + 2\pi\left(\frac{ab}{e}\right)\arcsin(e)$$

—CONTINUED—

20. —CONTINUED—

(d) $\dfrac{x^2}{9} + \dfrac{y^2}{4} = 1, a = 3, b = 2, c = \sqrt{5}, e = \dfrac{\sqrt{5}}{3}$

Prolate spheroid: $V = \dfrac{4\pi}{3}(3)(4) = 16\pi$

$$S = 2\pi(4) + 2\pi\left[\dfrac{(3)(2)}{\sqrt{5}/3}\right]\arcsin\left(\dfrac{\sqrt{5}}{3}\right) = 8\pi + \left(\dfrac{36\pi}{\sqrt{5}}\right)\arcsin\left(\dfrac{\sqrt{5}}{3}\right) \approx 67.673$$

Oblate spheroid: $V = \dfrac{4\pi}{3}(9)2 = 24\pi$

$$S = 18\pi + \pi\left(\dfrac{4}{\sqrt{5}/3}\right)\ln\left[\dfrac{1 + \left(\sqrt{5}/3\right)}{1 - \left(\sqrt{5}/3\right)}\right] = 18\pi + \dfrac{12\pi}{\sqrt{5}}\ln\left(\dfrac{3 + \sqrt{5}}{3 - \sqrt{5}}\right) \approx 89.001$$

21. $x = 1 + 4t$

$y = 2 - 3t$

(a) $\dfrac{dy}{dx} = -\dfrac{3}{4}$

No horizontal tangents

(b) $t = \dfrac{x - 1}{4}$

$y = 2 - \dfrac{3}{4}(x - 1) = \dfrac{-3x + 11}{4}$

(c)

22. $x = t + 4$

$y = t^2$

(a) $\dfrac{dy}{dx} = \dfrac{2t}{1} = 2t = 0$ when $t = 0$.

Point of horizontal tangency: $(4, 0)$

(b) $t = x - 4$

$y = (x - 4)^2$

(c)

23. $x = \dfrac{1}{t}$

$y = 2t + 3$

(a) $\dfrac{dy}{dx} = \dfrac{2}{-1/t^2} = -2t^2$

No horizontal tangents $\ (t \neq 0)$

(b) $t = \dfrac{1}{x}$

$y = \dfrac{2}{x} + 3$

(c)

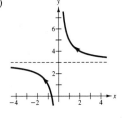

24. $x = \dfrac{1}{t}$

$y = t^2$

(a) $\dfrac{dy}{dx} = \dfrac{2t}{-1/t^2} = -2t^3$

No horizontal tangents $\ (t \neq 0)$

(b) $t = \dfrac{1}{x}$

$y = \dfrac{1}{x^2}$

(c)

25. $x = \dfrac{1}{2t + 1}$

$y = \dfrac{1}{t^2 - 2t}$

(a) $\dfrac{dy}{dx} = \dfrac{\dfrac{-(2t - 2)}{(t^2 - 2t)^2}}{\dfrac{-2}{(2t + 1)^2}} = \dfrac{(t - 1)(2t + 1)^2}{t^2(t - 2)^2} = 0$ when $t = 1$.

Point of horizontal tangency: $\left(\dfrac{1}{3}, -1\right)$

(b) $2t + 1 = \dfrac{1}{x} \Rightarrow t = \dfrac{1}{2}\left(\dfrac{1}{x} - 1\right)$

$y = \dfrac{1}{\dfrac{1}{2}\left(\dfrac{1 - x}{x}\right)\left[\dfrac{1}{2}\left(\dfrac{1 - x}{x}\right) - 2\right]}$

$\quad = \dfrac{4x^2}{(1 - x)^2 - 4x(1 - x)} = \dfrac{4x^2}{(5x - 1)(x - 1)}$

(c)

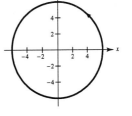

26. $x = 2t - 1$

$y = \dfrac{1}{t^2 - 2t}$

(a) $\dfrac{dy}{dx} = \dfrac{-(t^2 - 2t)^{-2}(2t - 2)}{2}$

$\quad = \dfrac{1 - t}{t^2(t - 2)^2} = 0$ when $t = 1$.

Point of horizontal tangency: $(1, -1)$

(b) $t = \dfrac{x + 1}{2}$

$y = \dfrac{1}{[(x + 1)/2]^2 - 2[(x + 1)/2]} = \dfrac{4}{(x - 3)(x + 1)}$

(c)

27. $x = 3 + 2\cos\theta$

$y = 2 + 5\sin\theta$

(a) $\dfrac{dy}{dx} = \dfrac{5\cos\theta}{-2\sin\theta} = -2.5\cot\theta = 0$ when $\theta = \dfrac{\pi}{2}, \dfrac{3\pi}{2}$.

Points of horizontal tangency: $(3, 7), (3, -3)$

(b) $\dfrac{(x - 3)^2}{4} + \dfrac{(y - 2)^2}{25} = 1$

(c)

28. $x = 6\cos\theta$

$y = 6\sin\theta$

(a) $\dfrac{dy}{dx} = \dfrac{6\cos\theta}{-6\sin\theta} = -\cot\theta = 0$ when $\theta = \dfrac{\pi}{2}, \dfrac{3\pi}{2}$.

Points of horizontal tangency: $(0, 6), (0, -6)$

(b) $\left(\dfrac{x}{6}\right)^2 + \left(\dfrac{y}{6}\right)^2 = 1$

$\quad\quad x^2 + y^2 = 36$

(c)

29. $x = \cos^3\theta$

$y = 4\sin^3\theta$

(a) $\dfrac{dy}{dx} = \dfrac{12\sin^2\theta\cos\theta}{3\cos^2\theta(-\sin\theta)} = \dfrac{-4\sin\theta}{\cos\theta} = -4\tan\theta = 0$ when $\theta = 0, \pi$.

But, $\dfrac{dy}{dt} = \dfrac{dx}{dt} = 0$ at $\theta = 0, \pi$. Hence no points of horizontal tangency.

(b) $x^{2/3} + \left(\dfrac{y}{4}\right)^{2/3} = 1$

(c)

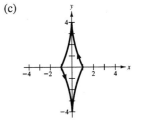

30. $x = e^t$

$y = e^{-t}$

(a) $\dfrac{dy}{dx} = \dfrac{-e^{-t}}{e^t} = -\dfrac{1}{e^{2t}} = -\dfrac{1}{x^2}$

No horizontal tangents

(b) $t = \ln x$

$y = e^{-\ln x} = e^{\ln(1/x)} = \dfrac{1}{x}, x > 0$

(c)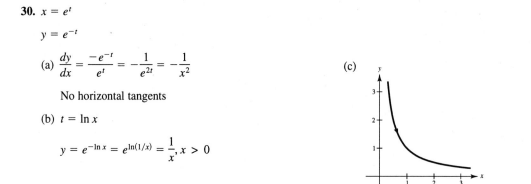

31. $x = \cot \theta$

$y = \sin 2\theta = 2 \sin \theta \cos \theta$

(a), (c)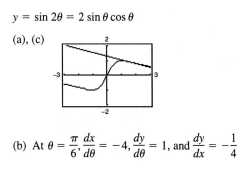

(b) At $\theta = \dfrac{\pi}{6}, \dfrac{dx}{d\theta} = -4, \dfrac{dy}{d\theta} = 1$, and $\dfrac{dy}{dx} = -\dfrac{1}{4}$

32. $x = 2\theta - \sin \theta$

$y = 2 - \cos \theta$

(a), (c)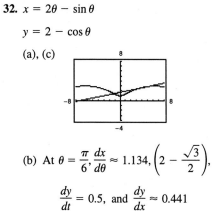

(b) At $\theta = \dfrac{\pi}{6}, \dfrac{dx}{d\theta} \approx 1.134, \left(2 - \dfrac{\sqrt{3}}{2}\right)$,

$\dfrac{dy}{dt} = 0.5$, and $\dfrac{dy}{dx} \approx 0.441$

33. $x = 3 + (3 - (-2))t = 3 + 5t$

$y = 2 + (2 - 6)t = 2 - 4t$

(other answers possible)

34. $(x - h)^2 + (y - k)^2 = r^2$

$(x - 5)^2 + (y - 3)^2 = 2^2 = 4$

35. $\dfrac{(x + 3)^2}{16} + \dfrac{(y - 4)^2}{9} = 1$

Let $\dfrac{(x + 3)^2}{16} = \cos^2 \theta$ and $\dfrac{(y - 4)^2}{9} = \sin^2 \theta$.

Then $x = -3 + 4 \cos \theta$ and $y = 4 + 3 \sin \theta$.

36. $a = 4, c = 5, b^2 = c^2 - a^2 = 9, \dfrac{y^2}{16} - \dfrac{x^2}{9} = 1$

Let $\dfrac{y^2}{16} = \sec^2 \theta$ and $\dfrac{x^2}{9} = \tan^2 \theta$.

Then $x = 3 \tan \theta$ and $y = 4 \sec \theta$.

37. $x = \cos 3\theta + 5 \cos \theta$

$y = \sin 3\theta + 5 \sin \theta$

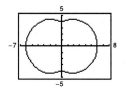

38. $x = (a - b) \cos t + b \cos \left(\dfrac{a - b}{b} t \right)$

$y = (a - b) \sin t - b \sin \left(\dfrac{a - b}{b} t \right)$

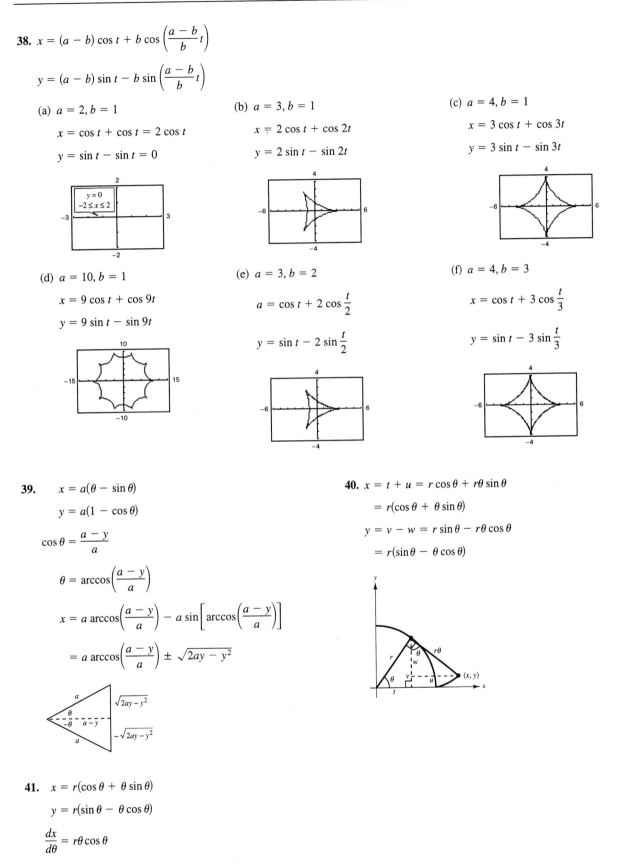

(a) $a = 2, b = 1$

$x = \cos t + \cos t = 2 \cos t$

$y = \sin t - \sin t = 0$

(b) $a = 3, b = 1$

$x = 2 \cos t + \cos 2t$

$y = 2 \sin t - \sin 2t$

(c) $a = 4, b = 1$

$x = 3 \cos t + \cos 3t$

$y = 3 \sin t - \sin 3t$

(d) $a = 10, b = 1$

$x = 9 \cos t + \cos 9t$

$y = 9 \sin t - \sin 9t$

(e) $a = 3, b = 2$

$a = \cos t + 2 \cos \dfrac{t}{2}$

$y = \sin t - 2 \sin \dfrac{t}{2}$

(f) $a = 4, b = 3$

$x = \cos t + 3 \cos \dfrac{t}{3}$

$y = \sin t - 3 \sin \dfrac{t}{3}$

39. $x = a(\theta - \sin \theta)$

$y = a(1 - \cos \theta)$

$\cos \theta = \dfrac{a - y}{a}$

$\theta = \arccos\left(\dfrac{a - y}{a} \right)$

$x = a \arccos\left(\dfrac{a - y}{a} \right) - a \sin\left[\arccos\left(\dfrac{a - y}{a} \right) \right]$

$= a \arccos\left(\dfrac{a - y}{a} \right) \pm \sqrt{2ay - y^2}$

40. $x = t + u = r \cos \theta + r\theta \sin \theta$

$= r(\cos \theta + \theta \sin \theta)$

$y = v - w = r \sin \theta - r\theta \cos \theta$

$= r(\sin \theta - \theta \cos \theta)$

41. $x = r(\cos \theta + \theta \sin \theta)$

$y = r(\sin \theta - \theta \cos \theta)$

$\dfrac{dx}{d\theta} = r\theta \cos \theta$

$\dfrac{dy}{d\theta} = r\theta \sin \theta$

$s = r \displaystyle\int_0^\pi \sqrt{\theta^2 \cos^2 \theta + \theta^2 \sin^2 \theta}\, d\theta = r \int_0^\pi \theta\, d\theta = \dfrac{r}{2}\left[\theta^2 \right]_0^\pi = \dfrac{1}{2}\pi^2 r$

42. $x = 6 \cos \theta$

$y = 6 \sin \theta$

$\dfrac{dx}{d\theta} = -6 \sin \theta$

$\dfrac{dy}{d\theta} = 6 \cos \theta$

$s = \displaystyle\int_0^\pi \sqrt{36 \sin^2 \theta + 36 \cos^2 \theta}\, d\theta = \Big[6\theta\Big]_0^\pi = 6\pi$ (one-half circumference of circle)

43. $(x, y) = (4, -4)$

$r = \sqrt{4^2 + (-4)^2} = 4\sqrt{2}$

$\theta = 7\dfrac{\pi}{4}$

$(r, \theta) = \left(4\sqrt{2}, \dfrac{7\pi}{4}\right), \left(-4\sqrt{2}, \dfrac{3\pi}{4}\right)$

44. $(x, y) = (-1, 3)$

$r = \sqrt{(-1)^2 + 3^2} = \sqrt{10}$

$\theta = \arctan(-3) \approx 1.89 \ (108.43°)$

$(r, \theta) = \left(\sqrt{10},\, 1.89\right), \left(-\sqrt{10},\, 5.03\right)$

45. $r = 3 \cos \theta$

$r^2 = 3r \cos \theta$

$x^2 + y^2 = 3x$

$x^2 + y^2 - 3x = 0$

46. $r = 10$

$r^2 = 100$

$x^2 + y^2 = 100$

47. $r = -2(1 + \cos \theta)$

$r^2 = -2r(1 + \cos \theta)$

$x^2 + y^2 = -2\left(\pm\sqrt{x^2 + y^2}\right) - 2x$

$(x^2 + y^2 + 2x)^2 = 4(x^2 + y^2)$

48. $r = \dfrac{1}{2 - \cos \theta}$

$2r - r \cos \theta = 1$

$2\left(\pm\sqrt{x^2 + y^2}\right) - x = 1$

$4(x^2 + y^2) = (x + 1)^2$

$3x^2 + 4y^2 - 2x - 1 = 0$

49. $r^2 = \cos 2\theta = \cos^2 \theta - \sin^2 \theta$

$r^4 = r^2 \cos^2 \theta - r^2 \sin^2 \theta$

$(x^2 + y^2)^2 = x^2 - y^2$

50. $r = 4 \sec\left(\theta - \dfrac{\pi}{3}\right) = \dfrac{4}{\cos[\theta - (\pi/3)]}$

$= \dfrac{4}{(1/2)\cos \theta + \left(\sqrt{3}/2\right)\sin \theta}$

$r\left(\cos \theta + \sqrt{3} \sin \theta\right) = 8$

$x + \sqrt{3}\, y = 8$

51. $r = 4 \cos 2\theta \sec \theta$

$= 4(2\cos^2\theta - 1)\left(\dfrac{1}{\cos\theta}\right)$

$r\cos\theta = 8\cos^2\theta - 4$

$x = 8\left(\dfrac{x^2}{x^2 + y^2}\right) - 4$

$x^3 + xy^2 = 4x^2 - 4y^2$

$y^2 = x^2\left(\dfrac{4-x}{4+x}\right)$

52. $\theta = \dfrac{3\pi}{4}$

$\tan\theta = -1$

$\dfrac{y}{x} = -1$

$y = -x$

53. $(x^2 + y^2)^2 = ax^2 y$

$r^4 = a(r^2\cos^2\theta)(r\sin\theta)$

$r = a\cos^2\theta\sin\theta$

54. $x^2 + y^2 - 4x = 0$

$r^2 - 4r\cos\theta = 0$

$r = 4\cos\theta$

55. $x^2 + y^2 = a^2\left(\arctan\dfrac{y}{x}\right)^2$

$r^2 = a^2\theta^2$

56. $(x^2 + y^2)\left(\arctan\dfrac{y}{x}\right)^2 = a^2$

$r^2\theta^2 = a^2$

57. $r = 4$

Circle of radius 4

Centered at the pole

Symmetric to polar axis,

$\theta = \pi/2$, and pole

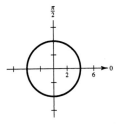

58. $\theta = \dfrac{\pi}{12}$

Line

59. $r = -\sec\theta = \dfrac{-1}{\cos\theta}$

$r\cos\theta = -1, x = -1$

Vertical line

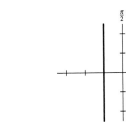

60. $r = 3\csc\theta, r\sin\theta = 3, y = 3$

Horizontal line

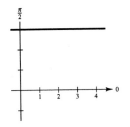

61. $r = -2(1 + \cos\theta)$

Cardioid

Symmetric to polar axis

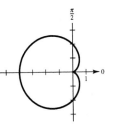

θ	0	$\dfrac{\pi}{3}$	$\dfrac{\pi}{2}$	$\dfrac{2\pi}{3}$	π
r	-4	-3	-2	-1	0

62. $r = 3 - 4\cos\theta$

Limaçon

Symmetric to polar axis

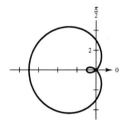

θ	0	$\dfrac{\pi}{3}$	$\dfrac{\pi}{2}$	$\dfrac{2\pi}{3}$	π
r	-1	1	3	5	7

63. $r = 4 - 3\cos\theta$

Limaçon

Symmetric to polar axis

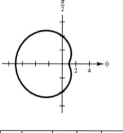

θ	0	$\dfrac{\pi}{3}$	$\dfrac{\pi}{2}$	$\dfrac{2\pi}{3}$	π
r	1	$\dfrac{5}{2}$	4	$\dfrac{11}{2}$	7

64. $r = 2\theta$

Spiral

Symmetric to $\theta = \pi/2$

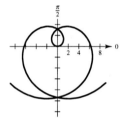

θ	0	$\dfrac{\pi}{4}$	$\dfrac{\pi}{2}$	$\dfrac{3\pi}{4}$	π	$\dfrac{5\pi}{4}$	$\dfrac{3\pi}{2}$
r	0	$\dfrac{\pi}{5}$	π	$\dfrac{3\pi}{2}$	2π	$\dfrac{5\pi}{2}$	3π

65. $r = -3\cos(2\theta)$

Rose curve with four petals

Symmetric to polar axis, $\theta = \dfrac{\pi}{2}$, and pole

Relative extrema: $(-3, 0), \left(3, \dfrac{\pi}{2}\right), (-3, \pi), \left(3, \dfrac{3\pi}{2}\right)$

Tangents at the pole: $\theta = \dfrac{\pi}{4}, \dfrac{3\pi}{4}$

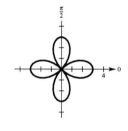

66. $r = \cos(5\theta)$

Rose curve with five petals

Symmetric to polar axis

Relative extrema: $(1, 0), \left(-1, \dfrac{\pi}{5}\right), \left(1, \dfrac{2\pi}{5}\right), \left(-1, \dfrac{3\pi}{5}\right), \left(1, \dfrac{4\pi}{5}\right)$

Tangents at the pole: $\theta = \dfrac{\pi}{10}, \dfrac{3\pi}{10}, \dfrac{\pi}{2}, \dfrac{7\pi}{10}, \dfrac{9\pi}{10}$

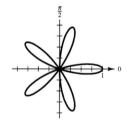

67. $r^2 = 4\sin^2(2\theta)$

$r = \pm 2\sin(2\theta)$

Rose curve with four petals

Symmetric to the polar axis, $\theta = \dfrac{\pi}{2}$, and pole

Relative extrema: $\left(\pm 2, \dfrac{\pi}{4}\right), \left(\pm 2, \dfrac{3\pi}{4}\right)$

Tangents at the pole: $\theta = 0, \dfrac{\pi}{2}$

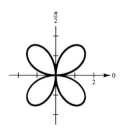

68. $r^2 = \cos(2\theta)$

Lemniscate

Symmetric to the polar axis

Relative extrema: $(\pm 1, 0)$

Tangents at the pole: $\theta = \dfrac{\pi}{4}, \dfrac{3\pi}{4}$

θ	0	$\dfrac{\pi}{6}$	$\dfrac{\pi}{4}$
r	± 1	$\pm\dfrac{\sqrt{2}}{2}$	0

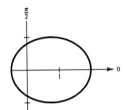

69. $r = \dfrac{2}{1 - \sin\theta}$

Parabola

Focus at the pole

Vertex: $\left(1, \dfrac{3\pi}{2}\right)$

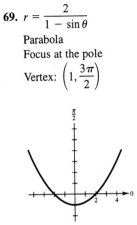

70. $r = \dfrac{4}{5 - 3\cos\theta} = \dfrac{4/5}{1 - (3/5)\cos\theta}$

Ellipse

Focus at the pole

$e = \dfrac{3}{5}$

Vertices: $(2, 0), \left(\dfrac{1}{2}, \pi\right)$

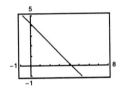

71. $r = \dfrac{3}{\cos[\theta - (\pi/4)]}$

Graph of $r = 3\sec\theta$ rotated through an angle of $\pi/4$

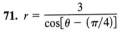

72. $r = 2\sin\theta\cos^2\theta$

Bifolium

Symmetric to $\theta = \pi/2$

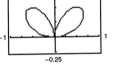

73. $r = 4\cos 2\theta \sec\theta$

Strophoid

Symmetric to the polar axis

$r \Rightarrow -\infty$ as $\theta \Rightarrow \dfrac{\pi^-}{2}$

$r \Rightarrow -\infty$ as $\theta \Rightarrow \dfrac{-\pi^+}{2}$

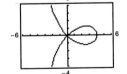

74. $r = 4(\sec\theta - \cos\theta)$

Semicubical parabola

Symmetric to the polar axis

$r \Rightarrow \infty$ as $\theta \Rightarrow \dfrac{\pi^-}{2}$

$r \Rightarrow \infty$ as $\theta \Rightarrow \dfrac{-\pi^+}{2}$

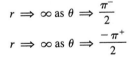
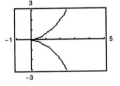

75. $r = 1 - 2\cos\theta$

(a) The graph has polar symmetry and the tangents at the pole are

$\theta = \dfrac{\pi}{3}, -\dfrac{\pi}{3}$.

—CONTINUED—

75. —CONTINUED—

(b) $\dfrac{dy}{dx} = \dfrac{2\sin^2\theta + (1 - 2\cos\theta)\cos\theta}{2\sin\theta\cos\theta - (1 - 2\cos\theta)\sin\theta}$

Horizontal tangents: $-4\cos^2\theta + \cos\theta + 2 = 0$, $\cos\theta = \dfrac{-1 \pm \sqrt{1 + 32}}{-8} = \dfrac{1 \pm \sqrt{33}}{8}$

When $\cos\theta = \dfrac{1 \pm \sqrt{33}}{8}$, $r = 1 - 2\left(\dfrac{1 + \sqrt{33}}{8}\right) = \dfrac{3 \mp \sqrt{33}}{4}$,

$\left[\dfrac{3 - \sqrt{33}}{4}, \arccos\left(\dfrac{1 + \sqrt{33}}{8}\right)\right] \approx (-0.686, 0.568)$

$\left[\dfrac{3 - \sqrt{33}}{4}, -\arccos\left(\dfrac{1 + \sqrt{33}}{8}\right)\right] \approx (-0.686, -0.568)$

$\left[\dfrac{3 + \sqrt{33}}{4}, \arccos\left(\dfrac{1 - \sqrt{33}}{8}\right)\right] \approx (2.186, 2.206)$

$\left[\dfrac{3 + \sqrt{33}}{4}, -\arccos\left(\dfrac{1 - \sqrt{33}}{8}\right)\right] \approx (2.186, -2.206)$.

Vertical tangents:

$\sin\theta(4\cos\theta - 1) = 0$, $\sin\theta = 0$, $\cos\theta = \dfrac{1}{4}$,

$\theta = 0, \pi, \theta = \pm\arccos\left(\dfrac{1}{4}\right), (-1, 0), (3, \pi)$

$\left(\dfrac{1}{2}, \pm\arccos\dfrac{1}{4}\right) \approx (0.5, \pm1.318)$

(c)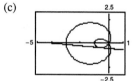

76. $r^2 = 4\sin(2\theta)$

(a) $2r\left(\dfrac{dr}{d\theta}\right) = 8\cos(2\theta)$

$\dfrac{dr}{d\theta} = \dfrac{4\cos(2\theta)}{r}$

Tangents at the pole: $\theta = 0, \dfrac{\pi}{2}$

(c)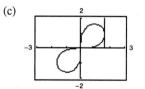

(b) $\dfrac{dy}{dx} = \dfrac{r\cos\theta + [(4\cos 2\theta\sin\theta)/r]}{-r\sin\theta + [(4\cos 2\theta\cos\theta)/r]}$

$= \dfrac{\cos(2\theta)\sin\theta + \sin(2\theta)\cos\theta}{\cos(2\theta)\cos\theta - \sin(2\theta)\sin\theta}$

Horizontal tangents:

$\dfrac{dy}{dx} = 0$ when $\cos(2\theta)\sin\theta + \sin(2\theta)\cos\theta = 0$,

$\tan\theta = -\tan(2\theta)$, $\theta = 0, \dfrac{\pi}{3}$, $(0, 0)$, $\left(\pm\sqrt{2\sqrt{3}}, \dfrac{\pi}{3}\right)$

Vertical tangents when $\cos 2\theta\cos\theta - \sin 2\theta\sin\theta = 0$:

$\tan 2\theta\tan\theta = 1$, $\theta = 0, \dfrac{\pi}{6}$, $(0, 0)$, $\left(\pm\sqrt{2\sqrt{3}}, \dfrac{\pi}{6}\right)$

77. $r = 1 + \cos\theta, r = 1 - \cos\theta$

The points $(1, \pi/2)$ and $(1, 3\pi/2)$ are the two points of intersection (other than the pole). The slope of the graph of $r = 1 + \cos\theta$ is

$$m_1 = \frac{dy}{dx} = \frac{r'\sin\theta + r\cos\theta}{r'\cos\theta - r\sin\theta} = \frac{-\sin^2\theta + \cos\theta(1 + \cos\theta)}{-\sin\theta\cos\theta - \sin\theta(1 + \cos\theta)}.$$

At $(1, \pi/2), m_1 = -1/-1 = 1$ and at $(1, 3\pi/2), m_1 = -1/1 = -1$. The slope of the graph of $r = 1 - \cos\theta$ is

$$m_2 = \frac{dy}{dx} = \frac{\sin^2\theta + \cos\theta(1 - \cos\theta)}{\sin\theta\cos\theta - \sin\theta(1 - \cos\theta)}.$$

At $(1, \pi/2), m_2 = 1/-1 = -1$ and at $(1, 3\pi/2), m_2 = 1/1 = 1$. In both cases, $m_1 = -1/m_2$ and we conclude that the graphs are orthogonal at $(1, \pi/2)$ and $(1, 3\pi/2)$.

78. $r = a\sin\theta, r = a\cos\theta$

The points of intersection are $\left(a/\sqrt{2}, \pi/4\right)$ and $(0, 0)$. For $r = a\sin\theta$,

$$m_1 = \frac{dy}{dx} = \frac{a\cos\theta\sin\theta + a\sin\theta\cos\theta}{a\cos^2\theta - a\sin^2\theta} = \frac{2\sin\theta\cos\theta}{\cos 2\theta}.$$

At $\left(a/\sqrt{2}, \pi/4\right), m_1$ is undefined and at $(0, 0), m_1 = 0$. For $r = a\cos\theta$,

$$m_2 = \frac{dy}{dx} = \frac{-a\sin^2\theta + a\cos^2\theta}{-a\sin\theta\cos\theta - a\cos\theta\sin\theta} = \frac{\cos 2\theta}{-2\sin\theta\cos\theta}.$$

At $\left(a/\sqrt{2}, \pi/4\right), m_2 = 0$ and at $(0, 0), m_2$ is undefined. Therefore, the graphs are orthogonal at $\left(a/\sqrt{2}, \pi/4\right)$ and $(0, 0)$.

79. Circle: $r = 3\sin\theta$

$$\frac{dy}{dx} = \frac{3\cos\theta\sin\theta + 3\sin\theta\cos\theta}{3\cos\theta\cos\theta - 3\sin\theta\sin\theta} = \frac{\sin 2\theta}{\cos^2\theta - \sin^2\theta} = \tan 2\theta \text{ at } \theta = \frac{\pi}{6}, \frac{dy}{dx} = \sqrt{3}$$

Limaçon: $r = 4 - 5\sin\theta$

$$\frac{dy}{dx} = \frac{-5\cos\theta\sin\theta + (4 - 5\sin\theta)\cos\theta}{-5\cos\theta\cos\theta - (4 - 5\sin\theta)\sin\theta} \text{ at } \theta = \frac{\pi}{6}, \frac{dy}{dx} = \frac{\sqrt{3}}{9}$$

Let α be the angle between the curves:

$$\tan\alpha = \frac{\sqrt{3} - \left(\sqrt{3}/9\right)}{1 + (1/3)} = \frac{2\sqrt{3}}{3}.$$

Therefore, $\alpha = \arctan\left(\frac{2\sqrt{3}}{3}\right) \approx 49.1°$.

80. False. There are an infinite number of polar coordinate representations of a point. For example, the point $(x, y) = (1, 0)$ has polar representations $(r, \theta) = (1, 0), (1, 2\pi), (-1, \pi)$, etc.

81. $r = 2 + \cos\theta$

$$A = 2\left[\frac{1}{2}\int_0^\pi (2 + \cos\theta)^2 \, d\theta\right] \approx 14.14 \quad \left(\frac{9\pi}{2}\right)$$

82. $r = 5(1 - \sin\theta)$

$$A = 2\left[\frac{1}{2}\int_{\pi/2}^{3\pi/2} [5(1 - \sin\theta)]^2 \, d\theta\right] \approx 117.81 \quad \left(75\frac{\pi}{2}\right)$$

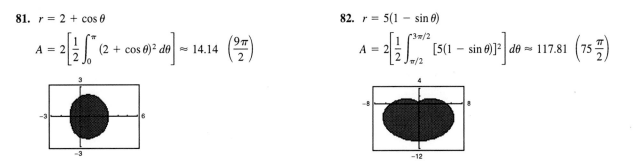

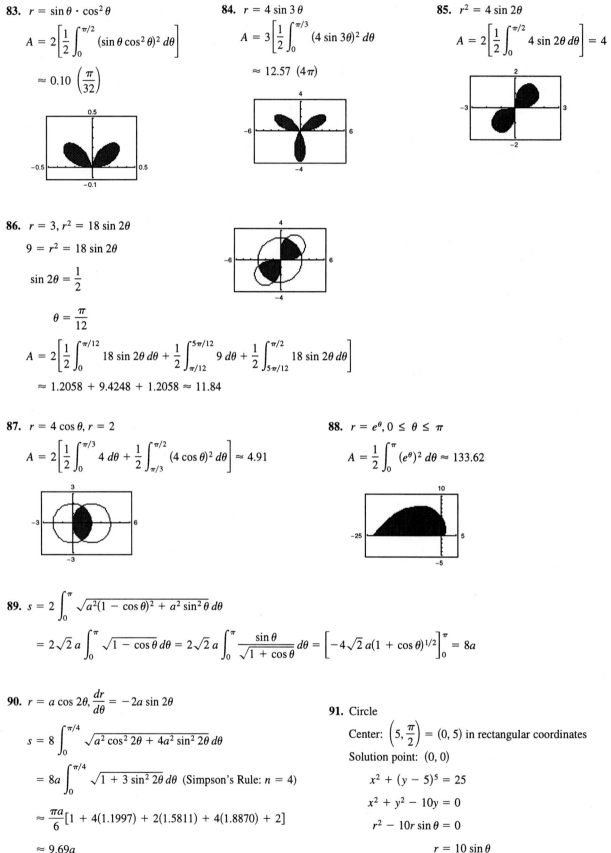

83. $r = \sin\theta \cdot \cos^2\theta$

$$A = 2\left[\frac{1}{2}\int_0^{\pi/2}(\sin\theta\cos^2\theta)^2\,d\theta\right]$$

$$\approx 0.10 \left(\frac{\pi}{32}\right)$$

84. $r = 4\sin 3\theta$

$$A = 3\left[\frac{1}{2}\int_0^{\pi/3}(4\sin 3\theta)^2\,d\theta\right]$$

$$\approx 12.57 \ (4\pi)$$

85. $r^2 = 4\sin 2\theta$

$$A = 2\left[\frac{1}{2}\int_0^{\pi/2}4\sin 2\theta\,d\theta\right] = 4$$

86. $r = 3, r^2 = 18\sin 2\theta$

$$9 = r^2 = 18\sin 2\theta$$

$$\sin 2\theta = \frac{1}{2}$$

$$\theta = \frac{\pi}{12}$$

$$A = 2\left[\frac{1}{2}\int_0^{\pi/12}18\sin 2\theta\,d\theta + \frac{1}{2}\int_{\pi/12}^{5\pi/12}9\,d\theta + \frac{1}{2}\int_{5\pi/12}^{\pi/2}18\sin 2\theta\,d\theta\right]$$

$$\approx 1.2058 + 9.4248 + 1.2058 \approx 11.84$$

87. $r = 4\cos\theta, r = 2$

$$A = 2\left[\frac{1}{2}\int_0^{\pi/3}4\,d\theta + \frac{1}{2}\int_{\pi/3}^{\pi/2}(4\cos\theta)^2\,d\theta\right] \approx 4.91$$

88. $r = e^\theta, 0 \le \theta \le \pi$

$$A = \frac{1}{2}\int_0^\pi(e^\theta)^2\,d\theta \approx 133.62$$

89. $s = 2\displaystyle\int_0^\pi\sqrt{a^2(1-\cos\theta)^2 + a^2\sin^2\theta}\,d\theta$

$$= 2\sqrt{2}\,a\int_0^\pi\sqrt{1-\cos\theta}\,d\theta = 2\sqrt{2}\,a\int_0^\pi\frac{\sin\theta}{\sqrt{1+\cos\theta}}\,d\theta = \left[-4\sqrt{2}\,a(1+\cos\theta)^{1/2}\right]_0^\pi = 8a$$

90. $r = a\cos 2\theta, \dfrac{dr}{d\theta} = -2a\sin 2\theta$

$$s = 8\int_0^{\pi/4}\sqrt{a^2\cos^2 2\theta + 4a^2\sin^2 2\theta}\,d\theta$$

$$= 8a\int_0^{\pi/4}\sqrt{1+3\sin^2 2\theta}\,d\theta \ \text{(Simpson's Rule: } n = 4)$$

$$\approx \frac{\pi a}{6}[1 + 4(1.1997) + 2(1.5811) + 4(1.8870) + 2]$$

$$\approx 9.69a$$

91. Circle

Center: $\left(5, \dfrac{\pi}{2}\right) = (0, 5)$ in rectangular coordinates

Solution point: $(0, 0)$

$$x^2 + (y - 5)^5 = 25$$

$$x^2 + y^2 - 10y = 0$$

$$r^2 - 10r\sin\theta = 0$$

$$r = 10\sin\theta$$

92. Line

Slope: $\sqrt{3}$

Solution point: $(0, 0)$

$y = \sqrt{3}\,x, r \sin \theta = \sqrt{3}\,r \cos \theta,$

$\tan \theta = \sqrt{3}, \theta = \dfrac{\pi}{3}$

93. Parabola

Vertex: $(2, \pi)$

Focus: $(0, 0)$

$e = 1, d = 4$

$r = \dfrac{4}{1 - \cos \theta}$

94. Parabola

Vertex: $\left(2, \dfrac{\pi}{2}\right)$

Focus: $(0, 0)$

$e = 1, d = 4$

$r = \dfrac{4}{1 + \sin \theta}$

95. Ellipse

Vertices: $(5, 0), (1, \pi)$

Focus: $(0, 0)$

$a = 3, c = 2, e = \dfrac{2}{3}, d = \dfrac{5}{2}$

$r = \dfrac{\left(\dfrac{2}{3}\right)\left(\dfrac{5}{2}\right)}{1 - \left(\dfrac{2}{3}\right)\cos \theta} = \dfrac{5}{3 - 2 \cos \theta}$

96. Hyperbola

Vertices: $(1, 0), (7, 0)$

Focus: $(0, 0)$

$a = 3, c = 4, e = \dfrac{4}{3}, d = \dfrac{7}{4}$

$r = \dfrac{\left(\dfrac{4}{3}\right)\left(\dfrac{7}{4}\right)}{1 + \left(\dfrac{4}{3}\right)\cos \theta} = \dfrac{7}{3 + 4 \cos \theta}$